中国生物多样性红色名录：脊椎动物

China's Red List of Biodiversity: Vertebrates

主编　蒋志刚

Chief Editor: Zhigang Jiang

第一卷　哺乳动物（下册）

Volume Ⅰ, Mammals (Ⅲ)

主编　蒋志刚

Chief Editor: Zhigang Jiang

副主编　吴　毅　刘少英　蒋学龙
周开亚　胡慧建

Vice-Chief Editors: Yi Wu　Shaoying Liu　Xuelong Jiang
Kaiya Zhou　Huijian Hu

科学出版社

北京

内 容 简 介

本书是“中国生物多样性红色名录：脊椎动物”的“第一卷　哺乳动物”，全书分为总论和各论两部分。总论介绍了哺乳动物的演化与现状、中国哺乳动物多样性与保护现状、本红色名录评估对象的分类系统、中国哺乳动物编目（2021）、中国哺乳动物分布格局和受保护状况；介绍了红色名录评估过程、依据的评估等级和标准，还介绍了咨询专家、评估队伍，以及建立数据库和开展初评、通讯评审、形成评估报告的过程；总结分析了评估结果，介绍了中国哺乳动物濒危状况，分析了野外灭绝的和区域灭绝的物种，分析了受威胁物种比例、哺乳动物的分布和濒危原因等，并将评估结果与《IUCN 受威胁物种红色名录》（2020-2）进行了比较分析；最后，分析了中国哺乳动物保护成效与远景。各论是图书的主体，对评估的 700 种中国哺乳动物物种的相关信息，即分类地位、评估信息、地理分布、种群状况、生境与生态系统、威胁因子、保护级别与保护行动及相关文献进行了详细叙述。

本书适合从事野生动物研究与保护的科研人员参考，适合自然保护区、环境保护、进出口对外贸易、检验检疫等相关各级行政管理部门作为行使管理职能的参照资料，适合作为高年级研究生教学参考书，适合国内大中型图书馆馆藏。

审图号：**GS (2020) 2858号**

图书在版编目（CIP）数据

中国生物多样性红色名录. 脊椎动物. 第一卷，哺乳动物. 下册 = China's Red List of Biodiversity: Vertebrates, Volume I, Mammals (III)：汉英对照 / 蒋志刚主编. —北京：科学出版社，2021.3

国家出版基金项目

ISBN 978-7-03-065664-3

Ⅰ. ①中…　Ⅱ. ①蒋…　Ⅲ. ①珍稀动物–中国–名录–汉、英 ②珍稀植物–中国–名录–汉、英 ③哺乳动物纲–中国–名录–汉、英　Ⅳ. ①Q958.52-62 ②Q948.52-62 ③Q959.808

中国版本图书馆CIP数据核字（2020）第131618号

责任编辑：马　俊　孙　青　郝晨扬 / 责任校对：严　娜

责任印制：肖　兴 / 排版设计：北京鑫诚文化传播有限公司

科学出版社 出版

北京东黄城根北街16号

邮政编码：100717

http: //www. sciencep. com

中国科学院印刷厂 印刷

科学出版社发行　各地新华书店经销

*

2021年3月第　一　版　开本：889×1194　1/16

2021年3月第一次印刷　印张：108 1/4

字数：3 117 000

定价：1428.00元（全三册）

（如有印装质量问题，我社负责调换）

China's Red List of Biodiversity: Vertebrates, Volume I, Mammals (III)

Chief Editor: Zhigang Jiang
Vice-Chief Editors: Yi Wu Shaoying Liu Xuelong Jiang Kaiya Zhou Huijian Hu

Abstract

This book is the first volume of the "*China's Red List of Biodiversity: Vertebrates*", *i.e.* "*Volume I, Mammals*", and is divided into two parts of "General Introduction" and "Species Monograph". General Introduction contains evolution and status of mammals, diversity and conservation status of mammals in China, taxonomic system of this red list, inventory of China's mammals (2021), the distribution patterns and conservation status of China's mammals; the evaluation process, categories and criteria that evaluation refers to, building the database, evaluation teams and how the evaluation teams to establish database, to carry out preliminary evaluation, to review the preliminary results by correspondence and to formulize the evaluation report; analyzes and summarizes the evaluation results, introduces the status of endangered mammal, analyzes the species extinct in the wild and regionally extinct, analyzes the proportion of threatened mammal species in different groups of mammals, in different habitats and its provincial distribution and the threats to endangered mammals, and compares the evaluation results with the *IUCN Red List of Endangered Species* (2020-2). Finally, the implications and prospects of mammal conservation in China are discussed. Species Monograph is the main part of this book, this part elaborates the detailed evaluating information of 700 Chinese mammal species, including taxonomic status, assessment information, geographical distribution, population situation, habitat and ecosystem, threats factor, protection category and conservation action, citations and references.

This book can be used as a reference for wild animal protect staffs and researchers, and can be used as an important data for decisions of government management department, *e.g.* nature reserve, environment protect, overseas trade, inspection and quarantine. This book is appropriate for senior grade graduated students in school, and appropriate to be collected by large- or medium-sized library.

ISBN 978-7-03-065664-3

Acknowledgement
Illustrations (with mark of "Lynx Edicions") by Toni Liobet from: Wilson, D. E., Lacher, T. E. Jr & Mittermeier, R. A. eds. (2009-2018). *Handbook of the Mammals of the World.* Volumes 1 to 8. Lynx Edicions, Barcelona.

《中国生物多样性红色名录：脊椎动物》

编 委 会

顾 问

陈宜瑜　郑光美　张亚平　金鉴明　马建章　曹文宣

主 编

蒋志刚

副主编

江建平　王跃招　张 鹗　张雁云

编委会成员

蔡 波　曹 亮　车 静　陈小勇　陈晓虹　丁 平
董 路　胡慧建　胡军华　计 翔　江建平　蒋学龙
蒋志刚　李 成　李春旺　李家堂　李丕鹏　梁 伟
刘 阳　刘少英　卢 欣　马 鸣　马 勇　马志军
饶定齐　史海涛　王 斌　王剑伟　王英永　王跃招
吴 华　吴 毅　吴孝兵　谢 锋　杨道德　杨晓君
曾晓茂　张 鹗　张 洁　张保卫　张雁云　张正旺
赵亚辉　周 放　周开亚

责任编辑

马 俊　李 迪　郝晨扬　孙 青

Editorial Committee of China's Red List of Biodiversity: Vertebrates

序 一

地球进入了一个崭新的地质纪元——人类世（Anthropocene），而地球上的人口仍呈指数增长。人类社会进入了全球化、信息化时代。人类的生态足迹日益扩大，人类对自然资源的消耗、对环境的污染达到了一个前所未有的水平，人类的影响已经遍及地球各个角落，导致了全球变化，影响了地球生物圈的结构与功能，危及了许多野生动植物的生存，造成全球范围的生物多样性危机，影响了人类社会的可持续发展。

中国是一个生物多样性大国，是地球上生物多样性最丰富的国家之一。根据《中国生物物种名录》（2019），中国已经记载生物达 106,509 个物种及种下单元，其中物种 94,260 个，种下单元 12,249 个。中国南北纬度跨度大，海拔跨度也大。中国还有多种气候类型、多样生境类型，栖息着丰富的高等生物。这些生物物种是国家重要的战略资源，是社会经济可持续发展中不可替代的物质基础。

濒危物种红色名录已经成为重要的生物多样性保护研究工具。目前，《世界自然保护联盟受威胁物种红色名录》（《IUCN 受威胁物种红色名录》）评估了 98,500 多个物种，发现其中 32,000 多个面临灭绝威胁，包括 41% 的两栖动物、34% 的针叶树、33% 的造礁珊瑚、26% 的哺乳动物和 14% 的鸟类。然而，《IUCN 受威胁物种红色名录》对物种的生存状况的评估是基于全球资料所做的，并不代表物种在各个分布国的生存状况。国家是生物多样性保护的主体，各国须开展自己的濒危物种红色名录研究，对其生物物种的生存状况进行评估。在某种程度上，可以说濒危物种红色名录研究反映了一个国家生物多样性综合研究的能力。

早在 20 世纪 90 年代，中国即引入 IUCN 受威胁物种红色名录等级标准开展了物种濒危状况评估工作，如 1991 年，我国学者发表了《中国植物红皮书》。1998 年，国家环境保护总局联合国家濒危物种科学委员会发表了《中国濒危动物红皮书 • 鱼类》《中国濒危动物红皮书 • 两栖类和爬行类》《中国濒危动物红皮书 • 鸟类》《中国濒危动物红皮书 • 兽类》等著作。另外，2004 年和 2009 年，相关领域专家开展了不同类群物种的濒危状况评估工作，先后发表了《中国物种红色名录

（第一卷）红色名录》和《中国物种红色名录（第二卷）脊椎动物》。

物种生存状况是变化的，于是，IUCN 每年定期更新 IUCN 红色名录。IUCN 红色名录并不反映一个跨越国家分布的物种在一个分布国家的生存状况。国家是濒危物种的管理主体，各国需要应用国际标准进行红色名录评估。鉴于此，为全面评估中国野生脊椎动物濒危状况，中国研究人员于 2013 年启动了"中国生物多样性红色名录——脊椎动物卷"的物种评估和报告编制工作。这次评估组织全国鱼类、两栖类、爬行类、鸟类与哺乳类专家收集数据，采用综合分析和专家评估相结合的方法，依据中国鱼类、两栖类、爬行类、鸟类和哺乳类野生种群与生境现状，利用 IUCN 红色名录标准第 3.1 版，编制了"中国生物多样性红色名录——脊椎动物卷"。该卷红色名录于 2015 年 5 月 6 日通过环境保护部和中国科学院的联合验收，并于 5 月 22 日以环境保护部、中国科学院 2015 年第 32 号公告形式正式发布。

2016 年以来，中国脊椎动物红色名录工作组组织中国研究人员再次厘定了中国脊椎动物多样性，重新评估了中国脊椎动物的生存状况。完成了《中国生物多样性红色名录：脊椎动物》(2021)。本次脊椎动物红色名录评估发现中国脊椎动物生存状况严峻，中国脊椎动物的灭绝风险高于世界平均水平。中国脊椎动物哺乳类、鸟类、两栖类、爬行类和鱼类等各个类群都发现了野外灭绝或区域灭绝的物种，有许多物种处于极危、濒危和易危的受威胁状态。

中国正处于发展期，人口众多，地貌复杂，区域发展程度差异大。如何拯救这些濒危物种是中国生物多样性保护面临的一项艰巨任务。中国政府十分重视生物多样性保护，缔结了《生物多样性公约》《濒危野生动植物种国际贸易公约》《关于特别是作为水禽栖息地的国际重要湿地公约》（简称《湿地公约》）等国际公约，并积极主动履约。中国大力开展了以自然保护区为主体、以国家公园为龙头的保护地建设。目前，我国建立的各类保护地已达 1.18 万处，面积占国土面积的 18% 以上。其中有 474 个国家级自然保护区。自然保护区保护了 90.5% 的陆地生态系统类型、85% 的野生动植物种类、65% 的高等植物群落。中国的森林覆盖率逐年增加，为濒危物种的种群与栖息地恢复奠定了基础。

生物多样性研究既是一项综合性研究，也是一项组合型研究。生物物种编目与受威胁状态评估需要不同学科的联合研究。为了生物多样性科学研究与保护，来自不同学科的学者走到一起，完成中国生物多样性红色名录研究。这项研究是中国整体生物多样性研究的重要组成部分。《中国生物多样性红色名录：脊椎动物》各卷在《生物多样性公约》第 15 次缔约方大会即将在中国召开之前出版发行，为中国生物多样性保护提供了基础数据，为监测中国生物多样性现状、开展阶段性 IUCN 红色名录指数研究积累了参数，也是中国保护生物学研究成果的展示。

中国科学院院士
国家自然科学基金委员会前主任
国家濒危物种科学委员会主任
2020 年 6 月 26 日

Foreword I

The earth has entered a new geological epoch, the Anthropocene, while the human population on the earth is still increasing exponentially. Human society has entered the era of globalization and information. The human ecological footprint is enlarging while the human consumption of natural resources increases; environmental pollution created by human being has reached an unprecedented level. Impact of human reaches throughout all corners of the earth, causes the global change and affects the structure and function of the earth's biosphere, threatens the survival of many wild animals and plants, causing a global biodiversity crisis, consequently, influences the sustainable development of human society.

China is a country with great biodiversity and one of the countries with the richest biodiversity on the earth. According to the *Species Catalogue of China* (2019), China has already recorded 106,509 species and subspecies taxa, including 94,260 species and 12,249 subspecies. The territory of China spans a large latitude and has huge elevation differences. China also has a variety of climate types, diverse habitat types, rich niches of higher organisms. These biological species are important strategic resources of the country and irreplaceable material basis for sustainable social and economic development.

The red list of endangered species has become an important tool for biodiversity conservation research. Currently, the *IUCN Red List of Threatened Species* has assessed more than 98,500 species and found that more than 32,000 of them are threatened with extinction, including 41% of amphibians, 34% of conifers, 33% of reef-building corals, 26% of mammals and 14% of birds. However, the *IUCN Red List of Threatened Species* is based on global data and does necessarily not represent the status of species in each range country. A country is the main sovereign body of biodiversity conservation, and each country needs to carry out the red list study of its endangered species to assess the survival status of its biological species. To some extent, red list study reflects the comprehensive research capacity of biodiversity in a country.

The red list of endangered species has become an important tool for biodiversity conservation research. As early as in the 1990s, China introduced the IUCN red list criteria for threatened species to carry out the assessment of status of endangered species. For example, in 1991, Chinese scholars published the *Red Data Book of Chinese Plants*. In 1998, Environmental Protection Administration, together with the National Scientific Committee on Endangered Species, published such books as *China Red Data Book of Endangered Animals: Pisces*; *China Red Data Book of Endangered Animals: Amphibia & Reptilia*; *China Red Data Book of Endangered Animals: Aves*; *China Red Data Book of Endangered Animals: Mammalia*. In addition, in 2004 and 2009, experts in related fields carried out the assessment of the endangered status of different taxa of species, and published the *China Species Red List Vol I Red List* and the *China Species Red List Vol II Vertebrates*.

Status of species is changing, therefore the IUCN updates its red list every year. However, the IUCN red list does not reflect the status of particular species in a range country if a species distributes in multi-countries. Countries are the main management bodies of endangered species; thus, each country needs to apply international standards for red list assessment. Therefore, in order to comprehensively assess the endangered status of wild vertebrates in China, Chinese researchers launched the compilation of the "China's Red List of Biodiversity: Volume of Vertebrates" in 2013. During that evaluation, by adopting the combination of comprehensive analysis and expert evaluation method, the experts were coordinated to collect data for assessing wild population and habitat status of China's fishes, amphibians, reptiles, birds and mammals and compiled the *China's Red List of Biodiversity: Volume of Vertebrates*. The red list was approved by Ministry of Environmental Protection and the Chinese Academy of Sciences on May 6, 2015, and was officially released on May 22 in the form of Announcement No. 32 of 2015 by the Ministry of Environmental Protection and Chinese Academy of Sciences.

Since 2016, the China's vertebrate red list working group has organized Chinese researchers to reassess the diversity and the status of Chinese vertebrates. The working group completed the *China's Red List of Biodiversity: Vertebrates* (2021). The assessment found that China's vertebrate survival situation is still grim, and the risk of extinction of Chinese vertebrates is higher than the world average. Various groups of vertebrates, including mammals, birds, amphibians, reptiles and fishes in China have found species of Extinct in the Wild or Regionally Extinct, and many species are in a state of Critically Endangered, Endangered and Vulnerable.

China is in the process of rapid development. China has the largest human population, complex landforms and huge differences in regional development levels. How to save these endangered species is a difficult task for China's biodiversity conservation. The Chinese government attaches great importance to the protection of biological diversity, and has signed and actively implemented international conventions such as the *Convention on Biological Diversity*, the *Convention on International Trade in Endangered Species of Wild Fauna and Flora*, and the *Convention on Wetlands of International Importance, Especially as Waterfowl Habitats* (*Convention on Wetlands* for short). China has vigorously established protected areas with the nature reserves as the main body and national parks as the leading part. At present, China has set up 11,800 protected areas of various types, accounting for more than 18% of the country's land area. There are 474 national nature reserves. The nature reserves protect 90.5% of terrestrial ecosystem types, 85% of wildlife species, and 65% of higher plant communities. On the other hand, China's forest coverage is increasing year by year, laying a sound foundation for the population and habitat restoration of endangered species.

Biodiversity research is not only a comprehensive research, but also a combined study. The inventory of biological species and the assessment of threatened status require joint efforts of different disciplines. For the scientific research and conservation of biodiversity, scholars from different disciplines came together to complete the red list of China's biodiversity. The study is an important part of China's overall biodiversity research. All volumes of *China's Red List of Biodiversity*: *Vertebrates* are published before the fifteenth meeting of the Conference of the Parties to the *Convention on Biological Diversity* which will be held in Kunming, China. The set of books will provide basic data for the biodiversity conservation in China, for monitoring the current situation of biological diversity, for accumulating parameters to conduct periodic IUCN red list index research, and is also an outcome of the Chinese conservation biology research.

Yiyu Chen

Member of the Chinese Academy of Sciences

Former Director of National Natural Science Foundation of China

Director of Endangered Species Scientific Commission, P. R. China

June 26, 2020

序　二

1948 年，在法国枫丹白露举行的一次由 23 个政府、126 个国家组织和 8 个国际组织参与的国际会议上，世界自然保护联盟（International Union for the Protection of Nature，IUPN）成立了。当时，这个组织没有财政来源、没有长期预算，甚至没有永久雇员，但是 IUPN 成为世界政府与非政府组织（Governmental and Nongovernmental Organization，GONGO）的发端。世界自然保护联盟成立后的第一个重大举措是在 1950 年建立了“生存服务（Survival Service）机构”。“生存服务机构”利用当时筹集到的 2,500 美元，召集全球的科学家、志愿者为全球濒危物种撰写评估报告，要求各国政府保护其境内的濒危物种。1964 年，世界自然保护联盟正式发布《濒危物种红皮书》（*Endangered Species Red Book*）。今天，世界自然保护联盟已经完全改变了它自己，包括其名称也改变为 International Union for Conservation of Nature（IUCN）。IUCN 已经成为联合国的观察员、世界范围内主要保护组织。IUCN“生存服务机构”已经演化为物种存续委员会（Species Survival Commission，SSC），IUCN《濒危物种红皮书》也演化为《IUCN 受威胁物种红色名录》。现在，IUCN 物种存续委员会每年发布《IUCN 受威胁物种红色名录》。《IUCN 受威胁物种红色名录》已发展成为世界上关于动物、植物和真菌物种全球保护状况最全面的信息源。

世界各国是生物多样性保护的主体。一个物种的 IUCN 红色名录等级并不一定等同于其在一个国家红色名录中的等级，除非这一物种是该国特有的物种。于是，世界各国也在制定各自的濒危物种红色名录。通过濒危物种红色名录的研究，各国对其境内分布的植物与动物物种的分布、生存状况和保护状况进行调查，然后，对物种的生存状况进行全面评估。因此，濒危物种红色名录是一份物种及其分布的清单，是对物种生存状况、保护状况的客观评估，是生物多样性健康状况的一个重要指标。

濒危物种红色名录被各国政府、自然保护地与野生动植物主管部门、与保护有关的非政府组织、研究人员、自然资源规划人员、教育机构使用。红色名录为生物多样性保护和政策变化提供了信息和促进行动的有力工具，对保护我们赖以生存的自然资源至关重要。通过网络应用，濒危物种红色名录也成为濒危物种信息库，成为保护工作者与研究人员的工具。

中国是 IUCN 的成员，中国也是联合国《生物多样性公约》的最早缔约方之一。中国一直走在生物多样性保护的前沿。中国还是世界上生物多样性最丰富的国家之一，有 7,300 余种脊椎动物，约占全球脊椎动物总数的 11%。中国动物区系组成复杂，空间分布格局差异显著，起源古老，拥有生物演化系统中的各种类群，如有“活化石”之称的大熊猫（*Ailuropoda melanoleuca*）、白鳖豚（*Lipotes vexillifer*）和扬子鳄（*Alligator sinensis*）等。此外，中国还是许多家养动物的起源中心。中国也是生物多样性受威胁最严重的国家之一。人类活动造成的资源过度利用、生境丧失与退化、环境污染及气候变化等因素导致脊椎动物多样性受到严重的威胁。

近年来，党中央和国务院高度重视生物多样性保护工作，将生物多样性保护上升为国家战略，发布了《中国生物多样性保护战略与行动计划（2011－2030 年）》，建立了生物物种资源保护部际联席会议制度，成立了中国生物多样性保护国家委员会，制定和实施了一系列生物多样性保护规划和计划，取得了积极进展。然而，中国生物多样性下降的总体趋势尚未得到有效遏制，保护形势依然严峻，特别是由于目前对中国物种受威胁状况缺乏全面的了解，影响了生物多样性的有效保护。因此，评估物种的受威胁状况，制定红色名录，从而提出针对性的保护策略，对于推动实施《中国生物多样性保护战略与行动计划（2011－2030 年）》和生态文明建设具有重要意义。

到目前，在中国科学家的努力下，中国哺乳动物、鸟类、两栖动物、爬行动物、淡水鱼类都得到了全面的评估。除了评估新发现的物种，中国濒危脊椎动物红色名录还重新评估了一些现存物种的状况，如大熊猫、藏羚（*Pantholops hodgsonii*）等物种，由于中国的保护努力，这些物种的中国濒危脊椎动物红色名录濒危等级下降。然而，中国生物多样性濒危局面仍然严峻。

尽管中国受威胁物种的比例很高，但中国政府正在加强生态环境保护，加强自然保护区、国家公园、世界遗产地及其他类型的保护地建设，加强荒漠化治理、湿地恢复、植树造林，努力扭转或至少制止生物多样性的下降。《生物多样性公约》第 15 次缔约方大会即将在中国昆明召开之际，中国科学家发表最新版《中国生物多样性红色名录：脊椎动物 第一卷 哺乳动物》、《中国生物多样性红色名录：脊椎动物 第二卷 鸟类》、《中国生物多样性红色名录：脊椎动物 第三卷 爬行动物》、《中国生物多样性红色名录：脊椎动物 第四卷 两栖动物》和《中国生物多样性红色名录：脊椎动物 第五卷 淡水鱼类》，全面更新了中国脊椎动物生存状况与种群和栖息地保护状况，从而为确定哪些物种须有针对性地努力恢复，为确定须保护的关键种群和栖息地提供了依据，有助于鉴别未来的濒危脊椎动物保护重点。这套图书的出版是中国自然保护史上的一件大事。

IUCN 主席

2020 年 6 月 26 日

Foreword II

The International Union for the Protection of Nature (IUPN) was established in Fontainebleau of France in 1948 at an international conference that 23 governments, 126 national organizations and 8 international organizations participated in. At that time, the organization had no financial resources, no long-term budget or not even a permanent employee, but it marked the born of the first world Governmental and Nongovernmental Organization (GONGO). Its first move was to establish the "Survival Service" in 1950. Using the $2,500 it raised at the time, the Survival Services called on scientists and volunteers from all around the world to write assessments of the world's endangered species, asking governments to protect those species within their borders. In 1964, the IUCN officially released the *Endangered Species Red Book*. Today, the International Union for Conservation of Nature (IUCN) has completely changed itself, including its name. The IUCN has become an observer of the United Nations and a leading conservation organization worldwide. IUCN Survival Service has evolved into the Species Survival Commission (SSC), the IUCN *Endangered Species Red Book* has been expanded into the website of *IUCN Red List of Threatened Species*, which is now renewed annually by the IUCN Species Survival Committee. The *IUCN Red List of Threatened Species* is the world's most comprehensive source of information on the status of global conservation of animal, plant and fungal species.

Sovereignty countries in the world are the main body of biodiversity protection. The status of a species in IUCN red list is not the affirmatively same in a country's red list except that the species is an endemic species in that country; countries around the world are also developing their own red lists of endangered species. Through the study of the red list of endangered species, countries conduct surveys on the distribution, survival and conservation status of plant and animal species in their territory and then conduct a comprehensive assessment of the survival status of species. Therefore, the red list of endangered species is a list of species and their distribution, an objective assessment of the survival and conservation status of species, and an important indicator of the health status of biodiversity. The red list is used by governments, natural protected areas and wildlife authorities, conservation NGOs, researchers, natural resource planners and educational institutions. It provides information and powerful tools for promoting action on biodiversity conservation and policy formation and is critical to safely guarding the natural resources on which we depend. Through the internet, the red list of endangered species has also become an information base of endangered species and a tool for conservation workers and researchers.

China is a member of the IUCN and one of the earliest parties to the UN *Convention on Biological Diversity*. China has been at the forefront of biodiversity conservation. China is also one of the most biodiverse countries in the world, with more than 7,300 vertebrate species, accounting for about 11% of the total number of vertebrates in the world. China has complex fauna composition, significant differences in spatial distribution pattern, ancient origins and various groups in the biological evolution system, such as Giant Panda (*Ailuropoda melanoleuca*), Baiji (*Lipotes vexillifer*) and Yangtze Alligator (*Alligator sinensis*). In addition, China is the origin center of many domestic animals and plants. China is also one of the countries where biodiversity is most threatened. Due to the overuse of resources, habitat loss and degradation caused by human activities, environmental pollution, climate change and other factors, vertebrate diversity is seriously threatened.

During recent years, the CPC Central Committee and the State Council attach great importance to the protection of biodiversity. Biodiversity conservation is announced as the national strategy, *China's Biodiversity Conservation Strategy and Action Plan* (*2011-2030*) is issued, the Joint Inter-Ministerial Meeting for Biological Species Resources Protection is regularly held, the China National Committee for Biodiversity Conservation has been set up, a series of biological diversity protection programs and plans have been formulated and implemented, and positive progress in the field has been made. However, the overall trend of biodiversity decline in China has not been effectively stopped, and the conservation situation is still pressing, especially, lack of comprehensive understanding of threatened species in China has hindered the effective conservation of biodiversity. Therefore, it is of great significance to assess the threatened status of species and to formulate the red list of endangered species, and to propose targeted conservation strategies for promoting the implementation of *China's Biodiversity Conservation Strategy and Action Plan* (*2011-2030*) and the construction of ecological civilization.

Now, the status of Chinese mammals, birds, reptiles, amphibians and freshwater fishes have all been comprehensively assessed by Chinese scientists. In addition to assessing newly discovered species, China's red list of endangered vertebrates has also reassessed the status of some species, including the Giant Panda and Tibetan Antelope (*Pantholops hodgsonii*), whose status on the red list of endangered vertebrates in China has been downgraded due to conservation efforts. However, the overall situation of endangered biodiversity in China is still serious.

Despite the high proportion of threatened species in China, the Chinese government is strengthening ecological protection, stepping up the construction of nature reserves, national parks, World Heritage Sites and other protected areas, working on desertification control, wetland restoration and afforestation, and trying to reverse or at least stop the trend of biodiversity decline. On the occasion that the fifteenth meeting of the Conference of the Parties to the *Convention on Biological Diversity*, which will be held in Kunming, China, Chinese scientists published the latest edition of the "*China's Red List of Biodiversity: Vertebrates, Volume I, Mammals*", "*China's Red List of Biodiversity: Vertebrates, Volume Ⅱ, Birds*", "*China's Red List of Biodiversity: Vertebrates, Volume Ⅲ, Reptiles*", "*China's Red List of Biodiversity: Vertebrates, Volume Ⅳ, Amphibians*" and "*China's Red List of Biodiversity: Vertebrates, Volume Ⅴ, Freshwater Fishes*". This set of books comprehensively update the survival status, population and habitat protection of vertebrate, determine the recovery efforts needed for the targeted species, identify key populations and habitats that need to be protected, thus provide a basis for identifying future priorities for endangered vertebrates conservation. Publication of these books is an important event in the history of Chinese nature conservation.

Xinsheng Zhang

President of IUCN,

the International Union for Conservation of Nature

June 26, 2020

总前言

物种的濒危现状和濒危机制是保护生物学的核心研究内容，其研究目标是评估人类对生物多样性的影响，提出防止物种灭绝及保护的策略，通过保护生物物种的种群和栖息地，避免物种受到灭绝的威胁。保护生物学研究既关注全球性问题，又具有鲜明的地域特色。中国具有世界上最多的人口，国土面积为世界第三，监测和评估其生物多样性、保护濒危物种，将为中国实现可持续发展提供科学支撑。中国研究人员通过濒危物种红色名录研究，量化了物种灭绝风险，预警了潜在的生态危机，为中国履行《生物多样性公约》等提供科技支撑。

世界自然保护联盟（International Union for Conservation of Nature，IUCN）成立后的第一个重大举措是在1950年建立了“生存服务（Survival Service）机构”。“生存服务机构”利用当时募集的2,500美元，召集科学家、志愿者评估全球濒危物种灭绝风险，发表有灭绝风险的物种研究报告，呼吁各国政府保护其境内的濒危物种，这是《IUCN受威胁物种红色名录》的发端。直到1964年，世界自然保护联盟才正式发布《濒危物种红皮书》。今天，世界自然保护联盟完全改变了它自己，“生存服务机构”已经演化成为物种存续委员会（Species Survival Commission，SSC）。IUCN《濒危物种红皮书》已经演变为网络版的《IUCN受威胁物种红色名录》。物种存续委员会从不定期发布IUCN濒危物种红皮书发展到现在每年发布更新的《IUCN受威胁物种红色名录》。

《IUCN受威胁物种红色名录》是世界生物多样性健康状况的重要指标。它是世界上最全面的一份动物、植物和真菌物种濒危状况清单。濒危物种红色名录基于物种种群数量、种群数量下降速率、生境破碎程度、生境面积及下降速率、预测灭绝概率等指标估测物种灭绝概率。《IUCN受威胁物种红色名录》（2020-2）发现地球上的32,000多个物种面临着灭绝的风险，占所有被评估物种的27%。其中，41%的两栖类、26%的哺乳类、34% 的针叶树、14% 的鸟类、30% 的鲨鱼、33% 的造礁珊瑚，以及 28% 的特定甲壳类物种面临灭绝风险。《IUCN 受威胁物种红色名录》为保护我们赖以生存的自然资源提供了至关重要的信息。

项目背景

自 1980 年以来，中国经济步入高速发展期。目前，中国已经成为世界第二大经济体。在人口增长、经济发展、全球变化的背景下，中国的生物多样性正面临着前所未有的城镇化、乡村和社会基础设施建设及全球变化的压力，野生生物的生存受到威胁。许多证据显示地球上的生物正面临生物进化中的第六次大灭绝。保护濒危物种是生物多样性保护的核心问题。评估物种濒危等级是生物多样性监测与保护的迫切需要。

虽然《IUCN 受威胁物种红色名录》没有国际法和国家法律的效力，但它是专家对全部物种生存状况的评估，它不仅限于评估濒危物种和明星物种，而是最大限度地涵盖了已知的物种，它不仅仅指导世界范围的濒危物种保护，也是指导生物多样性研究的有用工具。《IUCN 受威胁物种红色名录》对于政府间组织和非政府组织的保护决策及各国自然与自然保护法律法规的制定都产生了重要影响。

综上所述，濒危物种红色名录是物种灭绝风险的测度，IUCN 定期更新其濒危物种红色名录，预警全球物种的生存危机。同时，各国也开展了本国濒危物种红色名录研究。那么，既然已经有《IUCN 受威胁物种红色名录》，为什么还要开展国家濒危物种红色名录研究？

《IUCN 受威胁物种红色名录》与国家濒危物种红色名录都是物种灭绝风险的测度，前者是全球性评估，后者则是依国别的研究，两者的研究空间尺度不同。《IUCN 受威胁物种红色名录》预警了全球物种的濒危状况，为全球生物多样性研究提供了大数据；各国红色名录则确定了各国物种受威胁状况，填补了前者的知识空缺，两份红色名录互为补充。

基于如下原因，应当重视依国别的濒危物种红色名录。①国家是濒危物种保护的行为主体，物种在一个国家的生存状况是确定其保护级别、开展濒危物种保育的依据。②对于仅分布于一个国家的特有物种来说，其按国别的濒危物种红色名录等级即是其全球濒危等级。③《IUCN 受威胁物种红色名录》只提供了全球范围的物种濒危信息，并没有评估每一个国家所有物种的生存状况，特别是一些特有物种和跨越国境分布的物种。一些物种跨越国界分布，全球的生存状况并不反映其在个别国家的生存状况，一些全球无危的物种在其边缘分布区的国家里却是极度濒危的或受威胁的物种。世界各国的物种濒危状况有待各国科学家的研究。对于跨国境分布的物种来说，依国别的濒危物种红色名录等级则确定了该物种在本国的生存状况。④结合《IUCN 受威胁物种红色名录》，依国别的濒危物种红色名录为建立跨国保护地、保护迁徙物种的栖息地与跨国迁徙洄游通道提供依据。⑤依国别的濒危物种红色名录所特有的“区域灭绝”等级，反映了一个物种边缘种群在该国的区域灭绝，对于一个国家来说，事关重大；恢复“区域灭绝”物种是该物种原分布国家重新引入的相关保育工作的重点。⑥物种濒危状况是不断变化的。近年来，新种、新记录不断被发现。随着人们对生命世界认识的深入，脊椎动物分类系统也发生了变化。依国别的濒危物种红色名录提供了该国物种编目、分类、分布和生存状况的最新信息（蒋志刚等，2020）。

国家红色名录的重要性在许多情况下被忽视了。在研究报告和科普作品中，对国家濒危物种红色名录重视不够。论及物种濒危属性时，作者通常言必《IUCN 受威胁物种红色名录》濒危等级而不提其国家级的红色名录濒危等级。目前正值全球新型冠状病毒肺炎大流行，人们正在重新审视人与野生动物的关系。我国将修订有关野生动物保护与防疫法规和法律、重点保护野生物种名录，防控新的人与野生动物共患疾病再次暴发。对于确定《国家重点保护野生动物名录》而言，物种受威胁程度是物种列为国家重点保护野生物种的特征之一。重视依国别的红色名录有特别的意义。于是，生态环境部（原环境保护部）与中国科学院联合开展了中国生物多样性红色名录研究。

中国动物学家掌握了中国动物分布和生存状况的第一手资料，有必要组织全国淡水鱼类、两栖类、爬行类、鸟类与哺乳类专家及时更新中国脊椎动物分类系统，提供中国脊椎动物多样性的全面、完整的信息；有必要应用统一的国际物种濒危等级标准评估物种生存状况。在国家层面，定期组织全国淡水鱼类、两栖类、爬行类、鸟类与哺乳类专家应用 IUCN 受威胁物种红色名录等级标准和 IUCN 区域受威胁物种红色名录标准，全面评估更新的中国脊椎动物生物多样性红色名录，提供与国际红色名录研究可对比的结果，为红色名录指数的研究积累数据。

经过系统评审制定的中国生物多样性红色名录，由国家权威机构发布。中国生物多样性红色名录淡水鱼类、两栖类、爬行类、鸟类与哺乳类各卷将为监测中国生物多样性现状、为开展阶段性 IUCN 红色名录指数研究和履行《生物多样性公约》提供数据。

中国在 1998 年首次出版了《中国濒危动物红皮书》，2004 年，又出版了《中国物种红色名录》，2009 年，环境保护部组织开展了“中国陆栖脊椎动物物种濒危等级评估”。时隔多年，有必要重新全面评估中国生物多样性的濒危状况。于是，环境保护部委托中国科学院组织有关专家开展了“中国生物多样性红色名录——脊椎动物卷”的研究。在环境保护部和中国科学院的领导下，我们依据 IUCN 受威胁物种红色名录等级标准和 IUCN 区域受威胁物种红色名录标准，全面评估了中国哺乳动物生存状况。

2015 年，环境保护部与中国科学院联合发布了“中国生物多样性红色名录——脊椎动物卷”。现在，历时 6 年，我们全面编研、更新、丰富了此名录，形成了此 2021 版的《中国生物多样性红色名录：脊椎动物》。

项目目标

通过脊椎动物各类群的研究，收集整理中国脊椎动物现有物种种群、生境研究数据、资源监测数据，充实数据库；组织专家，采用综合分析和专家评估相结合的方法，依据中国脊椎动物野生种群与生境现状，利用 IUCN 受威胁物种红色名录等级标准第 3.1 版和 IUCN 区域受威胁物种红色名录标准第 4.0 版综合评价中国脊椎动物濒危状况，编制 2021 版《中国生物多样性红色名录：脊椎动物》。全面评价中国脊椎动物的灭绝风险，对中国濒危物种保护及时提供基础信息。

编研过程

2013 年 5 月 16 日，在中国科学院动物研究所召开了研究启动会。项目聘请陈宜瑜院士、郑光美院士、张亚平院士、金鉴明院士、马建章院士、曹文宣院士为咨询专家，并成立了哺乳类、鸟类、

爬行类、两栖类、淡水鱼类课题组。各课题组就评估程序和规范展开了研讨，对典型物种进行了评估并听取了专家委员会的意见。会后总结了专家意见，完善了中国脊椎动物红色名录评估程序和规范。

针对哺乳类、鸟类、爬行类、两栖类和淡水鱼类分别建立了工作组、核心专家组和咨询专家组。工作组负责按照预定的红色名录判定规程开展工作，工作包括资料收集与整理、红色名录初步评定、与通讯评审专家联络及通讯评估结果汇总。核心专家组对红色名录评估的方法、标准使用、数据来源等重要科学问题进行界定，讨论审核有关物种的受威胁等级。工作组在全国范围遴选咨询专家，建立咨询专家库。咨询专家参加了红色名录的通讯评审和会议评审。评审结束后，工作组按照统一格式，整理每个物种包含的信息，形成最终的物种评估说明书。物种评估说明书的内容包括物种的学名、中文名、评估受威胁等级及 IUCN 红色名录等级。“中国生物多样性红色名录——脊椎动物卷”于 2015 年 5 月 6 日通过环境保护部和中国科学院的联合验收，并于 5 月 22 日以环境保护部、中国科学院 2015 年第 32 号公告形式发布。

红色名录评估的信息来源主要有研究积累、标本数据、文献数据和专家咨询。项目各课题组相关研究团队是工作在中国淡水鱼类、两栖类、爬行类、鸟类和哺乳类研究一线的研究团队，在数十年的研究中积累了大量的科学数据，各分卷主持人还是国家濒危物种科学机构， 以及淡水鱼类、两栖类、爬行类、鸟类和哺乳类学术团体的骨干，所在单位是有关动物物种分类、标本收藏、研究的信息交换所，并各自建立了数据库。各分卷主持人还主持或参与了国家有关物种资源本底调查、科学评估、自然保护区生物多样性考察及相关的保护政策制定。

实践意义

《中国生物多样性红色名录：脊椎动物》的出版是一项重大的系统工程。这次生物多样性红色名录评估是迄今评估对象最广、涉及信息最全、参与专家人数最多的一次评估。通过 2015 版红色名录研究，我们更新了中国脊椎动物编目。中国有 2,854 种陆生脊椎动物，其中，有 407 种两栖类，402 种爬行类，1,372 种鸟类，673 种哺乳类。在 2021 版《中国生物多样性红色名录：脊椎动物》的编研中，我们再次更新了中国脊椎动物分类系统和编目。中国有 3,147 种陆生脊椎动物，其中，有 475 种两栖类，527 种爬行类，1,445 种鸟类，700 种哺乳类，比 2015 年的统计数据增加了 293 种。我们发现，中国是全球哺乳动物物种数最多的国家。中国陆生脊椎动物中，特有种超过 20%。我们还分析了中国脊椎动物的分布格局和特有类群，探讨了其濒危种类的空间分布规律。

中国濒危脊椎物种濒危模式与分布格局在我国生物多样性和生态系统保护中具有指导意义，这将为我国重点保护物种确定、国土空间开发和生态功能区的划分及各类保护地规划设计提供重要参考依据。也是确定中国物种多样性保护热点的依据之一。我们发现，中国脊椎动物生存危机依然严重，中国濒危脊椎物种的分布格局不均衡。物种的空间分布是一种立体格局，除了水平纬度上的物种分布格局，我们也需要物种多样性和濒危种类的垂直分布格局，这些格局对物种多样性保护具有重要参考价值。我们发现，高海拔地区受威胁哺乳动物的比例比低海拔地区高，高海拔地区的濒危物种应受到更多的关注。

展望与致谢

《中国生物多样性红色名录》的编制和发布为生物多样性保护政策和规划的制定提供了科学依据，发挥了中国科学家作为中国《生物多样性公约》履约“智库”的功能，同时，为开展生物多样性科学基础研究积累了基础数据，更新了脊椎动物分类系统与编目，为公众参与生物多样性保护创造了必要条件。《中国生物多样性红色名录》的编制是贯彻实施《中国生物多样性保护战略与行动计划（2011—2030 年）》和积极履行《生物多样性公约》的具体行动。通过《中国生物多样性红色名录》的编制，中国在生物多样性评价方面已经在全球先行一步，使我国在履行《生物多样性公约》方面走在世界的前列。

本项目得到了生态环境部（原环境保护部）、中国科学院、国家林业与草原局（原国家林业局）、中国科学院大学、科学出版社的关怀、指导和大力支持；得到了国家出版基金的大力支持。课题组还得到了如下项目的资助：中国科学院战略性先导科技专项（A 类）“地球大数据科学工程”（项目编号：XDA19050204）、国家重点研发计划项目（项目编号：2016YFC0503303）、国家科技基础性工作专项（项目编号：2013FY110300）的资助。在此谨致感谢！

蒋志刚
中国科学院动物研究所研究员
中国科学院大学岗位教授
国家濒危物种科学委员会前常务副主任

2020 年 6 月 6 日

Series' Preface

The current status and threats to species are the key issues in conservation biology. The primary goal of putting forward an endangered species red list is to evaluate human impact on biodiversity, identifying key threats and preventing species extinction by protecting the populations and habitats of threatened species. Conservation research not only pays attention to global issues, but also must focus attention on regional and national problems. China has the largest human population and the third largest terrestrial area in the world. Monitoring and evaluating the country's biodiversity and protecting endangered species will provide scientific support for China's sustainable development. Therefore, Chinese researchers are working to quantify the risk of extinctions through studies related to the red list of threatened species, providing early warning about potential ecological hazards, thus offering scientific support for implementation of the *Convention on Biological Diversity* in China.

The first major move for the International Union for Conservation of Nature (IUCN) after its establishment was to launch the Survival Service in 1950. Using the $2,500 raised at that time, the Survival Service called on scientists and volunteers to dedicate their expertise and time to assess the extinction risk of globally threatened species. The Survival Service then publicized their research reports on species at risk of extinction and called on governments to protect endangered species within their borders. Such an act marked the beginning of the *IUCN Red List of Threatened Species*. However, it was not until 1964 that IUCN officially published its first *Endangered Species Red Book*. Today, IUCN has completely changed itself, the Survival Service has been renamed as the Species Survival Commission (SSC). The IUCN

Endangered Species Red Book has evolved into the online version of *IUCN Red List of Threatened Species*. The Species Survival Commission refreshes and revises the *IUCN Red List of Threatened Species* periodically, and updates the *IUCN Red List of Threatened Species* annually.

The *IUCN Red List of Threatened Species* is an important indicator of the health of the world's biodiversity. It is the world's most comprehensive list of rare and threatened animal, plant and fungal species. The *IUCN Red List of Threatened Species* estimates the extinction probability of species based on population size, population decline rate, degree of habitat fragmentation, rate of decline of habitat area and other indicators. The *IUCN Red List of Threatened Species* (2020-2) estimated that more than 32,000 species on the earth are at risk of extinction, accounting for 27% of all assessed species globally. 41% of amphibians, 26% of mammals, 34% of conifers, 14% of birds, 30% of sharks, 33% of corals, and 28% of certain crustaceans are presently at risk of extinction. The *IUCN Red List of Threatened Species* provides vital information for protecting the biodiversity and natural resources on which we all collectively depend.

The Background

Since the 1980s, China has embarked on a fast track of socioeconomic development. China has become the world's second largest economy. Against a backdrop of population growth, rapid economic development and many global changes, China's biodiversity is under unprecedented pressure from urbanization, infrastructure development and a wide range of other factors, and the survival of wildlife is under threat. Ample evidence shows that life on the earth is facing its Sixth Mass Extinction in its long evolutionary history. Protecting endangered species is the core issue for biodiversity conservation. Thus, assessing the endangerment level of species is the primary and most urgent need that biodiversity monitoring and protection measures seek to address.

Though the *IUCN Red List of Threatened Species* does not possess the power of international or national laws, it is an expert assessment of the survival status of all species, not only endangered or charismatic species, but all known species to the greatest extent possible. It thus serves as a most useful tool not only to guide worldwide protection of endangered species but also for the study of biodiversity. The *IUCN Red List of Threatened Species* has significant impact on the conservation decisions of intergovernmental and non-governmental organizations as well as for the formulation of national laws and regulations regarding wildlife and nature conservation.

As stated above, the red list of threatened species provides a measure of the risk of extinction of species. IUCN regularly updates its global red list of endangered species in order to raise public awareness of the global status of wildlife and the species survival crisis. At the same time, countries also conduct national-level studies on the status of endangered species. However, since there is already an *IUCN Red List of Threatened Species*, the question may arise, why bother to conduct research at country level to produce national red lists of endangered species?

Both the *IUCN Red List of Threatened Species* and country red lists of threatened species assess species' risk of extinction, with the former being global in scope while the latter are regional assessments. The

IUCN Red List of Threatened Species alerts the world to the status of endangered species, and also serves as a database of global biodiversity. Country red lists, on the other hand, ascertain the status of species in particular countries, filling knowledge gaps in the former. The two lists are thus complementary to each other.

Country-level red lists should be given greater attention for at least the following reasons: (i) A sovereign country is the main authority for taking conservation action in regard to wildlife species within its boundaries, based on the level of endangerment (conservation status) of the species; (ii) For endemic species in a country, the country red list status constitutes its global status; (iii) The *IUCN Red List of Threatened Species* provides only the information on species at risk worldwide and does not assess the status of all species in each country, especially endemic species and those species with transboundary distribution. Some species are distributed across national boundaries and the global conservation status does not entirely reflect the survival status of the species in any particular country. Some global non-threatened species are critically endangered or threatened in the countries where they have peripheral ranges. The endangered status of species in different countries of the world thus remains to be studied by scientists in relation to specific countries. For species whose ranges cross national borders, the country's red list status reflects the survival status of the species in the country; (iv) Combined with the global *IUCN Red List of Threatened Species*, country red lists provide a basis from which to consider the establishment of transnational protected areas, the protection of important habitats for migratory species, and the protection of international migration corridors; (v) The category "Regionally Extinct" is unique to country (regional) red lists of endangered species as it refers only to a subset of the broader geographic distribution of the species, yet the national status is still indicative of the species' overall risk of extinction, this matters a lot for a country, and the restoration of "regionally extinct" species is the focus of conservation efforts for reintroduction in countries where the species originated; (vi) Country red lists provide updated information about endangered species with national inventories as well as with national reviews of classification, geographic distribution, and status of species at national level, which are also relevant for global species descriptions and assessments (Jiang *et al.*, 2020).

Despite these benefits, the significance of country-level red lists is often overlooked. Following onset of the global COVID-19 pandemic, however, people's outlook has been changing in regard to the relationship between people and wildlife. Consequently, China is amending its national laws on wildlife protection, epidemic prevention, and the list of state key protected wild species, in order to better prevent and control emerging zoonoses. The status of wildlife species included in China's red list of threatened species should be one of the defining elements for identifying and updating species on the *List of State Key Protected Wild Animal Species* in China. It is therefore critical to duly recognize the significance of the country red list at this special moment in time. For this purpose, the Ministry of Ecology and Environment (former Ministry of Environmental Protection) and the Chinese Academy of Sciences have jointly launched China's biodiversity red list.

Chinese zoologists have obtained first-hand information on the distribution and living status of animals in China. It is necessary to organize national experts on freshwater fishes, amphibians, reptiles, birds and

mammals to update the taxonomy of vertebrates in China in a timely manner and to provide systematic and comprehensive information on the diversity of vertebrates in China. It is necessary to apply standard international criteria for threatened species to assess the status of species. At the national level, it is necessary to coordinate national experts on freshwater fishes, amphibians, reptiles, birds and mammals to apply the IUCN red list criteria for threatened species and the IUCN regional red list criteria for threatened species, and through this process also to comprehensively update the *China's Red List of Biodiversity: Volume of Vertebrates* and thus to provide a country red list that is comparable to international red lists, and to enable index studies of red lists.

The red list of China's biodiversity, which has been systematically reviewed and formulated, shall be issued by the state authorities. The volumes of freshwater fishes, amphibians, reptiles, birds and mammals of the red list of china's biodiversity will provide data for the implementation of the *Convention on Biological Diversity*, for monitoring the state of biodiversity in China, as well as for conducting periodic IUCN red list index studies in the country.

China firstly published its *China Red Data Book of Endangered Animals* in 1998, followed by the *China Species Red List* in 2004 and in 2009, the Ministry of Environmental Protection coordinated the assessment and publishing of the "Assessment of the Red List of Endangered Species of Terrestrial Vertebrates in China", which is a multi-year project for the comprehensive re-assessment of the threatened status of China's biodiversity. Therefore, the Ministry of Environmental Protection entrusts the Chinese Academy of Sciences to organize experts to carry out research on the *China's Red List of Biodiversity: Volume of Vertebrates*. Under the leadership of the Ministry of Environmental Protection and the Chinese Academy of Sciences, we have conducted a comprehensive assessment of the living status of the vertebrates in China based on the IUCN red list criteria for threatened species and the IUCN regional red list criteria for endangered species.

In 2015, the Ministry of Environmental Protection and the Chinese Academy of Sciences jointly released the *China's Red List of Biodiversity*: *Volume of Vertebrates*. Now, six years on, we have thoroughly updated and compiled the series of books of *China's Red List of Biodiversity: Vertebrates* (2021).

The Goal

Through the study and preparation for each volume of vertebrates in China's biodiversity red list, we collected and sorted existing information on the population and habitat status of vertebrates in China and completed the database. We systematically and comprehensively evaluated the status of vertebrates in China, using the *IUCN Red List Categories and Criteria* (version 3.1) and the *Guidelines for Application of IUCN Red List Criteria at Regional and National Levels* (version 4.0) based on the status of the species' wild population and habitat. Combined with the empirical analysis and expert evaluation, we compiled the *China's Red List of Biodiversity: Vertebrates* (2021), which is a comprehensive assessment of extinction risk of China's vertebrates, providing the basic information pertinent for the protection of endangered species over the coming years.

The Assessment

The project launch meeting was held at the Institute of Zoology, Chinese Academy of Sciences on May 16, 2013. Academicians Yiyu Chen, Guangmei Zheng, Yaping Zhang, Jianming Jin, Jianzhang Ma and Wenxuan Cao were invited to participate in the meeting as consulting experts. Mammals, birds, reptiles, amphibians and freshwater fishes research groups were formed at the meeting. Each research group held a discussion on evaluation procedures and norms, assessed the typical species of their own taxonomic group, and consulted the opinion of the expert committee. After the meeting, all experts' opinions were summarized and the evaluation procedures and norms of the red list of vertebrates in China were finalized.

Working groups, core expert groups and communication expert groups were formed for each research group, focused respectively on mammals, birds, reptiles, amphibians and fresh water fishes. The working groups were responsible for carrying out work in accordance with the red list category assessment procedures, including data collection and classification, preliminary red list category evaluation, liaison with experts by correspondence, and providing summaries and communicating evaluation results. The core expert group defined the methods, standards, data sources and other important scientific issues of the red list assessment, discussed and reviewed the status of species. The working group selected consulting experts nationwide and established a database of consulting experts. Each consulting expert participated in the red list evaluation and conference review. After the review, the information about every species was sorted and summarized in a unified format to provide the final species evaluation specifications. The species description and assessment includes scientific name, Chinese name, threat level assessment and IUCN red list category criteria. The "*China's Red List of Biodiversity: Volume of Vertebrates*" was jointly approved by the Ministry of Environmental Protection and the Chinese Academy of Sciences on May 6, 2015, and officially released on May 22, 2015, through the Announcement No. 32 of the Ministry of Environmental Protection and the Chinese Academy of Sciences, on International Biodiversity Day of 2015.

The information sources for the evaluation of vertebrates in the red list of China's biodiversity included published and unpublished literature, specimen data, and expert consultation. Experts from the red list working groups for freshwater fishes, amphibians, reptiles, birds and mammals are experts who have accumulated a large amount of scientific data over decades. The coordinators of working groups are people from state endangered species scientific authorities and established academics from scientific communities focused on freshwater fishes, amphibians, reptiles, birds and mammals from across the country. The research institutions are the centers of taxonomy, specimen collections, and databases for animal species. The principal scientists of freshwater fishes, amphibians, reptiles, birds and mammals also often coordinated or participated in background investigations on national species resources, scientific assessments, biodiversity investigations of nature reserves, and the formulation of relevant government conservation policies.

The significance

The publishing of the *China's Red List of Biodiversity*: *Vertebrates* (2021) is a major systematic project. The

biodiversity red list assessment covered the widest range of subjects, providing the most complete information and involving the largest number of experts so far in the country. During the process of developing the 2015 edition of the red list, we updated the Chinese vertebrate inventory. On this basis, it was found that China has 2,854 terrestrial vertebrates, of which 407 are amphibians, 402 are reptiles, 1,372 are birds and 673 are mammals. In the preparation and research of the 2021 edition, we have once again updated the classification system and produced an updated inventory of vertebrates in China. Altogether there are 3,147 terrestrial vertebrates in China, among which there are 475 species of amphibians, 527 reptiles, 1,445 birds and 700 mammals. A further 293 vertebrate species were assessed for preparing the 2021 edition. China is now found to have the largest number of mammal species in the world. Among land vertebrates in China, more than 20% are endemic. We also analyzed the distribution pattern and endemic groups of vertebrates in China and discussed the spatial distribution pattern of endangered species.

The conservation status and distribution patterns of threatened vertebrates in China that are shared in this book provide an important reference for identification of key protected vertebrate species, planning and development of national strategic land use blueprints, the design of ecological functional zones and various protected sites, which are of great significance for biodiversity and ecosystem conservation in China. This information is also one of the criteria for determining hotspot locations for species diversity in China. We have found that the survival crisis of vertebrates is still present in China and the distribution pattern of endangered vertebrates remains uneven. The spatial distribution of species presents a three-dimensional pattern, including their geographic distribution (two dimensions) as well as vertical or elevational dimension where species including threatened species are generally situated. In particular, we found that the proportions of threatened mammals in different families and orders were greater at higher altitudes than those found at lower altitudes, and additionally we found that endangered species that live at higher altitudes often should receive more attention from the public as well as from the government.

The Outlook and Appreciation

The compiling and publishing of *China's Red List of Biodiversity* provides a scientific basis for biodiversity conservation planning and policy in China, based on the long-standing work of Chinese scientists, who constitute a *de facto* "think-tank" for research and implementation of the *Convention on Biological Diversity* in China. At the same time, the red list study has updated the vertebrate taxonomy and inventory in China, accumulated data for basic zoological research in biodiversity science both nationally and globally, and created the necessary conditions for public participation in biodiversity conservation. Producing the *China's Red List of Biodiversity* has been a concrete action in the implementation of *China's Biodiversity Conservation Strategy and Action Plan* (*2011-2030*) and has also been a key step in implementing the *Convention on Biological Diversity* in its territory. Through the compilation of the *China's Red List of Biodiversity*, China has demonstrated its leading role in the global assessment of the current status of biodiversity.

The project that enabled development and publication of this red list book received guidance and support from the Ministry of Ecology and Environment (former Ministry of Environmental Protection), the Chinese

Academy of Sciences, the National Forestry and Grassland Administration (former National Forestry Administration), the University of Chinese Academy of Sciences, and the Science Press (Beijing). The project received funding from the National Publication Foundation, and each individual research group also received support through the following projects: "Earth Big-Data Scientific Project" (XDA19050204) of the Strategic Leading Science and Technology Project, Chinese Academy of Sciences (Category A); National Key Research and Development Project (2016YFC0503303); Basic Science Special Project of the Ministry of Science and Technology of China (2013FY110300). We express our most sincere gratitude to all of these governmental bodies, institutions and funding agencies for their many different forms of support.

Zhigang Jiang, Ph.D.

Professor of Institute of Zoology, Chinese Academy of Sciences

Professor of University of Chinese Academy of Sciences

Former Executive Director of the Endangered Species Scientific Commission, P. R. China

前 言

中国哺乳动物区系有鲜明的特色：中国是世界上哺乳动物种类最多的国家之一，也是特有哺乳动物丰富的国家。进入 21 世纪后，中国哺乳动物研究得到了长足的发展。由于种种原因，中国的哺乳动物编目工作一直到 21 世纪初才由王应祥先生完成。王应祥先生在《中国哺乳动物种和亚种分类名录与分布大全》一书报道中国有哺乳动物 603 种。从 2008 年起，我们开始评估中国濒危哺乳动物生存状况。我们首先补充了新种与新记录种，删去了无效种，采用最新的哺乳动物分类系统，开展了中国哺乳动物编目研究。我们 2015 年报道了中国有哺乳动物 12 目 55 科 245 属 673 种（蒋志刚等，2015a）。2017 年，我们再次更新了分类系统，补充了新种与新记录种，删去了无效种，更新了这一数据，记录哺乳动物 13 目 56 科 248 属 693 种（蒋志刚等，2017）。在本次《中国生物多样性红色名录：脊椎动物　第一卷　哺乳动物》(2021) 的评估中，经过查遗补缺，再次更新中国哺乳动物为 13 目 56 科 248 属 700 种。我们共评估了除智人外的 700 种哺乳动物。

本书为中英文双语，面向全球读者。为了方便读者提纲挈领，掌握本红色名录研究方法和中国哺乳动物的生存与保护状况。本书在前面部分有介绍中国哺乳动物生存状况评估的“总论”。在总论中，介绍了哺乳动物的演化与现状，中国哺乳动物多样性与保护现状，本次红色名录评估对象的分类系统、分布格局、受保护状况；介绍了中国哺乳动物红色名录的评估过程、依据的评估等级标准，还介绍了咨询专家顾问、评估队伍，以及评估团队建立数据库和开展初步评定、通讯评审、形成评估报告的过程。最后，总结分析了本次红色名录评估结果，比较了《中国生物多样性红色名录：脊椎动物　第一卷　哺乳动物》2021 版与 2015 版的异同，介绍了中国哺乳动物濒危状况，分析了野外灭绝的和区域灭绝的物种，分析了不同哺乳动物类群、不同生境的受威胁哺乳动物物种比例及其省区哺乳动物的分布和濒危原因，并将本次评估结果与《IUCN 受威胁物种红色名录》(2020-2)的评估结果进行了比较分析。最后，分析展望了中国哺乳动物保护成效与远景。

本书“各论”中的物种编排顺序基本按照 IUCN 受威胁物种红色名录濒危等级：“极危”“濒危”“易危”“近危”“无危”“野外灭绝”“区域灭绝”（本书特有）“数据缺乏”排列。“各论”编排列出了每一物种中文名 Chinese Name 和其他物种的分类信息如目 Order、科 Family、学名 Scientific Name、命名人 Species Authority、英文名 English Name(s)、同物异名 Synonym(s)、种下单元评估 Infra-specific Taxa Assessed，并配有彩绘插图或照片。各论还列出了评估信息 Assessment Information、红色名录等级 Red List Category（评估标准版本，已列在物种标题上）、评估年份 Year Assessed、评定人 Assessor(s)、审定人 Reviewer(s) 和其他贡献人 Other Contributor(s)。并对评估对象进行了评估理由 Justification、地理分布 Geographical Distribution（国内分布 Domestic Distribution 和世界分布 World Distribution)，以及是否特有种的分布标注 Distribution Note 描述；还给出了国内分布图 Map of Domestic Distribution 和种群数量 Population Size、种群趋势 Population Trend、所在生境与生态系统 Habitat(s) and Ecosystem(s)，以及威胁 Threat(s)、保护级别与保护行动 Protection Category and Conservation Action(s)［国家重点保护野生动物等级 (2021) Category of National Key Protected Wild Animals (2021)、IUCN 红色名录 (2020-2) IUCN Red List (2020-2)、CITES 附录 (2019) CITES Appendix (2019)，是否开展了“保护行动 Conservation Action(s)”］。最后，列出了相关文献 Relevant References。书末附有检索表和参考文献。

在本书的编研过程中，有关工作得到了中国科学院、生态环境部（原环境保护部）、国家林业和草原局（原国家林业局）、中国科学院大学、国家濒危物种科学委员会、中国科学院动物研究所的精心指导、大力支持与帮助。在本书的编辑出版过程中，得到了科学出版社的细心帮助，得到国家出版基金的资助。在本书的编研过程中，我们执行了原环境保护部生物多样性专项：中国脊椎动物红色名录研究（2012—2018 年），国家科技基础性工作专项（项目编号：2013FY110300)、国家重点研发计划项目（项目编号：2016YFC0503303)、中国科学院战略性先导科技专项（A 类；项目编号：XDA19050204)、“美丽中国”生态文明建设科技工程项目（项目编号：XDA23100203)、全国第二次陆生野生动物资源调查项目、国家自然科学基金项目等项目。

本书的编研历时数年，由于研究团队知识与时间有限，我们在编研过程中深深领会到“吾生也有涯，而知也无涯。以有涯随无涯，殆已”。保护生物学是一门处理危机的学科，濒危物种红色名录是生物多样性预警报告。为了及时拯救濒危物种，尽管缺点、疏漏在所难免，我们仍将这本图书呈示于世，以期有关方面及时采取行动，保护人类赖以生存的基础。希望有关专家、读者对本书存在的问题不吝指正。

蒋志刚

2020 年 6 月 6 日

Preface

China has distinct characteristics in regard to its mammalian fauna: China has amongst the richest mammal diversity in the world, and it also has one of the greatest levels of endemism of mammal species globally. Since the beginning of the new millennium, mammal research in China has made significant progress. For various reasons, the inventory of mammals in China was not completed until the early 21st century. In his book *Taxonomy and Distribution of Mammal Species and Subspecies in China*, Professor Yingxiang Wang reported that there were 603 mammal species in China. Following this, since 2008, we have been systematically assessing the survival status of threatened mammals in China. First, we have added new species and new records of species, deleted invalid species, and adopted the latest mammal taxonomic system to carry out the inventory research of mammals in China. By 2015, we reported 673 species from 245 genera, 55 families and 12 orders of mammals in China (Jiang *et al*., 2015a). In 2017, we renewed the classification system, added several more species and new records of species, deleted additional invalid species, and through this process updated the data and inventory, recording in total 693 species from 248 genera, 56 families, 13 orders of mammals (Jiang *et al*., 2017). In this assessment, in the Mammals Volume of *China's Red List of Biodiversity*: *Vertebrates* (2021), we now recognize a total of 700 species of mammals (in 248 genera, 56 families, 13 orders) in the national inventory and evaluated 700 species except the *Homo sapiens*.

This book is bilingual in both Chinese and English, and is intended for a global audience, aiming to introduce readers to the basic concepts and grasp research methods concerning the red list of threatened species, and more specifically, to main findings from long-term research about the living conditions and conservation status of mammals in China. This book is preceded by a General Introduction regarding the assessment of the survival or conservation status of mammals in China. In the General Introduction, the evolution and current state of mammals are introduced including their diversity, their taxonomy, and their distribution patterns, along with their conservation or red list status. The General Introduction also describes the evaluation process that was adopted for the mammals of China's red list of biodiversity, including the evaluation criteria and consulting experts, and how the evaluation team established the mammal database, carried out preliminary evaluation, reviewed preliminary results by correspondence, and formalized the evaluation report. Finally, the evaluation results are analyzed and summarized for the readers. The authors compared the Mammal Volume of *China's Red List of Biodiversity: Vertebrates* 2021 edition with the 2015 edition, analyzed the status of threatened mammals, including special note of species that are Extinct in the Wild and Regionally Extinct, as well as the proportion of threatened species in different mammal groups, in different habitats and range provinces, and the threats that are faced by endangered mammals. A comparison of the results of mammals in *China's Red List of Biodiversity: Vertebrates* (2021) with the *IUCN Red List of Threatened Species* (2020-2) also is provided. Finally, the implications and prospects for mammal conservation in China are discussed.

In the book, the species are arranged in "Species Monograph" according to their status in the *IUCN Red List of Threatened Species*: "Critically Endangered", "Endangered", "Vulnerable", "Near Threatened", "Not Threatened", "Extinct in the Wild", "Regionally Extinct" (unique only in this book), and "Data Deficient". In each "Species Monograph", with colored illustrations or photographs, the Chinese Name and taxonomy of the species such as: Order, Family, Scientific Name, Species Authority, English Name(s), Synonym(s), Infra-specific Taxa Assessed of the species are given. In each "Species Monograph", Assessment Information, Red List Category (evaluation criteria version, listed in the title), Year Assessed, Assessor(s), Reviewer(s) and Other Contributor(s) are listed, plus the assessment Justification, Domestic Distribution in China and World Distribution, as well as the Distribution Note of whether the Species is Endemic or not. Map of Domestic Distribution, Population Size and Trend, Habitat(s) and Ecosystem(s) of species are also given. And more, "Species Monograph" presents the Threats, Protection Category and Conservation Action taken such as Category of National Key Protected Wild Animals (2021), IUCN Red List (2020-2), CITES Appendix (2019), and Whether there is a "Conservation Action(s)" for species. Finally, Relevant References are listed. At the end of the book, Index and Reference are listed.

During the compilation and research of the book, the work has been carefully guided, supported and helped by the Chinese Academy of Sciences, the Ministry of Ecology and Environment (formerly the Ministry of Environmental Protection), the State Forestry and Grassland Administration (formerly the State Forestry Administration), the University of Chinese Academy of Sciences, the National Endangered Species Scientific Commission, and the Institute of Zoology of the Chinese Academy of Sciences. During the process of editing and publishing this book, we have received help from Science Press in Beijing, and financial support from the National Publication Foundation. During the process of research for the book, we performed the biodiversity special project: China's Vertebrate Red List of 2012–2018 of the former Ministry of Environmental Protection,

Basic Science Special Project of the Ministry of Science and Technology of China (2013FY110300), National Key Research and Development Project (2016YFC0503303), Strategic Leading Science and Technology Project, Chinese Academy of Sciences (Category A, XDA19050204), Science and Technology Project of Ecological Civilization Construction of "The Beautiful China" (XDA23100203), as well as the special projects of the Second National Survey on Terrestrial Wild Animals and projects of the Natural Science Foundation, *etc*.

Although the compilation of this book took several years, due to the limited knowledge and time of the research team, we deeply understood an ancient proverb "My life is limited, and the knowledge is limitless. To use the limited time to explore the boundlessness, it will be exhausted", during the compilation and research process. Conservation biology is a crisis management discipline, and the red list of endangered species is an early-warning report. In order to save endangered species in a timely manner, despite the inevitable shortcomings, we present this book to the world so that action can be taken to protect the very foundation on which humanity depends. I hope that experts and readers will not hesitate to point out the problems in this book.

Zhigang Jiang

June 6, 2020

目 录 Contents

各论（下册） Species Monograph (III)

区域灭绝 RE

野外灭绝 EW

脊椎动物
Vertebrates

各 论（下册）
Species Monograph (III)

洮州绒鼠

Caryomys eva

无危 LC

数据缺乏 DD	无危 LC	近危 NT	易危 VU	濒危 EN	极危 CR	区域灭绝 RE	野外灭绝 EW	灭绝 EX

分类地位 Taxonomic Status

动物界 Animalia	脊索动物门 Chordata	哺乳纲 Mammalia	啮齿目 Rodentia	仓鼠科 Cricetidae

学 名 **Scientific Name**	*Caryomys eva*
命名人 **Species Authority**	Thomas, 1911
英文名 **English Name(s)**	Eva's Red-backed Vole
同物异名 **Synonym(s)**	Eva's Vole; Gansu Vole; Ganzu Vole; Taozhou Vole; *Caryomys alcinous* (Thomas, 1911); *aquilus* (G. M. Allen, 1912)
种下单元评估 **Infra-specific Taxa Assessed**	无 / None

评估信息 Assessment Information

评估年份 **Year Assessed**	2020
评定人 **Assessor(s)**	刘少英、蒋志刚 / Shaoying Liu, Zhigang Jiang
审定人 **Reviewer(s)**	马勇、路纪琪、鲍毅新 / Yong Ma, Jiqi Lu, Yixin Bao
其他贡献人 **Other Contributor(s)**	李立立、丁晨晨 / Lili Li, Chenchen Ding

理由 **Justification:** 洮州绒鼠的种群大。因此，列为无危等级 / Eva's Red-backed Vole has large populations. Thus, it is listed as Least Concern

地理分布 Geographical Distribution

国内分布 Domestic Distribution

陕西、宁夏、甘肃、四川 / Shaanxi, Ningxia, Gansu, Sichuan

世界分布 World Distribution

中国 / China

分布标注 Distribution Note

特有种 / Endemic

国内分布图
Map of Domestic Distribution

种群 Population

种群数量 Population Size	数量多 / Abundant
种群趋势 Population Trend	未知 / Unknown

生境与生态系统 Habitat (s) and Ecosystem (s)

生　　境 Habitat(s)	森林 / Forest
生态系统 Ecosystem(s)	森林生态系统 / Forest Ecosystem

威胁 Threat (s)

主要威胁 Major Threat(s)	无 / None

保护级别与保护行动 Protection Category and Conservation Action (s)

国家重点保护野生动物等级 (2021) Category of National Key Protected Wild Animals (2021)	无 / NA
IUCN 红色名录 (2020-2) IUCN Red List (2020-2)	无危 / LC
CITES 附录 (2019) CITES Appendix (2019)	无 / NA
保护行动 Conservation Action(s)	无 / None

相关文献 Relevant References

Burgin *et al.*, 2018; Jiang *et al.* (蒋志刚等), 2017; Wilson *et al.*, 2016; Zheng *et al.* (郑智民等), 2012; Liu *et al.*, 2012a; Wang and Xie (汪松和解焱), 2004; Luo (罗泽珣), 2000; Kaneko (金子之史), 1992

苛岚绒鼠

Caryomys inez

无危 LC

数据缺乏 DD	无危 LC	近危 NT	易危 VU	濒危 EN	极危 CR	区域灭绝 RE	野外灭绝 EW	灭绝 EX

分类地位 Taxonomic Status

动物界 Animalia	脊索动物门 Chordata	哺乳纲 Mammalia	啮齿目 Rodentia	仓鼠科 Cricetidae

学名 Scientific Name	*Caryomys inez*
命名人 Species Authority	Thomas, 1908
英文名 English Name(s)	Inez's Red-backed Vole
同物异名 Synonym(s)	Inez's Vole; Kolan Vole; *Caryomys nux* (Thomas, 1908)
种下单元评估 Infra-specific Taxa Assessed	无 / None

评估信息 Assessment Information

评估年份 Year Assessed	2020
评定人 Assessor(s)	刘少英、蒋志刚 / Shaoying Liu, Zhigang Jiang
审定人 Reviewer(s)	马勇、路纪琪、鲍毅新 / Yong Ma, Jiqi Lu, Yixin Bao
其他贡献人 Other Contributor(s)	李立立、丁晨晨 / Lili Li, Chenchen Ding

理由 Justification: 苛岚绒鼠发生区大，种群大。因此，列为无危等级 / Inez's Red-backed Vole has large area of occurrence and populations. Thus, it is listed as Least Concern

地理分布 Geographical Distribution

国内分布 Domestic Distribution

陕西、山西、甘肃、宁夏 / Shaanxi, Shanxi, Gansu, Ningxia

世界分布 World Distribution

中国 / China

分布标注 Distribution Note

特有种 / Endemic

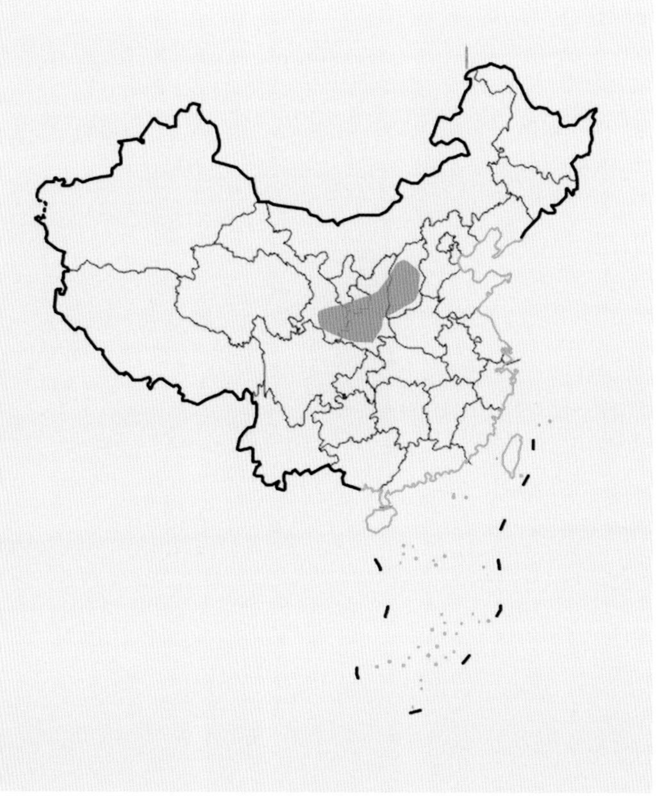

国内分布图
Map of Domestic Distribution

种群 Population

种群数量 Population Size	数量多 / Abundant
种群趋势 Population Trend	未知 / Unknown

生境与生态系统 Habitat (s) and Ecosystem (s)

生　　境 Habitat(s)	森林、溪流边 / Forest, Near Stream
生态系统 Ecosystem(s)	森林生态系统、湖泊河流生态系统 / Forest Ecosystem, Lake and River Ecosystem

威胁 Threat (s)

主要威胁 Major Threat(s)	无 / None

保护级别与保护行动 Protection Category and Conservation Action (s)

国家重点保护野生动物等级 (2021) Category of National Key Protected Wild Animals (2021)	无 / NA
IUCN 红色名录 (2020-2) IUCN Red List (2020-2)	无危 / LC
CITES 附录 (2019) CITES Appendix (2019)	无 / NA
保护行动 Conservation Action(s)	无 / None

相关文献 Relevant References

Burgin *et al*., 2018; Jiang *et al*. (蒋志刚等), 2015a; Wilson *et al*., 2017; Liu *et al*., 2012a; Zheng *et al*. (郑智民等), 2012; Smith *et al*. (史密斯等), 2009; Wang (王应祥), 2003; Luo (罗泽珣), 2000; Kaneko, 1996

戈壁阿尔泰高山䶄

Alticola barakshin

无危 LC

数据缺乏 DD	无危 LC	近危 NT	易危 VU	濒危 EN	极危 CR	区域灭绝 RE	野外灭绝 EW	灭绝 EX

分类地位 Taxonomic Status

动物界 Animalia	脊索动物门 Chordata	哺乳纲 Mammalia	啮齿目 Rodentia	仓鼠科 Cricetidae

学　名 Scientific Name	*Alticola barakshin*
命名人 Species Authority	Bannikov, 1947
英文名 English Name(s)	Gobi Altai Mountain Vole
同物异名 Synonym(s)	无 / None
种下单元评估 Infra-specific Taxa Assessed	无 / None

评估信息 Assessment Information

评估年份 Year Assessed	2020
评定人 Assessor(s)	刘少英、蒋志刚 / Shaoying Liu, Zhigang Jiang
审定人 Reviewer(s)	马勇、路纪琪、鲍毅新 / Yong Ma, Jiqi Lu, Yixin Bao
其他贡献人 Other Contributor(s)	李立立、丁晨晨 / Lili Li, Chenchen Ding

理由 Justification: 戈壁阿尔泰高山䶄的种群数量较多。因此，列为无危等级 / Gobi Altai Mountain Vole has large populations. Thus, it is listed as Least Concern

地理分布 Geographical Distribution

国内分布 Domestic Distribution

新疆 / Xinjiang

世界分布 World Distribution

中国、蒙古、俄罗斯 / China, Mongolia, Russia

分布标注 Distribution Note

非特有种 / Non-endemic

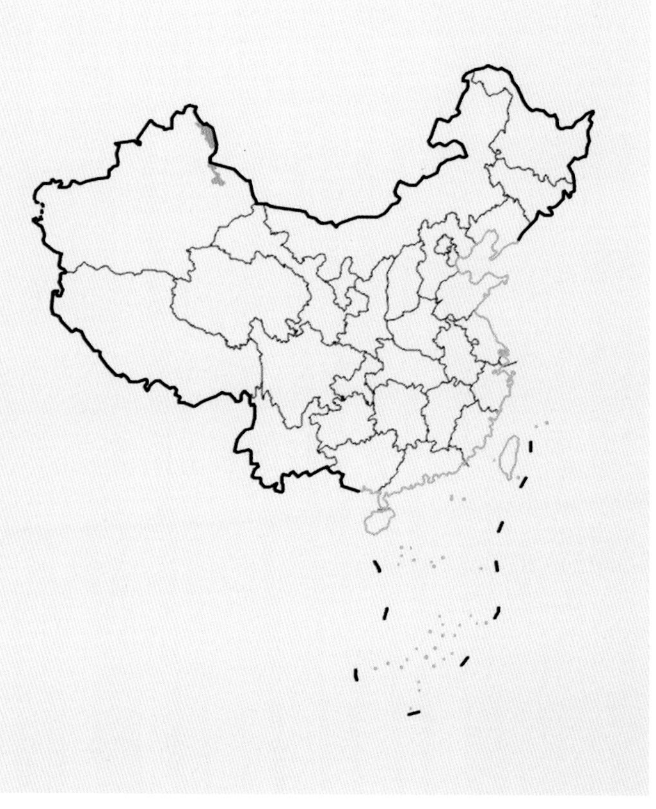

国内分布图
Map of Domestic Distribution

种群 Population

种群数量 **Population Size**	数量多 / Abundant
种群趋势 **Population Trend**	未知 / Unknown

生境与生态系统 Habitat (s) and Ecosystem (s)

生　　境 **Habitat(s)**	灌丛 / Shrubland
生态系统 **Ecosystem(s)**	灌丛生态系统 / Shrubland Ecosystem

威胁 Threat (s)

主要威胁 **Major Threat(s)**	无 / None

保护级别与保护行动 Protection Category and Conservation Action (s)

国家重点保护野生动物等级 **(2021) Category of National Key Protected Wild Animals (2021)**	无 / NA
IUCN 红色名录 **(2020-2) IUCN Red List (2020-2)**	无危 / LC
CITES 附录 **(2019) CITES Appendix (2019)**	无 / NA
保护行动 **Conservation Action(s)**	无 / None

相关文献 Relevant References

Burgin *et al.*, 2018; Jiang *et al.* (蒋志刚等), 2017; Wilson *et al.*, 2017; Zheng *et al.* (郑智民等), 2012; Smith *et al.* (史密斯等), 2009; Pan *et al.* (潘清华等), 2007; Luo (罗泽珣), 2000; Hou (侯兰新), 1995

中国生物多样性红色名录 China's Red List of Biodiversity

大耳高山䶄

Alticola macrotis

无危 LC

数据缺乏 DD	无危 LC	近危 NT	易危 VU	濒危 EN	极危 CR	区域灭绝 RE	野外灭绝 EW	灭绝 EX

分类地位 Taxonomic Status

动物界 Animalia	脊索动物门 Chordata	哺乳纲 Mammalia	啮齿目 Rodentia	仓鼠科 Cricetidae

学名 Scientific Name	*Alticola macrotis*
命名人 Species Authority	Radde, 1862
英文名 English Name(s)	Large-eared Vole
同物异名 Synonym(s)	*Alticola altaica* (Vinogradov, 1933); *A. fetisovi* (Galkina and Epifantseva, 1986); *A. macrotis* (Portenko, 1963) subsp. *vicina*; *A. vinogradovi* (Rasorenova, 1933)
种下单元评估 Infra-specific Taxa Assessed	无 / None

评估信息 Assessment Information

评估年份 Year Assessed	2020
评定人 Assessor(s)	刘少英、蒋志刚 / Shaoying Liu, Zhigang Jiang
审定人 Reviewer(s)	马勇、路纪琪、鲍毅新 / Yong Ma, Jiqi Lu, Yixin Bao
其他贡献人 Other Contributor(s)	李立立、丁晨晨 / Lili Li, Chenchen Ding

理由 **Justification:** 大耳高山䶄的种群数量较多。因此，列为无危等级 / Large-eared Vole has large populations. Thus, it is listed as Least Concern

地理分布 Geographical Distribution

国内分布 Domestic Distribution

新疆 / Xinjiang

世界分布 World Distribution

中国、哈萨克斯坦、蒙古、俄罗斯 / China, Kazakhstan, Mongolia, Russia

分布标注 Distribution Note

非特有种 / Non-endemic

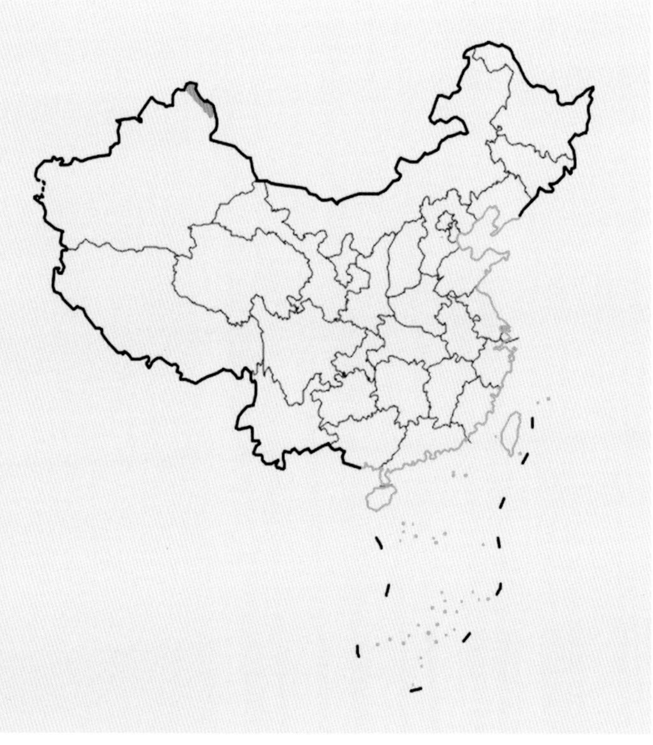

国内分布图
Map of Domestic Distribution

种群 Population

种群数量 **Population Size**	较多 / Relatively abundant
种群趋势 **Population Trend**	未知 / Unknown

生境与生态系统 Habitat (s) and Ecosystem (s)

生　　境 **Habitat(s)**	泰加林、针阔混交林 / Taiga Forest, Coniferous and Broad-leaved Mixed Forest
生态系统 **Ecosystem(s)**	森林生态系统 / Forest Ecosystem

威胁 Threat (s)

主要威胁 **Major Threat(s)**	无 / None

保护级别与保护行动 Protection Category and Conservation Action (s)

国家重点保护野生动物等级 **(2021) Category of National Key Protected Wild Animals (2021)**	无 / NA
IUCN 红色名录 **(2020-2) IUCN Red List (2020-2)**	无危 / LC
CITES 附录 **(2019) CITES Appendix (2019)**	无 / NA
保护行动 **Conservation Action(s)**	无 / None

相关文献 Relevant References

Burgin *et al*., 2018; Jiang *et al.* (蒋志刚等), 2017; Wilson *et al*., 2017; Zheng *et al.* (郑智民等), 2012; Wang (王应祥), 2003; Luo (罗泽珣), 2000; Zheng and Li (郑生武和李保国), 1999

蒋志刚 绘　Drawn by Zhigang Jiang

蒙古高山䶄
Alticola semicanus

无危 LC

数据缺乏 DD	无危 LC	近危 NT	易危 VU	濒危 EN	极危 CR	区域灭绝 RE	野外灭绝 EW	灭绝 EX

分类地位 Taxonomic Status

动物界 Animalia	脊索动物门 Chordata	哺乳纲 Mammalia	啮齿目 Rodentia	仓鼠科 Cricetidae

学　名 Scientific Name	*Alticola semicanus*
命名人 Species Authority	G. M. Allen, 1924
英文名 English Name(s)	Mongolian Silver Vole
同物异名 Synonym(s)	Mongolian Mountain Vole; *Alticola alleni* (Argyropulo, 1933)
种下单元评估 Infra-specific Taxa Assessed	无 / None

评估信息 Assessment Information

评估年份 Year Assessed	2020
评定人 Assessor(s)	刘少英、蒋志刚 / Shaoying Liu, Zhigang Jiang
审定人 Reviewer(s)	马勇、路纪琪、鲍毅新 / Yong Ma, Jiqi Lu, Yixin Bao
其他贡献人 Other Contributor(s)	李立立、丁晨晨 / Lili Li, Chenchen Ding

理由 Justification: 蒙古高山䶄的种群数量较多。因此，列为无危等级 / Mongolian Silver Vole has large populations. Thus, it is listed as Least Concern

地理分布 Geographical Distribution

国内分布 Domestic Distribution

内蒙古 / Inner Mongolia (Nei Mongol)

世界分布 World Distribution

中国、蒙古、俄罗斯 / China, Mongolia, Russia

分布标注 Distribution Note

非特有种 / Non-endemic

国内分布图
Map of Domestic Distribution

种群 Population

种群数量 Population Size	较多 / Relatively abundant
种群趋势 Population Trend	未知 / Unknown

生境与生态系统 Habitat (s) and Ecosystem (s)

生　　境 Habitat(s)	草地、内陆岩石区域 / Grassland, Inland Rocky Area
生态系统 Ecosystem(s)	草地生态系统 / Grassland Ecosystem

威胁 Threat (s)

主要威胁 Major Threat(s)	无 / None

保护级别与保护行动 Protection Category and Conservation Action (s)

国家重点保护野生动物等级 (2021) Category of National Key Protected Wild Animals (2021)	无 / NA
IUCN 红色名录 (2020-2) IUCN Red List (2020-2)	无危 / LC
CITES 附录 (2019) CITES Appendix (2019)	无 / NA
保护行动 Conservation Action(s)	无 / None

相关文献 Relevant References

Burgin *et al*., 2018; Jiang *et al.* (蒋志刚等), 2017; Wilson *et al*., 2017; Zheng *et al.* (郑智民等), 2012; Smith *et al.* (史密斯等), 2009; Wang (王应祥), 2003; Luo (罗泽珣), 2000

蒙古高山䶄 *Alticola semicanus*

扁颅高山䶄

Alticola strelzowi

无危 LC

数据缺乏 DD	无危 LC	近危 NT	易危 VU	濒危 EN	极危 CR	区域灭绝 RE	野外灭绝 EW	灭绝 EX

分类地位 Taxonomic Status

动物界 Animalia	脊索动物门 Chordata	哺乳纲 Mammalia	啮齿目 Rodentia	仓鼠科 Cricetidae
学　名 Scientific Name		*Alticola strelzowi*		
命名人 Species Authority		Kastschenko, 1899		
英文名 English Name(s)		Flat-headed Vole		
同物异名 Synonym(s)		Strelzow's Mountain Vole; *Alticola strelzowi* (Ognev, 1944) subsp. *depressus*; *A. strelzowi* (Kastschenko, 1901) subsp. *desertorum*		
种下单元评估 Infra-specific Taxa Assessed		无 / None		

评估信息 Assessment Information

评估年份 Year Assessed	2020
评　定　人 Assessor(s)	刘少英、蒋志刚 / Shaoying Liu, Zhigang Jiang
审　定　人 Reviewer(s)	马勇、路纪琪、鲍毅新 / Yong Ma, Jiqi Lu, Yixin Bao
其他贡献人 Other Contributor(s)	李立立、丁晨晨 / Lili Li, Chenchen Ding

理由 Justification: 扁颅高山䶄的种群数量较多。因此，列为无危等级 / Flat-headed Vole has large populations. Thus, it is listed as Least Concern

地理分布 Geographical Distribution

国内分布 Domestic Distribution

新疆 / Xinjiang

世界分布 World Distribution

中国；中亚 / China; Central Asia

分布标注 Distribution Note

非特有种 / Non-endemic

国内分布图
Map of Domestic Distribution

种群 Population

种群数量 Population Size	数量多 / Abundant
种群趋势 Population Trend	未知 / Unknown

生境与生态系统 Habitat (s) and Ecosystem (s)

生　　境 Habitat(s)	内陆岩石区域 / Inland Rocky Area
生态系统 Ecosystem(s)	森林生态系统 / Forest Ecosystem

威胁 Threat (s)

主要威胁 Major Threat(s)	无 / None

保护级别与保护行动 Protection Category and Conservation Action (s)

国家重点保护野生动物等级 (2021) Category of National Key Protected Wild Animals (2021)	无 / NA
IUCN红色名录(2020-2) IUCN Red List (2020-2)	无危 / LC
CITES 附录 (2019) CITES Appendix (2019)	无 / NA
保护行动 Conservation Action(s)	无 / None

相关文献 Relevant References

Burgin *et al*., 2018; Jiang *et al.* (蒋志刚等), 2017; Wilson *et al*., 2017; Zheng *et al.* (郑智民等), 2012; Smith *et al.* (史密斯等), 2009; Wang (王应祥), 2003; Luo (罗泽珣), 2000; Zheng and Li (郑生武和李保国), 1999

扁颅高山䶄 *Alticola strelzowi*　　方红霞 绘 Drawn by Hongxia Fang

草原兔尾鼠

Lagurus lagurus

无危 LC

数据缺乏 DD	无危 LC	近危 NT	易危 VU	濒危 EN	极危 CR	区域灭绝 RE	野外灭绝 EW	灭绝 EX

分类地位 Taxonomic Status

动物界 Animalia	脊索动物门 Chordata	哺乳纲 Mammalia	啮齿目 Rodentia	仓鼠科 Cricetidae

学　名 Scientific Name	*Lagurus lagurus*
命名人 Species Authority	Pallas, 1773
英文名 English Name(s)	Steppe Lemming
同物异名 Synonym(s)	Steppe Vole; *Lagurus abacanicus* (Serebrennikov, 1929); *agressus* (Serebrennikov, 1929); *altorum* (Thomas, 1912); *migratorius* (Gloger, 1841); *occidentalis* (Migulin, (转下页)
种下单元评估 Infra-specific Taxa Assessed	无 / None

评估信息 Assessment Information

评估年份 Year Assessed	2020
评定人 Assessor(s)	刘少英、蒋志刚 / Shaoying Liu, Zhigang Jiang
审定人 Reviewer(s)	马勇、路纪琪、鲍毅新 / Yong Ma, Jiqi Lu, Yixin Bao
其他贡献人 Other Contributor(s)	李立立、丁晨晨 / Lili Li, Chenchen Ding

理由 Justification: 草原兔尾鼠的种群数量多。因此，列为无危等级 / Steppe Lemming has large populations. Thus, it is listed as Least Concern

地理分布 Geographical Distribution

国内分布 Domestic Distribution

新疆 / Xinjiang

世界分布 World Distribution

中国；中亚、东欧 / China; Central Asia, East Europe

分布标注 Distribution Note

非特有种 / Non-endemic

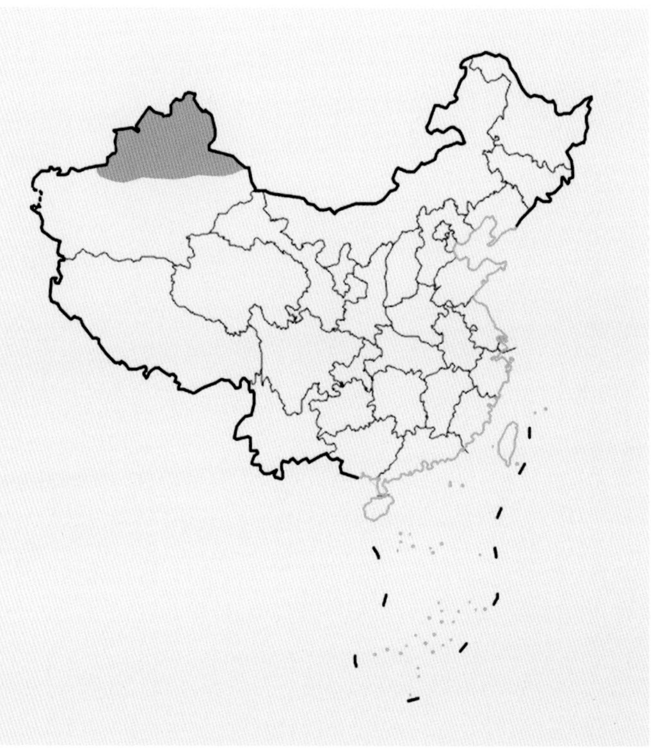

国内分布图
Map of Domestic Distribution

种群 Population

种群数量 Population Size	数量多 / Abundant
种群趋势 Population Trend	稳定 / Stable

生境与生态系统 Habitat (s) and Ecosystem (s)

生　　境 Habitat(s)	草地、半荒漠、耕地、沟渠、灌丛 / Grassland, Semi-desert, Arable Land, Irrigation Canal and Ditch, Shrubland
生态系统 Ecosystem(s)	灌丛生态系统、草地生态系统、农田生态系统、荒漠生态系统、湿地生态系统 / Shrubland Ecosystem, Grassland Ecosystem, Cropland Ecosystem, Desert Ecosystem, Wetland Ecosystem

威胁 Threat (s)

主要威胁 Major Threat(s)	无 / None

保护级别与保护行动 Protection Category and Conservation Action (s)

国家重点保护野生动物等级 (2021) Category of National Key Protected Wild Animals (2021)	无 / NA
IUCN 红色名录 (2020-2) IUCN Red List (2020-2)	无危 / LC
CITES 附录 (2019) CITES Appendix (2019)	无 / NA
保护行动 Conservation Action(s)	无 / None

相关文献 Relevant References

Burgin *et al.*, 2018; Jiang *et al.* (蒋志刚等), 2017; Wilson *et al.*, 2016; Zheng *et al.* (郑智民等), 2012; Smith *et al.* (史密斯等), 2009; Zhang *et al.* (张渝疆等), 2004; Wang (王应祥), 2003; Luo (罗泽珣), 2000

(接上页)
1938); *saturatus* (Ognev, 1950)

草原兔尾鼠 *Lagurus lagurus*

水鼾

Arvicola amphibius

无危 LC

数据缺乏 DD	无危 LC	近危 NT	易危 VU	濒危 EN	极危 CR	区域灭绝 RE	野外灭绝 EW	灭绝 EX

分类地位 Taxonomic Status

动物界 Animalia	脊索动物门 Chordata	哺乳纲 Mammalia	啮齿目 Rodentia	仓鼠科 Cricetidae

学　名 Scientific Name	*Arvicola amphibius*
命名人 Species Authority	Linnaeus, 1758
英文名 English Name(s)	European Water Vole
同物异名 Synonym(s)	Northern Water Vole
种下单元评估 Infra-specific Taxa Assessed	无 / None

评估信息 Assessment Information

评估年份 Year Assessed	2020
评定人 Assessor(s)	刘少英、蒋志刚 / Shaoying Liu, Zhigang Jiang
审定人 Reviewer(s)	马勇、路纪琪、鲍毅新 / Yong Ma, Jiqi Lu, Yixin Bao
其他贡献人 Other Contributor(s)	李立立、丁晨晨 / Lili Li, Chenchen Ding

理由 **Justification:** 水鼾的种群数量多。因此，列为无危等级 / European Water Vole has large populations. Thus, it is listed as Least Concern

地理分布 Geographical Distribution

国内分布 Domestic Distribution

新疆 / Xinjiang

世界分布 World Distribution

亚洲、欧洲 / Asia, Europe

分布标注 Distribution Note

非特有种 / Non-endemic

国内分布图
Map of Domestic Distribution

种群 Population

种群数量 **Population Size**	数量多 / Abundant
种群趋势 **Population Trend**	稳定 / Stable

生境与生态系统 Habitat (s) and Ecosystem (s)

生　　境 **Habitat(s)**	草地、淡水湖、江河、沼泽、耕地 / Grassland, Freshwater Lake, River, Swamp, Arable Land
生态系统 **Ecosystem(s)**	草地生态系统、湿地生态系统、湖泊河流生态系统、农田生态系统 / Grassland Ecosystem, Wetland Ecosystem, Lake and River Ecosystem, Cropland Ecosystem

威胁 Threat (s)

主要威胁 **Major Threat(s)**	无 / None

保护级别与保护行动 Protection Category and Conservation Action (s)

国家重点保护野生动物等级 **(2021) Category of National Key Protected Wild Animals (2021)**	无 / NA
IUCN 红色名录 **(2020-2) IUCN Red List (2020-2)**	无危 / LC
CITES 附录 **(2019) CITES Appendix (2019)**	无 / NA
保护行动 **Conservation Action(s)**	无 / None

相关文献 Relevant References

Burgin *et al*., 2018; Jiang *et al.* (蒋志刚等), 2017; Wilson *et al*., 2016; Zheng *et al.* (郑智民等), 2012; Wilson and Reeder, 2005; Wang (王应祥), 2003; Luo (罗泽珣), 2000

水䶄 *Arvicola amphibius*

云南松田鼠

Neodon forresti

无危 LC

数据缺乏 DD	无危 LC	近危 NT	易危 VU	濒危 EN	极危 CR	区域灭绝 RE	野外灭绝 EW	灭绝 EX

分类地位 Taxonomic Status

动物界 Animalia	脊索动物门 Chordata	哺乳纲 Mammalia	啮齿目 Rodentia	仓鼠科 Cricetidae
学名 Scientific Name		*Neodon forresti*		
命名人 Species Authority		Hinton, 1923		
英文名 English Name(s)		Forrest's Mountain Vole		
同物异名 Synonym(s)		无 / None		
种下单元评估 Infra-specific Taxa Assessed		无 / None		

评估信息 Assessment Information

评估年份 Year Assessed	2020
评定人 Assessor(s)	刘少英、蒋志刚 / Shaoying Liu, Zhigang Jiang
审定人 Reviewer(s)	马勇、鲍毅新、路纪琪 / Yong Ma, Yixin Bao, Jiqi Lu
其他贡献人 Other Contributor(s)	李立立、丁晨晨 / Lili Li, Chenchen Ding

理由 Justification: 云南松田鼠的种群数量较多。因此，列为无危等级 / Forrest's Mountain Vole has large populations. Thus, it is listed as Least Concern

地理分布 Geographical Distribution

国内分布 Domestic Distribution

云南 / Yunnan

世界分布 World Distribution

中国、缅甸 / China, Myanmar

分布标注 Distribution Note

非特有种 / Non-endemic

国内分布图
Map of Domestic Distribution

种群 Population

种群数量 Population Size	数量多 / Abundant
种群趋势 Population Trend	未知 / Unknown

生境与生态系统 Habitat (s) and Ecosystem (s)

生　　境 Habitat(s)	高海拔草地 / High Altitude Grassland
生态系统 Ecosystem(s)	草地生态系统 / Grassland Ecosystem

威胁 Threat (s)

主要威胁 Major Threat(s)	无 / None

保护级别与保护行动 Protection Category and Conservation Action (s)

国家重点保护野生动物等级 (2021) Category of National Key Protected Wild Animals (2021)	无 / NA
IUCN 红色名录 (2020-2) IUCN Red List (2020-2)	无危 / LC
CITES 附录 (2019) CITES Appendix (2019)	无 / NA
保护行动 Conservation Action(s)	无 / None

相关文献 Relevant References

Burgin *et al*., 2018; Jiang *et al.* (蒋志刚等), 2017; Wilson *et al*., 2016; Smith *et al.* (史密斯等), 2009; Pan *et al.* (潘清华等), 2007; Luo (罗泽珣), 2000; Zhang (张荣祖), 1997

中国生物多样性红色名录
China's Red List of Biodiversity

青海松田鼠

Neodon fuscus

无危 LC

数据缺乏 DD	无危 LC	近危 NT	易危 VU	濒危 EN	极危 CR	区域灭绝 RE	野外灭绝 EW	灭绝 EX

分类地位 Taxonomic Status

动物界 Animalia	脊索动物门 Chordata	哺乳纲 Mammalia	啮齿目 Rodentia	仓鼠科 Cricetidae
学名 Scientific Name		*Neodon fuscus*		
命名人 Species Authority		Büchner, 1889		
英文名 English Name(s)		Plateau Pine Vole		
同物异名 Synonym(s)		无 / None		
种下单元评估 Infra-specific Taxa Assessed		无 / None		

评估信息 Assessment Information

评估年份 Year Assessed	2020
评定人 Assessor(s)	刘少英、蒋志刚 / Shaoying Liu, Zhigang Jiang
审定人 Reviewer(s)	马勇、鲍毅新 / Yong Ma, Yixin Bao
其他贡献人 Other Contributor(s)	李立立、丁晨晨 / Lili Li, Chenchen Ding

理由 Justification: 青海松田鼠的种群大。因此，列为无危等级 / Plateau Pine Vole has large populations. Thus, it is listed as Least Concern

地理分布 Geographical Distribution

国内分布 Domestic Distribution

青海 / Qinghai

世界分布 World Distribution

中国 / China

分布标注 Distribution Note

特有种 / Endemic

国内分布图
Map of Domestic Distribution

种群 Population

种群数量 Population Size	数量多 / Abundant
种群趋势 Population Trend	未知 / Unknown

生境与生态系统 Habitat (s) and Ecosystem (s)

生　境 **Habitat(s)**	高寒草甸 / Alpine Meadow
生态系统 **Ecosystem(s)**	草地生态系统 / Grassland Ecosystem

威胁 Threat (s)

主要威胁 **Major Threat(s)**	无 / None

保护级别与保护行动 Protection Category and Conservation Action (s)

国家重点保护野生动物等级 **(2021) Category of National Key Protected Wild Animals (2021)**	无 / NA
IUCN 红色名录 **(2020-2) IUCN Red List (2020-2)**	无危 / LC
CITES 附录 **(2019) CITES Appendix (2019)**	无 / NA
保护行动 **Conservation Action(s)**	无 / None

相关文献 Relevant References

Burgin *et al*., 2018; Jiang *et al.* (蒋志刚等), 2017; Liu *et al*., 2017, 2012a; Wilson *et al*., 2016; Zheng *et al.* (郑智民等), 2012; Smith *et al.* (史密斯等), 2009; Wang (王应祥), 2003; Luo (罗泽珣), 2000; Li *et al.* (李德浩等), 1989

中国生物多样性红色名录
China's Red List of Biodiversity

高原松田鼠

Neodon irene

无危 LC

数据缺乏 DD	无危 LC	近危 NT	易危 VU	濒危 EN	极危 CR	区域灭绝 RE	野外灭绝 EW	灭绝 EX

分类地位 Taxonomic Status

动物界 Animalia	脊索动物门 Chordata	哺乳纲 Mammalia	啮齿目 Rodentia	仓鼠科 Cricetidae

学名 Scientific Name	*Neodon irene*
命名人 Species Authority	Thomas, 1911
英文名 English Name(s)	Irene's Mountain Vole
同物异名 Synonym(s)	Chinese Scrub Vole; *Neodon oniscus* (Thomas, 1911)
种下单元评估 Infra-specific Taxa Assessed	无 / None

评估信息 Assessment Information

评估年份 Year Assessed	2020
评定人 Assessor(s)	刘少英、蒋志刚 / Shaoying Liu, Zhigang Jiang
审定人 Reviewer(s)	马勇、鲍毅新 / Yong Ma, Yixin Bao
其他贡献人 Other Contributor(s)	李立立、丁晨晨 / Lili Li, Chenchen Ding

理由 Justification: 高原松田鼠的种群数量较多。因此，列为无危等级 / Irene's Mountain Vole has large populations. Thus, it is listed as Least Concern

地理分布 Geographical Distribution

国内分布 Domestic Distribution

甘肃、青海、四川、云南、西藏 / Gansu, Qinghai, Sichuan, Yunnan, Tibet (Xizang)

世界分布 World Distribution

中国 / China

分布标注 Distribution Note

特有种 / Endemic

国内分布图
Map of Domestic Distribution

种群 Population

种群数量 Population Size	数量多 / Abundant
种群趋势 Population Trend	未知 / Unknown

生境与生态系统 Habitat (s) and Ecosystem (s)

生　　境 Habitat(s)	高海拔草地、灌丛 / High Altitude Grassland, Shrubland
生态系统 Ecosystem(s)	灌丛生态系统、草地生态系统 / Shrubland Ecosystem, Grassland Ecosystem

威胁 Threat (s)

主要威胁 Major Threat(s)	无 / None

保护级别与保护行动 Protection Category and Conservation Action (s)

国家重点保护野生动物等级 (2021) Category of National Key Protected Wild Animals (2021)	无 / NA
IUCN 红色名录 (2020-2) IUCN Red List (2020-2)	无危 / LC
CITES 附录 (2019) CITES Appendix (2019)	无 / NA
保护行动 Conservation Action(s)	无 / None

相关文献 Relevant References

Burgin *et al.*, 2018; Jiang *et al.* (蒋志刚等), 2017; Liu *et al.*, 2017, 2012a; Wilson *et al.*, 2016; Zheng *et al.* (郑智民等), 2012; Luo (罗泽珣), 2000; Honacki *et al.*, 1982

高原松田鼠 *Neodon irene*

白尾松田鼠

Neodon leucurus

无危 LC

数据缺乏 DD	无危 LC	近危 NT	易危 VU	濒危 EN	极危 CR	区域灭绝 RE	野外灭绝 EW	灭绝 EX

分类地位 Taxonomic Status

动物界 Animalia	脊索动物门 Chordata	哺乳纲 Mammalia	啮齿目 Rodentia	仓鼠科 Cricetidae

学名 Scientific Name	*Neodon leucurus*
命名人 Species Authority	Blyth, 1863
英文名 English Name(s)	Blyth's Mountain Vole
同物异名 Synonym(s)	*Neodon blythi* (Blanford, 1875); *everesti* (Thomas and Hinton, 1922); *petulans* (Wroughton, 1911); *strauchi* (Büchner, 1889); *tsaidamensis* (Satunin, 1903); *waltoni* (转下页)
种下单元评估 Infra-specific Taxa Assessed	无 / None

评估信息 Assessment Information

评估年份 Year Assessed	2020
评定人 Assessor(s)	刘少英、蒋志刚 / Shaoying Liu, Zhigang Jiang
审定人 Reviewer(s)	马勇、鲍毅新 / Yong Ma, Yixin Bao
其他贡献人 Other Contributor(s)	李立立、丁晨晨 / Lili Li, Chenchen Ding

理由 Justification: 白尾松田鼠的种群大。因此，列为无危等级 / Blyth's Mountain Vole has large populations. Thus, it is listed as Least Concern

地理分布 Geographical Distribution

国内分布 Domestic Distribution

青海、西藏、新疆 / Qinghai, Tibet (Xizang), Xinjiang

世界分布 World Distribution

中国、印度、尼泊尔 / China, India, Nepal

分布标注 Distribution Note

非特有种 / Non-endemic

国内分布图
Map of Domestic Distribution

种群 Population

种群数量 **Population Size**	数量多 / Abundant
种群趋势 **Population Trend**	未知 / Unknown

生境与生态系统 Habitat (s) and Ecosystem (s)

生　　境 **Habitat(s)**	草地 / Grassland
生态系统 **Ecosystem(s)**	草地生态系统 / Grassland Ecosystem

威胁 Threat (s)

主要威胁 **Major Threat(s)**	无 / None

保护级别与保护行动 Protection Category and Conservation Action (s)

国家重点保护野生动物等级 **(2021) Category of National Key Protected Wild Animals (2021)**	无 / NA
IUCN 红色名录 **(2020-2) IUCN Red List (2020-2)**	无危 / LC
CITES 附录 **(2019) CITES Appendix (2019)**	无 / NA
保护行动 **Conservation Action(s)**	无 / None

相关文献 Relevant References

Burgin *et al.*, 2018; Jiang *et al.* (蒋志刚等), 2017; Liu *et al.*, 2017, 2012a; Wilson *et al.*, 2016; Zheng *et al.* (郑智民等), 2012; Smith *et al.* (史密斯等), 2009; Luo (罗泽珣), 2000; Zheng and Wang (郑昌琳和汪松), 1980

(接上页)
(Bonhote, 1905); *zadoensis* (Zheng and Wang, 1980)

白尾松田鼠 *Neodon leucurus*

墨脱松田鼠

Neodon medogensis

无危 LC

数据缺乏 DD	无危 LC	近危 NT	易危 VU	濒危 EN	极危 CR	区域灭绝 RE	野外灭绝 EW	灭绝 EX

分类地位 Taxonomic Status

动物界 Animalia	脊索动物门 Chordata	哺乳纲 Mammalia	啮齿目 Rodentia	仓鼠科 Cricetidae

学　名 Scientific Name	*Neodon medogensis*
命名人 Species Authority	Liu *et al.*, 2017
英文名 English Name(s)	Mêdog Mountain Vole
同物异名 Synonym(s)	*Neodon fidelis* (Hinton, 1923)
种下单元评估 Infra-specific Taxa Assessed	无 / None

评估信息 Assessment Information

评估年份 Year Assessed	2020
评定人 Assessor(s)	刘少英、蒋志刚 / Shaoying Liu, Zhigang Jiang
审定人 Reviewer(s)	马勇、鲍毅新 / Yong Ma, Yixin Bao
其他贡献人 Other Contributor(s)	李立立、丁晨晨 / Lili Li, Chenchen Ding

理由 Justification: 墨脱松田鼠的种群数量较多。因此，列为无危等级 / Mêdog Mountain Vole has large populations. Thus, it is listed as Least Concern

地理分布 Geographical Distribution

国内分布 Domestic Distribution

西藏 / Tibet (Xizang)

世界分布 World Distribution

中国；不详 / China; Unknown

分布标注 Distribution Note

非特有种 / Non-endemic

国内分布图
Map of Domestic Distribution

种群 Population

种群数量 Population Size	较多 / Relatively abundant
种群趋势 Population Trend	未知 / Unknown

生境与生态系统 Habitat (s) and Ecosystem (s)

生　　境 Habitat(s)	森林、灌丛 / Forest, Shrubland
生态系统 Ecosystem(s)	森林生态系统、灌丛生态系统 / Forest Ecosystem, Shrubland Ecosystem

威胁 Threat (s)

主要威胁 Major Threat(s)	无 / None

保护级别与保护行动 Protection Category and Conservation Action (s)

国家重点保护野生动物等级 (2021) Category of National Key Protected Wild Animals (2021)	无 / NA
IUCN 红色名录 (2020-2) IUCN Red List (2020-2)	无危 / LC
CITES 附录 (2019) CITES Appendix (2019)	无 / NA
保护行动 Conservation Action(s)	无 / None

相关文献 Relevant References

Burgin *et al.*, 2018; Jiang *et al.* (蒋志刚等), 2017; Liu *et al.* (刘少英等), 2017

中国生物多样性红色名录
China's Red List of Biodiversity

锡金松田鼠

Neodon sikimensis

无危 LC

数据缺乏 DD	无危 LC	近危 NT	易危 VU	濒危 EN	极危 CR	区域灭绝 RE	野外灭绝 EW	灭绝 EX

分类地位 Taxonomic Status

动物界 Animalia	脊索动物门 Chordata	哺乳纲 Mammalia	啮齿目 Rodentia	仓鼠科 Cricetidae

学名 Scientific Name	*Neodon sikimensis*
命名人 Species Authority	Horsfield, 1841
英文名 English Name(s)	Sikkim Mountain Vole
同物异名 Synonym(s)	*Arvicola thricolis* (Gray, 1863); *Microtus sikimensis* (Horsfield, 1841); *M. sikimensis* (Hodgson, 1849); *Pitymys sikimensis* (Hodgson, 1849) subsp. *sikimensis*
种下单元评估 Infra-specific Taxa Assessed	无 / None

评估信息 Assessment Information

评估年份 Year Assessed	2020
评定人 Assessor(s)	刘少英、蒋志刚 / Shaoying Liu, Zhigang Jiang
审定人 Reviewer(s)	马勇、鲍毅新 / Yong Ma, Yixin Bao
其他贡献人 Other Contributor(s)	李立立、丁晨晨 / Lili Li, Chenchen Ding

理由 Justification: 锡金松田鼠的种群数量较多。因此，列为无危等级 / Sikkim Mountain Vole has large populations. Thus, it is listed as Least Concern

地理分布 Geographical Distribution

国内分布 Domestic Distribution
西藏 / Tibet (Xizang)
世界分布 World Distribution
中国；南亚 / China; South Asia
分布标注 Distribution Note
非特有种 / Non-endemic

国内分布图
Map of Domestic Distribution

种群 Population

种群数量 Population Size	数量多 / Abundant
种群趋势 Population Trend	未知 / Unknown

生境与生态系统 Habitat (s) and Ecosystem (s)

生　　境 Habitat(s)	高海拔草地、泰加林 / Alpine Grassland, Taiga Forest
生态系统 Ecosystem(s)	森林生态系统、草地生态系统 / Forest Ecosystem, Grassland Ecosystem

威胁 Threat (s)

主要威胁 Major Threat(s)	无 / None

保护级别与保护行动 Protection Category and Conservation Action (s)

国家重点保护野生动物等级 (2021) Category of National Key Protected Wild Animals (2021)	无 / NA
IUCN 红色名录 (2020-2) IUCN Red List (2020-2)	无危 / LC
CITES 附录 (2019) CITES Appendix (2019)	无 / NA
保护行动 Conservation Action(s)	无 / None

相关文献 Relevant References

Burgin *et al.*, 2018; Jiang *et al.* (蒋志刚等), 2017; Liu *et al.*, 2017, 2012a; Wilson *et al.*, 2016; Tu *et al.* (涂飞云等), 2014; Zheng *et al.* (郑智民等), 2012; Wang (王应祥), 2003; Luo (罗泽珣), 2000

东方田鼠

Alexandromys fortis

无危 LC

数据缺乏 DD	无危 LC	近危 NT	易危 VU	濒危 EN	极危 CR	区域灭绝 RE	野外灭绝 EW	灭绝 EX

分类地位 Taxonomic Status

动物界 Animalia	脊索动物门 Chordata	哺乳纲 Mammalia	啮齿目 Rodentia	仓鼠科 Cricetidae

学　名 Scientific Name	*Alexandromys fortis*
命名人 Species Authority	Büchner, 1889
英文名 English Name(s)	Reed Vole
同物异名 Synonym(s)	*Alexandromys calamorum* (Thomas, 1902); *dolichocephalus* (Mori, 1930); *fujianensis* (Hong, 1981); *michnoi* (Kastschenko, 1910); *pelliceus* (Thomas, 1911); *superus* (Thomas, (转下页)
种下单元评估 Infra-specific Taxa Assessed	无 / None

评估信息 Assessment Information

评估年份 Year Assessed	2020
评定人 Assessor(s)	刘少英、蒋志刚 / Shaoying Liu, Zhigang Jiang
审定人 Reviewer(s)	马勇、鲍毅新 / Yong Ma, Yixin Bao
其他贡献人 Other Contributor(s)	李立立、丁晨晨 / Lili Li, Chenchen Ding

理由 Justification: 东方田鼠在其分布区为常见种，种群大。因此，列为无危等级 / Reed Vole is a common species in its distribution area with large populations. Thus, it is listed as Least Concern

地理分布 Geographical Distribution

国内分布 Domestic Distribution

内蒙古、山东、宁夏、陕西、贵州、广西、安徽、江苏、浙江、湖南、江西、福建 / Inner Mongolia (Nei Mongol), Shandong, Ningxia, Shaanxi, Guizhou, Guangxi, Anhui, Jiangsu, Zhejiang, Hunan, Jiangxi, Fujiang

世界分布 World Distribution

中国；东北亚 / China; Northeast Asia

分布标注 Distribution Note

非特有种 / Non-endemic

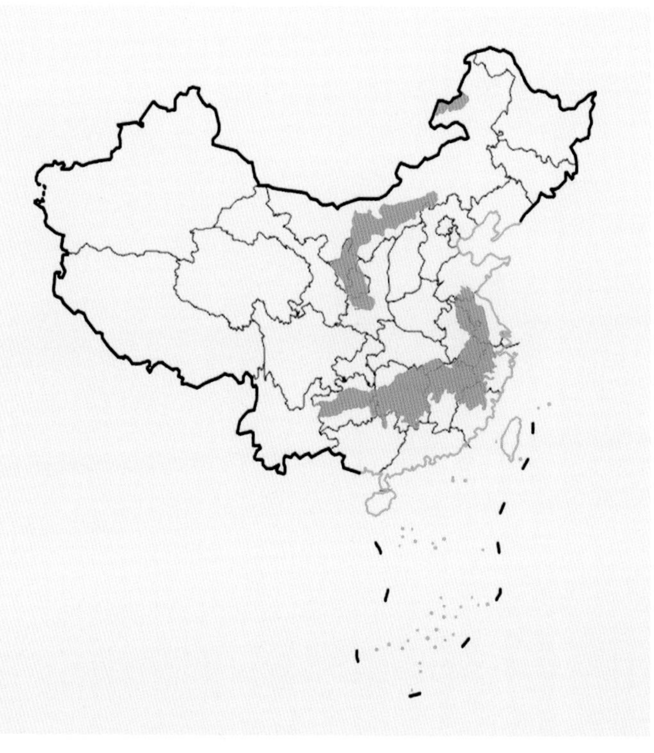

国内分布图
Map of Domestic Distribution

种群 Population

种群数量 Population Size	数量多 / Abundant
种群趋势 Population Trend	未知 / Unknown

生境与生态系统 Habitat (s) and Ecosystem (s)

生　　境 Habitat(s)	淡水湖、江河、溪流边、农田、沼泽、森林 / Freshwater Lake, River, Near Stream, Cropland, Swamp, Forest
生态系统 Ecosystem(s)	森林生态系统、农田生态系统、湖泊河流生态系统、湿地生态系统 / Forest Ecosystem, Cropland Ecosystem, Lake and River Ecosystem, Wetland Ecosystem

威胁 Threat (s)

主要威胁 Major Threat(s)	无 / None

保护级别与保护行动 Protection Category and Conservation Action (s)

国家重点保护野生动物等级 (2021) Category of National Key Protected Wild Animals (2021)	无 / NA
IUCN红色名录(2020-2) IUCN Red List (2020-2)	无危 / LC
CITES 附录 (2019) CITES Appendix (2019)	无 / NA
保护行动 Conservation Action(s)	无 / None

相关文献 Relevant References

Burgin *et al*., 2018; Jiang *et al*. (蒋志刚等), 2017; Liu *et al*., 2017; Wilson *et al*., 2016; Luo (罗泽珣), 2000

(接上页)
1911); *uliginosus* (James and Johnson, 1955)

东方田鼠 *Alexandromys fortis*

柴达木根田鼠

Alexandromys limnophilus

无危 LC

数据缺乏 DD	无危 LC	近危 NT	易危 VU	濒危 EN	极危 CR	区域灭绝 RE	野外灭绝 EW	灭绝 EX

分类地位 Taxonomic Status

动物界 Animalia	脊索动物门 Chordata	哺乳纲 Mammalia	啮齿目 Rodentia	仓鼠科 Cricetidae

学名 Scientific Name	*Alexandromys limnophilus*
命名人 Species Authority	Büchner, 1889
英文名 English Name(s)	Lacustrine Vole
同物异名 Synonym(s)	*Alexandromys flaviventris* (Satunin, 1903); *malcolmi* (Thomas, 1911); *malygini* (Courant *et al.*, 1999)
种下单元评估 Infra-specific Taxa Assessed	无 / None

评估信息 Assessment Information

评估年份 Year Assessed	2020
评定人 Assessor(s)	刘少英、蒋志刚 / Shaoying Liu, Zhigang Jiang
审定人 Reviewer(s)	马勇、鲍毅新 / Yong Ma, Yixin Bao
其他贡献人 Other Contributor(s)	李立立、丁晨晨 / Lili Li, Chenchen Ding

理由 **Justification:** 柴达木根田鼠的种群数量多。因此，列为无危等级 / Lacustrine Vole has large populations. Thus, it is listed as Least Concern

地理分布 Geographical Distribution

国内分布 Domestic Distribution

四川、青海、甘肃、陕西、宁夏、内蒙古、新疆 / Sichuan, Qinghai, Gansu, Shaanxi, Ningxia, Inner Mongolia (Nei Mongol), Xinjiang

世界分布 World Distribution

中国、蒙古 / China, Mongolia

分布标注 Distribution Note

非特有种 / Non-endemic

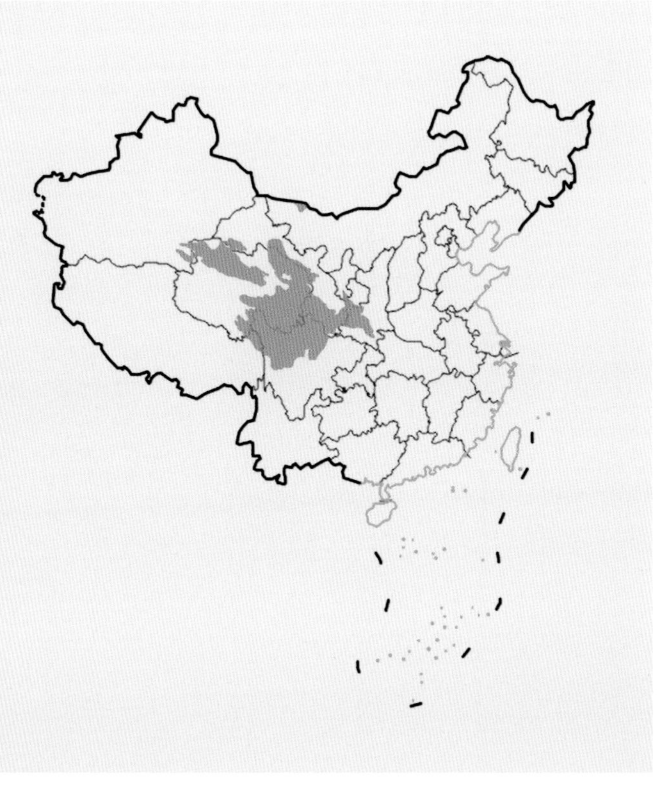

国内分布图
Map of Domestic Distribution

种群 Population

种群数量 Population Size	数量多 / Abundant
种群趋势 Population Trend	未知 / Unknown

生境与生态系统 Habitat (s) and Ecosystem (s)

生　　境 Habitat(s)	盐碱荒漠、高山草甸 / Saline Desert, Alpine Meadow
生态系统 Ecosystem(s)	草地生态系统、荒漠生态系统 / Grassland Ecosystem, Desert Ecosystem

威胁 Threat (s)

主要威胁 Major Threat(s)	无 / None

保护级别与保护行动 Protection Category and Conservation Action (s)

国家重点保护野生动物等级 (2021) Category of National Key Protected Wild Animals (2021)	无 / NA
IUCN 红色名录 (2020-2) IUCN Red List (2020-2)	无危 / LC
CITES 附录 (2019) CITES Appendix (2019)	无 / NA
保护行动 Conservation Action(s)	无 / None

相关文献 Relevant References

Burgin *et al*., 2018; Jiang *et al.* (蒋志刚等), 2017; Liu *et al*., 2017, 2012a; Wilson *et al*., 2016; Zheng *et al.* (郑智民等), 2012; Wang and Xie (汪松和解焱), 2004; Luo (罗泽珣), 2000

莫氏田鼠

Alexandromys maximowiczii

无危 LC

数据缺乏 DD	无危 LC	近危 NT	易危 VU	濒危 EN	极危 CR	区域灭绝 RE	野外灭绝 EW	灭绝 EX

分类地位 Taxonomic Status

动物界 Animalia	脊索动物门 Chordata	哺乳纲 Mammalia	啮齿目 Rodentia	仓鼠科 Cricetidae

学名 Scientific Name	*Alexandromys maximowiczii*
命名人 Species Authority	Schrenk, 1859
英文名 English Name(s)	Maximowicz's Vole
同物异名 Synonym(s)	*Microtus michnoi* (Kastschenko, 1913) subsp. *ungurensis*; *gromovi* (Vorontsov, Boeskorov, Ljapunova and Revin, 1988)
种下单元评估 Infra-specific Taxa Assessed	无 / None

评估信息 Assessment Information

评估年份 Year Assessed	2020
评定人 Assessor(s)	刘少英、蒋志刚 / Shaoying Liu, Zhigang Jiang
审定人 Reviewer(s)	马勇、鲍毅新 / Yong Ma, Yixin Bao
其他贡献人 Other Contributor(s)	李立立、丁晨晨 / Lili Li, Chenchen Ding

理由 Justification: 莫氏田鼠的种群数量多。因此，列为无危等级 / Maximowicz's Vole has large populations. Thus, it is listed as Least Concern

地理分布 Geographical Distribution

国内分布 Domestic Distribution

黑龙江、吉林、内蒙古、河北、陕西 / Heilongjiang, Jilin, Inner Mongolia (Nei Mongol), Hebei, Shaanxi

世界分布 World Distribution

中国、蒙古、俄罗斯 / China, Mongolia, Russia

分布标注 Distribution Note

非特有种 / Non-endemic

国内分布图
Map of Domestic Distribution

种群 Population

种群数量 Population Size	数量多 / Abundant
种群趋势 Population Trend	未知 / Unknown

生境与生态系统 Habitat (s) and Ecosystem (s)

生　　境 Habitat(s)	江河岸 / River Bank
生态系统 Ecosystem(s)	湖泊河流生态系统 / Lake and River Ecosystem

威胁 Threat (s)

主要威胁 Major Threat(s)	无 / None

保护级别与保护行动 Protection Category and Conservation Action (s)

国家重点保护野生动物等级 (2021) Category of National Key Protected Wild Animals (2021)	无 / NA
IUCN 红色名录 (2020-2) IUCN Red List (2020-2)	无危 / LC
CITES 附录 (2019) CITES Appendix (2019)	无 / NA
保护行动 Conservation Action(s)	无 / None

相关文献 Relevant References

Burgin *et al*., 2018; Jiang *et al.* (蒋志刚等), 2017; Liu *et al*., 2017, 2012a; Wilson *et al*., 2016; Zheng *et al.* (郑智民等), 2012; Luo (罗泽珣), 2000

莫氏田鼠 *Alexandromys maximowiczii*

根田鼠

Alexandromys oeconomus

无危 LC

数据缺乏 DD | 无危 LC | 近危 NT | 易危 VU | 濒危 EN | 极危 CR | 区域灭绝 RE | 野外灭绝 EW | 灭绝 EX

分类地位 Taxonomic Status

动 物 界 Animalia	脊索动物门 Chordata	哺 乳 纲 Mammalia	啮 齿 目 Rodentia	仓 鼠 科 Cricetidae
学　名 Scientific Name		*Alexandromys oeconomus*		
命 名 人 Species Authority		Pallas, 1776		
英 文 名 English Name(s)		Root Vole		
同物异名 Synonym(s)		Northern Vole; Tundra Vole; *Alexandromys altaicus* (Ognev, 1944); *amakensis* (Murie, 1930); *anikini* (Egorin, 1939); *arenicola* (de Sélys Longchamps, 1841); *dauricus* (转下页)		
种下单元评估 Infra-specific Taxa Assessed		无 / None		

评估信息 Assessment Information

评 估 年 份 Year Assessed	2020
评　定　人 Assessor(s)	刘少英、蒋志刚 / Shaoying Liu, Zhigang Jiang
审　定　人 Reviewer(s)	马勇、鲍毅新 / Yong Ma, Yixin Bao
其他贡献人 Other Contributor(s)	李立立、丁晨晨 / Lili Li, Chenchen Ding

理由 Justification: 根田鼠在其分布区为常见种，种群数量大。因此，列为无危等级 / Root Vole is a common species with large populations in its range. Thus, it is listed as Least Concern

地理分布 Geographical Distribution

国内分布 Domestic Distribution

新疆 / Xinjiang

世界分布 World Distribution

亚洲、欧洲、北美洲 / Asia, Europe, North America

分布标注 Distribution Note

非特有种 / Non-endemic

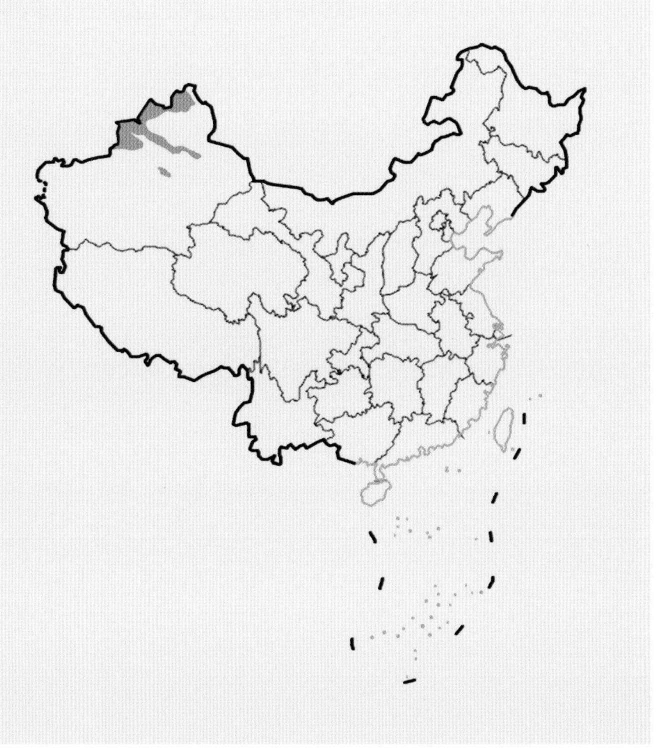

国内分布图
Map of Domestic Distribution

种群 Population

种群数量 Population Size	数量多 / Abundant
种群趋势 Population Trend	稳定 / Stable

生境与生态系统 Habitat (s) and Ecosystem (s)

生　　境 Habitat(s)	淡水湖、溪流边、沼泽、灌丛 / Freshwater Lake, Near Stream, Swamp, Shrubland
生态系统 Ecosystem(s)	灌丛生态系统、湖泊河流生态系统、湿地生态系统 / Shrubland Ecosystem, Lake and River Ecosystem, Wetland Ecosystem

威胁 Threat (s)

主要威胁 Major Threat(s)	无 / None

保护级别与保护行动 Protection Category and Conservation Action (s)

国家重点保护野生动物等级 (2021) Category of National Key Protected Wild Animals (2021)	无 / NA
IUCN 红色名录 (2020-2) IUCN Red List (2020-2)	无危 / LC
CITES 附录 (2019) CITES Appendix (2019)	无 / NA
保护行动 Conservation Action(s)	无 / None

相关文献 Relevant References

Burgin *et al.*, 2018; Jiang *et al.* (蒋志刚等), 2017; Liu *et al.*, 2017, 2012a; Wilson *et al.*, 2016; Zheng *et al.* (郑智民等), 2012; Smith *et al.* (史密斯等), 2009; Cui *et al.* (崔庆虎等), 2005; Sun *et al.* (孙平等), 2005; Luo (罗泽珣), 2000

(接上页)

(Kastschenko, 1910); *elymocetes* (Osgood, 1906); *endoecus* (Osgood, 1909); *epiratticeps* (Young, 1934); *finmarchicus* (Siivonen, 1967); *flaviventris* (Satunin, 1903); *gilmorei* (Setzer, 1952); *hahlovi* (Skalon, 1935); *innuitus* (Merriam, 1900); *kadiacensis* (Merriam, 1897); *kamtschatica* (Pallas, 1779); *karaginensis* (Kostenko, 1984); *karaginensis* (Kostenko, 1989); *kharanurensis* (Courant *et al.*, 1999); *kjusjurensis* (Koljuschev, 1935); *koreni* (G. M. Allen, 1914); *macfarlani* (Merriam, 1900); *medius* (Nilsson, 1844); *mehelyi* (Ehik, 1928); *montiumcaelestinum* (Ognev, 1944); *naumovi* (Stroganov, 1936); *operarius* (Nelson, 1893); *ouralensis* (Poliakov, 1881); *petshorae* (Ognev, 1944); *popofensis* (Merriam, 1900); *punukensis* (Hall and Gilmore, 1932); *ratticeps* (Keyserling and Blasius, 1841); *shantaricus* (Ognev, 1929); *sitkensis* (Merriam, 1897); *stimmingi* (Nehring, 1899); *suntaricus* (Dukelski, 1928); *tschuktschorum* (Miller, 1899); *uchidae* (Kuroda, 1924); *unalascensis* (Merriam, 1897); *uralensis* (Poljakov, 1881); *yakutatensis* (Merriam, 1900)

根田鼠 *Alexandromys oeconomus*

中国生物多样性红色名录 China's Red List of Biodiversity

狭颅田鼠

Microtus gregalis

无危 LC

数据缺乏 DD	无危 LC	近危 NT	易危 VU	濒危 EN	极危 CR	区域灭绝 RE	野外灭绝 EW	灭绝 EX

分类地位 Taxonomic Status

动 物 界 Animalia	脊索动物门 Chordata	哺 乳 纲 Mammalia	啮 齿 目 Rodentia	仓 鼠 科 Cricetidae
学 名 Scientific Name		*Microtus gregalis*		
命 名 人 Species Authority		Pallas, 1779		
英 文 名 English Name(s)		Narrow-headed Vole		
同物异名 Synonym(s)		*Microtus angelicus* (Hinton, 1910); *angustus* (Thomas, 1908); *brevicauda* (Kastschenko, 1901); *buturlini* (Ognev, 1922); *dolguschini* (Afanasiev, 1939); *dukelskiae* (转下页)		
种下单元评估 Infra-specific Taxa Assessed		无 / None		

评估信息 Assessment Information

评 估 年 份 Year Assessed	2020
评 定 人 Assessor(s)	刘少英、蒋志刚 / Shaoying Liu, Zhigang Jiang
审 定 人 Reviewer(s)	马勇、鲍毅新 / Yong Ma, Yixin Bao
其他贡献人 Other Contributor(s)	李立立、丁晨晨 / Lili Li, Chenchen Ding

理由 Justification: 狭颅田鼠分布较广，种群数量大。因此，列为无危等级 / Narrow-headed Vole is a widespread species with large populations. Thus, it is listed as Least Concern

地理分布 Geographical Distribution

国内分布 Domestic Distribution

黑龙江、内蒙古、河北、新疆 / Heilongjiang, Inner Mongolia (Nei Mongol), Hebei, Xinjiang

世界分布 World Distribution

中国；中亚、东亚 / China; Central Asia, East Asia

分布标注 Distribution Note

非特有种 / Non-endemic

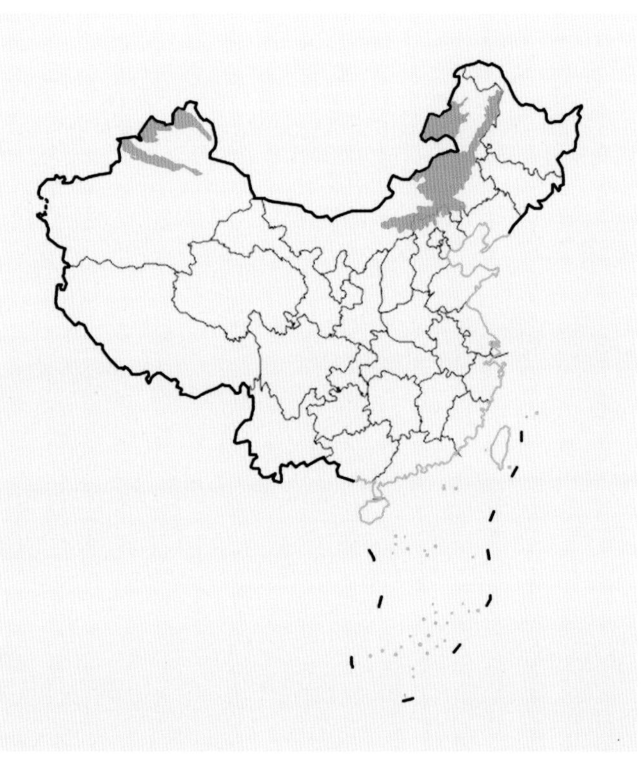

国内分布图 **Map of Domestic Distribution**

种群 Population

种群数量 Population Size	未知 / Unknown
种群趋势 Population Trend	稳定 / Stable

生境与生态系统 Habitat (s) and Ecosystem (s)

生　　境 Habitat(s)	干旱草地、草甸 / Arid Grassland, Meadow
生态系统 Ecosystem(s)	草地生态系统 / Grassland Ecosystem

威胁 Threat (s)

主要威胁 Major Threat(s)	无 / None

保护级别与保护行动 Protection Category and Conservation Action (s)

国家重点保护野生动物等级 (2021) Category of National Key Protected Wild Animals (2021)	无 / NA
IUCN 红色名录 (2020-2) IUCN Red List (2020-2)	无危 / LC
CITES 附录 (2019) CITES Appendix (2019)	无 / NA
保护行动 Conservation Action(s)	无 / None

相关文献 Relevant References

Burgin *et al*., 2018; Jiang *et al.* (蒋志刚等), 2017; Wilson *et al*., 2016; Zheng *et al.* (郑智民等), 2012; Smith *et al.* (史密斯等), 2009; Wang (王应祥), 2003; Luo (罗泽珣), 2000

(接上页)
(Ognev, 1950); *egorovi* (Baranov and Feigin, 1980); *eversmanni* (Poljakov, 1881); *kossogolicus* (Ognev, 1923); *kriogenicus* (Rekovets, 1978); *major* (Ognev, 1923); *montosus* (Argyropulo, 1932); *nordenskioldi* (Poljakov, 1881); *pallasii* (Kastschenko, 1901); *raei* (Poljakov, 1881); *ravidulus* (Miller, 1899); *sirtalaensis* (Yung, 1966); *slowzowi* (Poljakov, 1881); *talassicus* (Heptner, 1948); *tarbagataicus* (Ognev, 1944); *tianschanicus* (Buchner, 1889); *tundrae* (Ognev, 1944); *unguiculatus* (Vinogradov, 1935); *zachvatkini* (Heptner, 1945)

狭颅田鼠 *Microtus gregalis* 　　蒋卫 摄 By Wei Jiang

伊犁田鼠
Microtus ilaeus

无危 LC

数据缺乏 DD	无危 LC	近危 NT	易危 VU	濒危 EN	极危 CR	区域灭绝 RE	野外灭绝 EW	灭绝 EX

分类地位 Taxonomic Status

动物界 Animalia	脊索动物门 Chordata	哺乳纲 Mammalia	啮齿目 Rodentia	仓鼠科 Cricetidae

学　名 **Scientific Name**	*Microtus ilaeus*
命名人 **Species Authority**	Thomas, 1912
英文名 **English Name(s)**	Kazakhstan Vole
同物异名 **Synonym(s)**	Tien Shan Vole; *Microtus igromovi* (Meir *et al.*, 1996); *ileos* (Vinogradov, 1930); *innae* (Ognev, 1950); *kirgisorum* (Ognev, 1950)
种下单元评估 **Infra-specific Taxa Assessed**	无 / None

评估信息 Assessment Information

评估年份 **Year Assessed**	2020
评定人 **Assessor(s)**	刘少英、蒋志刚 / Shaoying Liu, Zhigang Jiang
审定人 **Reviewer(s)**	马勇、鲍毅新 / Yong Ma, Yixin Bao
其他贡献人 **Other Contributor(s)**	李立立、丁晨晨 / Lili Li, Chenchen Ding

理由 **Justification:** 伊犁田鼠在其分布区常见，种群数量大。因此，列为无危等级 / Kazakhstan Vole is a common species with large populations in its areas of occurrence. Thus, it is listed as Least Concern

地理分布 Geographical Distribution

国内分布 Domestic Distribution

新疆 / Xinjiang

世界分布 World Distribution

阿富汗、中国、哈萨克斯坦、吉尔吉斯斯坦、塔吉克斯坦、乌兹别克斯坦 / Afghanistan, China, Kazakhstan, Kyrgyzstan, Tajikistan, Uzbekistan

分布标注 Distribution Note

非特有种 / Non-endemic

国内分布图
Map of Domestic Distribution

种群 Population

种群数量 Population Size	数量多 / Abundant
种群趋势 Population Trend	未知 / Unknown

生境与生态系统 Habitat (s) and Ecosystem (s)

生　　境 Habitat(s)	森林、灌丛、草地 / Forest, Shrubland, Grassland
生态系统 Ecosystem(s)	森林生态系统、灌丛生态系统、草地生态系统 / Forest Ecosystem, Shrubland Ecosystem, Grassland Ecosystem

威胁 Threat (s)

主要威胁 Major Threat(s)	无 / None

保护级别与保护行动 Protection Category and Conservation Action (s)

国家重点保护野生动物等级 (2021) Category of National Key Protected Wild Animals (2021)	无 / NA
IUCN 红色名录 (2020-2) IUCN Red List (2020-2)	无危 / LC
CITES 附录 (2019) CITES Appendix (2019)	无 / NA
保护行动 Conservation Action(s)	无 / None

相关文献 Relevant References

Burgin *et al*., 2018; Hou (侯兰新), 2000; Wilson *et al*., 2016; Zheng *et al.* (郑智民等), 2012; Smith *et al.* (史密斯等), 2009; Wang (王应祥), 2003; Luo (罗泽珣), 2000

伊犁田鼠 *Microtus ilaeus*

帕米尔田鼠

Microtus juldaschi

无危 LC

数据缺乏 DD	无危 LC	近危 NT	易危 VU	濒危 EN	极危 CR	区域灭绝 RE	野外灭绝 EW	灭绝 EX

分类地位 Taxonomic Status

动物界 Animalia	脊索动物门 Chordata	哺乳纲 Mammalia	啮齿目 Rodentia	仓鼠科 Cricetidae

学　名 Scientific Name	*Microtus juldaschi*
命名人 Species Authority	Severtzov, 1879
英文名 English Name(s)	Juniper Mountain Vole
同物异名 Synonym(s)	Juniper Vole; *Neodon carruthersi* (Thomas, 1909); *pamirensis* (Miller, 1899); *thalassensis* (Sludsky, 1988); *yuldaschi* (Severtzov, 1879)
种下单元评估 Infra-specific Taxa Assessed	无 / None

评估信息 Assessment Information

评估年份 Year Assessed	2020
评定人 Assessor(s)	刘少英、蒋志刚 / Shaoying Liu, Zhigang Jiang
审定人 Reviewer(s)	马勇、鲍毅新 / Yong Ma, Yixin Bao
其他贡献人 Other Contributor(s)	李立立、丁晨晨 / Lili Li, Chenchen Ding

理由 Justification: 帕米尔田鼠的种群数量较多。因此，列为无危等级 / Juniper Mountain Vole has large populations. Thus, it is listed as Least Concern

地理分布 Geographical Distribution

国内分布 Domestic Distribution

西藏、新疆 / Tibet (Xizang), Xinjiang

世界分布 World Distribution

智利、中国；中亚 / Chile, China; Central Asia

分布标注 Distribution Note

非特有种 / Non-endemic

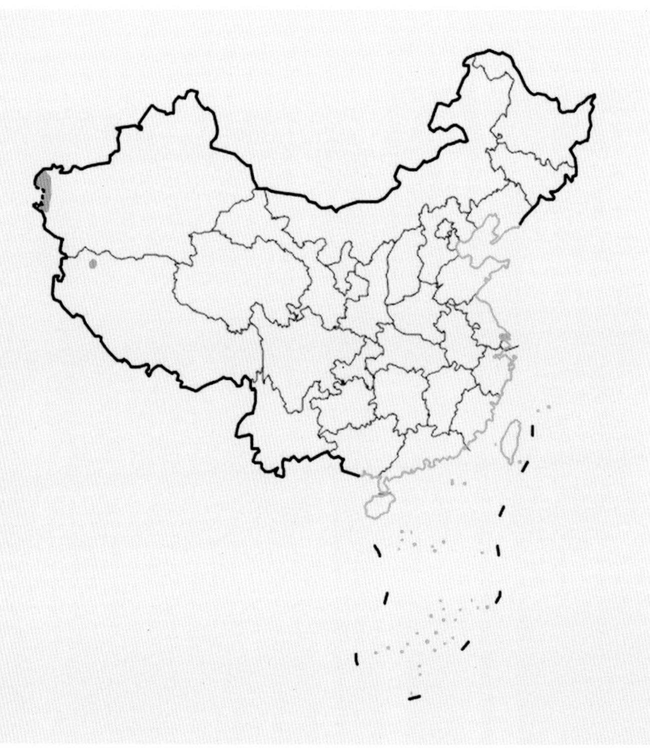

国内分布图
Map of Domestic Distribution

种群 Population

种群数量 Population Size	数量多 / Abundant
种群趋势 Population Trend	未知 / Unknown

生境与生态系统 Habitat (s) and Ecosystem (s)

生　　境 Habitat(s)	高海拔草地、灌丛 / High Altitude Grassland, Shrubland
生态系统 Ecosystem(s)	灌丛生态系统、草地生态系统 / Shrubland Ecosystem, Grassland Ecosystem

威胁 Threat (s)

主要威胁 Major Threat(s)	无 / None

保护级别与保护行动 Protection Category and Conservation Action (s)

国家重点保护野生动物等级 (2021) Category of National Key Protected Wild Animals (2021)	无 / NA
IUCN 红色名录 (2020-2) IUCN Red List (2020-2)	无危 / LC
CITES 附录 (2019) CITES Appendix (2019)	无 / NA
保护行动 Conservation Action(s)	无 / None

相关文献 Relevant References

Burgin *et al.*, 2018; Jiang *et al.* (蒋志刚等), 2017; Wilson *et al.*, 2016; Zheng *et al.* (郑智民等), 2012; Smith *et al.* (史密斯等), 2009; Wang (王应祥), 2003; Luo (罗泽珣), 2000; Corbet and Hill, 1991

帕米尔田鼠 *Microtus juldaschi*　　蒋志刚 绘　Drawn by Zhigang Jiang

蒙古田鼠
Microtus mongolicus

无危 LC

数据缺乏 DD	无危 LC	近危 NT	易危 VU	濒危 EN	极危 CR	区域灭绝 RE	野外灭绝 EW	灭绝 EX

分类地位 Taxonomic Status

动物界 Animalia	脊索动物门 Chordata	哺乳纲 Mammalia	啮齿目 Rodentia	仓鼠科 Cricetidae
学名 Scientific Name		*Microtus mongolicus*		
命名人 Species Authority		Radde, 1861		
英文名 English Name(s)		Mongolian Vole		
同物异名 Synonym(s)		*Microtus arvalis* (Fetisov, 1941) subsp. *baicalensis*; *poljakovi* (Kastschenko, 1901); *xerophilus* (Skalon, 1936)		
种下单元评估 Infra-specific Taxa Assessed		无 / None		

评估信息 Assessment Information

评估年份 Year Assessed	2020
评定人 Assessor(s)	刘少英、蒋志刚 / Shaoying Liu, Zhigang Jiang
审定人 Reviewer(s)	马勇、鲍毅新 / Yong Ma, Yixin Bao
其他贡献人 Other Contributor(s)	李立立、丁晨晨 / Lili Li, Chenchen Ding

理由 Justification: 蒙古田鼠在其生境中常见，种群数量大。因此，列为无危等级 / Mongolian Vole is a common species in its habitats and has large populations. Thus, it is listed as Least Concern

地理分布 Geographical Distribution

国内分布 Domestic Distribution

内蒙古、黑龙江、吉林 / Inner Mongolia (Nei Mongol), Heilongjiang, Jilin

世界分布 World Distribution

中国、蒙古、俄罗斯 / China, Mongolia, Russia

分布标注 Distribution Note

非特有种 / Non-endemic

国内分布图
Map of Domestic Distribution

种群 Population

种群数量 Population Size	种群数量大 / Large populations
种群趋势 Population Trend	稳定 / Stable

生境与生态系统 Habitat (s) and Ecosystem (s)

生　　境 Habitat(s)	森林、灌丛、草地 / Forest, Shrubland, Grassland
生态系统 Ecosystem(s)	森林生态系统、草地生态系统 / Forest Ecosystem, Grassland Ecosystem

威胁 Threat (s)

主要威胁 Major Threat(s)	无 / None

保护级别与保护行动 Protection Category and Conservation Action (s)

国家重点保护野生动物等级 (2021) Category of National Key Protected Wild Animals (2021)	无 / NA
IUCN 红色名录 (2020-2) IUCN Red List (2020-2)	无危 / LC
CITES 附录 (2019) CITES Appendix (2019)	无 / NA
保护行动 Conservation Action(s)	无 / None

相关文献 Relevant References

Burgin *et al*., 2018; Jiang *et al.* (蒋志刚等), 2017; Wilson *et al*., 2016; Zheng *et al.* (郑智民等), 2012; Smith *et al.* (史密斯等), 2009; Wang (王应祥), 2003; Luo (罗泽珣), 2000

蒙古田鼠 *Microtus mongolicus*　　方红霞 绘　Drawn by Hongxia Fang

社田鼠

Microtus socialis

无危 LC

数据缺乏 DD	无危 LC	近危 NT	易危 VU	濒危 EN	极危 CR	区域灭绝 RE	野外灭绝 EW	灭绝 EX

分类地位 Taxonomic Status

动物界 Animalia	脊索动物门 Chordata	哺乳纲 Mammalia	啮齿目 Rodentia	仓鼠科 Cricetidae
学名 Scientific Name		*Microtus socialis*		
命名人 Species Authority		Pallas, 1773		
英文名 English Name(s)		Social Vole		
同物异名 Synonym(s)		*Microtus aristovi* (Golenishchev, 2002); *astrachanensis* (Erxleben, 1777); *binominatus* (Ellerman, 1941); *bogdoensis* (Wang and Ma, 1982); *gravesi* (Goodwin, 1934); (转下页)		
种下单元评估 Infra-specific Taxa Assessed		无 / None		

评估信息 Assessment Information

评估年份 Year Assessed	2020
评定人 Assessor(s)	刘少英、蒋志刚 / Shaoying Liu, Zhigang Jiang
审定人 Reviewer(s)	马勇、鲍毅新 / Yong Ma, Yixin Bao
其他贡献人 Other Contributor(s)	李立立、丁晨晨 / Lili Li, Chenchen Ding

理由 Justification: 社田鼠在其生境中常见，种群数量大。因此，列为无危等级 / Social Vole is a common species with large populations in its habitat. Thus, it is listed as Least Concern

地理分布 Geographical Distribution

国内分布 Domestic Distribution

新疆 / Xinjiang

世界分布 World Distribution

中国；中亚、西亚 / China; Central Asia, West Asia

分布标注 Distribution Note

非特有种 / Non-endemic

国内分布图
Map of Domestic Distribution

种群 Population

种群数量 Population Size	种群数量大 / Large populations
种群趋势 Population Trend	稳定 / Stable

生境与生态系统 Habitat (s) and Ecosystem (s)

生　　境 Habitat(s)	草原、半荒漠 / Grassland, Semi-desert
生态系统 Ecosystem(s)	草地生态系统、荒漠荒原生态系统 / Grassland Ecosystem, Desert Ecosystem

威胁 Threat (s)

主要威胁 Major Threat(s)	无 / None

保护级别与保护行动 Protection Category and Conservation Action (s)

国家重点保护野生动物等级 (2021) Category of National Key Protected Wild Animals (2021)	无 / NA
IUCN 红色名录 (2020-2) IUCN Red List (2020-2)	无危 / LC
CITES 附录 (2019) CITES Appendix (2019)	无 / NA
保护行动 Conservation Action(s)	无 / None

相关文献 Relevant References

Burgin *et al.*, 2018; Jiang *et al.* (蒋志刚等), 2017; Wilson *et al.*, 2016; Zheng *et al.* (郑智民等), 2012; Luo (罗泽珣), 2000; Shayrave and Vashkent (沙依拉吾和武什肯), 1996

(接上页)
nikolajevi (Ognev, 1950); *parvus* (Satunin, 1901); *satunini* (Ognev, 1924); *syriacus* (Brants, 1827); *zaitsevi* (Golenishchev, 2002)

社田鼠 *Microtus socialis*

布氏田鼠
Lasiopodomys brandtii

无危 LC

数据缺乏 DD	无危 LC	近危 NT	易危 VU	濒危 EN	极危 CR	区域灭绝 RE	野外灭绝 EW	灭绝 EX

分类地位 Taxonomic Status

动物界 Animalia	脊索动物门 Chordata	哺乳纲 Mammalia	啮齿目 Rodentia	仓鼠科 Cricetidae

学　名 Scientific Name	*Lasiopodomys brandtii*
命名人 Species Authority	Radde, 1861
英文名 English Name(s)	Brandt's Vole
同物异名 Synonym(s)	Steppe Vole; *Lasiopodomys brandtii* (Bannikov, 1948) subsp. *hangaicus*; *Microtus brandtii* (Kastschenko, 1912) subsp. *aga*; *M. warringtoni* (Miller, 1913)
种下单元评估 Infra-specific Taxa Assessed	无 / None

评估信息 Assessment Information

评估年份 Year Assessed	2020
评定人 Assessor(s)	刘少英、蒋志刚 / Shaoying Liu, Zhigang Jiang
审定人 Reviewer(s)	马勇、鲍毅新 / Yong Ma, Yixin Bao
其他贡献人 Other Contributor(s)	李立立、丁晨晨 / Lili Li, Chenchen Ding

理由 Justification: 布氏田鼠分布广，种群数量大。因此，列为无危等级 / Brandt's Vole is a widespread species with large populations. Thus, it is listed as Least Concern

地理分布 Geographical Distribution

国内分布 Domestic Distribution

黑龙江、吉林、内蒙古、河北 / Heilongjiang, Jilin, Inner Mongolia (Nei Mongol), Hebei

世界分布 World Distribution

中国、蒙古、俄罗斯 / China, Mongolia, Russia

分布标注 Distribution Note

非特有种 / Non-endemic

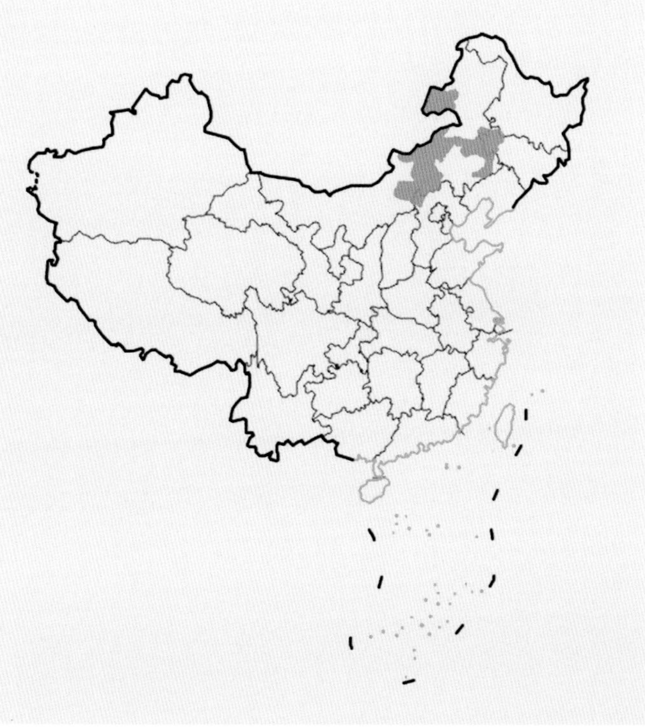

国内分布图
Map of Domestic Distribution

种群 Population

种群数量 **Population Size**	野外常见 / Common in field
种群趋势 **Population Trend**	稳定 / Stable

生境与生态系统 Habitat (s) and Ecosystem (s)

生　　境 **Habitat(s)**	干旱草地 / Dry Steppe
生态系统 **Ecosystem(s)**	草地生态系统 / Grassland Ecosystem

威胁 Threat (s)

主要威胁 **Major Threat(s)**	无 / None

保护级别与保护行动 Protection Category and Conservation Action (s)

国家重点保护野生动物等级 **(2021) Category of National Key Protected Wild Animals (2021)**	无 / NA
IUCN 红色名录 **(2020-2) IUCN Red List (2020-2)**	无危 / LC
CITES 附录 **(2019) CITES Appendix (2019)**	无 / NA
保护行动 **Conservation Action(s)**	无 / None

相关文献 Relevant References

Burgin *et al.*, 2018; Jiang *et al.* (蒋志刚等), 2017; Wilson *et al.*, 2016; Zheng *et al.* (郑智民等), 2012; Smith *et al.* (史密斯等), 2009; Luo (罗泽珣), 2000; Shou (寿振黄), 1962

棕色田鼠

Lasiopodomys mandarinus

无危 LC

数据缺乏 DD	无危 LC	近危 NT	易危 VU	濒危 EN	极危 CR	区域灭绝 RE	野外灭绝 EW	灭绝 EX

分类地位 Taxonomic Status

动 物 界 Animalia	脊索动物门 Chordata	哺 乳 纲 Mammalia	啮 齿 目 Rodentia	仓 鼠 科 Cricetidae

学 名 Scientific Name	*Lasiopodomys mandarinus*
命 名 人 Species Authority	Milne-Edwards, 1871
英 文 名 English Name(s)	Mandarin Vole
同物异名 Synonym(s)	*Lasiopodomys faeceus* (G. M. Allen, 1924); *jeholensis* (Mori, 1939); *johannes* (Thomas, 1910); *kishidai* (Mori, 1930); *mandrianus* (Miller, 1896); *pullus* (Miller, 1911); (转下页)
种下单元评估 Infra-specific Taxa Assessed	无 / None

评估信息 Assessment Information

评 估 年 份 Year Assessed	2020
评 定 人 Assessor(s)	刘少英、蒋志刚 / Shaoying Liu, Zhigang Jiang
审 定 人 Reviewer(s)	马勇、鲍毅新 / Yong Ma, Yixin Bao
其他贡献人 Other Contributor(s)	李立立、丁晨晨 / Lili Li, Chenchen Ding

理由 Justification: 棕色田鼠分布广，种群数量大。因此，列为无危等级 / Mandarin Vole is a widespread species with large populations. Thus, it is listed as Least Concern

地理分布 Geographical Distribution

国内分布 Domestic Distribution

吉林、内蒙古、辽宁、河北、北京、山西、山东、河南、安徽、江苏、陕西 / Jilin, Inner Mongolia (Nei Mongol), Liaoning, Hebei, Beijing, Shanxi, Shandong, Henan, Anhui, Jiangsu, Shaanxi

世界分布 World Distribution

中国；东北亚 / China; Northeast Asia

分布标注 Distribution Note

非特有种 / Non-endemic

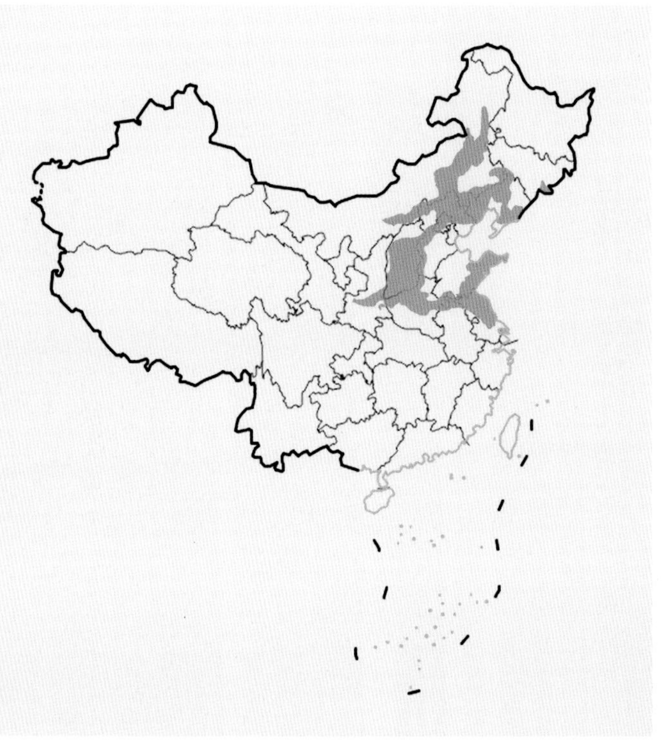

国内分布图
Map of Domestic Distribution

种群 Population

种群数量 **Population Size**	种群数量大 / Large populations
种群趋势 **Population Trend**	未知 / Unknown

生境与生态系统 Habitat (s) and Ecosystem (s)

生　　境 **Habitat(s)**	草地、溪流边、灌丛 / Grassland, Near Stream, Shrubland
生态系统 **Ecosystem(s)**	灌丛生态系统、草地生态系统、湖泊河流生态系统 / Shrubland Ecosystem, Grassland Ecosystem, Lake and River Ecosystem

威胁 Threat (s)

主要威胁 **Major Threat(s)**	无 / None

保护级别与保护行动 Protection Category and Conservation Action (s)

国家重点保护野生动物等级 **(2021) Category of National Key Protected Wild Animals (2021)**	无 / NA
IUCN 红色名录 **(2020-2) IUCN Red List (2020-2)**	无危 / LC
CITES 附录 **(2019) CITES Appendix (2019)**	无 / NA
保护行动 **Conservation Action(s)**	无 / None

相关文献 Relevant References

Burgin *et al*., 2018; Jiang *et al.* (蒋志刚等), 2017; Wilson *et al*., 2016; Zheng *et al.* (郑智民等), 2012; Luo (罗泽珣), 2000; Fan and Liu (樊龙锁和刘焕金), 1998; Wang and Xu (王廷正和许文贤), 1993

(接上页)
vinogradovi (Fetisov, 1936)

棕色田鼠 *Lasiopodomys mandarinus*

笔尾树鼠

Chiropodomys gliroides

无危 LC

数据缺乏 DD	无危 LC	近危 NT	易危 VU	濒危 EN	极危 CR	区域灭绝 RE	野外灭绝 EW	灭绝 EX

分类地位 Taxonomic Status

动物界 Animalia	脊索动物门 Chordata	哺乳纲 Mammalia	啮齿目 Rodentia	鼠科 Muridae

学名 **Scientific Name**	*Chiropodomys gliroides*
命名人 **Species Authority**	Blyth, 1856
英文名 **English Name(s)**	Indomalayan Pencil-tailed Tree Mouse
同物异名 **Synonym(s)**	Pencil-tailed Tree Mouse; *Chiropodomys ana* (Thomas and Wroughton, 1909); *jingdongensis* (Wu and Deng, 1984); *niadis* (Miller, 1903); *peguensis* (Blyth, 1859); *penicillatus* (Peters, 1868)
种下单元评估 **Infra-specific Taxa Assessed**	无 / None

评估信息 Assessment Information

评估年份 **Year Assessed**	2020
评定人 **Assessor(s)**	刘少英、蒋志刚 / Shaoying Liu, Zhigang Jiang
审定人 **Reviewer(s)**	马勇、鲍毅新 / Yong Ma, Yixin Bao
其他贡献人 **Other Contributor(s)**	李立立、丁晨晨 / Lili Li, Chenchen Ding

理由 **Justification:** 笔尾树鼠分布广，种群数量大。因此，列为无危等级 / Indomalayan Pencil-tailed Tree Mouse is a widespread species with large populations. Thus, it is listed as Least Concern

地理分布 Geographical Distribution

国内分布 Domestic Distribution

云南、广西、海南 / Yunnan, Guangxi, Hainan

世界分布 World Distribution

中国；南亚、东南亚 / China; South Asia, Southeast Asia

分布标注 Distribution Note

非特有种 / Non-endemic

国内分布图
Map of Domestic Distribution

种群 Population

种群数量 Population Size	数量多 / Abundant
种群趋势 Population Trend	稳定 / Stable

生境与生态系统 Habitat (s) and Ecosystem (s)

生　　境 Habitat(s)	森林、次生林 / Forest, Secondary Forest
生态系统 Ecosystem(s)	森林生态系统 / Forest Ecosystem

威胁 Threat (s)

主要威胁 Major Threat(s)	无 / None

保护级别与保护行动 Protection Category and Conservation Action (s)

国家重点保护野生动物等级 (2021) Category of National Key Protected Wild Animals (2021)	无 / NA
IUCN 红色名录 (2020-2) IUCN Red List (2020-2)	无危 / LC
CITES 附录 (2019) CITES Appendix (2019)	无 / NA
保护行动 Conservation Action(s)	无 / None

相关文献 Relevant References

Burgin *et al.*, 2018; Jiang *et al.* (蒋志刚等), 2017; Wilson *et al.*, 2016; Zheng *et al.* (郑智民等), 2012; Smith *et al.* (史密斯等) , 2009; Wang (王应祥), 2003

笔尾树鼠 *Chiropodomys gliroides*

© Lynx Edicions

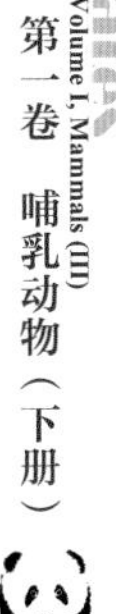

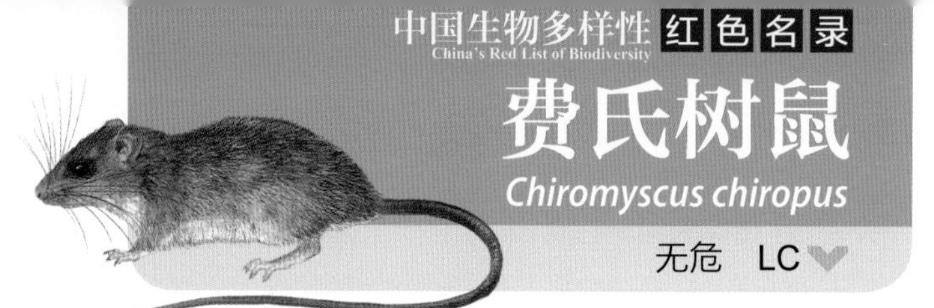

费氏树鼠
Chiromyscus chiropus

无危 LC

数据缺乏 DD	无危 LC	近危 NT	易危 VU	濒危 EN	极危 CR	区域灭绝 RE	野外灭绝 EW	灭绝 EX

分类地位 Taxonomic Status

动物界 Animalia	脊索动物门 Chordata	哺乳纲 Mammalia	啮齿目 Rodentia	鼠科 Muridae

学　　名 **Scientific Name**	*Chiromyscus chiropus*
命 名 人 **Species Authority**	Thomas, 1891
英 文 名 **English Name(s)**	Indochinese Chiromyscus
同物异名 **Synonym(s)**	Fea's Tree Rat
种下单元评估 **Infra-specific Taxa Assessed**	无 / None

评估信息 Assessment Information

评 估 年 份 **Year Assessed**	2020
评　定　人 **Assessor(s)**	刘少英、蒋志刚 / Shaoying Liu, Zhigang Jiang
审　定　人 **Reviewer(s)**	马勇、鲍毅新 / Yong Ma, Yixin Bao
其他贡献人 **Other Contributor(s)**	李立立、丁晨晨 / Lili Li, Chenchen Ding

理由 **Justification:** 费氏树鼠的种群数量大。因此，列为无危等级 / Indochinese Chiromyscus has large populations. Thus, it is listed as Least Concern

地理分布 Geographical Distribution

国内分布 Domestic Distribution

云南 / Yunnan

世界分布 World Distribution

中国；东南亚 / China; Southeast Asia

分布标注 Distribution Note

非特有种 / Non-endemic

国内分布图
Map of Domestic Distribution

种群 Population

种群数量 **Population Size**	种群数量大 / Large populations
种群趋势 **Population Trend**	稳定 / Stable

生境与生态系统 Habitat (s) and Ecosystem (s)

生　　境 **Habitat(s)**	森林 / Forest
生态系统 **Ecosystem(s)**	森林生态系统 / Forest Ecosystem

威胁 Threat (s)

主要威胁 **Major Threat(s)**	无 / None

保护级别与保护行动 Protection Category and Conservation Action (s)

国家重点保护野生动物等级 **(2021) Category of National Key Protected Wild Animals (2021)**	无 / NA
IUCN 红色名录 **(2020-2) IUCN Red List (2020-2)**	无危 / LC
CITES 附录 **(2019) CITES Appendix (2019)**	无 / NA
保护行动 **Conservation Action(s)**	无 / None

相关文献 Relevant References

Burgin *et al.*, 2018; Jiang *et al.* (蒋志刚等), 2017; Wilson *et al.*, 2016; Zheng *et al.* (郑智民等), 2012; Smith *et al.* (史密斯等), 2009; Pan *et al.* (潘清华等), 2007; Wang (王应祥), 2003

费氏树鼠 *Chiromyscus chiropus*

南洋鼠

Chiromyscus langbianis

无危 LC

数据缺乏 DD	无危 LC	近危 NT	易危 VU	濒危 EN	极危 CR	区域灭绝 RE	野外灭绝 EW	灭绝 EX

分类地位 Taxonomic Status

动物界 Animalia	脊索动物门 Chordata	哺乳纲 Mammalia	啮齿目 Rodentia	鼠科 Muridae

学名 Scientific Name	*Chiromyscus langbianis*
命名人 Species Authority	Robinson and Kloss, 1922
英文名 English Name(s)	Indochinese Arboreal Niviventer
同物异名 Synonym(s)	无 / None
种下单元评估 Infra-specific Taxa Assessed	无 / None

评估信息 Assessment Information

评估年份 Year Assessed	2020
评定人 Assessor(s)	刘少英、蒋志刚 / Shaoying Liu, Zhigang Jiang
审定人 Reviewer(s)	马勇、鲍毅新 / Yong Ma, Yixin Bao
其他贡献人 Other Contributor(s)	李立立、丁晨晨 / Lili Li, Chenchen Ding

理由 Justification: 南洋鼠在其分布区内种群数量大。因此，列为无危等级 / Indochinese Arboreal Niviventer has large populations within its range. Thus, it is listed as Least Concern

地理分布 Geographical Distribution

国内分布 Domestic Distribution

云南 / Yunnan

世界分布 World Distribution

中国；东南亚 / China; Southeast Asia

分布标注 Distribution Note

非特有种 / Non-endemic

国内分布图
Map of Domestic Distribution

种群 Population

种群数量 Population Size	种群数量大 / Large populations
种群趋势 Population Trend	未知 / Unknown

生境与生态系统 Habitat (s) and Ecosystem (s)

生　　境 Habitat(s)	森林 / Forest
生态系统 Ecosystem(s)	森林生态系统 / Forest Ecosystem

威胁 Threat (s)

主要威胁 Major Threat(s)	无 / None

保护级别与保护行动 Protection Category and Conservation Action (s)

国家重点保护野生动物等级 (2021) Category of National Key Protected Wild Animals (2021)	无 / NA
IUCN 红色名录 (2020-2) IUCN Red List (2020-2)	无危 / LC
CITES 附录 (2019) CITES Appendix (2019)	无 / NA
保护行动 Conservation Action(s)	无 / None

相关文献 Relevant References

Burgin *et al*., 2018; Jiang *et al.* (蒋志刚等), 2017; Wilson *et al*., 2016; Smith *et al.* (史密斯等), 2009

红耳巢鼠

Micromys erythrotis

无危 LC

数据缺乏 DD	无危 LC	近危 NT	易危 VU	濒危 EN	极危 CR	区域灭绝 RE	野外灭绝 EW	灭绝 EX

分类地位 Taxonomic Status

动物界 Animalia	脊索动物门 Chordata	哺乳纲 Mammalia	啮齿目 Rodentia	鼠科 Muridae

学名 Scientific Name	*Micromys erythrotis*
命名人 Species Authority	Blyth, 1855
英文名 English Name(s)	Red-eared Harvest Mouse
同物异名 Synonym(s)	*Micromys fidelis* (Hinton, 1923)
种下单元评估 Infra-specific Taxa Assessed	无 / None

评估信息 Assessment Information

评估年份 Year Assessed	2020
评定人 Assessor(s)	刘少英、蒋志刚 / Shaoying Liu, Zhigang Jiang
审定人 Reviewer(s)	马勇、鲍毅新 / Yong Ma, Yixin Bao
其他贡献人 Other Contributor(s)	李立立、丁晨晨 / Lili Li, Chenchen Ding

理由 Justification: 红耳巢鼠分布广，种群数量大。因此，列为无危等级 / Red-eared Harvest Mouse is a widespread species with large populations. Thus, it is listed as Least Concern

地理分布 Geographical Distribution

国内分布 Domestic Distribution

西藏、云南、四川、重庆、贵州、广西、福建 / Tibet (Xizang), Yunnan, Sichuan, Chongqing, Guizhou, Guangxi, Fujian

世界分布 World Distribution

中国；南亚、东南亚 / China; South Asia, Southeast Asia

分布标注 Distribution Note

非特有种 / Non-endemic

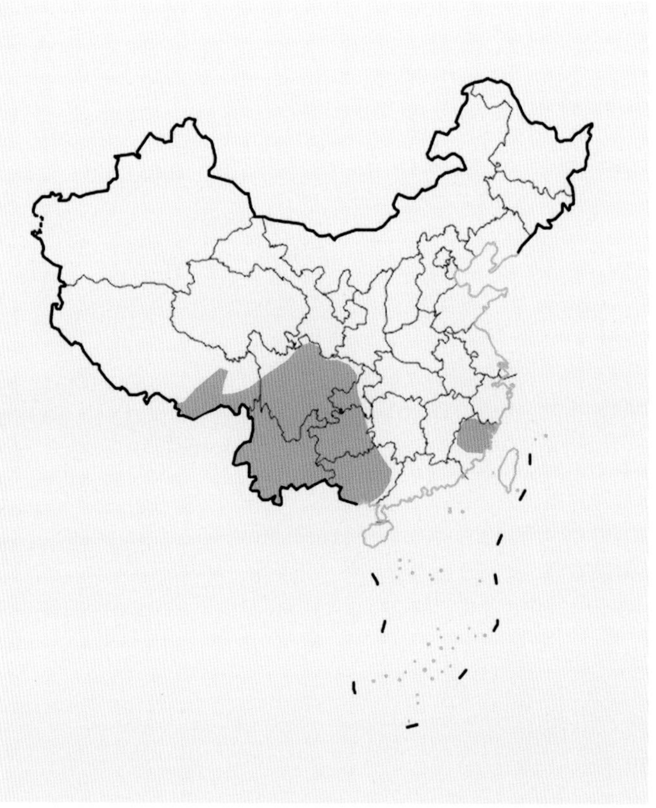

国内分布图
Map of Domestic Distribution

种群 Population

种群数量 **Population Size**	种群数量大 / Large populations
种群趋势 **Population Trend**	未知 / Unknown

生境与生态系统 Habitat (s) and Ecosystem (s)

生　　境 **Habitat(s)**	耕地、竹林 / Arable Land, Bamboo Grove
生态系统 **Ecosystem(s)**	森林生态系统、农田生态系统 / Forest Ecosystem, Cropland Ecosystem

威胁 Threat (s)

主要威胁 **Major Threat(s)**	无 / None

保护级别与保护行动 Protection Category and Conservation Action (s)

国家重点保护野生动物等级 **(2021) Category of National Key Protected Wild Animals (2021)**	无 / NA
IUCN 红色名录 **(2020-2) IUCN Red List (2020-2)**	无危 / LC
CITES 附录 **(2019) CITES Appendix (2019)**	无 / NA
保护行动 **Conservation Action(s)**	无 / None

相关文献 Relevant References

Burgin *et al*., 2018; Jiang *et al.* (蒋志刚等), 2017; Wilson *et al*., 2016; Abramov *et al*., 2009

红耳巢鼠 *Micromys erythrotis*　　沈成 摄 By Cheng Shen

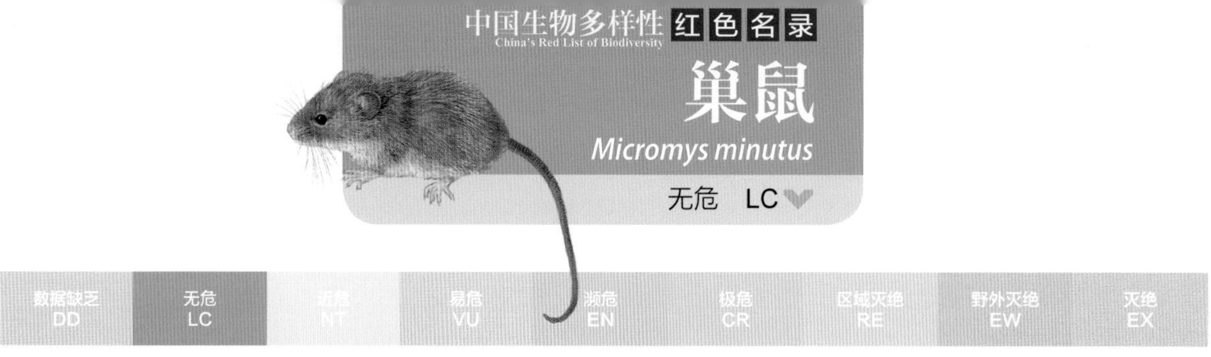

分类地位 Taxonomic Status

动 物 界 **Animalia**	脊索动物门 **Chordata**	哺 乳 纲 **Mammalia**	啮 齿 目 **Rodentia**	鼠 科 **Muridae**
学 名 **Scientific Name**		*Micromys minutus*		
命 名 人 **Species Authority**		Pallas, 1771		
英 文 名 **English Name(s)**		Eurasian Harvest Mouse		
同物异名 **Synonym(s)**		*Micromys agilis* (Dehne, 1841); *aokii* (Kuroda, 1922); *arundinaceus* (Petenyi, 1882); *arvensis* (Leach, 1816); *avenarius* (Wolf, 1794); *batarovi* (Kastschenko, (转下页)		
种下单元评估 **Infra-specific Taxa Assessed**		无 / None		

评估信息 Assessment Information

评 估 年 份 **Year Assessed**	2020
评 定 人 **Assessor(s)**	刘少英、蒋志刚 / Shaoying Liu, Zhigang Jiang
审 定 人 **Reviewer(s)**	马勇、鲍毅新 / Yong Ma, Yixin Bao
其他贡献人 **Other Contributor(s)**	李立立、丁晨晨 / Lili Li, Chenchen Ding

理由 **Justification:** 巢鼠分布广，种群数量大。因此，列为无危等级 / Eurasian Harvest Mouse is a widespread species with large populations. Thus, it is listed as Least Concern

地理分布 Geographical Distribution

国内分布 Domestic Distribution

黑龙江、吉林、辽宁、内蒙古、河北、陕西、甘肃、宁夏、四川、贵州、新疆、江苏、安徽、浙江、湖北、湖南、江西、广东、广西、福建、台湾、重庆 / Heilongjiang, Jilin, Liaoning, Inner Mongolia (Nei Mongol), Hebei, Shaanxi, Gansu, Ningxia, Sichuan, Guizhou, Xinjiang, Jiangsu, Anhui, Zhejiang, Hubei, Hunan, Jiangxi, Guangdong, Guangxi, Fujian, Taiwan, Chongqing

世界分布 World Distribution

亚洲、欧洲 / Asia, Europe

分布标注 Distribution Note

非特有种 / Non-endemic

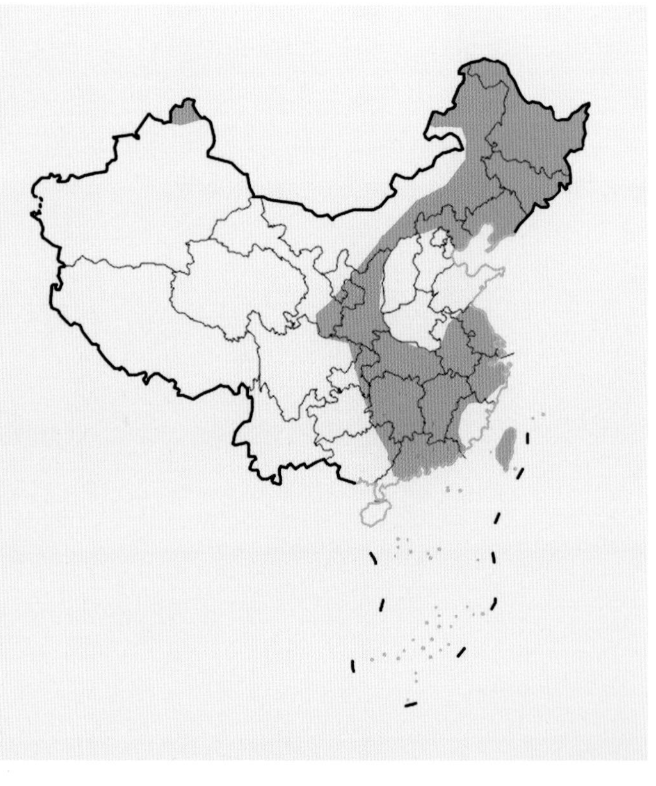

国内分布图
Map of Domestic Distribution

种群 Population

种群数量 Population Size	种群数量大 / Large populations
种群趋势 Population Trend	稳定 / Stable

生境与生态系统 Habitat (s) and Ecosystem (s)

生　　境 Habitat(s)	耕地、竹林 / Arable Land, Bamboo Grove
生态系统 Ecosystem(s)	森林生态系统、农田生态系统 / Forest Ecosystem, Cropland Ecosystem

威胁 Threat (s)

主要威胁 Major Threat(s)	无 / None

保护级别与保护行动 Protection Category and Conservation Action (s)

国家重点保护野生动物等级 (2021) Category of National Key Protected Wild Animals (2021)	无 / NA
IUCN 红色名录 (2020-2) IUCN Red List (2020-2)	无危 / LC
CITES 附录 (2019) CITES Appendix (2019)	无 / NA
保护行动 Conservation Action(s)	无 / None

相关文献 Relevant References

Burgin *et al*., 2018; Jiang *et al.* (蒋志刚等), 2017; Wilson *et al*., 2016; Zheng *et al.* (郑智民等), 2012; Smith *et al.* (史密斯等), 2009; Pan *et al.* (潘清华等), 2007; Wang (王应祥), 2003

(接上页)
1910); *berezowskii* (Argyropulo, 1929); *brauneri* (Martino, 1930); *campestris* (Desmarest, 1822); *danubialis* (Simonescu, 1971); *erythrotis* (Blyth, 1856); *fenniae* (Hilzheimer, 1911); *flavus* (Kerr, 1792); *hertigi* (Johnson and Jones, 1955); *hondonis* (Kuroda, 1933); *japonicus* (Thomas, 1906); *kastschenkoi* (Charlamagne, 1915); *kurodai* (Mori, 1942); *kytmanovi* (Kastschenko, 1910); *mehelyi* (Bolkay, 1925); *meridionalis* (Costa, 1844); *messorius* (Kerr, 1792); *minatus* (Schinz, 1840); *minimus* (White, 1789); *oryzivorus* (de Sélys-Longchamps, 1841); *parvulus* (Hermann, 1804); *pendulinus* (Hermann, 1804); *pianmaensis* (Peng, 1981); *pratensis* (Ockskay, 1831); *pumilus* (F. Cuvier, 1842); *pygmaeus* (Milne-Edwards, 1872); *sareptae* (Hilzheimer, 1911); *shenshiensis* (Li, Wu and Shao, 1965); *soricinus* (Hermann, 1780); *subobscurus* (Fritsche, 1934); *takasagoensis* (Tokuda, 1939); *takasagoensis* (Tokuda, 1941); *triticeus* (Boaert, 1785); *typicus* (Barrett-Hamilton, 1899); *ussuricus* (Barrett-Hamilton, 1899); *zhenjiangensis* (Huang, 1989)

巢鼠 *Micromys minutus*

黑线姬鼠
Apodemus agrarius

无危 LC

数据缺乏 DD | 无危 LC | 近危 NT | 易危 VU | 濒危 EN | 极危 CR | 区域灭绝 RE | 野外灭绝 EW | 灭绝 EX

分类地位 Taxonomic Status

动物界 Animalia	脊索动物门 Chordata	哺乳纲 Mammalia	啮齿目 Rodentia	鼠科 Muridae
学名 Scientific Name		*Apodemus agrarius*		
命名人 Species Authority		Pallas, 1771		
英文名 English Name(s)		Striped Field Mouse		
同物异名 Synonym(s)		*Apodemus albostriatus* (Bechstein, 1801); *caucasicus* (Kuznetzov, 1944); *chejuensis* (Johnson and Jones, 1955); *coreae* (Thomas, 1908); *gloveri* (Kuroda, 1939); *harti* (转下页)		
种下单元评估 Infra-specific Taxa Assessed		无 / None		

评估信息 Assessment Information

评估年份 Year Assessed	2020
评定人 Assessor(s)	刘少英、蒋志刚 / Shaoying Liu, Zhigang Jiang
审定人 Reviewer(s)	马勇、鲍毅新 / Yong Ma, Yixin Bao
其他贡献人 Other Contributor(s)	李立立、丁晨晨 / Lili Li, Chenchen Ding

理由 Justification: 黑线姬鼠分布广，种群数量大。因此，列为无危等级 / Striped Field Mouse is a widespread species with large populations. Thus, it is listed as Least Concern

地理分布 Geographical Distribution

国内分布 Domestic Distribution

黑龙江、吉林、辽宁、内蒙古、河北、北京、天津、山东、河南、山西、陕西、宁夏、甘肃、上海、江苏、安徽、浙江、江西、湖北、湖南、四川、贵州、云南、广西、广东、福建、台湾、新疆、重庆 / Heilongjiang, Jilin, Liaoning, Inner Mongolia (Nei Mongol), Hebei, Beijing, Tianjin, Shandong, Henan, Shanxi, Shaanxi, Ningxia, Gansu, Shanghai, Jiangsu, Anhui, Zhejiang, Jiangxi, Hubei, Hunan, Sichuan, Guizhou, Yunnan, Guangxi, Guangdong, Fujian, Taiwan, Xinjiang, Chongqing

世界分布 World Distribution

亚洲、欧洲 / Asia, Europe

分布标注 Distribution Note

非特有种 / Non-endemic

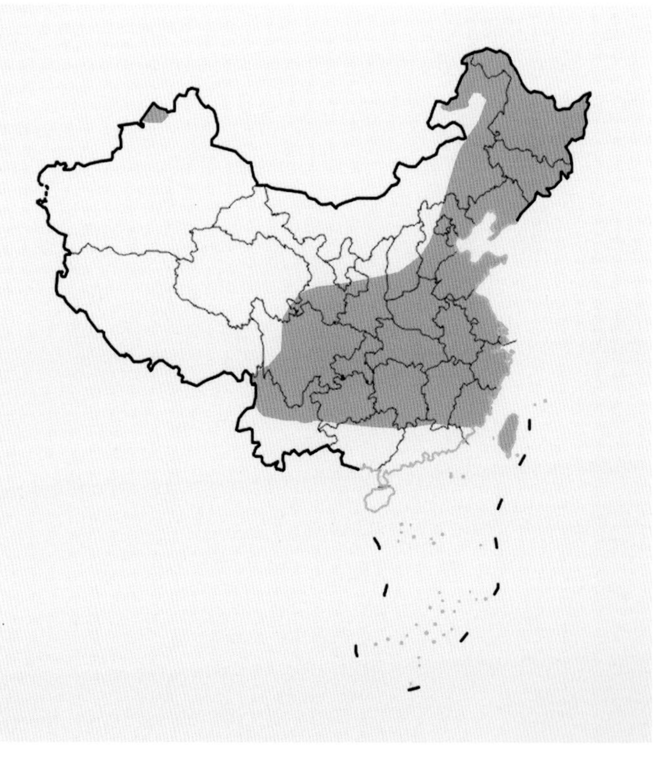

国内分布图
Map of Domestic Distribution

种群 Population

种群数量 Population Size	种群数量大 / Large populations
种群趋势 Population Trend	稳定 / Stable

生境与生态系统 Habitat (s) and Ecosystem (s)

生　　境 Habitat(s)	耕地、草地、森林、灌丛 / Arable Land, Grassland, Forest, Shrubland
生态系统 Ecosystem(s)	森林生态系统、草地生态系统、农田生态系统、灌丛生态系统 / Forest Ecosystem, Grassland Ecosystem, Cropland Ecosystem, Shrubland Ecosystem

威胁 Threat (s)

主要威胁 Major Threat(s)	无 / None

保护级别与保护行动 Protection Category and Conservation Action (s)

国家重点保护野生动物等级 (2021) Category of National Key Protected Wild Animals (2021)	无 / NA
IUCN 红色名录 (2020-2) IUCN Red List (2020-2)	无危 / LC
CITES 附录 (2019) CITES Appendix (2019)	无 / NA
保护行动 Conservation Action(s)	无 / None

相关文献 Relevant References

Burgin *et al*., 2018; Jiang *et al.* (蒋志刚等), 2017; Wilson *et al*., 2016; Zheng *et al.* (郑智民等), 2012; Yang *et al.* (杨再学等), 2007; Wang (王应祥), 2003

(接上页)
(Thomas, 1898); *henrici* (Lehmann, 1970); *insulaemus* (Tokuda, 1939); *insulaemus* (Tokuda, 1941); *istrianus* (Kryštufek, 1985); *kahmanni* (Malec and Storch, 1963); *karelicus* (Ehrström, 1914); *maculatus* (Bechstein, 1801); *mantchuricus* (Thomas, 1898); *nicolskii* (Charlemagne, 1933); *nikolskii* (Migouline, 1927); *ningpoensis* (Swinhoe, 1870); *ognevi* (Johansen, 1923); *pallescens* (Johnson and Jones, 1955); *pallidior* (Thomas, 1908); *pratensis* (Ockskay, 1831); *rubens* (Oken, 1816); *septentrionalis* (Ognev, 1924); *tianschanicus* (Ognev, 1940); *volgensis* (Kuznetzov, 1944)

黑线姬鼠 *Apodemus agrarius*

高山姬鼠

Apodemus chevrieri

无危 LC

数据缺乏 DD	无危 LC	近危 NT	易危 VU	濒危 EN	极危 CR	区域灭绝 RE	野外灭绝 EW	灭绝 EX

分类地位 Taxonomic Status

动物界 Animalia	脊索动物门 Chordata	哺乳纲 Mammalia	啮齿目 Rodentia	鼠科 Muridae

学名 Scientific Name	*Apodemus chevrieri*
命名人 Species Authority	Milne-Edwards, 1868
英文名 English Name(s)	Chevrier's Field Mouse
同物异名 Synonym(s)	*Apodemus fergussoni* (Thomas, 1911）
种下单元评估 Infra-specific Taxa Assessed	无 / None

评估信息 Assessment Information

评估年份 Year Assessed	2020
评定人 Assessor(s)	刘少英、蒋志刚 / Shaoying Liu, Zhigang Jiang
审定人 Reviewer(s)	马勇、鲍毅新 / Yong Ma, Yixin Bao
其他贡献人 Other Contributor(s)	李立立、丁晨晨 / Lili Li, Chenchen Ding

理由 Justification: 高山姬鼠分布广，种群数量大。因此，列为无危等级 / Chevrier's Field Mouse is a widespread species with large populations. Thus, it is listed as Least Concern

地理分布 Geographical Distribution

国内分布 Domestic Distribution

陕西、甘肃、四川、湖北、贵州、云南、重庆 / Shaanxi, Gansu, Sichuan, Hubei, Guizhou, Yunnan, Chongqing

世界分布 World Distribution

中国 / China

分布标注 Distribution Note

特有种 / Endemic

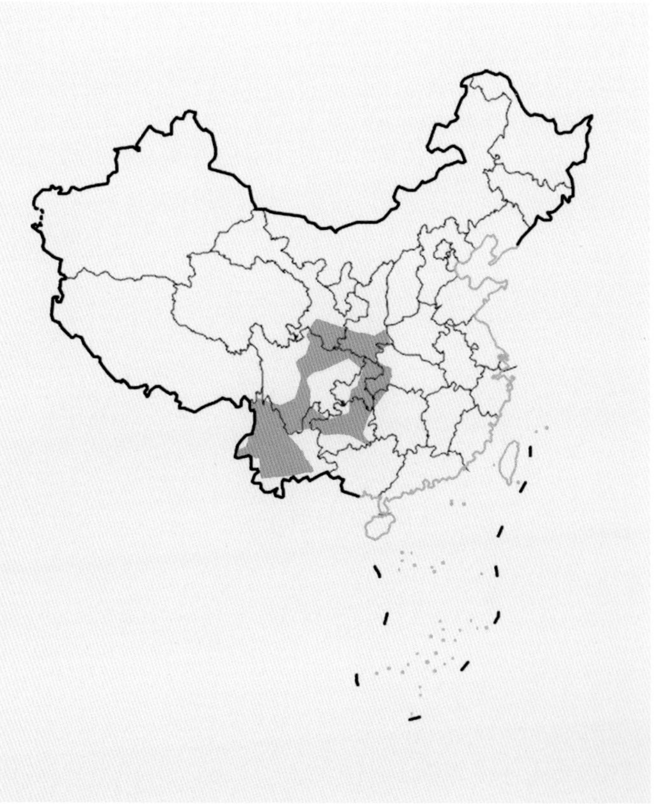

国内分布图
Map of Domestic Distribution

种群 Population

种群数量 Population Size	种群数量大 / Large populations
种群趋势 Population Trend	未知 / Unknown

生境与生态系统 Habitat (s) and Ecosystem (s)

生　　境 Habitat(s)	耕地、草地、森林、灌丛 / Arable Land, Grassland, Forest, Shrubland
生态系统 Ecosystem(s)	森林生态系统、草地生态系统、农田生态系统、灌丛生态系统 / Forest Ecosystem, Grassland Ecosystem, Cropland Ecosystem, Shrubland Ecosystem

威胁 Threat (s)

主要威胁 Major Threat(s)	无 / None

保护级别与保护行动 Protection Category and Conservation Action (s)

国家重点保护野生动物等级 (2021) Category of National Key Protected Wild Animals (2021)	无 / NA
IUCN 红色名录 (2020-2) IUCN Red List (2020-2)	无危 / LC
CITES 附录 (2019) CITES Appendix (2019)	无 / NA
保护行动 Conservation Action(s)	无 / None

相关文献 Relevant References

Burgin *et al.*, 2018; Jiang *et al.* (蒋志刚等), 2017; Wilson *et al.*, 2016; Li *et al.* (黎运喜等), 2012

高山姬鼠 *Apodemus chevrieri*　　© Lynx Edicions

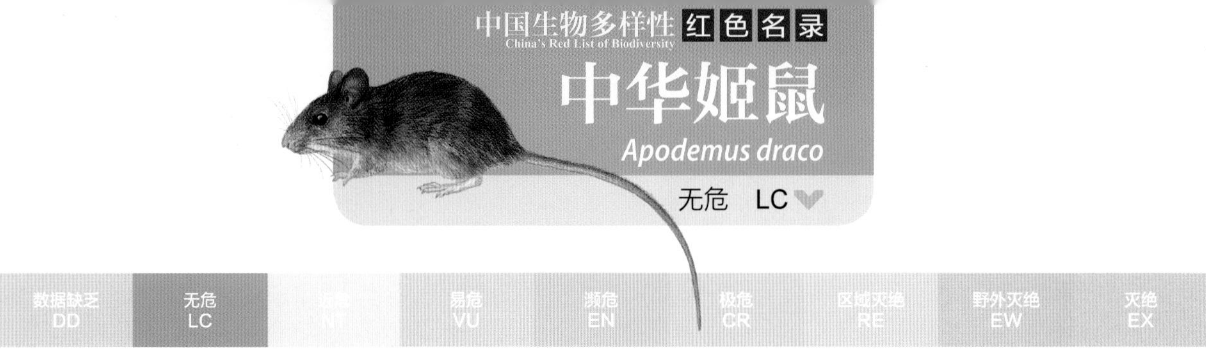

中华姬鼠

Apodemus draco

无危 LC

分类地位 Taxonomic Status

动物界 **Animalia**	脊索动物门 **Chordata**	哺乳纲 **Mammalia**	啮齿目 **Rodentia**	鼠科 **Muridae**
学名 **Scientific Name**		*Apodemus draco*		
命名人 **Species Authority**		Barrett-Hamilton, 1900		
英文名 **English Name(s)**		South China Field Mouse		
同物异名 **Synonym(s)**		*Apodemus argenteus* (Swinhoe, 1870); *badius* (Swinhoe, 1870); *ilex* (Thomas, 1922); *orestes* (Thomas, 1911)		
种下单元评估 **Infra-specific Taxa Assessed**		无 / None		

评估信息 Assessment Information

评估年份 **Year Assessed**	2020
评定人 **Assessor(s)**	刘少英、蒋志刚 / Shaoying Liu, Zhigang Jiang
审定人 **Reviewer(s)**	马勇、鲍毅新 / Yong Ma, Yixin Bao
其他贡献人 **Other Contributor(s)**	李立立、丁晨晨 / Lili Li, Chenchen Ding

理由 **Justification:** 中华姬鼠分布广，种群数量大。因此，列为无危等级 / South China Field Mouse is a widespread species with large populations. Thus, it is listed as Least Concern

地理分布 Geographical Distribution

国内分布 Domestic Distribution

河北、河南、宁夏、陕西、甘肃、四川、云南、青海、西藏、贵州、安徽、浙江、江西、湖北、湖南、广东、广西、福建、台湾、重庆、北京、天津、山西、山东、上海 / Hebei, Henan, Ningxia, Shaanxi, Gansu, Sichuan, Yunnan, Qinghai, Tibet (Xizang), Guizhou, Anhui, Zhejiang, Jiangxi, Hubei, Hunan, Guangdong, Guangxi, Fujian, Taiwan, Chongqing, Beijing, Tianjin, Shanxi, Shandong, Shanghai

世界分布 World Distribution

中国、缅甸、印度 / China, Myanmar, India

分布标注 Distribution Note

非特有种 / Non-endemic

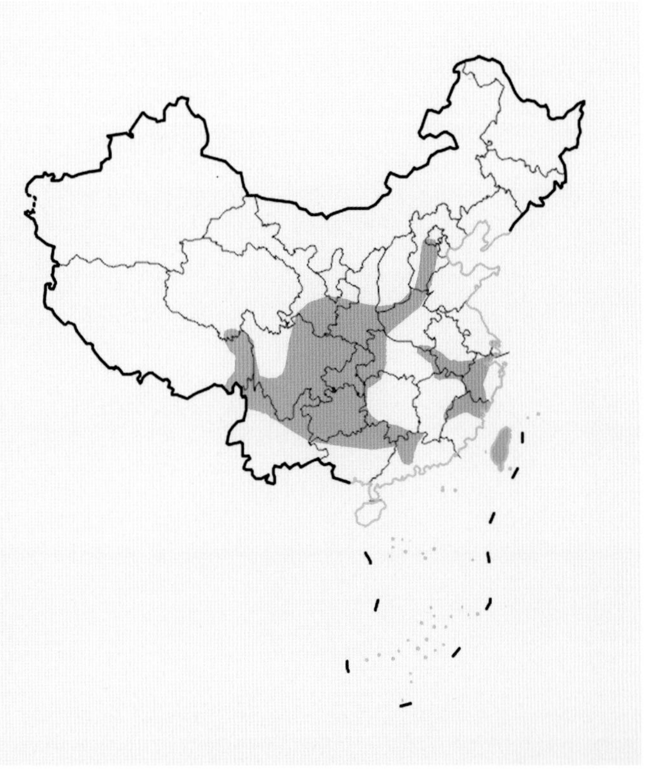

国内分布图
Map of Domestic Distribution

种群 Population

种群数量 Population Size	种群数量大 / Large populations
种群趋势 Population Trend	稳定 / Stable

生境与生态系统 Habitat (s) and Ecosystem (s)

生　　境 Habitat(s)	森林、灌丛 / Forest, Shrubland
生态系统 Ecosystem(s)	森林生态系统、灌丛生态系统 / Forest Ecosystem, Shrubland Ecosystem

威胁 Threat (s)

主要威胁 Major Threat(s)	无 / None

保护级别与保护行动 Protection Category and Conservation Action (s)

国家重点保护野生动物等级 (2021) Category of National Key Protected Wild Animals (2021)	无 / NA
IUCN 红色名录 (2020-2) IUCN Red List (2020-2)	无危 / LC
CITES 附录 (2019) CITES Appendix (2019)	无 / NA
保护行动 Conservation Action(s)	无 / None

相关文献 Relevant References

Burgin *et al*., 2018; Jiang *et al.* (蒋志刚等), 2017; Wilson *et al*., 2016; Li *et al.* (黎运喜等), 2012; Zheng *et al.* (郑智民等), 2012; Smith *et al.* (史密斯等), 2009

数据缺乏 DD	无危 LC	近危 NT	易危 VU	濒危 EN	极危 CR	区域灭绝 RE	野外灭绝 EW	灭绝 EX

分类地位 Taxonomic Status

动物界 Animalia	脊索动物门 Chordata	哺乳纲 Mammalia	啮齿目 Rodentia	鼠科 Muridae

学名 Scientific Name	*Apodemus ilex*
命名人 Species Authority	Thomas, 1922
英文名 English Name(s)	Mekong Field Mouse
同物异名 Synonym(s)	Lancangjiang Field Mouse
种下单元评估 Infra-specific Taxa Assessed	无 / None

评估信息 Assessment Information

评估年份 Year Assessed	2020
评定人 Assessor(s)	刘少英、蒋志刚 / Shaoying Liu, Zhigang Jiang
审定人 Reviewer(s)	马勇、鲍毅新 / Yong Ma, Yixin Bao
其他贡献人 Other Contributor(s)	李立立、丁晨晨 / Lili Li, Chenchen Ding

理由 Justification: 澜沧江姬鼠分布广，种群数量大。因此，列为无危等级 / Mekong Field Mouse is a widespread species with large populations. Thus, it is listed as Least Concern

地理分布 Geographical Distribution

国内分布 Domestic Distribution

云南 / Yunnan

世界分布 World Distribution

中国 / China

分布标注 Distribution Note

特有种 / Endemic

国内分布图
Map of Domestic Distribution

种群 Population

种群数量 **Population Size**	未知 / Unknown
种群趋势 **Population Trend**	未知 / Unknown

生境与生态系统 Habitat (s) and Ecosystem (s)

生　　境 **Habitat(s)**	森林 / Forest
生态系统 **Ecosystem(s)**	森林生态系统 / Forest Ecosystem

威胁 Threat (s)

主要威胁 **Major Threat(s)**	无 / None

保护级别与保护行动 Protection Category and Conservation Action (s)

国家重点保护野生动物等级 **(2021) Category of National Key Protected Wild Animals (2021)**	无 / NA
IUCN 红色名录 **(2020-2) IUCN Red List (2020-2)**	无危 / LC
CITES 附录 **(2019) CITES Appendix (2019)**	无 / NA
保护行动 **Conservation Action(s)**	无 / None

相关文献 Relevant References

Burgin *et al*., 2018; Jiang *et al.* (蒋志刚等), 2017; Wilson *et al*., 2016; Pan *et al.* (潘清华等), 2007

澜沧江姬鼠 *Apodemus ilex*　　沈成 摄　By Cheng Shen

大耳姬鼠

Apodemus latronum

无危 LC

数据缺乏 DD	无危 LC	近危 NT	易危 VU	濒危 EN	极危 CR	区域灭绝 RE	野外灭绝 EW	灭绝 EX

分类地位 Taxonomic Status

动物界 Animalia	脊索动物门 Chordata	哺乳纲 Mammalia	啮齿目 Rodentia	鼠科 Muridae

学名 Scientific Name	*Apodemus latronum*
命名人 Species Authority	Thomas, 1911
英文名 English Name(s)	Large-eared Field Mouse
同物异名 Synonym(s)	Sichuan Field Mouse
种下单元评估 Infra-specific Taxa Assessed	无 / None

评估信息 Assessment Information

评估年份 Year Assessed	2020
评定人 Assessor(s)	刘少英、蒋志刚 / Shaoying Liu, Zhigang Jiang
审定人 Reviewer(s)	马勇、鲍毅新 / Yong Ma, Yixin Bao
其他贡献人 Other Contributor(s)	李立立、丁晨晨 / Lili Li, Chenchen Ding

理由 **Justification:** 大耳姬鼠分布广，种群数量大。因此，列为无危等级 / Large-eared Field Mouse is a widespread species with large populations. Thus, it is listed as Least Concern

地理分布 Geographical Distribution

国内分布 Domestic Distribution

四川、青海、西藏、云南 / Sichuan, Qinghai, Tibet (Xizang), Yunnan

世界分布 World Distribution

中国、缅甸、印度 / China, Myanmar, India

分布标注 Distribution Note

非特有种 / Non-endemic

国内分布图
Map of Domestic Distribution

种群 Population

种群数量 Population Size	种群数量大 / Large populations
种群趋势 Population Trend	未知 / Unknown

生境与生态系统 Habitat (s) and Ecosystem (s)

生　　境 Habitat(s)	森林、草地、灌丛 / Forest, Grassland, Shrubland
生态系统 Ecosystem(s)	森林生态系统、草地生态系统、灌丛生态系统 / Forest Ecosystem, Grassland Ecosystem, Shrubland Ecosystem

威胁 Threat (s)

主要威胁 Major Threat(s)	无 / None

保护级别与保护行动 Protection Category and Conservation Action (s)

国家重点保护野生动物等级 (2021) Category of National Key Protected Wild Animals (2021)	无 / NA
IUCN 红色名录 (2020-2) IUCN Red List (2020-2)	无危 / LC
CITES 附录 (2019) CITES Appendix (2019)	无 / NA
保护行动 Conservation Action(s)	无 / None

相关文献 Relevant References

Burgin *et al*., 2018; Jiang *et al.* (蒋志刚等), 2017; Wilson *et al*., 2016; Zheng *et al.* (郑智民等), 2012; Fan *et al.* (范振鑫等), 2010; Wang (王应祥), 2003; Chen *et al.* (陈志平等), 1996

大林姬鼠

Apodemus peninsulae

无危 LC

分类地位 Taxonomic Status

动物界 Animalia	脊索动物门 Chordata	哺乳纲 Mammalia	啮齿目 Rodentia	鼠科 Muridae
学名 **Scientific Name**		*Apodemus peninsulae*		
命名人 **Species Authority**		Thomas, 1907		
英文名 **English Name(s)**		Korean Field Mouse		
同物异名 **Synonym(s)**		Korean Wood Mouse; *Apodemus giliacus* (Thomas, 1907); *major* (Rae, 1862); *majusculus* (Turov, 1924); *nigritalus* (Hollister, 1913); *praetor* (Miller, 1914); *qinghaiensis* (转下页)		
种下单元评估 **Infra-specific Taxa Assessed**		无 / None		

评估信息 Assessment Information

评估年份 **Year Assessed**	2020
评定人 **Assessor(s)**	刘少英、蒋志刚 / Shaoying Liu, Zhigang Jiang
审定人 **Reviewer(s)**	马勇、鲍毅新 / Yong Ma, Yixin Bao
其他贡献人 **Other Contributor(s)**	李立立、丁晨晨 / Lili Li, Chenchen Ding

理由 **Justification:** 大林姬鼠分布广，种群数量大。因此，列为无危等级 / Korean Field Mouse is a widespread species with large populations. Thus, it is listed as Least Concern

地理分布 Geographical Distribution

国内分布 Domestic Distribution

黑龙江、吉林、辽宁、内蒙古、河北、天津、北京、山东、河南、山西、陕西、甘肃、宁夏、青海、四川、西藏、云南、湖北 / Heilongjiang, Jilin, Liaoning, Inner Mongolia (Nei Mongol), Hebei, Tianjin, Beijing, Shandong, Henan, Shanxi, Shaanxi, Gansu, Ningxia, Qinghai, Sichuan, Tibet (Xizang), Yunnan, Hubei

世界分布 World Distribution

中国、日本、哈萨克斯坦、朝鲜、韩国、俄罗斯 / China, Japan, Kazakhstan, Korea (the Democratic People's Republic of), Korea (the Republic of), Russia

分布标注 Distribution Note

非特有种 / Non-endemic

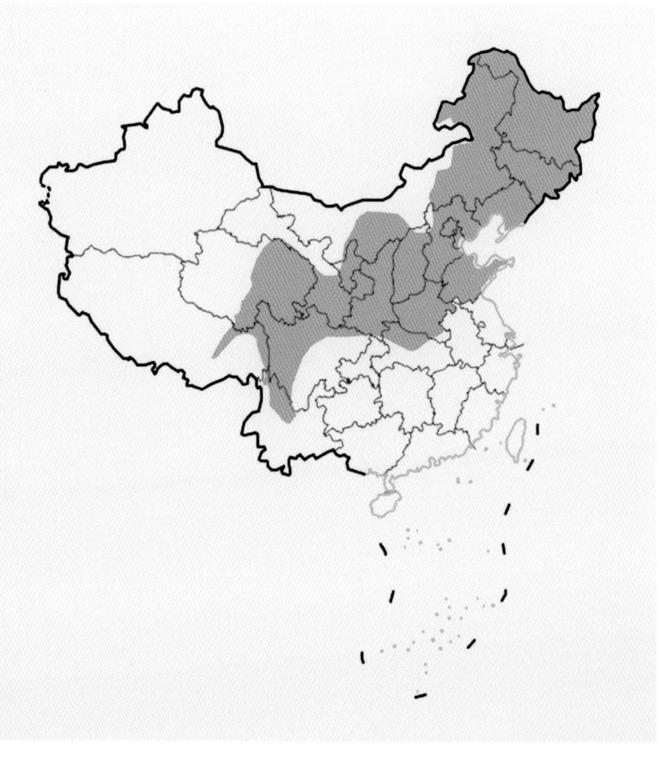

国内分布图
Map of Domestic Distribution

种群 Population

种群数量 Population Size	种群数量大 / Large populations
种群趋势 Population Trend	稳定 / Stable

生境与生态系统 Habitat (s) and Ecosystem (s)

生　　境 Habitat(s)	灌丛、森林 / Shrubland, Forest
生态系统 Ecosystem(s)	森林生态系统、灌丛生态系统 / Forest Ecosystem, Shrubland Ecosystem

威胁 Threat (s)

主要威胁 Major Threat(s)	无 / None

保护级别与保护行动 Protection Category and Conservation Action (s)

国家重点保护野生动物等级 (2021) Category of National Key Protected Wild Animals (2021)	无 / NA
IUCN 红色名录 (2020-2) IUCN Red List (2020-2)	无危 / LC
CITES 附录 (2019) CITES Appendix (2019)	无 / NA
保护行动 Conservation Action(s)	无 / None

相关文献 Relevant References

Burgin *et al*., 2018; Jiang *et al.* (蒋志刚等), 2017; Wilson *et al*., 2016; Zheng *et al.* (郑智民等), 2012; Fu *et al.* (付必谦等), 2008; Koh and Lee, 1994

(接上页)
(Feng, Zheng and Wu, 1983); *rufulus* (Dukelski, 1928); *sowerbyi* (Jones, 1956); *tscherga* (Kastchenko, 1899)

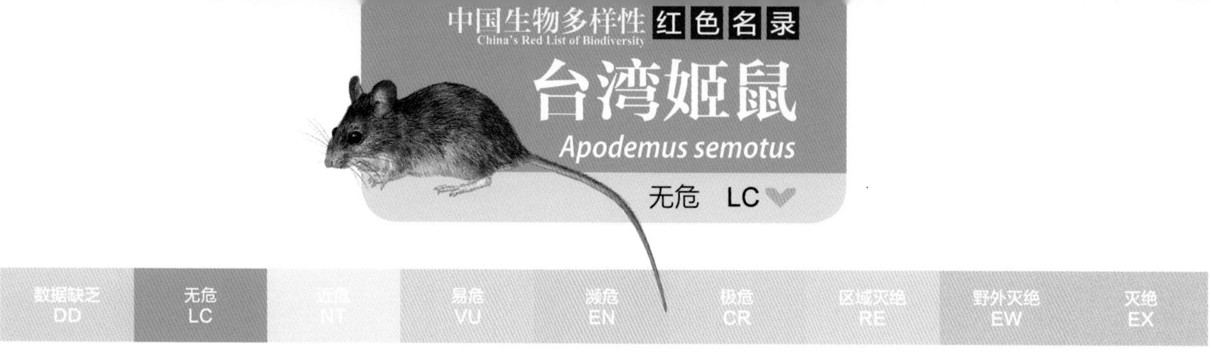

分类地位 Taxonomic Status

动物界 Animalia	脊索动物门 Chordata	哺乳纲 Mammalia	啮齿目 Rodentia	鼠科 Muridae

学名 Scientific Name	*Apodemus semotus*
命名人 Species Authority	Thomas, 1908
英文名 English Name(s)	Taiwan Field Mouse
同物异名 Synonym(s)	Formosan Wood Mouse
种下单元评估 Infra-specific Taxa Assessed	无 / None

评估信息 Assessment Information

评估年份 Year Assessed	2020
评定人 Assessor(s)	刘少英、蒋志刚 / Shaoying Liu, Zhigang Jiang
审定人 Reviewer(s)	马勇、鲍毅新 / Yong Ma, Yixin Bao
其他贡献人 Other Contributor(s)	李立立、丁晨晨 / Lili Li, Chenchen Ding

理由 Justification: 台湾姬鼠种群数量大。因此，列为无危等级 / Taiwan Field Mouse has large populations. Thus, it is listed as Least Concern

地理分布 Geographical Distribution

国内分布 Domestic Distribution

台湾 / Taiwan

世界分布 World Distribution

中国 / China

分布标注 Distribution Note

特有种 / Endemic

国内分布图
Map of Domestic Distribution

种群 Population

种群数量 Population Size	常见 / Common
种群趋势 Population Trend	未知 / Unknown

生境与生态系统 Habitat (s) and Ecosystem (s)

生　　境 Habitat(s)	草地、森林、竹林、灌丛 / Grassland, Forest, Bamboo Grove, Shrubland
生态系统 Ecosystem(s)	森林生态系统、灌丛生态系统、草地生态系统 / Forest Ecosystem, Shrubland Ecosystem, Grassland Ecosystem

威胁 Threat (s)

主要威胁 Major Threat(s)	无 / None

保护级别与保护行动 Protection Category and Conservation Action (s)

国家重点保护野生动物等级 (2021) Category of National Key Protected Wild Animals (2021)	无 / NA
IUCN 红色名录 (2020-2) IUCN Red List (2020-2)	无危 / LC
CITES 附录 (2019) CITES Appendix (2019)	无 / NA
保护行动 Conservation Action(s)	无 / None

相关文献 Relevant References

Burgin *et al*., 2018; Jiang *et al.* (蒋志刚等), 2017; Wilson *et al*., 2016; Zheng *et al.* (郑智民等), 2012; Liu *et al.* (刘晓明等), 2002; Adler, 1996

黑缘齿鼠

Rattus andamanensis

无危 LC

数据缺乏 DD	无危 LC	近危 NT	易危 VU	濒危 EN	极危 CR	区域灭绝 RE	野外灭绝 EW	灭绝 EX

分类地位 Taxonomic Status

动物界 Animalia	脊索动物门 Chordata	哺乳纲 Mammalia	啮齿目 Rodentia	鼠科 Muridae

学名 Scientific Name	*Rattus andamanensis*
命名人 Species Authority	Blyth, 1860
英文名 English Name(s)	Indochinese Forest Rat
同物异名 Synonym(s)	Sikkim Rat; *Rattus burrulus* (Miller, 1902); *flebilis* (Miller, 1902); *hainanicus* (G. M. Allen, 1925); *holchu* (Chaturvedi, 1965); *klumensis* (Kloss, 1916); *koratensis* (Kloss, (转下页)
种下单元评估 Infra-specific Taxa Assessed	无 / None

评估信息 Assessment Information

评估年份 Year Assessed	2020
评定人 Assessor(s)	刘少英、蒋志刚 / Shaoying Liu, Zhigang Jiang
审定人 Reviewer(s)	马勇、鲍毅新 / Yong Ma, Yixin Bao
其他贡献人 Other Contributor(s)	李立立、丁晨晨 / Lili Li, Chenchen Ding

理由 **Justification:** 黑缘齿鼠分布广，种群数量大。因此，列为无危等级 / Indochinese Forest Rat is a widespread species with large populations. Thus, it is listed as Least Concern

地理分布 Geographical Distribution

国内分布 Domestic Distribution

西藏、云南、贵州、四川、香港、海南、广西、广东 / Tibet (Xizang), Yunnan, Guizhou, Sichuan, Hong Kong, Hainan, Guangxi, Guangdong

世界分布 World Distribution

中国；南亚、东南亚 / China; South Asia, Southeast Asia

分布标注 Distribution Note

非特有种 / Non-endemic

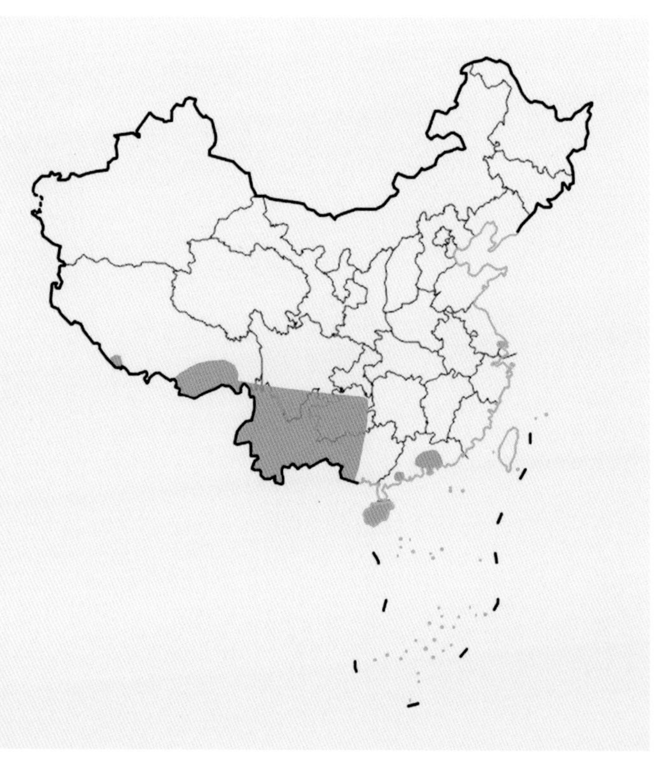

国内分布图
Map of Domestic Distribution

种群 Population

种群数量 Population Size	种群数量大 / Large populations
种群趋势 Population Trend	稳定 / Stable

生境与生态系统 Habitat (s) and Ecosystem (s)

生　　境 Habitat(s)	耕地、灌丛、人造建筑 / Arable Land, Shrubland, Building
生态系统 Ecosystem(s)	灌丛生态系统、农田生态系统、人类聚落生态系统 / Shrubland Ecosystem, Cropland Ecosystem, Human Settlement Ecosystem

威胁 Threat (s)

主要威胁 Major Threat(s)	无 / None

保护级别与保护行动 Protection Category and Conservation Action (s)

国家重点保护野生动物等级 (2021) Category of National Key Protected Wild Animals (2021)	无 / NA
IUCN 红色名录 (2020-2) IUCN Red List (2020-2)	无危 / LC
CITES 附录 (2019) CITES Appendix (2019)	无 / NA
保护行动 Conservation Action(s)	无 / None

相关文献 Relevant References

Burgin *et al*., 2018; Jiang *et al.* (蒋志刚等), 2017; Wilson *et al*., 2016; Smith *et al.* (史密斯等), 2009; Pan *et al.* (潘清华等), 2007; Wang and Xie (汪松和解焱), 2004; Wang (王应祥), 2003

(接上页)
1919); *kraensis* (Kloss, 1916); *remotus* (Robinson and Kloss, 1914); *sikkimensis* (Hinton, 1919); *yaoshanensis* (Shih, 1930)

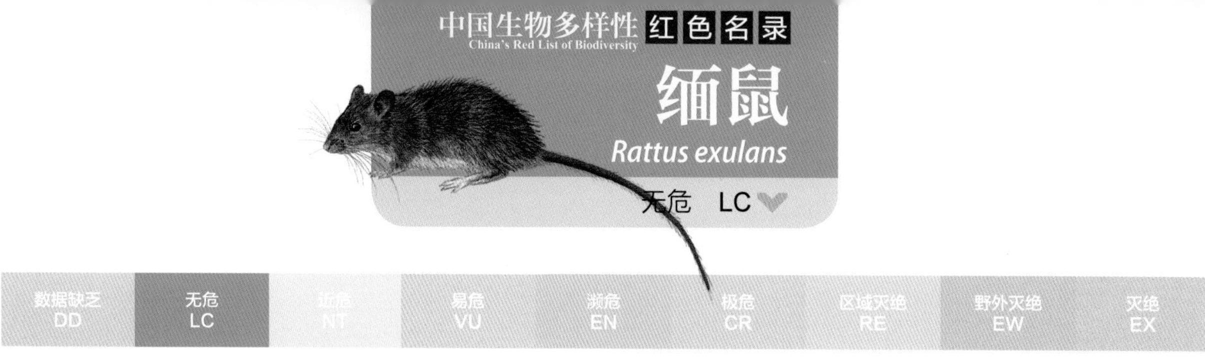

分类地位 Taxonomic Status

动 物 界 Animalia	脊索动物门 Chordata	哺 乳 纲 Mammalia	啮 齿 目 Rodentia	鼠 科 Muridae
学 名 Scientific Name		*Rattus exulans*		
命 名 人 Species Authority		Peale, 1848		
英 文 名 English Name(s)		Polynesian Rat		
同物异名 Synonym(s)		Pacific Rat; Little Rat; *Rattus aemuli* (Thomas, 1896); *aitape* (Troughton, 1937); *apicus* (Mearns, 1905); *basilanus* (Hollister, 1913); *bocourti* (Milne-Edwards, 1872); (转下页)		
种下单元评估 Infra-specific Taxa Assessed		无 / None		

评估信息 Assessment Information

评 估 年 份 Year Assessed	2020
评 定 人 Assessor(s)	刘少英、蒋志刚 / Shaoying Liu, Zhigang Jiang
审 定 人 Reviewer(s)	马勇、鲍毅新 / Yong Ma, Yixin Bao
其他贡献人 Other Contributor(s)	李立立、丁晨晨 / Lili Li, Chenchen Ding

理由 Justification: 缅鼠的种群数量大。因此，列为无危等级 / Polynesian Rat has large populations. Thus, it is listed as Least Concern

地理分布 Geographical Distribution

国内分布 Domestic Distribution

台湾；西沙群岛永兴岛 / Taiwan; Yongxing Island of Xisha Islands

世界分布 World Distribution

中国；南亚、东南亚、太平洋诸岛；北美洲 / China; South Asia, Southeast Asia, Pacific Ocean Islands; North America

分布标注 Distribution Note

非特有种 / Non-endemic

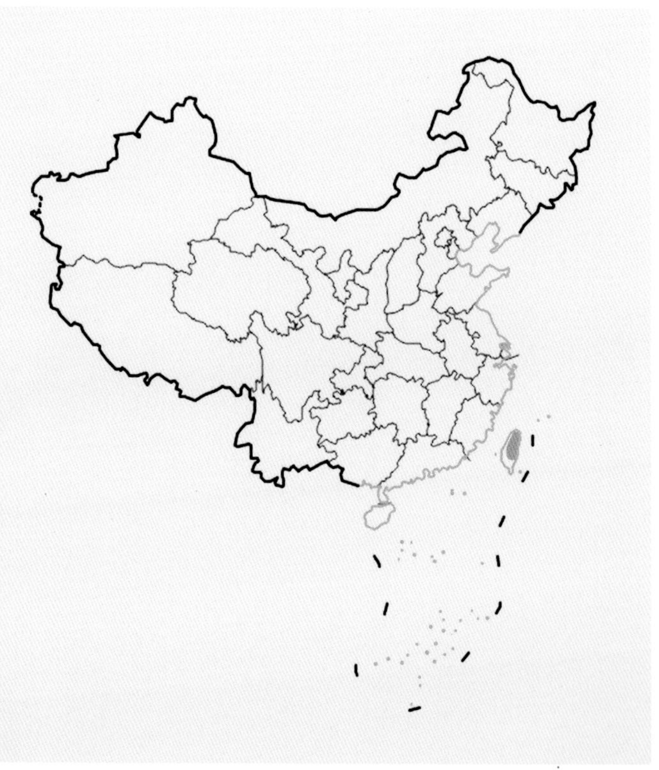

国内分布图
Map of Domestic Distribution

种群 Population

种群数量 Population Size	种群数量大 / Large populations
种群趋势 Population Trend	稳定 / Stable

生境与生态系统 Habitat (s) and Ecosystem (s)

生　　境 Habitat(s)	人工环境 / Artificial Environment
生态系统 Ecosystem(s)	人类聚落生态系统、农田生态系统 / Human Settlement Ecosystem, Cropland Ecosystem

威胁 Threat (s)

主要威胁 Major Threat(s)	无 / None

保护级别与保护行动 Protection Category and Conservation Action (s)

国家重点保护野生动物等级 (2021) Category of National Key Protected Wild Animals (2021)	无 / NA
IUCN 红色名录 (2020-2) IUCN Red List (2020-2)	无危 / LC
CITES 附录 (2019) CITES Appendix (2019)	无 / NA
保护行动 Conservation Action(s)	无 / None

相关文献 Relevant References

Burgin *et al*., 2018; Jiang *et al.* (蒋志刚等), 2017; Wilson *et al*., 2016; Zheng *et al.* (郑智民等), 2012; Smith *et al.* (史密斯等), 2009; Zhang (张荣祖), 1996

(接上页)
browni (Alston, 1877); *buruensis* (J. A. Allen, 1911); *calcis* (Hollister, 1911); *clabatus* (Lyon, 1906); *concolor* (Blyth, 1859); *echimyoides* (Ramsay, 1877); *ephippium* (Jentink, 1880); *equile* (Robinson and Kloss, 1927); *eurous* (Miller and Hollister, 1921); *gawae* (Troughton, 1845); *hawaiiensis* (Stone, 1917); *huegeli* (Thomas, 1880); *jessook* (Jentink, 1879); *lassacquerei* (Sody, 1933); *leucophaetus* (Hollister, 1913); *luteiventris* (J. A. Allen, 1910); *malengiensis* (Sody, 1941); *manoquarius* (Sody, 1934); *maorium* (Hutton, 1870); *mayonicus* (Hollister, 1913); *melanoderma* (Dieterlen, 1986); *meringgit* (Sody, 1941); *micronesiensis* (Tokuda, 1933); *negrinus* (Thomas, 1898); *obscurus* (Miller, 1900); *ornatulus* (Hollister, 1913); *otteni* (Kopstein, 1931); *pantarensis* (Mearns, 1905); *praecelsus* (Troughton, 1937); *pullus* (Miller, 1901); *querceti* (Hollister, 1911); *raveni* (Miller and Hollister, 1921); *rennelli* (Troughton, 1945); *schuitemakeri* (Sody, 1933); *solatus* (Kellogg, 1945); *stragulum* (Robinson and Kloss, 1916); *suffectus* (Troughton, 1937); *surdus* (Miller, 1903); *tibicen* (Troughton, 1937); *todayensis* (Mearns, 1905); *vigoratus* (Hollister, 1913); *vitiensis* (Peale, 1848); *vulcani* (Mearns, 1905); *wichmanni* (Jentink, 1890)

缅鼠 *Rattus exulans*

黄毛鼠

Rattus losea

无危 LC

数据缺乏 DD	无危 LC	近危 NT	易危 VU	濒危 EN	极危 CR	区域灭绝 RE	野外灭绝 EW	灭绝 EX

分类地位 Taxonomic Status

动物界 Animalia	脊索动物门 Chordata	哺乳纲 Mammalia	啮齿目 Rodentia	鼠科 Muridae

学名 Scientific Name	*Rattus losea*
命名人 Species Authority	Swinhoe, 1871
英文名 English Name(s)	Losea Rat
同物异名 Synonym(s)	Lesser Ricefield Rat; *Rattus exiguus* (Howell, 1927); *sakeratensis* (Gyldenstolpe, 1917)
种下单元评估 Infra-specific Taxa Assessed	无 / None

评估信息 Assessment Information

评估年份 Year Assessed	2020
评定人 Assessor(s)	刘少英、蒋志刚 / Shaoying Liu, Zhigang Jiang
审定人 Reviewer(s)	马勇、鲍毅新 / Yong Ma, Yixin Bao
其他贡献人 Other Contributor(s)	李立立、丁晨晨 / Lili Li, Chenchen Ding

理由 Justification: 黄毛鼠分布广，种群数量大。因此，列为无危等级 / Losea Rat is a widespread species with large populations. Thus, it is listed as Least Concern

地理分布 Geographical Distribution

国内分布 Domestic Distribution

贵州、安徽、福建、海南、浙江、江西、湖北、湖南、广东、广西、云南、台湾、香港 / Guizhou, Anhui, Fujian, Hainan, Zhejiang, Jiangxi, Hubei, Hunan, Guangdong, Guangxi, Yunnan, Taiwan, Hong Kong

世界分布 World Distribution

中国；东南亚 / China; Southeast Asia

分布标注 Distribution Note

非特有种 / Non-endemic

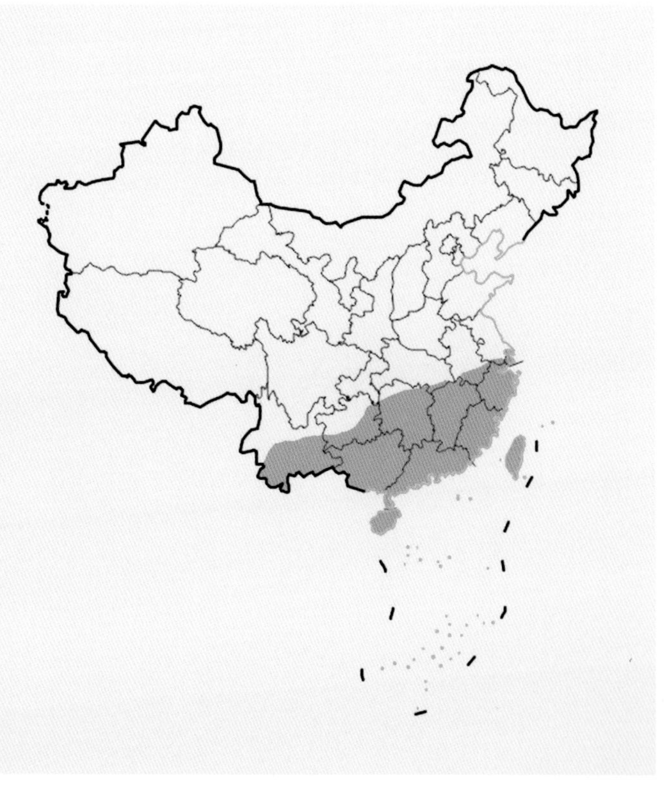

国内分布图
Map of Domestic Distribution

种群 Population

种群数量 **Population Size**	常见 / Common
种群趋势 **Population Trend**	上升 / Increasing

生境与生态系统 Habitat (s) and Ecosystem (s)

生　　境 **Habitat(s)**	草地、灌丛、红树林、耕地 / Grassland, Shrubland, Mangrove, Arable Land
生态系统 **Ecosystem(s)**	灌丛生态系统、草地生态系统、农田生态系统、湿地生态系统 / Shrubland Ecosystem, Grassland Ecosystem, Cropland Ecosystem, Wetland Ecosystem

威胁 Threat (s)

主要威胁 **Major Threat(s)**	无 / None

保护级别与保护行动 Protection Category and Conservation Action (s)

国家重点保护野生动物等级 **(2021) Category of National Key Protected Wild Animals (2021)**	无 / NA
IUCN 红色名录 **(2020-2) IUCN Red List (2020-2)**	无危 / LC
CITES 附录 **(2019) CITES Appendix (2019)**	无 / NA
保护行动 **Conservation Action(s)**	无 / None

相关文献 Relevant References

Burgin *et al*., 2018; Jiang *et al.* (蒋志刚等), 2017; Wilson *et al*., 2016; Zheng *et al.* (郑智民等), 2012; Smith *et al.* (史密斯等), 2009; Zhou *et al.* (周树武等), 2007; Feng *et al.* (冯志勇等), 1990

黄毛鼠 *Rattus losea*

大足鼠

Rattus nitidus

无危 LC

数据缺乏 DD	无危 LC	近危 NT	易危 VU	濒危 EN	极危 CR	区域灭绝 RE	野外灭绝 EW	灭绝 EX

分类地位 Taxonomic Status

动物界 Animalia	脊索动物门 Chordata	哺乳纲 Mammalia	啮齿目 Rodentia	鼠科 Muridae
学名 Scientific Name		*Rattus nitidus*		
命名人 Species Authority		Hodgson, 1845		
英文名 English Name(s)		Himalayan Field Rat		
同物异名 Synonym(s)		White-footed Indochinese Rat; *Rattus aequicaudalus* (Hodgson, 1849); *guhai* (Nath, 1952); *horeites* (Hodgson, 1845); *anuselae* (Thomas, 1920); *obsoletus* (Hinton, (转下页)		
种下单元评估 Infra-specific Taxa Assessed		无 / None		

评估信息 Assessment Information

评估年份 Year Assessed	2020
评定人 Assessor(s)	刘少英、蒋志刚 / Shaoying Liu, Zhigang Jiang
审定人 Reviewer(s)	马勇、鲍毅新 / Yong Ma, Yixin Bao
其他贡献人 Other Contributor(s)	李立立、丁晨晨 / Lili Li, Chenchen Ding

理由 Justification: 大足鼠分布广，种群数量大。因此，列为无危等级 / Himalayan Field Rat is a widespread species with large populations. Thus, it is listed as Least Concern

地理分布 Geographical Distribution

国内分布 Domestic Distribution

四川、贵州、云南、西藏、安徽、江苏、上海、浙江、湖南、江西、广东、广西、海南、福建、甘肃、陕西、重庆 / Sichuan, Guizhou, Yunnan, Tibet (Xizang), Anhui, Jiangsu, Shanghai, Zhejiang, Hunan, Jiangxi, Guangdong, Guangxi, Hainan, Fujian, Gansu, Shaanxi, Chongqing

世界分布 World Distribution

中国；南亚、东南亚 / China; South Asia, Southeast Asia

分布标注 Distribution Note

非特有种 / Non-endemic

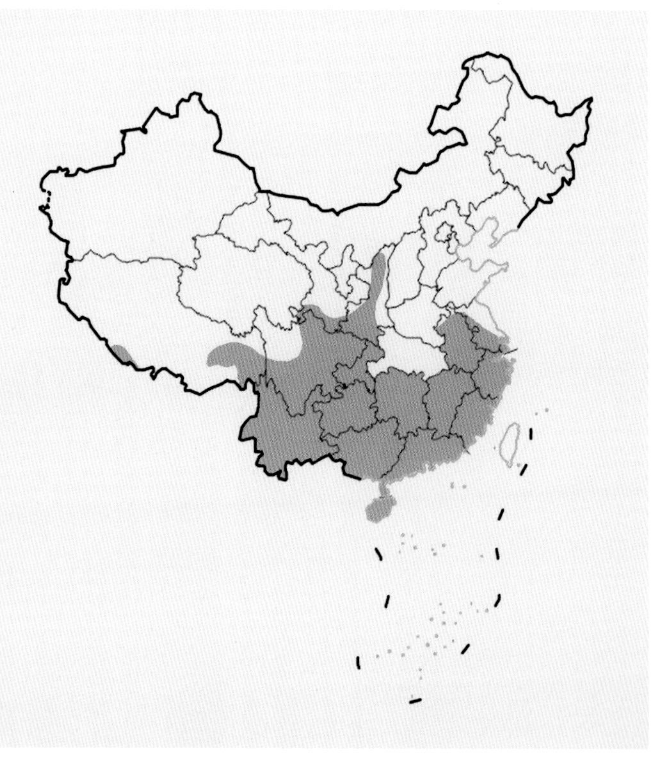

国内分布图
Map of Domestic Distribution

种群 Population

种群数量 Population Size	种群数量大 / Large populations
种群趋势 Population Trend	稳定 / Stable

生境与生态系统 Habitat (s) and Ecosystem (s)

生　　境 Habitat(s)	耕地、溪流边 / Arable Land, Near Stream
生态系统 Ecosystem(s)	农田生态系统、湖泊河流生态系统 / Cropland Ecosystem, Lake and River Ecosystem

威胁 Threat (s)

主要威胁 Major Threat(s)	无 / None

保护级别与保护行动 Protection Category and Conservation Action (s)

国家重点保护野生动物等级 (2021) Category of National Key Protected Wild Animals (2021)	无 / NA
IUCN 红色名录 (2020-2) IUCN Red List (2020-2)	无危 / LC
CITES 附录 (2019) CITES Appendix (2019)	无 / NA
保护行动 Conservation Action(s)	无 / None

相关文献 Relevant References

Burgin *et al*., 2018; Jiang *et al.* (蒋志刚等), 2017; Wilson *et al*., 2016; Zheng *et al.* (郑智民等), 2012; Wang (王红慷), 2008; Jiang *et al.* (蒋光藻等), 1999; Yang *et al.* (杨跃敏等), 1994

(接上页)
1919); *rahengis* (Kloss, 1919); *ruber* (Jentink, 1880); *rubricosa* (Anderson, 1879); *subditivus* (Miller and Hollister, 1921); *vanheurni* (Sody, 1933)

大足鼠 *Rattus nitidus*

数据缺乏 DD	无危 LC	近危 NT	易危 VU	濒危 EN	极危 CR	区域灭绝 RE	野外灭绝 EW	灭绝 EX

分类地位 Taxonomic Status

动物界 Animalia	脊索动物门 Chordata	哺乳纲 Mammalia	啮齿目 Rodentia	鼠科 Muridae
学名 **Scientific Name**		*Rattus norvegicus*		
命名人 **Species Authority**		Berkenhout, 1769		
英文名 **English Name(s)**		Brown Rat		
同物异名 **Synonym(s)**		Common Rat; Norwegian Rat; *Rattus aquaticus* (Rutty, 1772); *albinus* (Donaldson, 1912); *albus* (Hatai, 1907); *americanus* (de Kay, 1842); *caraco* (Pallas, 1778); （转下页）		
种下单元评估 **Infra-specific Taxa Assessed**		无 / None		

评估信息 Assessment Information

评估年份 **Year Assessed**	2020
评定人 **Assessor(s)**	刘少英、蒋志刚 / Shaoying Liu, Zhigang Jiang
审定人 **Reviewer(s)**	马勇、鲍毅新 / Yong Ma, Yixin Bao
其他贡献人 **Other Contributor(s)**	李立立、丁晨晨 / Lili Li, Chenchen Ding

理由 **Justification:** 褐家鼠分布广，种群数量大，其在中国的分布区仍在扩大。因此，列为无危等级 / Brown Rat is a widespread species with large populations. Furthermore, its areas of distribution is expanding in China. Thus, it is listed as Least Concern

地理分布 Geographical Distribution

国内分布 Domestic Distribution

黑龙江、吉林、辽宁、内蒙古、北京、天津、河北、山西、山东、河南、陕西、宁夏、甘肃、青海、新疆、四川、贵州、云南、广西、广东、海南、香港、上海、江苏、浙江、安徽、江西、湖南、湖北、福建、台湾、重庆 / Heilongjiang, Jilin, Liaoning, Inner Mongolia (Nei Mongol), Beijing, Tianjin, Hebei, Shanxi, Shandong, Henan, Shaanxi, Ningxia, Gansu, Qinghai, Xinjiang, Sichuan, Guizhou, Yunnan, Guangxi, Guangdong, Hainan, Hong Kong, Shanghai, Jiangsu, Zhejiang, Anhui, Jiangxi, Hunan, Hubei, Fujian, Taiwan, Chongqing

世界分布 World Distribution

亚洲、欧洲 / Asia, Europe

分布标注 Distribution Note

非特有种 / Non-endemic

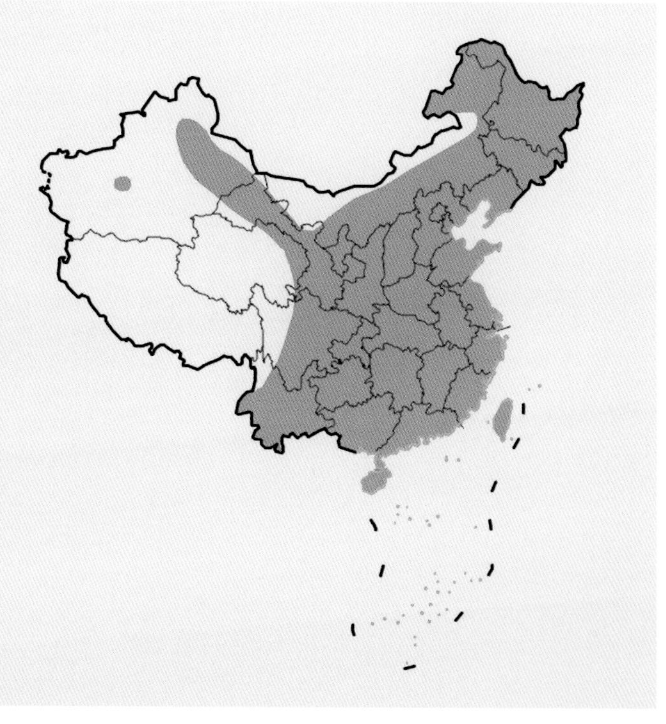

国内分布图
Map of Domestic Distribution

种群 Population

种群数量 Population Size	种群数量大 / Large populations
种群趋势 Population Trend	稳定 / Stable

生境与生态系统 Habitat (s) and Ecosystem (s)

生　境 Habitat(s)	人工环境 / Artificial Environment
生态系统 Ecosystem(s)	人类聚落生态系统、农田生态系统 / Human Settlement Ecosystem, Cropland Ecosystem

威胁 Threat (s)

主要威胁 Major Threat(s)	无 / None

保护级别与保护行动 Protection Category and Conservation Action (s)

国家重点保护野生动物等级 (2021) Category of National Key Protected Wild Animals (2021)	无 / NA
IUCN 红色名录 (2020-2) IUCN Red List (2020-2)	无危 / LC
CITES 附录 (2019) CITES Appendix (2019)	无 / NA
保护行动 Conservation Action(s)	无 / None

相关文献 Relevant References

Burgin *et al*., 2018; Jiang *et al.* (蒋志刚等), 2017; Wilson *et al*., 2016; Zhong *et al.* (钟宇等), 2014; Zheng *et al.* (郑智民等), 2012; Zhou *et al.* (周朝霞等), 2009; Li *et al.* (李世斌等), 1993

(接上页)
caspius (Oken, 1816); *cauquenensis* (Philippi, 1900); *decaryi* (Grandidier, 1934); *decumanoides* (Hodgson, 1841); *decumanus* (Pallas, 1779); *discolor* (Noack, 1918); *fossilis* (Ameghino, 1889); *fossor* (Walker, 1808); *griseipectus* (Milne-Edwards, 1872); *hibernicus* (Thompson, 1837); *hoffmanni* (Trouessart, 1904); *humiliatus* (Milne-Edwards, 1868); *hybridus* (Bechstein, 1800); *insolatus* (Howell, 1927); *javanus* (Hermann, 1804); *kurodobu* (Kuroda, 1953); *leucosternum* (Rüppell, 1842); *lutescens* (Gay, 1848); *magnirostris* (Mearns, 1905); *major* (Hoffmann, 1887); *maniculatus* (Wagner, 1848); *maurus* (Waterhouse, 1839); *migrans* (Zimmermann, 1777); *orii* (Kuroda, 1952); *otomoi* (Yamada, 1930); *ouangthomae* (Milne-Edwards, 1871); *plumbeus* (Milne-Edwards, 1874); *praestans* (Trouessart, 1904); *primarius* (Kastschenko, 1912); *shirokuma* (Kuroda, 1953); *simpsoni* (Philippi, 1900); *socer* (Miller, 1914); *sowerbyi* (Howell, 1928); *suffureoventris* (Kuroda, 1952); *surmulottus* (Severinus, 1779); *tamarensis* (Higgins and Petterd, 1883)

褐家鼠 *Rattus norvegicus*

黄胸鼠
Rattus tanezumi

无危 LC

数据缺乏 DD | 无危 LC | 近危 NT | 易危 VU | 濒危 EN | 极危 CR | 区域灭绝 RE | 野外灭绝 EW | 灭绝 EX

分类地位 Taxonomic Status

动物界 **Animalia**	脊索动物门 **Chordata**	哺乳纲 **Mammalia**	啮齿目 **Rodentia**	鼠科 **Muridae**
学名 **Scientific Name**		*Rattus tanezumi*		
命名人 **Species Authority**		Temminck, 1844		
英文名 **English Name(s)**		Oriental House Rat		
同物异名 **Synonym(s)**		Tanezumi Rat; Asian Rat; Asian House Rat; *Rattus alangensis* (Chasen, 1937); *amboinensis* (Laurie and Hill, 1954); *argyraceus* (Sody, 1941); *auroreus* (Sody, 1941); *barussanoides* (转下页)		
种下单元评估 **Infra-specific Taxa Assessed**		无 / None		

评估信息 Assessment Information

评估年份 **Year Assessed**	2020
评定人 **Assessor(s)**	刘少英、蒋志刚 / Shaoying Liu, Zhigang Jiang
审定人 **Reviewer(s)**	马勇、鲍毅新 / Yong Ma, Yixin Bao
其他贡献人 **Other Contributor(s)**	李立立、丁晨晨 / Lili Li, Chenchen Ding

理由 **Justification:** 黄胸鼠分布广，种群数量大。因此，列为无危等级 / Oriental House Rat is a widespread species with large populations. Thus, it is listed as Least Concern

地理分布 Geographical Distribution

国内分布 Domestic Distribution

河北、北京、天津、河南、陕西、四川、贵州、云南、西藏、安徽、江苏、上海、浙江、江西、湖南、湖北、广东、香港、海南、广西、福建、宁夏、甘肃、新疆、重庆、辽宁、山西、台湾 / Hebei, Beijing, Tianjin, Henan, Shaanxi, Sichuan, Guizhou, Yunnan, Tibet (Xizang), Anhui, Jiangsu, Shanghai, Zhejiang, Jiangxi, Hunan, Hubei, Guangdong, Hong Kong, Hainan, Guangxi, Fujian, Ningxia, Gansu, Xinjiang, Chongqing, Liaoning, Shanxi, Taiwan

世界分布 World Distribution

中国；中亚、南亚、东南亚 / China; Central Asia, South Asia, Southeast Asia

分布标注 Distribution Note

非特有种 / Non-endemic

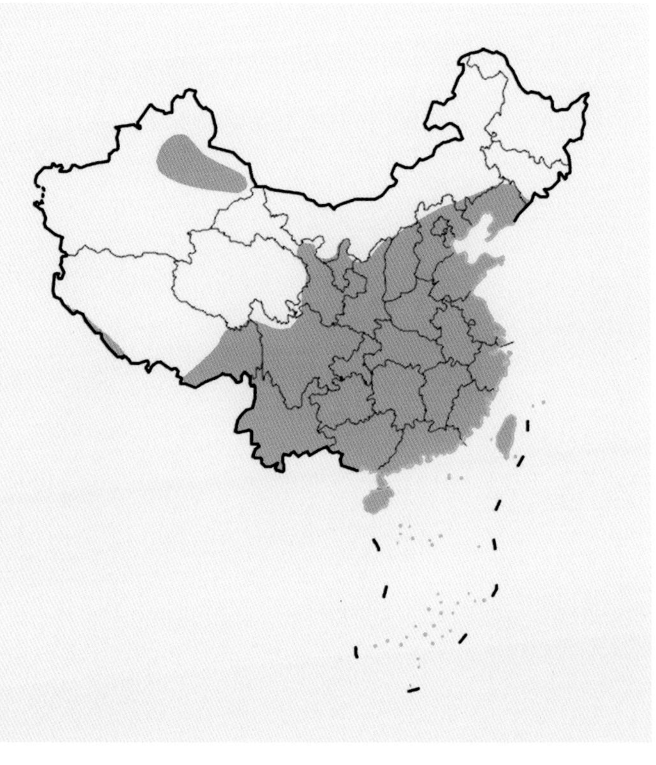

国内分布图
Map of Domestic Distribution

种群 Population

种群数量 Population Size	种群数量大 / Large populations
种群趋势 Population Trend	上升 / Increasing

生境与生态系统 Habitat (s) and Ecosystem (s)

生　　境 Habitat(s)	耕地 / Arable Land
生态系统 Ecosystem(s)	农田生态系统 / Cropland Ecosystem

威胁 Threat (s)

主要威胁 Major Threat(s)	无 / None

保护级别与保护行动 Protection Category and Conservation Action (s)

国家重点保护野生动物等级 (2021) Category of National Key Protected Wild Animals (2021)	无 / NA
IUCN 红色名录 (2020-2) IUCN Red List (2020-2)	无危 / LC
CITES 附录 (2019) CITES Appendix (2019)	无 / NA
保护行动 Conservation Action(s)	无 / None

相关文献 Relevant References

Burgin *et al.*, 2018; Jiang *et al.* (蒋志刚等), 2017; Wilson *et al.*, 2016; Hu *et al.* (胡秋波等), 2014; Li *et al.* (李秋阳等), 2013; Zheng *et al.* (郑智民等), 2012; Tong and Lu (仝磊和路红琪), 2010a; Wang (王应祥), 2003

(接上页)
(Sody, 1941); *benguetensis* (Hollister, 1913); *bhotia* (Hinton, 1918); *brevicaudus* (Kuroda, 1952); *brevicaudus* (Chakraborty, 1975); *brunneus* (Hodgson, 1845); *brunneusculus* (Hodgson, 1845); *bullocki* (Roonwal, 1948); *canna* (Swinhoe, 1871); *coloratus* (Hollister, 1913); *dammermani* (Thomas, 1921); *dentatus* (Miller, 1913); *diardii* (Jentink, 1880); *exsul* (Miller, 1913); *flavipectus* (Milne-Edwards, 1872); *fortunatus* (Miller, 1913); *gangutrianus* (Hinton, 1919); *germaini* (Milne-Edwards, 1872); *griseiventer* (Bonhote, 1903); *insulanus* (Miller, 1913); *kadanus* (Chasen, 1937); *kelleri* (Mearns, 1905); *khyensis* (Hinton, 1919); *kurokuma* (Kuroda, 1953); *kramensis* (Kloss, 1919); *lalolis* (Tate and Archbold, 1935); *lanensis* (Kloss, 1919); *lontaris* (Chasen, 1937); *macmillani* (Hinton, 1919); *makassarius* (Sody, 1941); *makensis* (Kloss, 1916); *mansorius* (Johnson, 1962); *masaretes* (Sody, 1937); *mesanis* (Kloss, 1919); *mindanensis* (Mearns, 1905); *moheius* (Chasen, 1937); *molliculus* (Robinson and Kloss, 1922); *moluccarius* (Sody, 1933); *neglectus* (Jentink, 1880); *nemoralis* (Blyth, 1851); *obiensis* (Sody, 1941); *ouangthomae* (Milne-Edwards, 1872); *palelae* (Miller and Hollister, 1921); *palembang* (Tate and Archbold, 1935); *panjius* (Chasen, 1937); *pannellus* (Miller, 1913); *pannosus* (Miller, 1900); *pelengensis* (Sody, 1941); *pipidonis* (Chasen, 1937); *poenitentiarii* (Kloss, 1915); *portus* (Kloss, 1915); *povolny* (Niethammer and Martens, 1975); *pulliventer* (Miller, 1902); *rangensis* (Kloss, 1916); *robiginosus* (Hollister, 1913); *robinsoni* (Chasen, 1940); *robustulus* (Blyth, 1859); *sakisimana* (Tokuda, 1939); *samati* (Sody, 1932); *santalum* (Sody, 1932); *sapoensis* (Sody, 1941); *satarae* (Hinton, 1918); *septicus* (Sody, 1933); *shirokuma* (Kuroda, 1953); *sladeni* (Anderson, 1879); *sumbae* (Sody, 1930); *tablasi* (Taylor, 1934); *talaudensis* (Sody, 1941); *tatkonensis* (Hinton, 1910); *thai* (Kloss, 1917); *tikos* (Hinton, 1919); *tistae* (Hinton, 1918); *toxi* (Sody, 1941); *turbidus* (Miller, 1913); *yeni* (Dao, 1960); *yunnanensis* (Anderson, 1879); *zamboangae* (Mearns, 1905)

黄胸鼠 *Rattus tanezumi*

数据缺乏 DD	无危 LC	近危 NT	易危 VU	濒危 EN	极危 CR	区域灭绝 RE	野外灭绝 EW	灭绝 EX

分类地位 Taxonomic Status

动物界 Animalia	脊索动物门 Chordata	哺乳纲 Mammalia	啮齿目 Rodentia	鼠科 Muridae
学名 Scientific Name		*Niviventer andersoni*		
命名人 Species Authority		Thomas, 1911		
英文名 English Name(s)		Anderson's Niviventer		
同物异名 Synonym(s)		Anderson's White-bellied Rat; *Niviventer lushuiensis* (Wu and Wang, 2002)		
种下单元评估 Infra-specific Taxa Assessed		无 / None		

评估信息 Assessment Information

评估年份 Year Assessed	2020
评定人 Assessor(s)	刘少英、蒋志刚 / Shaoying Liu, Zhigang Jiang
审定人 Reviewer(s)	马勇、鲍毅新 / Yong Ma, Yixin Bao
其他贡献人 Other Contributor(s)	李立立、丁晨晨 / Lili Li, Chenchen Ding

理由 Justification: 安氏白腹鼠分布广，种群数量大。因此，列为无危等级 / Anderson's Niviventer is a widespread species with large populations. Thus, it is listed as Least Concern

地理分布 Geographical Distribution

国内分布 Domestic Distribution

四川、云南、西藏、重庆、湖北、湖南、贵州、甘肃、陕西 / Sichuan, Yunnan, Tibet (Xizang), Chongqing, Hubei, Hunan, Guizhou, Gansu, Shaanxi

世界分布 World Distribution

中国 / China

分布标注 Distribution Note

特有种 / Endemic

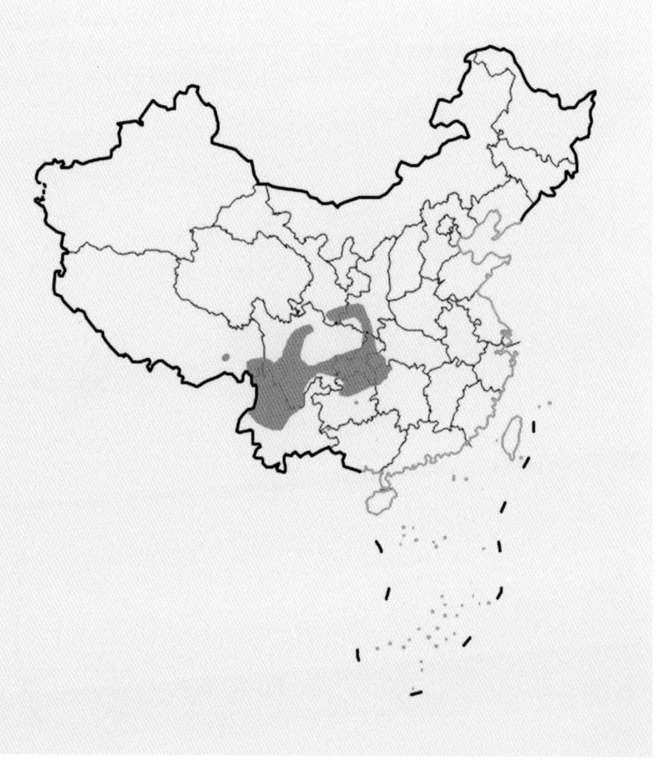

国内分布图
Map of Domestic Distribution

种群 Population

种群数量 **Population Size**	种群数量大 / Large populations
种群趋势 **Population Trend**	未知 / Unknown

生境与生态系统 Habitat (s) and Ecosystem (s)

生　　境 **Habitat(s)**	森林 / Forest
生态系统 **Ecosystem(s)**	森林生态系统 / Forest Ecosystem

威胁 Threat (s)

主要威胁 **Major Threat(s)**	无 / None

保护级别与保护行动 Protection Category and Conservation Action (s)

国家重点保护野生动物等级 **(2021) Category of National Key Protected Wild Animals (2021)**	无 / NA
IUCN 红色名录 **(2020-2) IUCN Red List (2020-2)**	无危 / LC
CITES 附录 **(2019) CITES Appendix (2019)**	无 / NA
保护行动 **Conservation Action(s)**	无 / None

相关文献 Relevant References

Li *et al.* (李飞虹等), 2020; Burgin *et al.*, 2018; Jiang *et al.* (蒋志刚等), 2017; Wilson *et al.*, 2016; Zheng *et al.* (郑智民等), 2012; Smith *et al.* (史密斯等), 2009; Feng *et al.* (冯祚建等), 1986; Musser and Chiu, 1979

安氏白腹鼠 *Niviventer andersoni*

北社鼠

Niviventer confucianus

无危 LC

数据缺乏 DD	无危 LC	近危 NT	易危 VU	濒危 EN	极危 CR	区域灭绝 RE	野外灭绝 EW	灭绝 EX

分类地位 Taxonomic Status

动物界 Animalia	脊索动物门 Chordata	哺乳纲 Mammalia	啮齿目 Rodentia	鼠科 Muridae

学名 Scientific Name	*Niviventer confucianus*
命名人 Species Authority	Milne-Edwards, 1871
英文名 English Name(s)	Confucian Niviventer
同物异名 Synonym(s)	Chinese White-bellied Rat; *Niviventer canorus* (Thomas, 1911); *chihliensis* (Thomas, 1917); *deqinensis* (Deng and Wang, 2000); *elegans* (Shih, 1931); *littoreus* (Cabrera, （转下页）
种下单元评估 Infra-specific Taxa Assessed	无 / None

评估信息 Assessment Information

评估年份 Year Assessed	2020
评定人 Assessor(s)	刘少英、蒋志刚 / Shaoying Liu, Zhigang Jiang
审定人 Reviewer(s)	马勇、鲍毅新 / Yong Ma, Yixin Bao
其他贡献人 Other Contributor(s)	李立立、丁晨晨 / Lili Li, Chenchen Ding

理由 Justification: 北社鼠分布广，种群数量大。因此，列为无危等级 / Confucian Niviventer is a widespread species with large populations. Thus, it is listed as Least Concern

地理分布 Geographical Distribution

国内分布 Domestic Distribution

山西、陕西、云南、浙江、北京、天津、河北、内蒙古、辽宁、上海、江苏、安徽、福建、江西、山东、河南、湖北、湖南、广东、广西、四川、贵州、西藏、甘肃、青海、宁夏、吉林、重庆 / Shanxi, Shaanxi, Yunnan, Zhejiang, Beijing, Tianjin, Hebei, Inner Mongolia (Nei Mongol), Liaoning, Shanghai, Jiangsu, Anhui, Fujian, Jiangxi, Shandong, Henan, Hubei, Hunan, Guangdong, Guangxi, Sichuan, Guizhou, Tibet (Xizang), Gansu, Qinghai, Ningxia, Jilin, Chongqing

世界分布 World Distribution

中国；东南亚 / China; Southeast Asia

分布标注 Distribution Note

非特有种 / Non-endemic

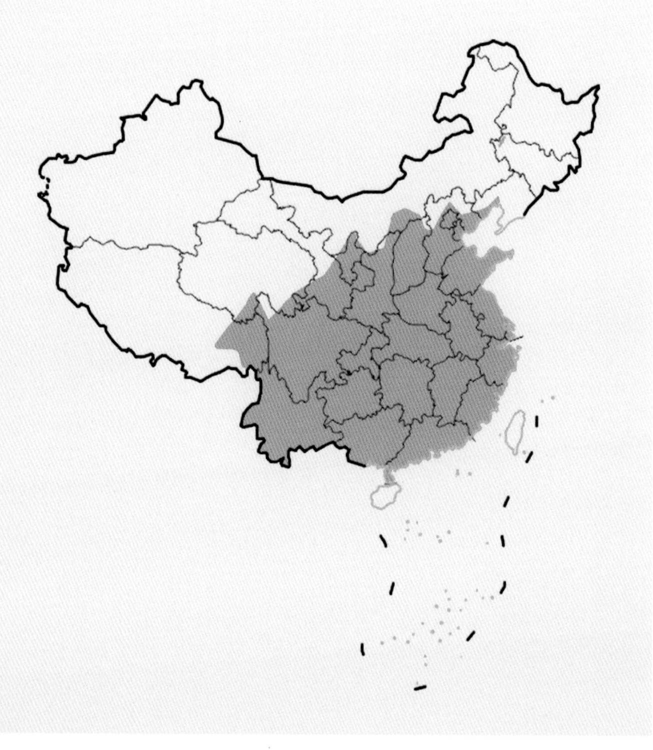

国内分布图
Map of Domestic Distribution

种群 Population

种群数量 Population Size	种群数量大 / Large populations
种群趋势 Population Trend	稳定 / Stable

生境与生态系统 Habitat (s) and Ecosystem (s)

生　　境 Habitat(s)	森林、耕地 / Forest, Arable Land
生态系统 Ecosystem(s)	森林生态系统、农田生态系统 / Forest Ecosystem, Cropland Ecosystem

威胁 Threat (s)

主要威胁 Major Threat(s)	无 / None

保护级别与保护行动 Protection Category and Conservation Action (s)

国家重点保护野生动物等级 (2021) Category of National Key Protected Wild Animals (2021)	无 / NA
IUCN 红色名录 (2020-2) IUCN Red List (2020-2)	无危 / LC
CITES 附录 (2019) CITES Appendix (2019)	无 / NA
保护行动 Conservation Action(s)	无 / None

相关文献 Relevant References

Burgin *et al*., 2018; Jiang *et al.* (蒋志刚等), 2017; Wilson *et al*., 2016; Ma *et al.* (马晓婷等), 2014; Peng and Guo (彭培英和郭宪国), 2014; Zhang *et al.* (张旭等), 2013; Zheng *et al.* (郑智民等), 2012

(接上页)
1922); *luticolor* (Thomas, 1908); *mentosus* (Thomas, 1916); *naoniuensis* (Zhang and Zhao, 1984); *sacer* (Thomas, 1908); *sinianus* (Shih, 1931); *yajiangensis* (Deng and Wang, 2000); *yushuensis* (Wang and Zheng, 1981); *zappeyi* (G. M. Allen, 1912)

北社鼠 *Niviventer confucianus*　　周佳俊 摄　By Jiajun Zhou

台湾白腹鼠

Niviventer coninga

无危 LC

数据缺乏 DD	无危 LC	近危 NT	易危 VU	濒危 EN	极危 CR	区域灭绝 RE	野外灭绝 EW	灭绝 EX

分类地位 Taxonomic Status

动 物 界 Animalia	脊索动物门 Chordata	哺 乳 纲 Mammalia	啮 齿 目 Rodentia	鼠 科 Muridae

学 名 Scientific Name	*Niviventer coninga*
命 名 人 Species Authority	Swinhoe, 1864
英 文 名 English Name(s)	Spiny Taiwan Niviventer
同物异名 Synonym(s)	Coxing's White-bellied Rat; *Niviventer coxinga* (Swinhoe, 1871); *coxingi* (Thomas, 1892); *coxingi* (Swinhoe, 1864)
种下单元评估 Infra-specific Taxa Assessed	无 / None

评估信息 Assessment Information

评 估 年 份 Year Assessed	2020
评 定 人 Assessor(s)	刘少英、蒋志刚 / Shaoying Liu, Zhigang Jiang
审 定 人 Reviewer(s)	马勇、鲍毅新 / Yong Ma, Yixin Bao
其他贡献人 Other Contributor(s)	李立立、丁晨晨 / Lili Li, Chenchen Ding

理由 Justification: 台湾白腹鼠分布广，种群数量大。因此，列为无危等级 / Spiny Taiwan Niviventer is a widespread species with large populations. Thus, it is listed as Least Concern

地理分布 Geographical Distribution

国内分布 Domestic Distribution

台湾 / Taiwan

世界分布 World Distribution

中国 / China

分布标注 Distribution Note

特有种 / Endemic

国内分布图
Map of Domestic Distribution

种群 Population

种群数量 Population Size	种群数量大 / Large populations
种群趋势 Population Trend	未知 / Unknown

生境与生态系统 Habitat (s) and Ecosystem (s)

生　　境 Habitat(s)	森林、灌丛 / Forest, Shrubland
生态系统 Ecosystem(s)	森林生态系统、灌丛生态系统 / Forest Ecosystem, Shrub Ecosystem

威胁 Threat (s)

主要威胁 Major Threat(s)	无 / None

保护级别与保护行动 Protection Category and Conservation Action (s)

国家重点保护野生动物等级 (2021) Category of National Key Protected Wild Animals (2021)	无 / NA
IUCN 红色名录 (2020-2) IUCN Red List (2020-2)	无危 / LC
CITES 附录 (2019) CITES Appendix (2019)	无 / NA
保护行动 Conservation Action(s)	无 / None

相关文献 Relevant References

Burgin *et al.*, 2018; Jiang *et al.* (蒋志刚等), 2017; Wilson *et al.*, 2016; Zheng *et al.* (郑智民等), 2012; Li *et al.* (李裕冬等), 2007; Wang (王应祥), 2003; Adler, 1996

灰腹鼠

Niviventer eha

无危 LC

数据缺乏 DD	无危 LC	近危 NT	易危 VU	濒危 EN	极危 CR	区域灭绝 RE	野外灭绝 EW	灭绝 EX

分类地位 Taxonomic Status

动物界 Animalia	脊索动物门 Chordata	哺乳纲 Mammalia	啮齿目 Rodentia	鼠科 Muridae
学名 Scientific Name		*Niviventer eha*		
命名人 Species Authority		Wroughton, 1916		
英文名 English Name(s)		Little Himalayan Rat		
同物异名 Synonym(s)		Smoke-bellied Niviventer; *Niviventer ninus* (Thomas, 1922)		
种下单元评估 Infra-specific Taxa Assessed		无 / None		

评估信息 Assessment Information

评估年份 Year Assessed	2020
评定人 Assessor(s)	刘少英、蒋志刚 / Shaoying Liu, Zhigang Jiang
审定人 Reviewer(s)	马勇、鲍毅新 / Yong Ma, Yixin Bao
其他贡献人 Other Contributor(s)	李立立、丁晨晨 / Lili Li, Chenchen Ding

理由 Justification: 灰腹鼠分布广，种群数量大。因此，列为无危等级 / Little Himalayan Rat is a widespread species with large populations. Thus, it is listed as Least Concern

地理分布 Geographical Distribution

国内分布 Domestic Distribution

云南、西藏、广西、贵州 / Yunnan, Tibet (Xizang), Guangxi, Guizhou

世界分布 World Distribution

中国；南亚 / China; South Asia

分布标注 Distribution Note

非特有种 / Non-endemic

国内分布图
Map of Domestic Distribution

种群 Population

种群数量 Population Size	种群数量大 / Large populations
种群趋势 Population Trend	稳定 / Stable

生境与生态系统 Habitat (s) and Ecosystem (s)

生　　境 Habitat(s)	泰加林、温带森林 / Taiga Forest, Temperate Forest
生态系统 Ecosystem(s)	森林生态系统 / Forest Ecosystem

威胁 Threat (s)

主要威胁 Major Threat(s)	无 / None

保护级别与保护行动 Protection Category and Conservation Action (s)

国家重点保护野生动物等级 (2021) Category of National Key Protected Wild Animals (2021)	无 / NA
IUCN 红色名录 (2020-2) IUCN Red List (2020-2)	无危 / LC
CITES 附录 (2019) CITES Appendix (2019)	无 / NA
保护行动 Conservation Action(s)	无 / None

相关文献 Relevant References

Burgin *et al*., 2018; Jiang *et al.* (蒋志刚等), 2017; Wilson *et al*., 2016; Zheng *et al.* (郑智民等), 2012; Jing *et al*., 2007; Pan *et al.* (潘清华等), 2007; Wang (王应祥), 2003

灰腹鼠 *Niviventer eha*

中国生物多样性红色名录
China's Red List of Biodiversity

川西白腹鼠

Niviventer excelsior

无危 LC

数据缺乏 DD	无危 LC	近危 NT	易危 VU	濒危 EN	极危 CR	区域灭绝 RE	野外灭绝 EW	灭绝 EX

分类地位 Taxonomic Status

动物界 Animalia	脊索动物门 Chordata	哺乳纲 Mammalia	啮齿目 Rodentia	鼠科 Muridae

学名 Scientific Name	*Niviventer excelsior*
命名人 Species Authority	Thomas, 1911
英文名 English Name(s)	Sichuan Niviventer
同物异名 Synonym(s)	Large White-bellied Rat; *Niviventer tengchongensis* (Deng and Wang, 2002)
种下单元评估 Infra-specific Taxa Assessed	无 / None

评估信息 Assessment Information

评估年份 Year Assessed	2020
评定人 Assessor(s)	刘少英、蒋志刚 / Shaoying Liu, Zhigang Jiang
审定人 Reviewer(s)	马勇、鲍毅新 / Yong Ma, Yixin Bao
其他贡献人 Other Contributor(s)	李立立、丁晨晨 / Lili Li, Chenchen Ding

理由 Justification: 川西白腹鼠分布广，种群数量大。因此，列为无危等级 / Sichuan Niviventer is a widespread species with large populations. Thus, it is listed as Least Concern

地理分布 Geographical Distribution

国内分布 Domestic Distribution

云南、四川、西藏 / Yunnan, Sichuan, Tibet (Xizang)

世界分布 World Distribution

中国 / China

分布标注 Distribution Note

特有种 / Endemic

国内分布图
Map of Domestic Distribution

种群 Population

种群数量 Population Size	种群数量大 / Large populations
种群趋势 Population Trend	未知 / Unknown

生境与生态系统 Habitat (s) and Ecosystem (s)

生　　境 Habitat(s)	热带和亚热带湿润山地森林 / Tropical and Subtropical Moist Montane Forest
生态系统 Ecosystem(s)	森林生态系统 / Forest Ecosystem

威胁 Threat (s)

主要威胁 Major Threat(s)	无 / None

保护级别与保护行动 Protection Category and Conservation Action (s)

国家重点保护野生动物等级 (2021) Category of National Key Protected Wild Animals (2021)	无 / NA
IUCN 红色名录 (2020-2) IUCN Red List (2020-2)	无危 / LC
CITES 附录 (2019) CITES Appendix (2019)	无 / NA
保护行动 Conservation Action(s)	无 / None

相关文献 Relevant References

Burgin *et al.*, 2018; Jiang *et al.* (蒋志刚等), 2017; Wilson *et al.*, 2016; Zheng *et al.* (郑智民等), 2012; Li *et al.* (李裕冬等), 2007; Liu *et al.* (刘少英等), 2005; Musser and Chiu, 1979

川西白腹鼠 *Niviventer excelsior*

针毛鼠

Niviventer fulvescens

无危 LC

数据缺乏 DD	无危 LC	近危 NT	易危 VU	濒危 EN	极危 CR	区域灭绝 RE	野外灭绝 EW	灭绝 EX

分类地位 Taxonomic Status

动物界 Animalia	脊索动物门 Chordata	哺乳纲 Mammalia	啮齿目 Rodentia	鼠科 Muridae

学名 Scientific Name	*Niviventer fulvescens*
命名人 Species Authority	Gray, 1847
英文名 English Name(s)	Chestnut White-bellied Rat
同物异名 Synonym(s)	Indomalayan Niviventer; *Niviventer baturus* (Sody, 1932); *besuki* (Sody, 1931); *blythi* (Kloss, 1917); *bukit* (Bonhote, 1903); *caudatior* (Hodgson, 1849); *cinnamomeus* (转下页)
种下单元评估 Infra-specific Taxa Assessed	无 / None

评估信息 Assessment Information

评估年份 Year Assessed	2020
评定人 Assessor(s)	刘少英、蒋志刚 / Shaoying Liu, Zhigang Jiang
审定人 Reviewer(s)	马勇、鲍毅新 / Yong Ma, Yixin Bao
其他贡献人 Other Contributor(s)	李立立、丁晨晨 / Lili Li, Chenchen Ding

理由 Justification: 针毛鼠分布广，种群数量大。因此，列为无危等级 / Chestnut White-bellied Rat is a widespread species with large populations. Thus, it is listed as Least Concern

地理分布 Geographical Distribution

国内分布 Domestic Distribution

西藏、云南、四川、贵州、重庆、湖南、湖北、广东、广西、海南、江西 / Tibet (Xizang), Yunnan, Sichuan, Guizhou, Chongqing, Hunan, Hubei, Guangdong, Guangxi, Hainan, Jiangxi

世界分布 World Distribution

中国；南亚、东南亚 / China; South Asia, Southeast Asia

分布标注 Distribution Note

非特有种 / Non-endemic

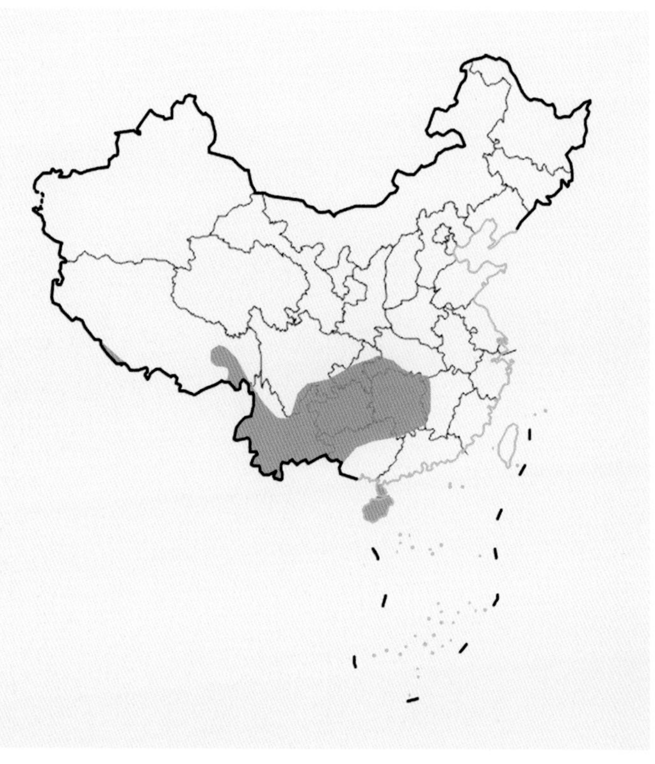

国内分布图
Map of Domestic Distribution

种群 Population

种群数量 Population Size	种群数量大 / Large populations
种群趋势 Population Trend	下降 / Decreasing

生境与生态系统 Habitat (s) and Ecosystem (s)

生　　境 Habitat(s)	森林、灌丛、竹林、耕地 / Forest, Shrubland, Bamboo Grove, Arable Land
生态系统 Ecosystem(s)	森林生态系统、灌丛生态系统、农田生态系统 / Forest Ecosystem, Shrubland Ecosystem, Cropland Ecosystem

威胁 Threat (s)

主要威胁 Major Threat(s)	无 / None

保护级别与保护行动 Protection Category and Conservation Action (s)

国家重点保护野生动物等级 (2021) Category of National Key Protected Wild Animals (2021)	无 / NA
IUCN 红色名录 (2020-2) IUCN Red List (2020-2)	无危 / LC
CITES 附录 (2019) CITES Appendix (2019)	无 / NA
保护行动 Conservation Action(s)	无 / None

相关文献 Relevant References

Burgin *et al*., 2018; Jiang *et al.* (蒋志刚等), 2017; Yang *et al.* (杨再学等), 2014; Huang *et al.* (黄辉等), 2013; Zheng *et al.* (郑智民等), 2012; Smith *et al.* (史密斯等), 2009

(接上页)
(Blyth, 1859); *condorensis* (Kloss, 1926); *flavipilis* (Shih, 1930); *gracilis* (Miller, 1913); *huang* (Bonhote, 1905); *jacobsoni* (Bartels, 1937); *jerdoni* (Blyth, 1863); *lepidus* (Miller, 1913); *lepturoides* (Sody, 1934); *lieftincki* (Chasen, 1939); *ling* (Bonhote, 1905); *marinus* (Kloss, 1916); *mekongis* (Robinson and Kloss, 1922); *minor* (Shih, 1930); *octomammis* (Gray, 1863); *orbus* (Robinson and Kloss, 1914); *pan* (Robinson and Kloss, 1914); *temmincki* (Kloss, 1921); *treubii* (Robinson and Kiloss, 1919); *vulpicolor* (G. M. Allen, 1926); *wongi* (Shih, 1931)

拟刺毛鼠
Niviventer huang

无危 LC

数据缺乏 DD	无危 LC	近危 NT	易危 VU	濒危 EN	极危 CR	区域灭绝 RE	野外灭绝 EW	灭绝 EX

分类地位 Taxonomic Status

动物界 Animalia	脊索动物门 Chordata	哺乳纲 Mammalia	啮齿目 Rodentia	鼠科 Muridae
学名 Scientific Name		*Niviventer huang*		
命名人 Species Authority		Bonhote, 1905		
英文名 English Name(s)		Eastern Spiny-haired Rat		
同物异名 Synonym(s)		无 / None		
种下单元评估 Infra-specific Taxa Assessed		无 / None		

评估信息 Assessment Information

评估年份 Year Assessed	2020
评定人 Assessor(s)	刘少英、蒋志刚 / Shaoying Liu, Zhigang Jiang
审定人 Reviewer(s)	马勇、鲍毅新 / Yong Ma, Yixin Bao
其他贡献人 Other Contributor(s)	李立立、丁晨晨 / Lili Li, Chenchen Ding

理由 Justification: 拟刺毛鼠分布广，种群数量大。因此，列为无危等级 / Eastern Spiny-haired Rat is a widespread species with large populations. Thus, it is listed as Least Concern

地理分布 Geographical Distribution

国内分布 Domestic Distribution

四川、陕西、重庆、安徽、浙江、江西、福建、广东、广西、香港、澳门 / Sichuan, Shaanxi, Chongqing, Anhui, Zhejiang, Jiangxi, Fujian, Guangdong, Guangxi, Hong Kong, Macao

世界分布 World Distribution

中国 / China

分布标注 Distribution Note

特有种 / Endemic

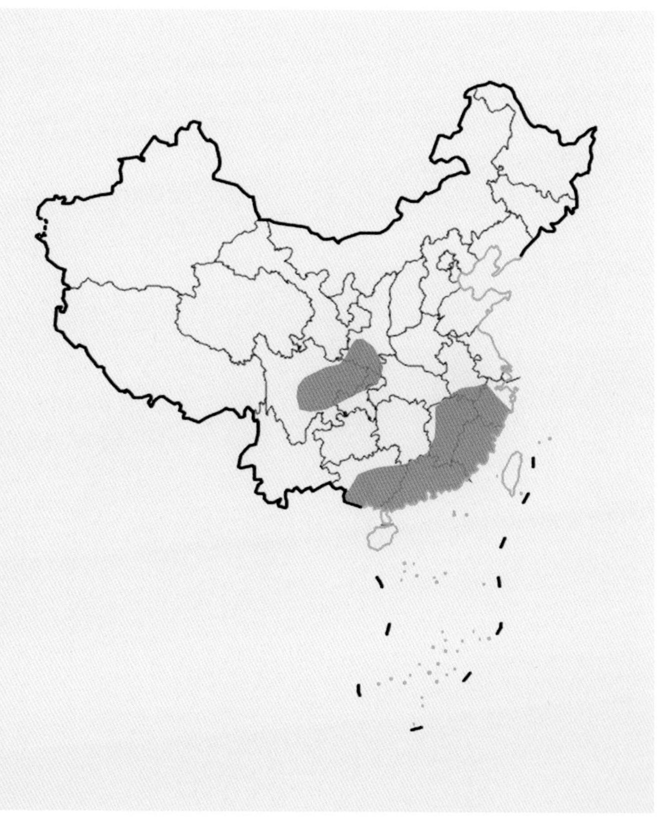

国内分布图
Map of Domestic Distribution

种群 Population

种群数量 **Population Size**	种群数量大 / Large populations
种群趋势 **Population Trend**	未知 / Unknown

生境与生态系统 Habitat (s) and Ecosystem (s)

生　境 **Habitat(s)**	水塘边草丛植被 / Grass Vegetation Near Pond
生态系统 **Ecosystem(s)**	草丛生态系统 / Grass Ecosystem

威胁 Threat (s)

主要威胁 **Major Threat(s)**	无 / None

保护级别与保护行动 Protection Category and Conservation Action (s)

国家重点保护野生动物等级 **(2021) Category of National Key Protected Wild Animals (2021)**	无 / NA
IUCN 红色名录 **(2020-2) IUCN Red List (2020-2)**	无危 / LC
CITES 附录 **(2019) CITES Appendix (2019)**	无 / NA
保护行动 **Conservation Action(s)**	无 / None

相关文献 Relevant References

Burgin *et al.*, 2018; Chen *et al.*, 2017b; He and Jiang, 2015; Jiang *et al.* (蒋志刚等), 2017

拟刺毛鼠 *Niviventer huang*

数据缺乏 DD	无危 LC	近危 NT	易危 VU	濒危 EN	极危 CR	区域灭绝 RE	野外灭绝 EW	灭绝 EX

分类地位 Taxonomic Status

动物界 Animalia	脊索动物门 Chordata	哺乳纲 Mammalia	啮齿目 Rodentia	鼠科 Muridae
学名 Scientific Name		*Niviventer lotipes*		
命名人 Species Authority		Allen, 1926		
英文名 English Name(s)		Hainan Niviventer		
同物异名 Synonym(s)		无 / None		
种下单元评估 Infra-specific Taxa Assessed		无 / None		

评估信息 Assessment Information

评估年份 Year Assessed	2020
评定人 Assessor(s)	刘少英、蒋志刚 / Shaoying Liu, Zhigang Jiang
审定人 Reviewer(s)	马勇、鲍毅新 / Yong Ma, Yixin Bao
其他贡献人 Other Contributor(s)	李立立、丁晨晨 / Lili Li, Chenchen Ding

理由 Justification: 海南白腹鼠分布广，种群数量大。因此，列为无危等级 / Hainan Niviventer is a widespread species with large populations. Thus, it is listed as Least Concern

地理分布 Geographical Distribution

国内分布 Domestic Distribution

海南 / Hainan

世界分布 World Distribution

中国 / China

分布标注 Distribution Note

特有种 / Endemic

国内分布图
Map of Domestic Distribution

种群 Population

种群数量 Population Size	种群数量大 / Large populations
种群趋势 Population Trend	未知 / Unknown

生境与生态系统 Habitat (s) and Ecosystem (s)

生　境 Habitat(s)	森林、耕地 / Forest, Arable Land
生态系统 Ecosystem(s)	森林生态系统、农田生态系统 / Forest Ecosystem, Cropland Ecosystem

威胁 Threat (s)

主要威胁 Major Threat(s)	无 / None

保护级别与保护行动 Protection Category and Conservation Action (s)

国家重点保护野生动物等级 (2021) Category of National Key Protected Wild Animals (2021)	无 / NA
IUCN 红色名录 (2020-2) IUCN Red List (2020-2)	无危 / LC
CITES 附录 (2019) CITES Appendix (2019)	无 / NA
保护行动 Conservation Action(s)	无 / None

相关文献 Relevant References

Burgin *et al*., 2018; Jiang *et al.* (蒋志刚等), 2017; Wilson *et al*., 2016; Li *et al*., 2008

红毛王鼠

Maxomys surifer

无危 LC

数据缺乏 DD	无危 LC	近危 NT	易危 VU	濒危 EN	极危 CR	区域灭绝 RE	野外灭绝 EW	灭绝 EX

分类地位 Taxonomic Status

动物界 Animalia	脊索动物门 Chordata	哺乳纲 Mammalia	啮齿目 Rodentia	鼠科 Muridae
学名 Scientific Name		*Maxomys surifer*		
命名人 Species Authority		Miller, 1900		
英文名 English Name(s)		Indomalayan Maxomys		
同物异名 Synonym(s)		Red Spiny Rat; *Maxomys anambae* (Miller, 1900); *antucus* (Lyon, 1916); *aoris* (Robinson, 1912); *banacus* (Lyon, 1916); *bandahara* (Robinson, 1921); *bentincanus* (转下页)		
种下单元评估 Infra-specific Taxa Assessed		无 / None		

评估信息 Assessment Information

评估年份 Year Assessed	2020
评定人 Assessor(s)	刘少英、蒋志刚 / Shaoying Liu, Zhigang Jiang
审定人 Reviewer(s)	马勇、鲍毅新 / Yong Ma, Yixin Bao
其他贡献人 Other Contributor(s)	李立立、丁晨晨 / Lili Li, Chenchen Ding

理由 **Justification:** 红毛王鼠的种群数量大。因此，列为无危等级 / Indomalayan Maxomys has large populations. Thus, it is listed as Least Concern

地理分布 Geographical Distribution

国内分布 Domestic Distribution

云南 / Yunnan

世界分布 World Distribution

中国；东南亚 / China; Southeast Asia

分布标注 Distribution Note

非特有种 / Non-endemic

国内分布图
Map of Domestic Distribution

种群 Population

种群数量 **Population Size**	种群数量大 / Large populations
种群趋势 **Population Trend**	下降 / Decreasing

生境与生态系统 Habitat (s) and Ecosystem (s)

生　境 **Habitat(s)**	热带湿润低地森林 / Tropical Moist Lowland Forest
生态系统 **Ecosystem(s)**	森林生态系统 / Forest Ecosystem

威胁 Threat (s)

主要威胁 **Major Threat(s)**	无 / None

保护级别与保护行动 Protection Category and Conservation Action (s)

国家重点保护野生动物等级 **(2021) Category of National Key Protected Wild Animals (2021)**	无 / NA
IUCN 红色名录 **(2020-2) IUCN Red List (2020-2)**	无危 / LC
CITES 附录 **(2019) CITES Appendix (2019)**	无 / NA
保护行动 **Conservation Action(s)**	无 / None

相关文献 Relevant References

Burgin *et al.*, 2018; Jiang *et al.* (蒋志刚等), 2017; Wilson *et al.*, 2016; Zheng *et al.* (郑智民等), 2012; Smith *et al.* (史密斯等), 2009; Wang (王应祥), 2003; Wu *et al.*, 1996

(接上页)
(Miller, 1903); *binominatus* (Kloss, 1915); *butangensis* (Miller, 1900); *carimatae* (Miller, 1906); *casensis* (Miller, 1903); *catellifer* (Miller, 1903); *changensis* (Kloss, 1916); *connectens* (Kloss, 1916); *domelicus* (Miller, 1903); *eclipsis* (Kloss, 1916); *finis* (Kloss, 1916); *flavidulus* (Miller, 1900); *flavigrandis* (Kloss, 1911); *grandis* (Kloss, 1911); *koratis* (Kloss, 1919); *kramis* (Kloss, 1919); *kutensis* (Kloss, 1916); *leonis* (Robinson and Kloss, 1911); *luteolus* (Miller, 1903); *mabalus* (Lyon, 1916); *manicalis* (Robinson and Kloss, 1914); *microdon* (Kloss, 1908); *muntia* (Chasen, 1940); *natunae* (Chasen, 1940); *pelagius* (Kloss, 1916); *pemangilis* (Robinson, 1912); *perflavus* (Lyon, 1911); *pidonis* (Chasen, 1940); *pinacus* (Lyon, 1916); *puket* (Chasen, 1940); *ravus* (Robinson and Kloss, 1916); *saturatus* (Lyon, 1911); *serutus* (Miller, 1906); *siarma* (Kloss, 1919); *solaris* (Sody, 1934); *spurcus* (Robinson and Kloss, 1914); *telibon* (Chasen, 1940); *tenebrosus* (Kloss, 1916); *ubecus* (Lyon, 1911); *umbridorsum* (Miller, 1903); *verbeeki* (Sody, 1930)

红毛王鼠 *Maxomys surifer*

大泡灰鼠

Berylmys berdmorei

无危 LC

数据缺乏 DD	无危 LC	近危 NT	易危 VU	濒危 EN	极危 CR	区域灭绝 RE	野外灭绝 EW	灭绝 EX

分类地位 Taxonomic Status

动物界 Animalia	脊索动物门 Chordata	哺乳纲 Mammalia	啮齿目 Rodentia	鼠科 Muridae

学名 Scientific Name	*Berylmys berdmorei*
命名人 Species Authority	Blyth, 1851
英文名 English Name(s)	Berdmore's Berylmy
同物异名 Synonym(s)	Small White-toothed Rat; *Berylmys magnus* (Kloss, 1916); *mullulus* (Thomas, 1916)
种下单元评估 Infra-specific Taxa Assessed	无 / None

评估信息 Assessment Information

评估年份 Year Assessed	2020
评定人 Assessor(s)	刘少英、蒋志刚 / Shaoying Liu, Zhigang Jiang
审定人 Reviewer(s)	马勇、鲍毅新 / Yong Ma, Yixin Bao
其他贡献人 Other Contributor(s)	李立立、丁晨晨 / Lili Li, Chenchen Ding

理由 **Justification:** 大泡灰鼠的种群数量大。因此，列为无危等级 / Berdmore's Berylmy has large populations. Thus, it is listed as Least Concern

地理分布 Geographical Distribution

国内分布 Domestic Distribution

云南 / Yunnan

世界分布 World Distribution

中国；东南亚 / China; Southeast Asia

分布标注 Distribution Note

非特有种 / Non-endemic

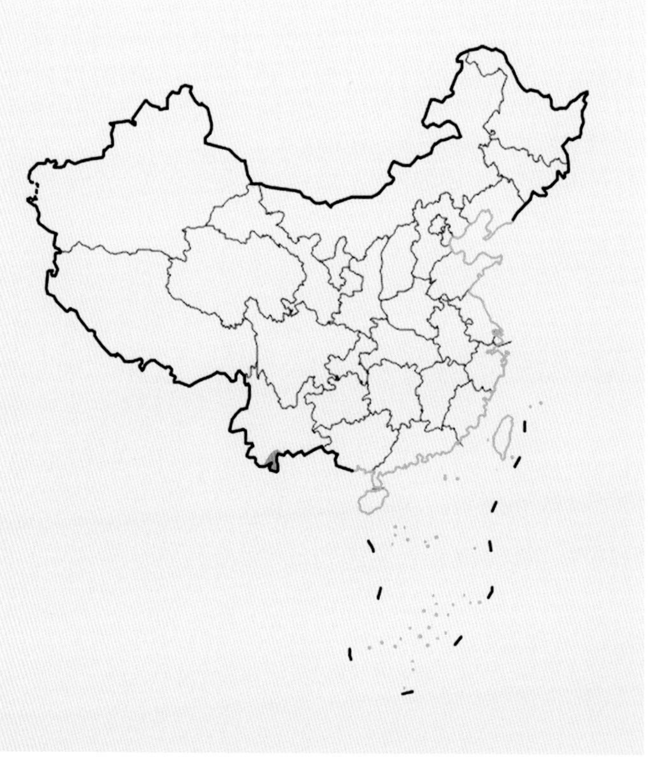

国内分布图
Map of Domestic Distribution

种群 Population

种群数量 Population Size	种群数量大 / Large populations
种群趋势 Population Trend	稳定 / Stable

生境与生态系统 Habitat (s) and Ecosystem (s)

生　　境 Habitat(s)	森林、沼泽 / Forest, Swamp
生态系统 Ecosystem(s)	森林生态系统、湿地生态系统 / Forest Ecosystem, Wetland Ecosystem

威胁 Threat (s)

主要威胁 Major Threat(s)	无 / None

保护级别与保护行动 Protection Category and Conservation Action (s)

国家重点保护野生动物等级 (2021) Category of National Key Protected Wild Animals (2021)	无 / NA
IUCN 红色名录 (2020-2) IUCN Red List (2020-2)	无危 / LC
CITES 附录 (2019) CITES Appendix (2019)	无 / NA
保护行动 Conservation Action(s)	无 / None

相关文献 Relevant References

Burgin *et al*., 2018; Jiang *et al.* (蒋志刚等), 2017; Wilson *et al*., 2016; Zheng *et al.* (郑智民等), 2012; Smith *et al.* (史密斯等), 2009; Xing *et al.* (邢雅俊等), 2008; Wang (王应祥), 2003

青毛巨鼠

Berylmys bowersi

无危 LC

数据缺乏 DD	无危 LC	近危 NT	易危 VU	濒危 EN	极危 CR	区域灭绝 RE	野外灭绝 EW	灭绝 EX

分类地位 Taxonomic Status

动物界 Animalia	脊索动物门 Chordata	哺乳纲 Mammalia	啮齿目 Rodentia	鼠科 Muridae

学名 Scientific Name	*Berylmys bowersi*
命名人 Species Authority	Anderson, 1879
英文名 English Name(s)	Bower's White-toothed Rat
同物异名 Synonym(s)	*Berylmys ferreocanus* (Miller, 1900); *kennethi* (Kloss, 1919); *lactiiventer* (Kloss, 1919); *latouchei* (Thomas, 1897); *totipes* (Dao, 1966); *wellsi* (Thomas, 1921)
种下单元评估 Infra-specific Taxa Assessed	无 / None

评估信息 Assessment Information

评估年份 Year Assessed	2020
评定人 Assessor(s)	刘少英、蒋志刚 / Shaoying Liu, Zhigang Jiang
审定人 Reviewer(s)	马勇、鲍毅新 / Yong Ma, Yixin Bao
其他贡献人 Other Contributor(s)	李立立、丁晨晨 / Lili Li, Chenchen Ding

理由 Justification: 青毛巨鼠的种群数量大。因此，列为无危等级 / Bower's White-toothed Rat has large populations. Thus, it is listed as Least Concern

地理分布 Geographical Distribution

国内分布 Domestic Distribution

浙江、福建、云南、广西、安徽、江西、湖南、湖北、广东、四川、重庆、贵州、西藏 / Zhejiang, Fujian, Yunnan, Guangxi, Anhui, Jiangxi, Hunan, Hubei, Guangdong, Sichuan, Chongqing, Guizhou, Tibet (Xizang)

世界分布 World Distribution

中国；南亚、东南亚 / China; South Asia, Southeast Asia

分布标注 Distribution Note

非特有种 / Non-endemic

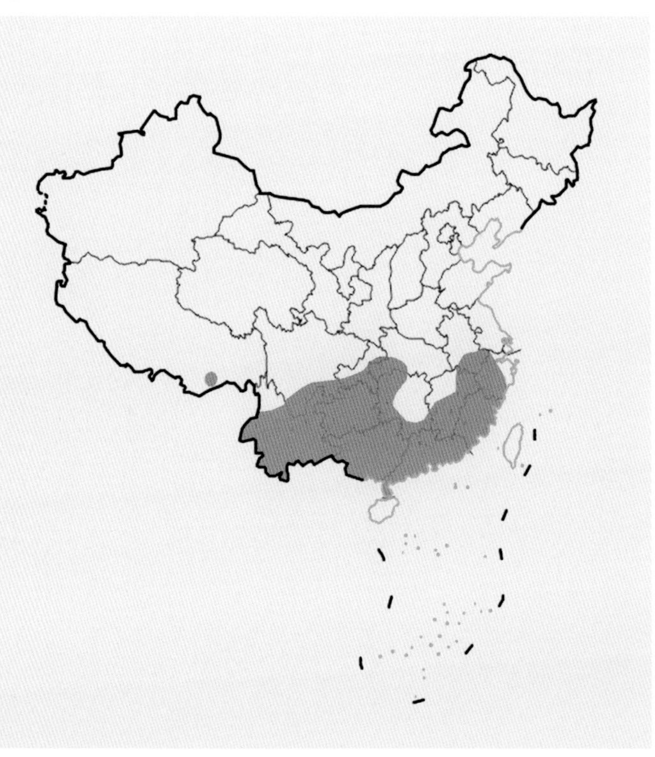

国内分布图
Map of Domestic Distribution

种群 Population

种群数量 Population Size	种群数量大 / Large populations
种群趋势 Population Trend	下降 / Decreasing

生境与生态系统 Habitat (s) and Ecosystem (s)

生　　境 Habitat(s)	森林、次生林、灌丛、耕地 / Forest, Secondary Forest, Shrubland, Arable Land
生态系统 Ecosystem(s)	森林生态系统、灌丛生态系统、农田生态系统 / Forest Ecosystem, Shrubland Ecosystem, Cropland Ecosystem

威胁 Threat (s)

主要威胁 Major Threat(s)	无 / None

保护级别与保护行动 Protection Category and Conservation Action (s)

国家重点保护野生动物等级 (2021) Category of National Key Protected Wild Animals (2021)	无 / NA
IUCN 红色名录 (2020-2) IUCN Red List (2020-2)	无危 / LC
CITES 附录 (2019) CITES Appendix (2019)	无 / NA
保护行动 Conservation Action(s)	无 / None

相关文献 Relevant References

Burgin *et al*., 2018; Jiang *et al.* (蒋志刚等), 2017; Wilson *et al*., 2016; Jiang *et al.* (江广华等), 2013; Zheng *et al.* (郑智民等), 2012; Pan *et al.* (潘清华等), 2007

青毛巨鼠 *Berylmys bowersi*　　By www.flickr.com

白腹巨鼠

Leopoldamys edwardsi

无危 LC

数据缺乏 DD	无危 LC	近危 NT	易危 VU	濒危 EN	极危 CR	区域灭绝 RE	野外灭绝 EW	灭绝 EX

分类地位 Taxonomic Status

动物界 Animalia	脊索动物门 Chordata	哺乳纲 Mammalia	啮齿目 Rodentia	鼠科 Muridae

学名 Scientific Name	*Leopoldamys edwardsi*
命名人 Species Authority	Thomas, 1882
英文名 English Name(s)	Edward's Rat
同物异名 Synonym(s)	Edward's Long-tailed Giant Rat; *Leopoldamys garonum* (Thomas, 1921); *gigas* (Satunin, 1903); *hainanensis* (Xu and Yu, 1985); *listeri* (Thomas, 1916); *melli* (Matschie, 1922)
种下单元评估 Infra-specific Taxa Assessed	无 / None

评估信息 Assessment Information

评估年份 Year Assessed	2020
评定人 Assessor(s)	刘少英、蒋志刚 / Shaoying Liu, Zhigang Jiang
审定人 Reviewer(s)	马勇、鲍毅新 / Yong Ma, Yixin Bao
其他贡献人 Other Contributor(s)	李立立、丁晨晨 / Lili Li, Chenchen Ding

理由 Justification: 白腹巨鼠分布广，种群数量大。因此，列为无危等级 / Edward's Rat is a widespread species with large populations. Thus, it is listed as Least Concern

地理分布 Geographical Distribution

国内分布 Domestic Distribution

西藏、云南、甘肃、贵州、广东、广西、海南、福建、浙江、重庆、湖北、湖南、四川、安徽、江西、陕西 / Tibet (Xizang), Yunnan, Gansu, Guizhou, Guangdong, Guangxi, Hainan, Fujian, Zhejiang, Chongqing, Hubei, Hunan, Sichuan, Anhui, Jiangxi, Shaanxi

世界分布 World Distribution

中国；南亚、东南亚 / China; South Asia, Southeast Asia

分布标注 Distribution Note

非特有种 / Non-endemic

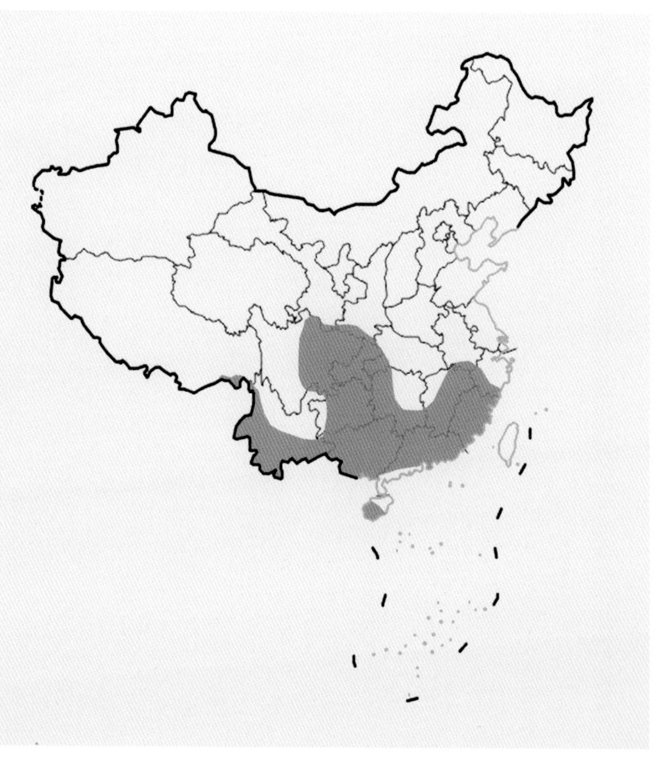

国内分布图
Map of Domestic Distribution

种群 Population

种群数量 Population Size	种群数量大 / Large populations
种群趋势 Population Trend	未知 / Unknown

生境与生态系统 Habitat (s) and Ecosystem (s)

生　　境 Habitat(s)	热带和亚热带湿润低地山地森林 / Tropical and Subtropical Moist Lowland Montane Forest
生态系统 Ecosystem(s)	森林生态系统 / Forest Ecosystem

威胁 Threat (s)

主要威胁 Major Threat(s)	无 / None

保护级别与保护行动 Protection Category and Conservation Action (s)

国家重点保护野生动物等级 (2021) Category of National Key Protected Wild Animals (2021)	无 / NA
IUCN 红色名录 (2020-2) IUCN Red List (2020-2)	无危 / LC
CITES 附录 (2019) CITES Appendix (2019)	无 / NA
保护行动 Conservation Action(s)	无 / None

相关文献 Relevant References

Burgin *et al*., 2018; Jiang *et al.* (蒋志刚等), 2017; Wilson *et al*., 2016; Zheng *et al.* (郑智民等), 2012; Liu *et al.* (刘鑫等), 2011; Lu and Liu (路纪琪和刘彬), 2008; Xia (夏武平), 1964

数据缺乏 DD	无危 LC	近危 NT	易危 VU	濒危 EN	极危 CR	区域灭绝 RE	野外灭绝 EW	灭绝 EX

分类地位 Taxonomic Status

动物界 Animalia	脊索动物门 Chordata	哺乳纲 Mammalia	啮齿目 Rodentia	鼠科 Muridae
学名 Scientific Name	*Mus booduga*			
命名人 Species Authority	Gray, 1837			
英文名 English Name(s)	Little Indian Field Mouse			
同物异名 Synonym(s)	无 / None			
种下单元评估 Infra-specific Taxa Assessed	无 / None			

评估信息 Assessment Information

评估年份 Year Assessed	2020
评定人 Assessor(s)	刘少英、蒋志刚 / Shaoying Liu, Zhigang Jiang
审定人 Reviewer(s)	马勇、鲍毅新 / Yong Ma, Yixin Bao
其他贡献人 Other Contributor(s)	李立立、丁晨晨 / Lili Li, Chenchen Ding

理由 Justification: 印度小鼠分布较广。因此，列为无危等级 / Little Indian Field Mouse is relatively widespread. Thus, it is listed as Least Concern

地理分布 Geographical Distribution

国内分布 Domestic Distribution

西藏 / Tibet (Xizang)

世界分布 World Distribution

中国；南亚 / China; South Asia

分布标注 Distribution Note

非特有种 / Non-endemic

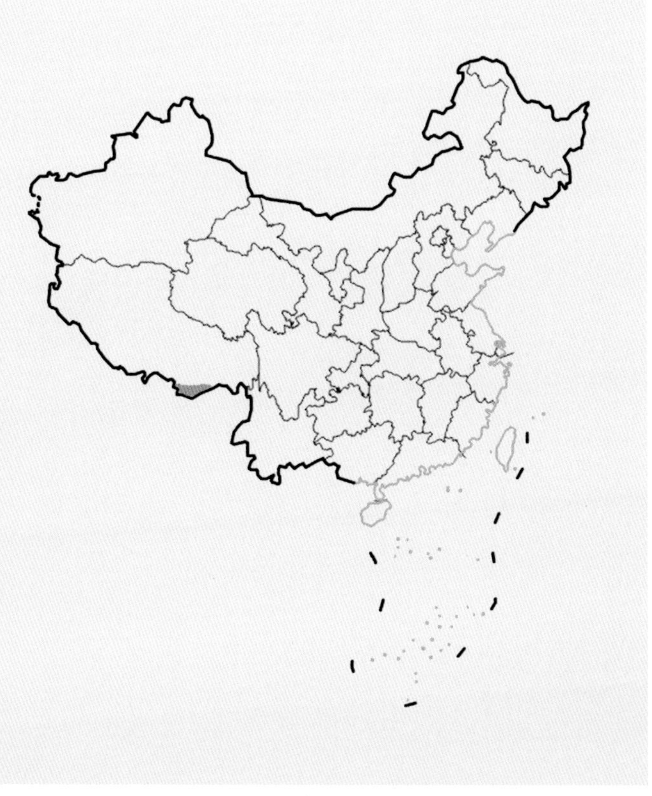

国内分布图
Map of Domestic Distribution

种群 Population

种群数量 Population Size	未知 / Unknown
种群趋势 Population Trend	未知 / Unknown

生境与生态系统 Habitat (s) and Ecosystem (s)

生　　境 Habitat(s)	海拔 400m 以下的多种生境 / Various habitats below 400m a.s.l.
生态系统 Ecosystem(s)	森林生态系统、灌丛生态系统、草地生态系统、农田生态系统 / Forest Ecosystem, Shrubland Ecosystem, Grassland Ecosystem, Cropland Ecosystem

威胁 Threat (s)

主要威胁 Major Threat(s)	无 / None

保护级别与保护行动 Protection Category and Conservation Action (s)

国家重点保护野生动物等级 (2021) Category of National Key Protected Wild Animals (2021)	无 / NA
IUCN 红色名录 (2020-2) IUCN Red List (2020-2)	无危 / LC
CITES 附录 (2019) CITES Appendix (2019)	无 / NA
保护行动 Conservation Action(s)	无 / None

相关文献 Relevant References

Burgin *et al*., 2018; Jiang *et al.* (蒋志刚等), 2017; Wilson *et al*., 2016; Choudhury, 2003

卡氏小鼠

Mus caroli

无危 LC

数据缺乏 DD	无危 LC	近危 NT	易危 VU	濒危 EN	极危 CR	区域灭绝 RE	野外灭绝 EW	灭绝 EX

分类地位 Taxonomic Status

动 物 界 Animalia	脊索动物门 Chordata	哺 乳 纲 Mammalia	啮 齿 目 Rodentia	鼠 科 Muridae

学 名 Scientific Name	*Mus caroli*
命 名 人 Species Authority	Bonhote, 1902
英 文 名 English Name(s)	Ryukyu Mouse
同物异名 Synonym(s)	*Mus boninensis* (Kishida, 1926); *boninensis* (Kuroda, 1930); *formosanus* (Kuroda, 1925); *kurilensis* (Kuroda, 1924); *ouwensi* (Kloss, 1921)
种下单元评估 Infra-specific Taxa Assessed	无 / None

评估信息 Assessment Information

评 估 年 份 Year Assessed	2020
评 定 人 Assessor(s)	刘少英、蒋志刚 / Shaoying Liu, Zhigang Jiang
审 定 人 Reviewer(s)	马勇、鲍毅新 / Yong Ma, Yixin Bao
其他贡献人 Other Contributor(s)	李立立、丁晨晨 / Lili Li, Chenchen Ding

理由 Justification: 卡氏小鼠分布广，种群数量大。因此，列为无危等级 / Ryukyu Mouse is a widespread species with large populations. Thus, it is listed as Least Concern

地理分布 Geographical Distribution

国内分布 Domestic Distribution

福建、台湾、贵州、广西、云南、广东、海南、香港 / Fujian, Taiwan, Guizhou, Guangxi, Yunnan, Guangdong, Hainan, Hong Kong

世界分布 World Distribution

中国；东南亚 / China; Southeast Asia

分布标注 Distribution Note

非特有种 / Non-endemic

国内分布图
Map of Domestic Distribution

种群 Population

种群数量 Population Size	种群数量大 / Large populations
种群趋势 Population Trend	稳定 / Stable

生境与生态系统 Habitat (s) and Ecosystem (s)

生　　境 Habitat(s)	耕地、草地、灌丛、次生林 / Arable Land, Grassland, Shrubland, Secondary Forest
生态系统 Ecosystem(s)	森林生态系统、灌丛生态系统、草地生态系统、农田生态系统 / Forest Ecosystem, Shrubland Ecosystem, Grassland Ecosystem, Cropland Ecosystem

威胁 Threat (s)

主要威胁 Major Threat(s)	无 / None

保护级别与保护行动 Protection Category and Conservation Action (s)

国家重点保护野生动物等级 (2021) Category of National Key Protected Wild Animals (2021)	无 / NA
IUCN 红色名录 (2020-2) IUCN Red List (2020-2)	无危 / LC
CITES 附录 (2019) CITES Appendix (2019)	无 / NA
保护行动 Conservation Action(s)	无 / None

相关文献 Relevant References

Burgin *et al*., 2018; Jiang *et al.* (蒋志刚等), 2017; Wilson *et al*., 2016; Peng and Guo (彭培英和郭宪国), 2014; Zheng *et al.* (郑智民等), 2012; Wu (吴爱国), 2002; Wu *et al.* (吴德林等), 1995

仔鹿小鼠

Mus cervicolor

无危 LC

数据缺乏 DD	无危 LC	近危 NT	易危 VU	濒危 EN	极危 CR	区域灭绝 RE	野外灭绝 EW	灭绝 EX

分类地位 Taxonomic Status

动物界 Animalia	脊索动物门 Chordata	哺乳纲 Mammalia	啮齿目 Rodentia	鼠科 Muridae

学名 Scientific Name	*Mus cervicolor*
命名人 Species Authority	Hodgson, 1845
英文名 English Name(s)	Fawn-colored Mouse
同物异名 Synonym(s)	*Mus annamensis* (Robinson and Kloss, 1922); *cunicularis* (Blyth, 1855); *imphalensis* (Roonwal, 1948)
种下单元评估 Infra-specific Taxa Assessed	无 / None

评估信息 Assessment Information

评估年份 Year Assessed	2020
评定人 Assessor(s)	刘少英、蒋志刚 / Shaoying Liu, Zhigang Jiang
审定人 Reviewer(s)	马勇、鲍毅新 / Yong Ma, Yixin Bao
其他贡献人 Other Contributor(s)	李立立、丁晨晨 / Lili Li, Chenchen Ding

理由 **Justification:** 仔鹿小鼠分布广，种群数量大。因此，列为无危等级 / Fawn-colored Mouse is a widespread species with large populations. Thus, it is listed as Least Concern

地理分布 Geographical Distribution

国内分布 Domestic Distribution

云南 / Yunnan

世界分布 World Distribution

中国；南亚、东南亚 / China; South Asia, Southeast Asia

分布标注 Distribution Note

非特有种 / Non-endemic

国内分布图
Map of Domestic Distribution

种群 Population

种群数量 **Population Size**	较多 / Relatively abundant
种群趋势 **Population Trend**	稳定 / Stable

生境与生态系统 Habitat (s) and Ecosystem (s)

生　　境 **Habitat(s)**	次生林、草地、灌丛、耕地 / Secondary Forest, Grassland, Shrubland, Arable Land
生态系统 **Ecosystem(s)**	森林生态系统、灌丛生态系统、草地生态系统、农田生态系统 / Forest Ecosystem, Shrubland Ecosystem, Grassland Ecosystem, Cropland Ecosystem

威胁 Threat (s)

主要威胁 **Major Threat(s)**	无 / None

保护级别与保护行动 Protection Category and Conservation Action (s)

国家重点保护野生动物等级 **(2021) Category of National Key Protected Wild Animals (2021)**	无 / NA
IUCN 红色名录 **(2020-2) IUCN Red List (2020-2)**	无危 / LC
CITES 附录 **(2019) CITES Appendix (2019)**	无 / NA
保护行动 **Conservation Action(s)**	无 / None

相关文献 Relevant References

Burgin *et al*., 2018; Jiang *et al.* (蒋志刚等), 2017; Wilson *et al*., 2016; Smith *et al.* (史密斯等), 2009; Wang (王应祥), 2003; Marshall, 1977

丛林小鼠

Mus cookii

无危 LC

数据缺乏 DD	无危 LC	近危 NT	易危 VU	濒危 EN	极危 CR	区域灭绝 RE	野外灭绝 EW	灭绝 EX

分类地位 Taxonomic Status

动物界 Animalia	脊索动物门 Chordata	哺乳纲 Mammalia	啮齿目 Rodentia	鼠科 Muridae

学名 Scientific Name	*Mus cookii*
命名人 Species Authority	Ryley, 1914
英文名 English Name(s)	Cook's Mouse
同物异名 Synonym(s)	*Mus darjilingensis* (Hodgson, 1849); *nagarum* (Thomas, 1921); *palnica* (Thomas, 1923); *rahengis* (Kloss, 1920); *thai* (Kloss, 1917)
种下单元评估 Infra-specific Taxa Assessed	无 / None

评估信息 Assessment Information

评估年份 Year Assessed	2020
评定人 Assessor(s)	刘少英、蒋志刚 / Shaoying Liu, Zhigang Jiang
审定人 Reviewer(s)	马勇、鲍毅新 / Yong Ma, Yixin Bao
其他贡献人 Other Contributor(s)	李立立、丁晨晨 / Lili Li, Chenchen Ding

理由 Justification: 丛林小鼠分布广，种群数量大。因此，列为无危等级 / Cook's Mouse is a widespread species with large populations. Thus, it is listed as Least Concern

地理分布 Geographical Distribution

国内分布 Domestic Distribution

云南 / Yunnan

世界分布 World Distribution

中国；南亚、东南亚 / China; South Asia, Southeast Asia

分布标注 Distribution Note

非特有种 / Non-endemic

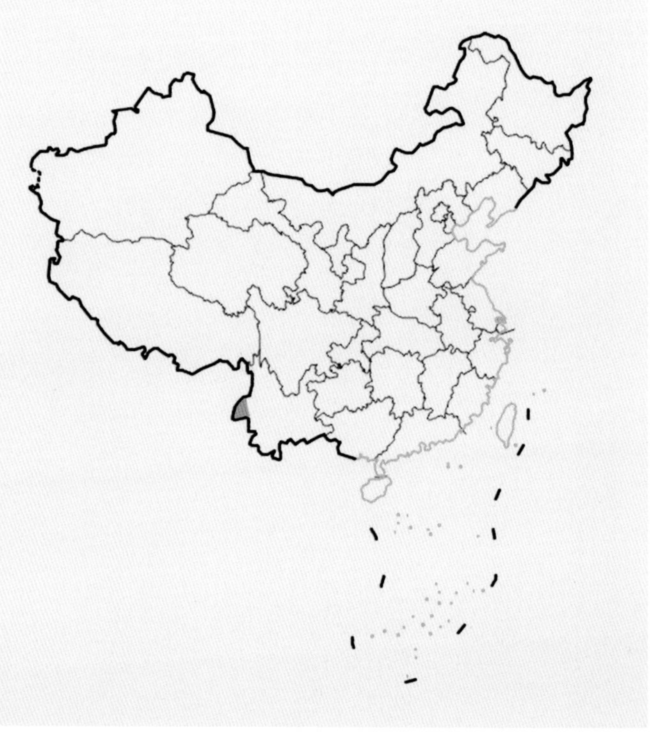

国内分布图
Map of Domestic Distribution

种群 Population

种群数量 **Population Size**	未知 / Unknown
种群趋势 **Population Trend**	稳定 / Stable

生境与生态系统 Habitat (s) and Ecosystem (s)

生　　境 **Habitat(s)**	耕地 / Arable Land
生态系统 **Ecosystem(s)**	农田生态系统 / Cropland Ecosystem

威胁 Threat (s)

主要威胁 **Major Threat(s)**	无 / None

保护级别与保护行动 Protection Category and Conservation Action (s)

国家重点保护野生动物等级 **(2021) Category of National Key Protected Wild Animals (2021)**	无 / NA
IUCN红色名录**(2020-2) IUCN Red List (2020-2)**	无危 / LC
CITES 附录 **(2019) CITES Appendix (2019)**	无 / NA
保护行动 **Conservation Action(s)**	无 / None

相关文献 Relevant References

Burgin *et al*., 2018; Jiang *et al.* (蒋志刚等), 2017; Wilson *et al*., 2016; Zheng *et al.* (郑智民等), 2012; Wang (王应祥), 2003; Marshall, 1977

数据缺乏 DD	无危 LC	近危 NT	易危 VU	濒危 EN	极危 CR	区域灭绝 RE	野外灭绝 EW	灭绝 EX

分类地位 Taxonomic Status

动物界 Animalia	脊索动物门 Chordata	哺乳纲 Mammalia	啮齿目 Rodentia	鼠科 Muridae
学名 Scientific Name		*Mus musculus*		
命名人 Species Authority		Linnaeus, 1758		
英文名 English Name(s)		House Mouse		
同物异名 Synonym(s)		*Mus albula* (Kishida, 1924); *cinereomaculatus* (Fitzinger, 1867); *molossinus* (Temminck, 1844); *nordmanni* (Keyserling and Blasius, 1840); *reboudi* (Loche, 1867); *tantillus* (转下页)		
种下单元评估 Infra-specific Taxa Assessed		无 / None		

评估信息 Assessment Information

评估年份 Year Assessed	2020
评定人 Assessor(s)	刘少英、蒋志刚 / Shaoying Liu, Zhigang Jiang
审定人 Reviewer(s)	马勇、鲍毅新 / Yong Ma, Yixin Bao
其他贡献人 Other Contributor(s)	李立立、丁晨晨 / Lili Li, Chenchen Ding

理由 **Justification:** 小家鼠分布广，种群数量大。因此，列为无危等级 / House Mouse is a widespread species with large populations. Thus, it is listed as Least Concern

地理分布 Geographical Distribution

国内分布 Domestic Distribution

黑龙江、吉林、辽宁、内蒙古、河北、北京、天津、山东、河南、山西、陕西、甘肃、宁夏、青海、四川、贵州、云南、西藏、上海、江苏、浙江、安徽、新疆、江西、湖北、湖南、广西、广东、海南、福建、台湾、重庆、香港、澳门 / Heilongjiang, Jilin, Liaoning, Inner Mongolia (Nei Mongol), Hebei, Beijing, Tianjin, Shandong, Henan, Shanxi, Shaanxi, Gansu, Ningxia, Qinghai, Sichuan, Guizhou, Yunnan, Tibet (Xizang), Shanghai, Jiangsu, Zhejiang, Anhui, Xinjiang, Jiangxi, Hubei, Hunan, Guangxi, Guangdong, Hainan, Fujian, Taiwan, Chongqing, Hong Kong, Macao

世界分布 World Distribution

全球分布 / Globally distributed

分布标注 Distribution Note

非特有种 / Non-endemic

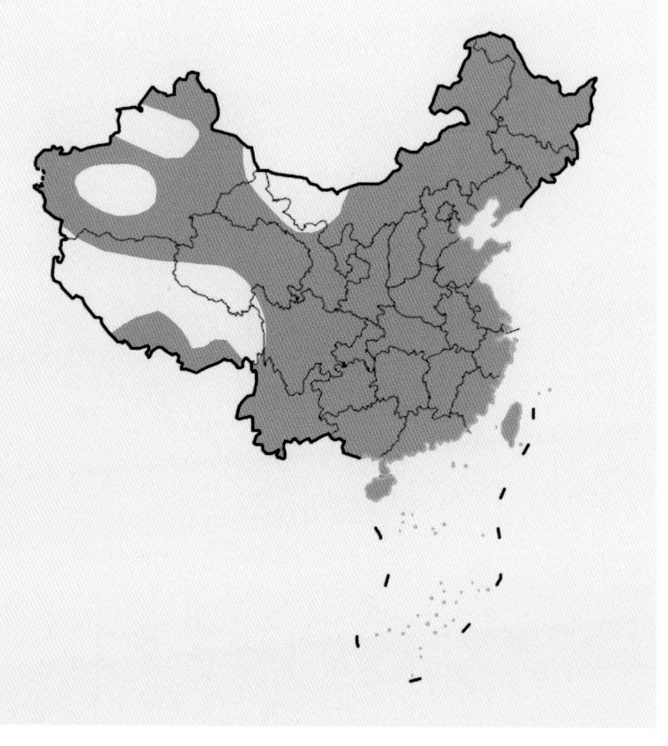

国内分布图 **Map of Domestic Distribution**

种群 Population

种群数量 Population Size	种群数量大 / Large populations
种群趋势 Population Trend	稳定 / Stable

生境与生态系统 Habitat (s) and Ecosystem (s)

生　　境 Habitat(s)	人工环境 / Human Settlement
生态系统 Ecosystem(s)	人类聚落生态系统、农田生态系统 / Human Settlement Ecosystem, Cropland Ecosystem

威胁 Threat (s)

主要威胁 Major Threat(s)	无 / None

保护级别与保护行动 Protection Category and Conservation Action (s)

国家重点保护野生动物等级 (2021) Category of National Key Protected Wild Animals (2021)	无 / NA
IUCN 红色名录 (2020-2) IUCN Red List (2020-2)	无危 / LC
CITES 附录 (2019) CITES Appendix (2019)	无 / NA
保护行动 Conservation Action(s)	无 / None

相关文献 Relevant References

Burgin *et al*., 2018; Jiang *et al.* (蒋志刚等), 2017; Wilson *et al*., 2016; Zhu *et al.* (朱琼蕊等), 2014; Zheng *et al.* (郑智民等), 2012; Macholán, 1999; Yan and Zhong (严志堂和钟明明), 1984

(接上页)
(G. M. Allen, 1927); *varius* (Fitzinger, 1867); *yonakuni* (Kuroda, 1924)

小家鼠 *Mus musculus*

锡金小鼠

Mus pahari

无危 LC

数据缺乏 DD	无危 LC	近危 NT	易危 VU	濒危 EN	极危 CR	区域灭绝 RE	野外灭绝 EW	灭绝 EX

分类地位 Taxonomic Status

动物界 Animalia	脊索动物门 Chordata	哺乳纲 Mammalia	啮齿目 Rodentia	鼠科 Muridae

学名 Scientific Name	*Mus pahari*
命名人 Species Authority	Thomas, 1916
英文名 English Name(s)	Gairdner's Shrew-mouse
同物异名 Synonym(s)	Indochinese Shrewlike Mouse; *Mus gairdneri* (Kloss, 1920); *jacksoniae* (Thomas, 1921); *meator* (G. M. Allen, 1927); *mocchauensis* (Dao, 1978)
种下单元评估 Infra-specific Taxa Assessed	无 / None

评估信息 Assessment Information

评估年份 Year Assessed	2020
评定人 Assessor(s)	刘少英、蒋志刚 / Shaoying Liu, Zhigang Jiang
审定人 Reviewer(s)	马勇、鲍毅新 / Yong Ma, Yixin Bao
其他贡献人 Other Contributor(s)	李立立、丁晨晨 / Lili Li, Chenchen Ding

理由 Justification: 锡金小鼠分布广，种群数量大。因此，列为无危等级 / Gairdner's Shrew-mouse is a widespread species with large populations. Thus, it is listed as Least Concern

地理分布 Geographical Distribution

国内分布 Domestic Distribution

四川、贵州、云南、西藏、广西 / Sichuan, Guizhou, Yunnan, Tibet (Xizang), Guangxi

世界分布 World Distribution

中国；南亚、东南亚 / China; South Asia, Southeast Asia

分布标注 Distribution Note

非特有种 / Non-endemic

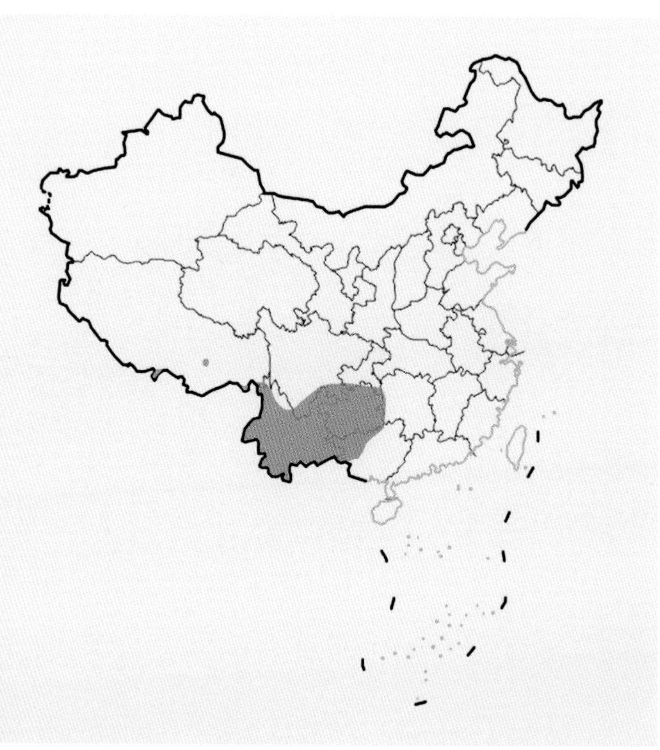

国内分布图
Map of Domestic Distribution

种群 Population

种群数量 Population Size	种群数量大 / Large populations
种群趋势 Population Trend	稳定 / Stable

生境与生态系统 Habitat (s) and Ecosystem (s)

生　境 Habitat(s)	森林 / Forest
生态系统 Ecosystem(s)	森林生态系统 / Forest Ecosystem

威胁 Threat (s)

主要威胁 Major Threat(s)	无 / None

保护级别与保护行动 Protection Category and Conservation Action (s)

国家重点保护野生动物等级 (2021) Category of National Key Protected Wild Animals (2021)	无 / NA
IUCN 红色名录 (2020-2) IUCN Red List (2020-2)	无危 / LC
CITES 附录 (2019) CITES Appendix (2019)	无 / NA
保护行动 Conservation Action(s)	无 / None

相关文献 Relevant References

Burgin *et al*., 2018; Jiang *et al.* (蒋志刚等), 2017; Wilson *et al*., 2016; Yang *et al.* (杨再学等), 2013; Pan *et al*. (潘会等), 2012; Zheng *et al.* (郑智民等), 2012; Wang (王应祥), 2003

小板齿鼠

Bandicota bengalensis

无危 LC

数据缺乏 DD	无危 LC	近危 NT	易危 VU	濒危 EN	极危 CR	区域灭绝 RE	野外灭绝 EW	灭绝 EX

分类地位 Taxonomic Status

动物界 Animalia	脊索动物门 Chordata	哺乳纲 Mammalia	啮齿目 Rodentia	鼠科 Muridae

学名 Scientific Name	*Bandicota bengalensis*
命名人 Species Authority	Gray, 1835
英文名 English Name(s)	Lesser Bandicoot Rat
同物异名 Synonym(s)	Indian Mole-rat, Sind Rice Rat
种下单元评估 Infra-specific Taxa Assessed	无 / None

评估信息 Assessment Information

评估年份 Year Assessed	2020
评定人 Assessor(s)	刘少英、蒋志刚 / Shaoying Liu, Zhigang Jiang
审定人 Reviewer(s)	马勇、鲍毅新 / Yong Ma, Yixin Bao
其他贡献人 Other Contributor(s)	李立立、丁晨晨 / Lili Li, Chenchen Ding

理由 Justification: 小板齿鼠分布广，种群大，在许多保护区内有分布，并对栖息地改变有一定的容忍度。因此，列为无危等级 / Lesser Bandicoot Rat is a widespread species with large populations, and is distributed in many protected area. It is an adaptable species, and could bear the habitats change in some degree. Thus, it is listed as Least Concern

地理分布 Geographical Distribution

国内分布 Domestic Distribution

西藏 / Tibet (Xizang)

世界分布 World Distribution

中国；南亚 / China; South Asia

分布标注 Distribution Note

非特有种 / Non-endemic

国内分布图
Map of Domestic Distribution

种群 Population

种群数量 **Population Size**	种群数量大 / Large populations
种群趋势 **Population Trend**	未知 / Unknown

生境与生态系统 Habitat (s) and Ecosystem (s)

生　　境 **Habitat(s)**	耕地、民居 / Arable Land, Human Dwelling
生态系统 **Ecosystem(s)**	农田生态系统、人类聚落生态系统 / Cropland Ecosystem, Human Settlement Ecosystem

威胁 Threat (s)

主要威胁 **Major Threat(s)**	无 / None

保护级别与保护行动 Protection Category and Conservation Action (s)

国家重点保护野生动物等级 **(2021) Category of National Key Protected Wild Animals (2021)**	无 / NA
IUCN 红色名录 **(2020-2) IUCN Red List (2020-2)**	无危 / LC
CITES 附录 **(2019) CITES Appendix (2019)**	无 / NA
保护行动 **Conservation Action(s)**	无 / None

相关文献 Relevant References

Burgin *et al*., 2018; Jiang *et al.* (蒋志刚等), 2017; Wilson *et al*., 2016; Choudhury, 2003

数据缺乏 DD	无危 LC	近危 NT	易危 VU	濒危 EN	极危 CR	区域灭绝 RE	野外灭绝 EW	灭绝 EX

分类地位 Taxonomic Status

动物界 Animalia	脊索动物门 Chordata	哺乳纲 Mammalia	啮齿目 Rodentia	鼠科 Muridae
学名 Scientific Name		*Bandicota indica*		
命名人 Species Authority		Bechstein, 1800		
英文名 English Name(s)		Greater Bandicoot Rat		
同物异名 Synonym(s)		*Bandicota bandicota* (Bechstein, 1800); *elliotanus* (Anderson, 1878); *eloquens* (Kishida, 1926); *gigantea* (Hardwicke, 1804); *jabouillei* (Thomas, 1927); *macropus* (Hodgson, (转下页)		
种下单元评估 Infra-specific Taxa Assessed		无 / None		

评估信息 Assessment Information

评估年份 Year Assessed	2020
评定人 Assessor(s)	刘少英、蒋志刚 / Shaoying Liu, Zhigang Jiang
审定人 Reviewer(s)	马勇、鲍毅新 / Yong Ma, Yixin Bao
其他贡献人 Other Contributor(s)	李立立、丁晨晨 / Lili Li, Chenchen Ding

理由 Justification: 板齿鼠分布较广，种群数量大。因此，列为无危等级 / Greater Bandicoot Rat is a widespread species with large populations. Thus, it is listed as Least Concern

地理分布 Geographical Distribution

国内分布 Domestic Distribution

四川、贵州、云南、广西、广东、福建、台湾、香港 / Sichuan, Guizhou, Yunnan, Guangxi, Guangdong, Fujian, Taiwan, Hong Kong

世界分布 World Distribution

中国；南亚、东南亚 / China; South Asia, Southeast Asia

分布标注 Distribution Note

非特有种 / Non-endemic

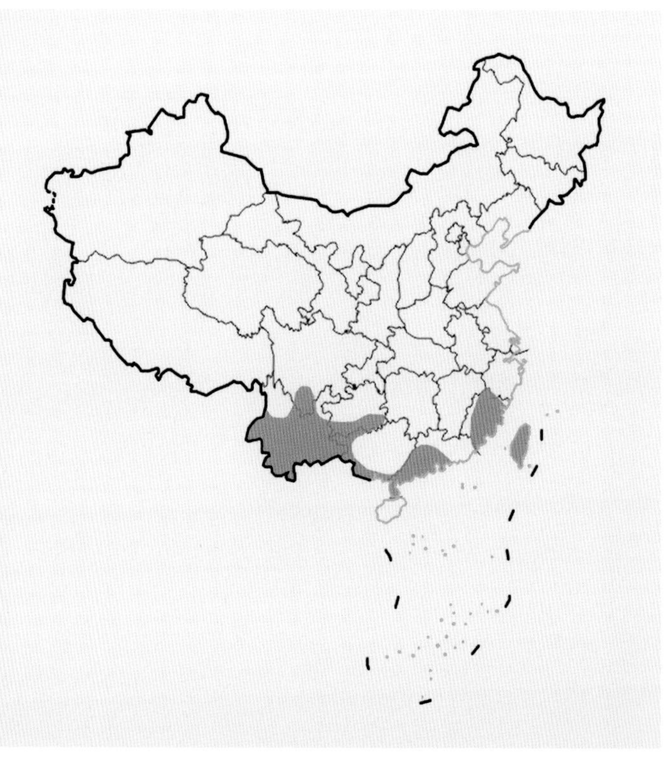

国内分布图
Map of Domestic Distribution

种群 Population

种群数量 Population Size	未知 / Unknown
种群趋势 Population Trend	上升 / Increasing

生境与生态系统 Habitat (s) and Ecosystem (s)

生　　境 Habitat(s)	城市、耕地、江河岸 / Urban Area, Arable Land, River Bank
生态系统 Ecosystem(s)	农田生态系统、湖泊河流生态系统、人类聚落生态系统 / Cropland Ecosystem, Lake and River Ecosystem, Human Settlement Ecosystem

威胁 Threat (s)

主要威胁 Major Threat(s)	无 / None

保护级别与保护行动 Protection Category and Conservation Action (s)

国家重点保护野生动物等级 (2021) Category of National Key Protected Wild Animals (2021)	无 / NA
IUCN 红色名录 (2020-2) IUCN Red List (2020-2)	无危 / LC
CITES 附录 (2019) CITES Appendix (2019)	无 / NA
保护行动 Conservation Action(s)	无 / None

相关文献 Relevant References

Burgin *et al*., 2018; Jiang *et al.* (蒋志刚等), 2017; Wilson *et al*., 2016; Smith *et al.* (史密斯等), 2009; Wu (吴爱国), 2001; Zhang *et al.* (张世炎等), 2012; Zheng *et al.* (郑智民等), 2012

(接上页)

1845); *malabarica* (Shaw, 1801); *maxima* (Pradhan, Mondal, Bhagwat and Agrawal, 1993); *mordax* (Thomas, 1916); *nemorivaga* (Hodgson, 1836); *perchal* (Shaw, 1801); *setifera* (Horsfield, 1824); *siamensis* (Kloss, 1919); *sonlaensis* (Dao, 1975); *taiwanus* (Tokuda, 1939)

板齿鼠 *Bandicota indica*

印度地鼠

Nesokia indica

无危 LC

数据缺乏 DD	无危 LC	近危 NT	易危 VU	濒危 EN	极危 CR	区域灭绝 RE	野外灭绝 EW	灭绝 EX

分类地位 Taxonomic Status

动物界 Animalia	脊索动物门 Chordata	哺乳纲 Mammalia	啮齿目 Rodentia	鼠科 Muridae
学名 Scientific Name		*Nesokia indica*		
命名人 Species Authority		Gray, 1830		
英文名 English Name(s)		Short-tailed Bandicoot Rat		
同物异名 Synonym(s)		*Nesokia bacheri* (Nehring, 1897); *bailwardi* (Thomas, 1907); *beaba* (Wroughton, 1908); *boettgeri* (Rae and Walter, 1889); *brachyura* (Büchner, 1889); *buxtoni* (Thomas, 1919); (转下页)		
种下单元评估 Infra-specific Taxa Assessed		无 / None		

评估信息 Assessment Information

评估年份 Year Assessed	2020
评定人 Assessor(s)	刘少英、蒋志刚 / Shaoying Liu, Zhigang Jiang
审定人 Reviewer(s)	胡慧建 / Huijian Hu
其他贡献人 Other Contributor(s)	李立立、丁晨晨 / Lili Li, Chenchen Ding

理由 Justification: 印度地鼠分布广，种群数量大。因此，列为无危等级 / Short-tailed Bandicoot Rat is a widespread species with large populations. Thus, it is listed as Least Concern

地理分布 Geographical Distribution

国内分布 Domestic Distribution

新疆 / Xinjiang

世界分布 World Distribution

阿富汗、孟加拉国、中国、埃及、印度、伊朗、伊拉克、以色列、约旦、巴基斯坦、巴勒斯坦、沙特阿拉伯、叙利亚、塔吉克斯坦、土库曼斯坦、乌兹别克斯坦 / Afghanistan, Bangladesh, China, Egypt, India, Iran, Iraq, Israel, Jordan, Pakistan, Palestine, Saudi Arabia, Syria Arab Republic, Tajikistan, Turkmenistan, Uzbekistan

分布标注 Distribution Note

非特有种 / Non-endemic

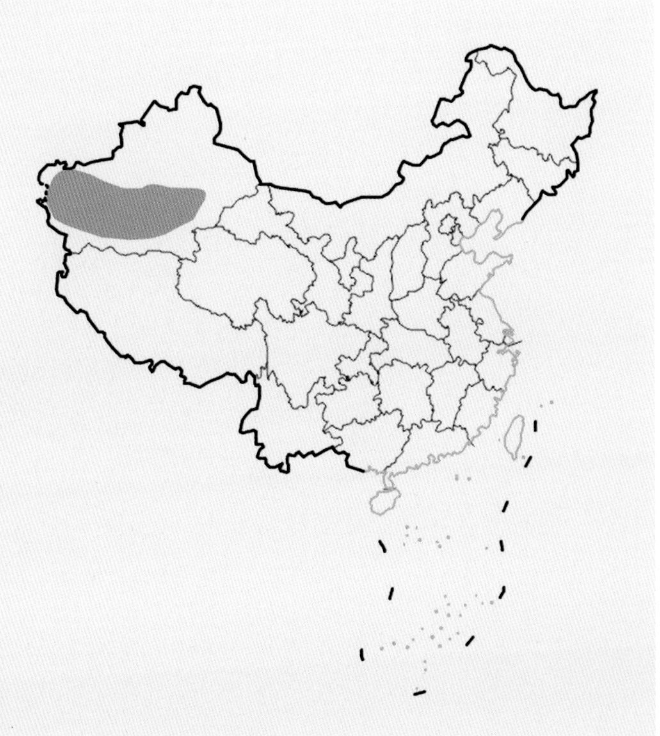

国内分布图
Map of Domestic Distribution

种群 Population

种群数量 Population Size	种群数量大 / Large populations
种群趋势 Population Trend	未知 / Unknown

生境与生态系统 Habitat (s) and Ecosystem (s)

生　　境 Habitat(s)	耕地 / Arable Land
生态系统 Ecosystem(s)	农田生态系统 / Cropland Ecosystem

威胁 Threat (s)

主要威胁 Major Threat(s)	无 / None

保护级别与保护行动 Protection Category and Conservation Action (s)

国家重点保护野生动物等级 (2021) Category of National Key Protected Wild Animals (2021)	无 / NA
IUCN 红色名录 (2020-2) IUCN Red List (2020-2)	无危 / LC
CITES 附录 (2019) CITES Appendix (2019)	无 / NA
保护行动 Conservation Action(s)	无 / None

相关文献 Relevant References

Burgin *et al*., 2018; Jiang *et al.* (蒋志刚等), 2017; Wilson *et al*., 2016; Choudhury, 2003

(接上页)
chitralensis (Schlitter and Setzer, 1973); *dukelskiana* (Heptner, 1928); *griffithi* (Horsfield, 1851); *hardwickei* (Gray, 1837); *huttoni* (Blyth, 1846); *indicus* (Peters, 1860); *insularis* (Goodwin, 1940); *legendrei* (Goodwin, 1939); *myosura* (Wagner, 1845); *satunini* (Nehring, 1899); *scullyi* (Wood-Mason, 1876); *suilla* (Thomas, 1907)

短耳沙鼠

Brachiones przewalskii

无危 LC

数据缺乏 DD	无危 LC	近危 NT	易危 VU	濒危 EN	极危 CR	区域灭绝 RE	野外灭绝 EW	灭绝 EX

分类地位 Taxonomic Status

动物界 Animalia	脊索动物门 Chordata	哺乳纲 Mammalia	啮齿目 Rodentia	鼠科 Muridae

学名 Scientific Name	*Brachiones przewalskii*
命名人 Species Authority	Büchner, 1889
英文名 English Name(s)	Przewalski's Jird
同物异名 Synonym(s)	Przewalski's Gerbil; *Brachiones arenicolor* (Miller, 1900); *callichrous* (Heptner, 1934)
种下单元评估 Infra-specific Taxa Assessed	无 / None

评估信息 Assessment Information

评估年份 Year Assessed	2020
评定人 Assessor(s)	刘少英、蒋志刚 / Shaoying Liu, Zhigang Jiang
审定人 Reviewer(s)	马勇、刘伟 / Yong Ma, Wei Liu
其他贡献人 Other Contributor(s)	李立立、丁晨晨 / Lili Li, Chenchen Ding

理由 Justification: 短耳沙鼠分布广，种群数量大。因此，列为无危等级 / Przewalski's Jird is a widespread species with large populations. Thus, it is listed as Least Concern

地理分布 Geographical Distribution

国内分布 Domestic Distribution

甘肃、新疆、内蒙古 / Gansu, Xinjiang, Inner Mongolia (Nei Mongol)

世界分布 World Distribution

中国 / China

分布标注 Distribution Note

特有种 / Endemic

国内分布图
Map of Domestic Distribution

种群 Population

种群数量 Population Size	种群数量大 / Large populations
种群趋势 Population Trend	未知 / Unknown

生境与生态系统 Habitat (s) and Ecosystem (s)

生　　境 Habitat(s)	灌丛 / Shrubland
生态系统 Ecosystem(s)	灌丛生态系统 / Shrubland Ecosystem

威胁 Threat (s)

主要威胁 Major Threat(s)	无 / None

保护级别与保护行动 Protection Category and Conservation Action (s)

国家重点保护野生动物等级 (2021) Category of National Key Protected Wild Animals (2021)	无 / NA
IUCN 红色名录 (2020-2) IUCN Red List (2020-2)	无危 / LC
CITES 附录 (2019) CITES Appendix (2019)	无 / NA
保护行动 Conservation Action(s)	无 / None

相关文献 Relevant References

Burgin *et al*., 2018; Jiang *et al.* (蒋志刚等), 2017; Wilson *et al*., 2016; Zheng *et al.* (郑智民等), 2012; Smith *et al.* (史密斯等), 2009; Zheng and Zhang (郑涛和张迎梅), 1990; Wang (王定国), 1988

红尾沙鼠

Meriones libycus

无危 LC

数据缺乏 DD	无危 LC	近危 NT	易危 VU	濒危 EN	极危 CR	区域灭绝 RE	野外灭绝 EW	灭绝 EX

分类地位 Taxonomic Status

动物界 Animalia	脊索动物门 Chordata	哺乳纲 Mammalia	啮齿目 Rodentia	鼠科 Muridae

学　名 Scientific Name	*Meriones libycus*
命名人 Species Authority	Lichtenstein, 1823
英文名 English Name(s)	Libyan Jird
同物异名 Synonym(s)	*Meriones afghanus* (Pavlinov and Rossolimo, 1987); *amplus* (Ranck, 1968); *aquilo* (Thomas, 1912); *caucasicus* (Satunin, 1896); *caucasius* (Brandt, 1855); *caudatus* (Thomas, (转下页)
种下单元评估 Infra-specific Taxa Assessed	无 / None

评估信息 Assessment Information

评估年份 Year Assessed	2020
评定人 Assessor(s)	刘少英、蒋志刚 / Shaoying Liu, Zhigang Jiang
审定人 Reviewer(s)	马勇、刘伟 / Yong Ma, Wei Liu
其他贡献人 Other Contributor(s)	李立立、丁晨晨 / Lili Li, Chenchen Ding

理由 Justification: 红尾沙鼠分布广，种群数量大。因此，列为无危等级 / Libyan Jird is a widespread species with large populations. Thus, it is listed as Least Concern

地理分布 Geographical Distribution

国内分布 Domestic Distribution

新疆 / Xinjiang

世界分布 World Distribution

中国；中亚、南亚、西亚、北非 / China; Central Asia, South Asia, West Asia, North Africa

分布标注 Distribution Note

非特有种 / Non-endemic

国内分布图
Map of Domestic Distribution

种群 Population

种群数量 Population Size	种群数量大 / Large populations
种群趋势 Population Trend	稳定 / Stable

生境与生态系统 Habitat (s) and Ecosystem (s)

生　　境 Habitat(s)	沙漠 / Desert
生态系统 Ecosystem(s)	荒漠生态系统 / Desert Ecosystem

威胁 Threat (s)

主要威胁 Major Threat(s)	无 / None

保护级别与保护行动 Protection Category and Conservation Action (s)

国家重点保护野生动物等级 (2021) Category of National Key Protected Wild Animals (2021)	无 / NA
IUCN 红色名录 (2020-2) IUCN Red List (2020-2)	无危 / LC
CITES 附录 (2019) CITES Appendix (2019)	无 / NA
保护行动 Conservation Action(s)	无 / None

相关文献 Relevant References

Burgin *et al*., 2018; Jiang *et al.* (蒋志刚等), 2017; Wilson *et al*., 2016; Zheng *et al.* (郑智民等), 2012; Li and Abdukadir (李俊和阿布力米提・阿不都卡迪尔), 2007; Turkan (买尔旦・吐尔干), 2006; Anwar *et al.* (艾尼瓦尔・铁木尔等), 1998

(接上页)
1919); *collium* (Severtzov, 1873); *confalonieri* (de Beaux, 1931); *edithae* (Cheesman and Hinton, 1924); *erythrourus* (Gray, 1842); *evelynae* (Cheesman and Hinton, 1924); *eversmanni* (Bogdanov, 1889); *farsi* (Schlitter and Setzer, 1973); *gaetulus* (Lataste, 1882); *guyonii* (Loche, 1867); *heptneri* (Argyropulo, 1936); *intermedius* (Gromov, 1952); *iranensis* (Goodwin, 1939); *luridus* (Ranck, 1968); *marginae* (Heptner, 1933); *mariae* (Cabrera, 1907); *maxeratis* (Heptner, 1933); *melanurus* (Rüppell, 1842); *oxianus* (Heptner, 1933); *renaultii* (Loche, 1867); *schousboeii* (Loche, 1867); *schwarzovi* (Toktosunov, 1977); *sogdianus* (Heptner, 1933); *syrius* (Thomas, 1919); *tuareg* (Thomas, 1925); *turfanensis* (Satunin, 1903)

红尾沙鼠 *Meriones libycus*

数据缺乏 DD	无危 LC	近危 NT	易危 VU	濒危 EN	极危 CR	区域灭绝 RE	野外灭绝 EW	灭绝 EX

分类地位 Taxonomic Status

动物界 Animalia	脊索动物门 Chordata	哺乳纲 Mammalia	啮齿目 Rodentia	鼠科 Muridae
学名 Scientific Name		*Meriones meridianus*		
命名人 Species Authority		Pallas, 1773		
英文名 English Name(s)		Mid-day Gerbil		
同物异名 Synonym(s)		*Meriones auceps* (Thomas, 1908); *brevicaudatus* (Milne-Edwards, 1867); *buechneri* (Thomas, 1909); *cryptorhinus* (Blanford, 1875); *fulvus* (Eversmann, 1848); (转下页)		
种下单元评估 Infra-specific Taxa Assessed		无 / None		

评估信息 Assessment Information

评估年份 Year Assessed	2020
评定人 Assessor(s)	刘少英、蒋志刚 / Shaoying Liu, Zhigang Jiang
审定人 Reviewer(s)	马勇、刘伟 / Yong Ma, Wei Liu
其他贡献人 Other Contributor(s)	李立立、丁晨晨 / Lili Li, Chenchen Ding

理由 **Justification:** 子午沙鼠分布较广，种群数量大。因此，列为无危等级 / Mid-day Gerbil is a widespread species with large populations. Thus, it is listed as Least Concern

地理分布 Geographical Distribution

国内分布 Domestic Distribution

内蒙古、河北、河南、陕西、山西、新疆、甘肃、宁夏、青海 / Inner Mongolia (Nei Mongol), Hebei, Henan, Shaanxi, Shanxi, Xinjiang, Gansu, Ningxia, Qinghai

世界分布 World Distribution

中国；中亚、西亚 / China; Central Asia, West Asia

分布标注 Distribution Note

非特有种 / Non-endemic

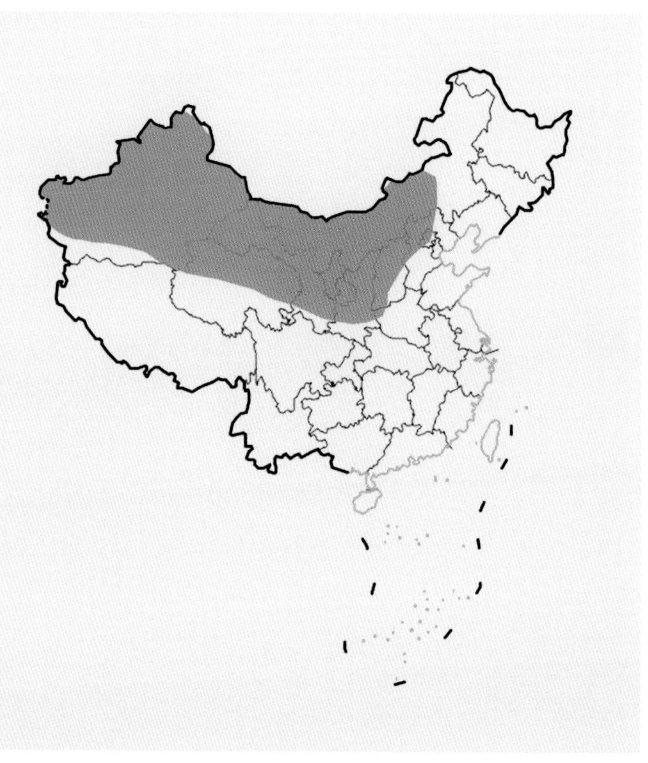

国内分布图
Map of Domestic Distribution

种群 Population

种群数量 Population Size	种群数量大 / Large populations
种群趋势 Population Trend	未知 / Unknown

生境与生态系统 Habitat (s) and Ecosystem (s)

生　　境 Habitat(s)	沙漠、灌丛 / Desert, Shrubland
生态系统 Ecosystem(s)	荒漠生态系统 / Desert Ecosystem

威胁 Threat (s)

主要威胁 Major Threat(s)	无 / None

保护级别与保护行动 Protection Category and Conservation Action (s)

国家重点保护野生动物等级 (2021) Category of National Key Protected Wild Animals (2021)	无 / NA
IUCN 红色名录 (2020-2) IUCN Red List (2020-2)	无危 / LC
CITES 附录 (2019) CITES Appendix (2019)	无 / NA
保护行动 Conservation Action(s)	无 / None

相关文献 Relevant References

Burgin *et al*., 2018; Jiang *et al.* (蒋志刚等), 2017; Wilson *et al*., 2016; Huang and Zhou (黄翔和周立志), 2012; Zheng *et al.* (郑智民等), 2012; E *et al.* (鄂晋等), 2009; Zhao *et al.* (赵天飙等), 2001

(接上页)
heptneri (Kuznetzov, 1944); *jei* (Wang, 1964); *karelini* (Kolossow, 1935); *lepturus* (Büchner, 1889); *littoralis* (Heptner, 1927); *massagetes* (Heptner, 1933); *muleiensis* (Wang, 1981); *nogaiorum* (Heptner, 1927); *penicilliger* (Heptner, 1933); *psammophilus* (Milne-Edwards, 1871); *roborowskii* (Büchner, 1889); *shitkovi* (Heptner, 1933); *tropini* (Kartavtseva and Korobitsyna, 1986); *urianchaicus* (Vinogradov, 1927); *uschtaganicus* (Rall, 1940); *zhitkovi* (Heptner, 1936)

子午沙鼠 *Meriones meridianus*

柽柳沙鼠

Meriones tamariscinus

无危 LC

数据缺乏 DD	无危 LC	近危 NT	易危 VU	濒危 EN	极危 CR	区域灭绝 RE	野外灭绝 EW	灭绝 EX

分类地位 Taxonomic Status

动物界 Animalia	脊索动物门 Chordata	哺乳纲 Mammalia	啮齿目 Rodentia	鼠科 Muridae

学名 Scientific Name	*Meriones tamariscinus*
命名人 Species Authority	Pallas, 1773
英文名 English Name(s)	Tamarisk Gerbil
同物异名 Synonym(s)	*Meriones ciscaucasicus* (Satunin, 1903); *collium* (Severtsov, 1873); *jaxartensis* (Ognev and Heptner, 1928); *kokandicus* (Heptner, 1933); *montanus* (Severtsov, 1873); (转下页)
种下单元评估 Infra-specific Taxa Assessed	无 / None

评估信息 Assessment Information

评估年份 Year Assessed	2020
评定人 Assessor(s)	刘少英、蒋志刚 / Shaoying Liu, Zhigang Jiang
审定人 Reviewer(s)	马勇、刘伟 / Yong Ma, Wei Liu
其他贡献人 Other Contributor(s)	李立立、丁晨晨 / Lili Li, Chenchen Ding

理由 **Justification:** 柽柳沙鼠分布较广，种群数量大。因此，列为无危等级 / Tamarisk Gerbil is a widespread species with large populations. Thus, it is listed as Least Concern

地理分布 Geographical Distribution

国内分布 Domestic Distribution

甘肃、新疆、内蒙古 / Gansu, Xinjiang, Inner Mongolia (Nei Mongol)

世界分布 World Distribution

中国；中亚 / China; Central Asia

分布标注 Distribution Note

非特有种 / Non-endemic

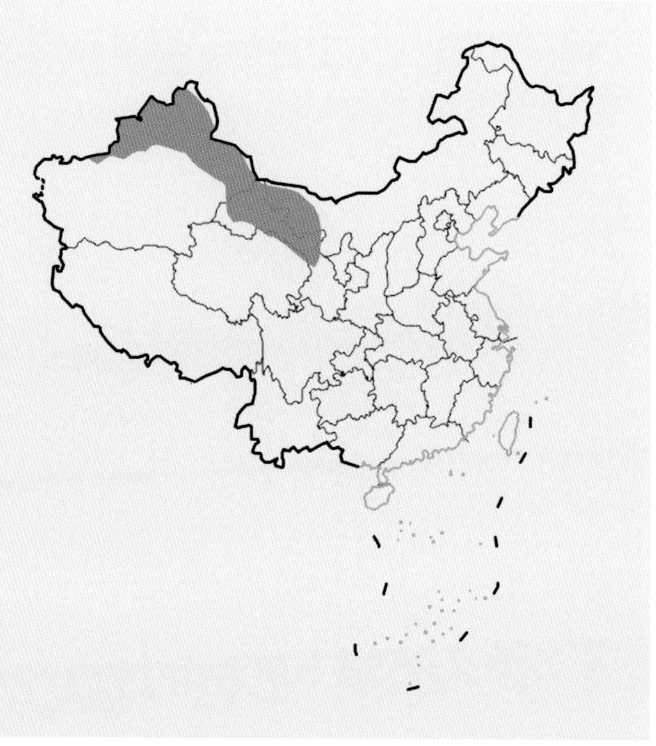

国内分布图
Map of Domestic Distribution

种群 Population

种群数量 Population Size	种群数量大 / Large populations
种群趋势 Population Trend	未知 / Unknown

生境与生态系统 Habitat (s) and Ecosystem (s)

生　　境 Habitat(s)	沙漠、半荒漠、灌丛、盐碱沼泽地 / Desert, Semi-desert, Shrubland; Saline, Brackish or Alkaline Marsh
生态系统 Ecosystem(s)	灌丛生态系统、荒漠生态系统 / Shrubland Ecosystem, Desert Ecosystem

威胁 Threat (s)

主要威胁 Major Threat(s)	无 / None

保护级别与保护行动 Protection Category and Conservation Action (s)

国家重点保护野生动物等级 (2021) Category of National Key Protected Wild Animals (2021)	无 / NA
IUCN 红色名录 (2020-2) IUCN Red List (2020-2)	无危 / LC
CITES 附录 (2019) CITES Appendix (2019)	无 / NA
保护行动 Conservation Action(s)	无 / None

相关文献 Relevant References

Burgin *et al*., 2018; Jiang *et al.* (蒋志刚等), 2017; Wilson *et al*., 2016; Zheng *et al.* (郑智民等), 2012; Gu *et al.* (谷登芝等), 2011; Smith *et al.* (史密斯等), 2009; Wang (王应祥), 2003

(接上页)
satschouensis (Satunin, 1903); *tamaricinus* (Pallas, 1779)

数据缺乏 DD	无危 LC	近危 NT	易危 VU	濒危 EN	极危 CR	区域灭绝 RE	野外灭绝 EW	灭绝 EX

分类地位 Taxonomic Status

动物界 **Animalia**	脊索动物门 **Chordata**	哺乳纲 **Mammalia**	啮齿目 **Rodentia**	鼠科 **Muridae**
学名 **Scientific Name**		*Meriones unguiculatus*		
命名人 **Species Authority**		Milne-Edwards, 1867		
英文名 **English Name(s)**		Mongolian Gerbil		
同物异名 **Synonym(s)**		*Meriones chihfengensis* (Mori, 1939); *M. koslovi* (Satunin, 1903); *M. kurauchii* (Mori, 1930); *Pallasiomys unguiculatus* (Heptner, 1949) subsp. *selenginus*		
种下单元评估 **Infra-specific Taxa Assessed**		无 / None		

评估信息 Assessment Information

评估年份 **Year Assessed**	2020
评定人 **Assessor(s)**	刘少英、蒋志刚 / Shaoying Liu, Zhigang Jiang
审定人 **Reviewer(s)**	马勇、刘伟 / Yong Ma, Wei Liu
其他贡献人 **Other Contributor(s)**	李立立、丁晨晨 / Lili Li, Chenchen Ding

理由 Justification: 长爪沙鼠分布广，种群数量大。因此，列为无危等级 / Mongolian Gerbil is a widespread species with large populations. Thus, it is listed as Least Concern

地理分布 Geographical Distribution

国内分布 Domestic Distribution

内蒙古、辽宁、山西、河北、陕西、宁夏、甘肃、吉林 / Inner Mongolia (Nei Mongol), Liaoning, Shanxi, Hebei, Shaanxi, Ningxia, Gansu, Jilin

世界分布 World Distribution

东亚 / East Asia

分布标注 Distribution Note

非特有种 / Non-endemic

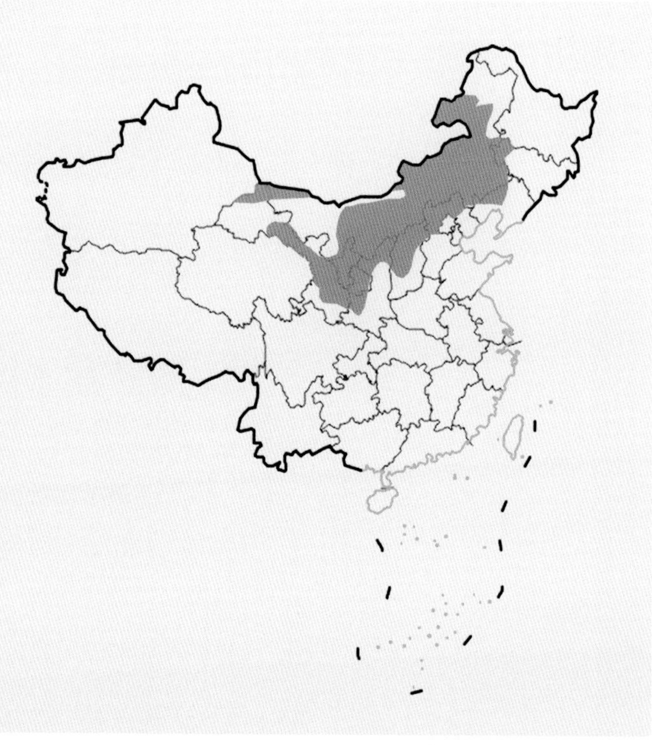

国内分布图
Map of Domestic Distribution

种群 Population

种群数量 Population Size	种群数量大 / Large populations
种群趋势 Population Trend	未知 / Unknown

生境与生态系统 Habitat (s) and Ecosystem (s)

生　　境 Habitat(s)	半荒漠、干旱草地 / Semi-desert, Arid Grassland
生态系统 Ecosystem(s)	草地生态系统、荒漠生态系统 / Grassland Ecosystem, Desert Ecosystem

威胁 Threat (s)

主要威胁 Major Threat(s)	无 / None

保护级别与保护行动 Protection Category and Conservation Action (s)

国家重点保护野生动物等级 (2021) Category of National Key Protected Wild Animals (2021)	无 / NA
IUCN 红色名录 (2020-2) IUCN Red List (2020-2)	无危 / LC
CITES 附录 (2019) CITES Appendix (2019)	无 / NA
保护行动 Conservation Action(s)	无 / None

相关文献 Relevant References

Burgin *et al.*, 2018; Jiang *et al.* (蒋志刚等), 2017; Wilson *et al.*, 2016; Zhang *et al.* (张晓东等), 2013; Zheng *et al.* (郑智民等), 2012; Ding *et al.* (丁贤明等), 2008; Huang *et al.* (黄继荣等), 2006

数据缺乏 DD	无危 LC	近危 NT	易危 VU	濒危 EN	极危 CR	区域灭绝 RE	野外灭绝 EW	灭绝 EX

分类地位 Taxonomic Status

动物界 **Animalia**	脊索动物门 **Chordata**	哺乳纲 **Mammalia**	啮齿目 **Rodentia**	鼠科 **Muridae**
学名 **Scientific Name**		*Rhombomys opimus*		
命名人 **Species Authority**		Lichtenstein, 1823		
英文名 **English Name(s)**		Great Gerbil		
同物异名 **Synonym(s)**		*Rhombomys alaschanicus* (Matschie, 1911); *dalversinicus* (Kashkarov, 1926); *fumicolor* (Heptner, 1933); *giganteus* (Buchner, 1889); *major* (Burdelov, 1989); *minor* (转下页)		
种下单元评估 **Infra-specific Taxa Assessed**		无 / None		

评估信息 Assessment Information

评估年份 **Year Assessed**	2020
评定人 **Assessor(s)**	刘少英、蒋志刚 / Shaoying Liu, Zhigang Jiang
审定人 **Reviewer(s)**	马勇、刘伟 / Yong Ma, Wei Liu
其他贡献人 **Other Contributor(s)**	李立立、丁晨晨 / Lili Li, Chenchen Ding

理由 **Justification:** 大沙鼠分布广，种群数量大。因此，列为无危等级 / Great Gerbil is a widespread species with large populations. Thus, it is listed as Least Concern

地理分布 Geographical Distribution

国内分布 Domestic Distribution

内蒙古、新疆、甘肃 / Inner Mongolia (Nei Mongol), Xinjiang, Gansu

世界分布 World Distribution

中国；中亚、西亚 / China; Central Asia, West Asia

分布标注 Distribution Note

非特有种 / Non-endemic

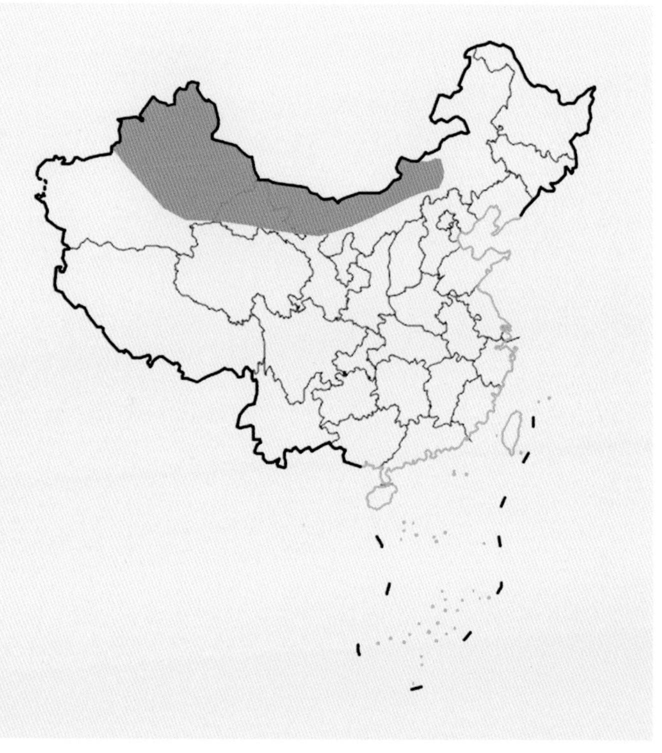

国内分布图
Map of Domestic Distribution

种群 Population

种群数量 Population Size	种群数量大 / Large populations
种群趋势 Population Trend	稳定 / Stable

生境与生态系统 Habitat (s) and Ecosystem (s)

生　　境 Habitat(s)	荒漠、半荒漠 / Desert, Semi-desert
生态系统 Ecosystem(s)	荒漠生态系统 / Desert Ecosystem

威胁 Threat (s)

主要威胁 Major Threat(s)	无 / None

保护级别与保护行动 Protection Category and Conservation Action (s)

国家重点保护野生动物等级 (2021) Category of National Key Protected Wild Animals (2021)	无 / NA
IUCN 红色名录 (2020-2) IUCN Red List (2020-2)	无危 / LC
CITES 附录 (2019) CITES Appendix (2019)	无 / NA
保护行动 Conservation Action(s)	无 / None

相关文献 Relevant References

Burgin *et al*., 2018; Jiang *et al.* (蒋志刚等), 2017; Wilson *et al*., 2016; Zheng *et al.* (郑智民等), 2012; Qiao *et al.* (乔洪海等), 2011; Zhao *et al.* (赵天飙等), 2005; Zhou *et al.* (周立志等), 2000

(接上页)
(Burdelov, 1989); *nigrescens* (Satunin, 1903); *pallidus* (Wagner, 1841); *pevzovi* (Heptner, 1939); *sargadensis* (Heptner, 1939); *sodalis* (Goodwin, 1939)

大沙鼠 *Rhombomys opimus*

沙巴猪尾鼠

Typhlomys chapensis

无危 LC

分类地位 Taxonomic Status

动 物 界 Animalia	脊索动物门 Chordata	哺 乳 纲 Mammalia	啮 齿 目 Rodentia	刺山鼠科 Platacanthomyidae

学　名 Scientific Name	*Typhlomys chapensis*
命 名 人 Species Authority	Milne-Edwards, 1877
英 文 名 English Name(s)	Sort-furred Tree Mouse
同物异名 Synonym(s)	*Typhlomys chapensis* (Osgood, 1932); *daloushanensis* (Wang and Li, 1996); *jingdongensis* (Wu and Wang, 1984); *guangxiensis* (Wang and Chen, 1996)
种下单元评估 Infra-specific Taxa Assessed	无 / None

评估信息 Assessment Information

评 估 年 份 Year Assessed	2020
评　定　人 Assessor(s)	刘少英、蒋志刚 / Shaoying Liu, Zhigang Jiang
审　定　人 Reviewer(s)	蒋学龙 / Xuelong Jiang
其他贡献人 Other Contributor(s)	李立立、丁晨晨 / Lili Li, Chenchen Ding

理由 Justification: 沙巴猪尾鼠分布广，种群数量大。因此，列为无危等级 / Sort-furred Tree Mouse is a widespread species with large populations. Thus, it is listed as Least Concern

地理分布 Geographical Distribution

国内分布 Domestic Distribution

云南 / Yunnan

世界分布 World Distribution

中国、越南 / China, Viet Nam

分布标注 Distribution Note

非特有种 / Non-endemic

国内分布图
Map of Domestic Distribution

种群 Population

种群数量 Population Size	未知 / Unknown
种群趋势 Population Trend	下降 / Decreasing

生境与生态系统 Habitat (s) and Ecosystem (s)

生　　境 Habitat(s)	亚热带湿润低地森林 / Subtropical Moist Lowland Forest
生态系统 Ecosystem(s)	森林生态系统 / Forest Ecosystem

威胁 Threat (s)

主要威胁 Major Threat(s)	无 / None

保护级别与保护行动 Protection Category and Conservation Action (s)

国家重点保护野生动物等级 (2021) Category of National Key Protected Wild Animals (2021)	无 / NA
IUCN 红色名录 (2020-2) IUCN Red List (2020-2)	无危 / LC
CITES 附录 (2019) CITES Appendix (2019)	无 / NA
保护行动 Conservation Action(s)	无 / None

相关文献 Relevant References

Burgin *et al*., 2018; Chen *et al*., 2017b; Jiang *et al*. (蒋志刚等), 2017; Wilson *et al*., 2016

银星竹鼠

Rhizomys pruinosus

无危 LC

数据缺乏 DD	无危 LC	近危 NT	易危 VU	濒危 EN	极危 CR	区域灭绝 RE	野外灭绝 EW	灭绝 EX

分类地位 Taxonomic Status

动物界 Animalia	脊索动物门 Chordata	哺乳纲 Mammalia	啮齿目 Rodentia	鼹型鼠科 Spalacidae

学名 Scientific Name	*Rhizomys pruinosus*
命名人 Species Authority	Blyth, 1851
英文名 English Name(s)	Hoary Bamboo Rat
同物异名 Synonym(s)	*Rhizomys latouchei* (Thomas, 1915); *pannosus* (Thomas, 1915); *prusianus* (Shih, 1930); *senex* (Thomas, 1915); *umbriceps* (Thomas, 1916)
种下单元评估 Infra-specific Taxa Assessed	无 / None

评估信息 Assessment Information

评估年份 Year Assessed	2020
评定人 Assessor(s)	刘少英、蒋志刚 / Shaoying Liu, Zhigang Jiang
审定人 Reviewer(s)	蒋学龙 / Xuelong Jiang
其他贡献人 Other Contributor(s)	李立立、丁晨晨 / Lili Li, Chenchen Ding

理由 Justification: 银星竹鼠分布广，种群数量大。因此，列为无危等级 / Hoary Bamboo Rat is a widespread species with large populations. Thus, it is listed as Least Concern

地理分布 Geographical Distribution

国内分布 Domestic Distribution

贵州、云南、四川、江西、湖南、广西、广东、福建 / Guizhou, Yunnan, Sichuan, Jiangxi, Hunan, Guangxi, Guangdong, Fujian

世界分布 World Distribution

中国；南亚、东南亚 / China; South Asia, Southeast Asia

分布标注 Distribution Note

非特有种 / Non-endemic

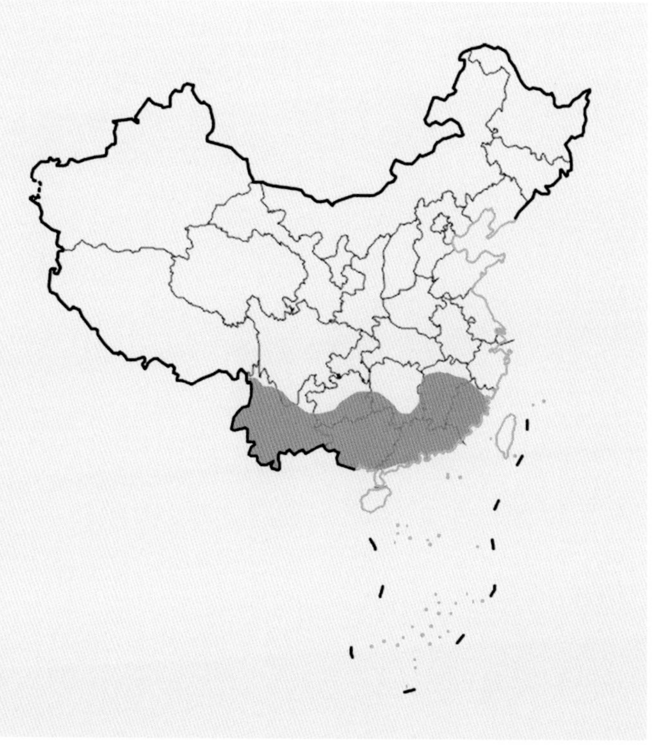

国内分布图
Map of Domestic Distribution

种群 Population

种群数量 **Population Size**	种群数量大 / Large populations
种群趋势 **Population Trend**	稳定 / Stable

生境与生态系统 Habitat (s) and Ecosystem (s)

生　　境 **Habitat(s)**	竹林、草地 / Bamboo Grove, Grassland
生态系统 **Ecosystem(s)**	竹林生态系统 / Bamboo Ecosystem

威胁 Threat (s)

主要威胁 **Major Threat(s)**	无 / None

保护级别与保护行动 Protection Category and Conservation Action (s)

国家重点保护野生动物等级 **(2021) Category of National Key Protected Wild Animals (2021)**	无 / NA
IUCN 红色名录 **(2020-2) IUCN Red List (2020-2)**	无危 / LC
CITES 附录 **(2019) CITES Appendix (2019)**	无 / NA
保护行动 **Conservation Action(s)**	无 / None

相关文献 Relevant References

Burgin *et al*., 2018; Jiang *et al.* (蒋志刚等), 2017; Wilson *et al*., 2016; Zheng *et al.* (郑智民等), 2012; Smith *et al.* (史密斯等), 2009; Wang (王应祥), 2003; Xu (徐龙辉), 1984

中华竹鼠

Rhizomys sinensis

无危 LC

数据缺乏 DD	无危 LC	近危 NT	易危 VU	濒危 EN	极危 CR	区域灭绝 RE	野外灭绝 EW	灭绝 EX

分类地位 Taxonomic Status

动物界 Animalia	脊索动物门 Chordata	哺乳纲 Mammalia	啮齿目 Rodentia	鼹型鼠科 Spalacidae
学　名 **Scientific Name**		*Rhizomys sinensis*		
命名人 **Species Authority**		Gray, 1831		
英文名 **English Name(s)**		Chinese Bamboo Rat		
同物异名 **Synonym(s)**		*Rhizomys chinensis* (Swinhoe, 1870); *davidi* (Thomas, 1911); *neowardi* (Wang, 2003); *pediculus* (Wang, 2003); *reductus* (Dao and Cao, 1990); *vestitus* (Milne-Edwards, 1871); (转下页)		
种下单元评估 **Infra-specific Taxa Assessed**		无 / None		

评估信息 Assessment Information

评估年份 **Year Assessed**	2020
评　定　人 **Assessor(s)**	刘少英、蒋志刚 / Shaoying Liu, Zhigang Jiang
审　定　人 **Reviewer(s)**	蒋学龙 / Xuelong Jiang
其他贡献人 **Other Contributor(s)**	李立立、丁晨晨 / Lili Li, Chenchen Ding

理由 **Justification:** 中华竹鼠分布广，种群数量大。因此，列为无危等级 / Chinese Bamboo Rat is a widespread species with large populations. Thus, it is listed as Least Concern

地理分布 Geographical Distribution

国内分布 Domestic Distribution

云南、四川、贵州、重庆、湖南、湖北、江西、浙江、福建、安徽、广东、广西、甘肃、陕西 / Yunnan, Sichuan, Guizhou, Chongqing, Hunan, Hubei, Jiangxi, Zhejiang, Fujian, Anhui, Guangdong, Guangxi, Gansu, Shaanxi

世界分布 World Distribution

中国、缅甸、越南 / China, Myanmar, Viet Nam

分布标注 Distribution Note

非特有种 / Non-endemic

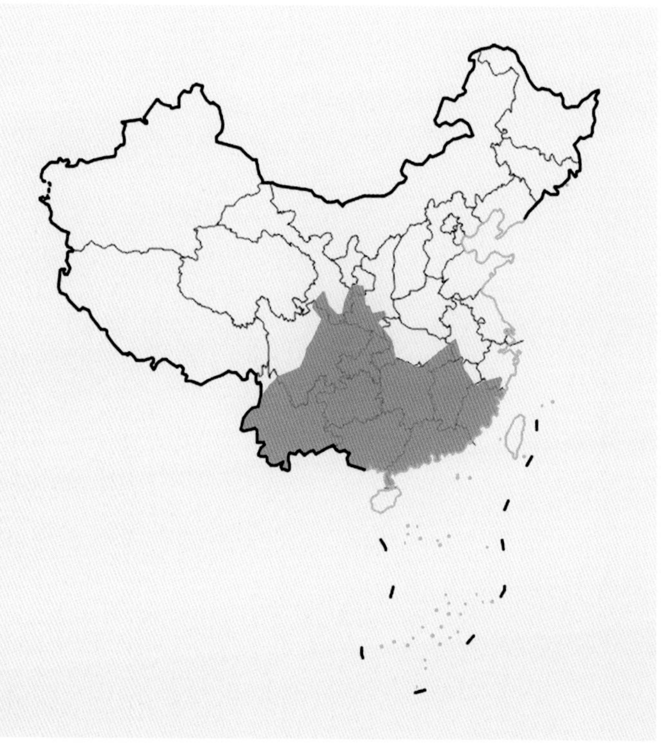

国内分布图
Map of Domestic Distribution

种群 Population

种群数量 Population Size	种群数量大 / Large populations
种群趋势 Population Trend	未知 / Unknown

生境与生态系统 Habitat (s) and Ecosystem (s)

生　　境 Habitat(s)	竹林、松林 / Bamboo Grove, Pine Forest
生态系统 Ecosystem(s)	森林生态系统 / Forest Ecosystem

威胁 Threat (s)

主要威胁 Major Threat(s)	无 / None

保护级别与保护行动 Protection Category and Conservation Action (s)

国家重点保护野生动物等级 (2021) Category of National Key Protected Wild Animals (2021)	无 / NA
IUCN 红色名录 (2020-2) IUCN Red List (2020-2)	无危 / LC
CITES 附录 (2019) CITES Appendix (2019)	无 / NA
保护行动 Conservation Action(s)	无 / None

相关文献 Relevant References

Burgin *et al*., 2018; Jiang *et al.* (蒋志刚等), 2017; Wilson *et al*., 2016; Zheng *et al.* (郑智民等), 2012; Smith *et al.* (史密斯等), 2009; Tang *et al.* (唐中海等), 2009; Wang (王应祥), 2003

(接上页)
wardi (Thomas, 1921)

大竹鼠

Rhizomys sumatrensis

无危 LC

数据缺乏 DD	无危 LC	近危 NT	易危 VU	濒危 EN	极危 CR	区域灭绝 RE	野外灭绝 EW	灭绝 EX

分类地位 Taxonomic Status

动物界 Animalia	脊索动物门 Chordata	哺乳纲 Mammalia	啮齿目 Rodentia	鼹型鼠科 Spalacidae

学名 Scientific Name	*Rhizomys sumatrensis*
命名人 Species Authority	Raffles, 1821
英文名 English Name(s)	Indomalayan Bamboo Rat
同物异名 Synonym(s)	Large Bamboo Rat; Sumatran Rat; Indomalayan Rat; *Rhizomys cinereus* (M'Clelland, 1842); *dekan* (Temminck, 1832); *erythrogenys* (Anderson, 1877); *insularis* (转下页)
种下单元评估 Infra-specific Taxa Assessed	无 / None

评估信息 Assessment Information

评估年份 Year Assessed	2020
评定人 Assessor(s)	刘少英、蒋志刚 / Shaoying Liu, Zhigang Jiang
审定人 Reviewer(s)	蒋学龙 / Xuelong Jiang
其他贡献人 Other Contributor(s)	李立立、丁晨晨 / Lili Li, Chenchen Ding

理由 Justification: 大竹鼠分布广，种群数量大。因此，列为无危等级 / Indomalayan Bamboo Rat is a widespread species with fairly large populations. Thus, it is listed as Least Concern

地理分布 Geographical Distribution

国内分布 Domestic Distribution	
云南 / Yunnan	
世界分布 World Distribution	
中国；东南亚 / China; Southeast Asia	
分布标注 Distribution Note	
非特有种 / Non-endemic	

国内分布图
Map of Domestic Distribution

种群 Population

种群数量 **Population Size**	种群数量大 / Large populations
种群趋势 **Population Trend**	未知 / Unknown

生境与生态系统 Habitat (s) and Ecosystem (s)

生　　境 **Habitat(s)**	竹林 / Bamboo Grove
生态系统 **Ecosystem(s)**	森林生态系统 / Forest Ecosystem

威胁 Threat (s)

主要威胁 **Major Threat(s)**	无 / None

保护级别与保护行动 Protection Category and Conservation Action (s)

国家重点保护野生动物等级 **(2021) Category of National Key Protected Wild Animals (2021)**	无 / NA
IUCN 红色名录 (2020-2) IUCN Red List (2020-2)	无危 / LC
CITES 附录 (2019) CITES Appendix (2019)	无 / NA
保护行动 **Conservation Action(s)**	无 / None

相关文献 Relevant References

Burgin *et al*., 2018; Jiang *et al.* (蒋志刚等), 2017; Wilson *et al*., 2016; Zheng *et al.* (郑智民等), 2012; Smith *et al.* (史密斯等), 2009; Shou and Zhang (寿振黄和张洁), 1958

(接上页)
(Thomas, 1915); *javanus* (Cuvier, 1829); *padangensis* (Brongersma, 1936)

大竹鼠 *Rhizomys sumatrensis*

中华鼢鼠

Eospalax fontanierii

无危 LC

数据缺乏 DD	无危 LC	近危 NT	易危 VU	濒危 EN	极危 CR	区域灭绝 RE	野外灭绝 EW	灭绝 EX

分类地位 Taxonomic Status

动 物 界 Animalia	脊索动物门 Chordata	哺 乳 纲 Mammalia	啮 齿 目 Rodentia	鼹型鼠科 Spalacidae

学 名 Scientific Name	*Eospalax fontanierii*
命 名 人 Species Authority	Milne-Edwards, 1867
英 文 名 English Name(s)	Chinese Zokor
同物异名 Synonym(s)	高原鼢鼠 (Plateau Zokor); *Eospalax baileyi* (Thomas, 1911); *cansus* (Lyon, 1907); *fontanus* (Thomas, 1912); *kukunoriensis* (Lönnberg, 1926); *rufescens* (J. A. Allen, 1909); (转下页)
种下单元评估 Infra-specific Taxa Assessed	无 / None

评估信息 Assessment Information

评 估 年 份 Year Assessed	2020
评 定 人 Assessor(s)	蒋志刚、刘少英 / Zhigang Jiang, Shaoying Liu
审 定 人 Reviewer(s)	李保国 / Baoguo Li
其他贡献人 Other Contributor(s)	李立立、丁晨晨 / Lili Li, Chenchen Ding

理由 Justification: 中华鼢鼠分布广，种群数量大。因此，列为无危等级 / Chinese Zokor is a widespread species with large populations. Thus, it is listed as Least Concern

地理分布 Geographical Distribution

国内分布 Domestic Distribution

内蒙古、河北、北京、山东、山西、河南、陕西、宁夏、甘肃、青海、四川 / Inner Mongolia (Nei Mongol), Hebei, Beijing, Shandong, Shanxi, Henan, Shaanxi, Ningxia, Gansu, Qinghai, Sichuan

世界分布 World Distribution

中国 / China

分布标注 Distribution Note

特有种 / Endemic

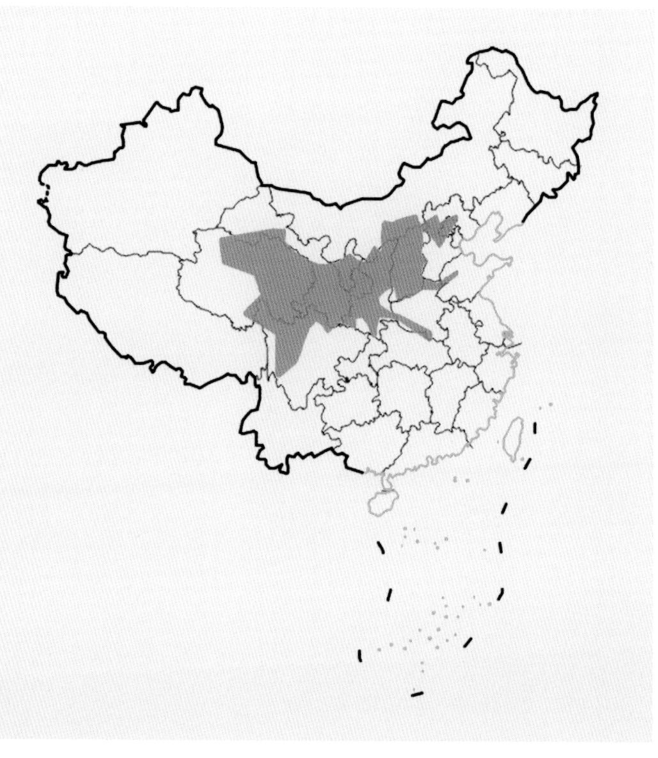

国内分布图
Map of Domestic Distribution

种群 Population

种群数量 Population Size	种群数量大 / Large populations
种群趋势 Population Trend	未知 / Unknown

生境与生态系统 Habitat (s) and Ecosystem (s)

生　　境 Habitat(s)	草地 / Grassland
生态系统 Ecosystem(s)	草地生态系统 / Grassland Ecosystem

威胁 Threat (s)

主要威胁 Major Threat(s)	无 / None

保护级别与保护行动 Protection Category and Conservation Action (s)

国家重点保护野生动物等级 (2021) Category of National Key Protected Wild Animals (2021)	无 / NA
IUCN 红色名录 (2020-2) IUCN Red List (2020-2)	无危 / LC
CITES 附录 (2019) CITES Appendix (2019)	无 / NA
保护行动 Conservation Action(s)	无 / None

相关文献 Relevant References

Burgin *et al.*, 2018; Jiang *et al.* (蒋志刚等), 2017; Wilson *et al.*, 2016; Zheng *et al.* (郑智民等), 2012; Zhang *et al.* (张阳等), 2011; Li and Chen (李保国和陈服官), 1989; Zhang and Yang (张孚允和杨若莉), 1980

(接上页)
shenseius (Thomas, 1911)

罗氏鼢鼠

Eospalax rothschildi

无危 LC

数据缺乏 DD	无危 LC	近危 NT	易危 VU	濒危 EN	极危 CR	区域灭绝 RE	野外灭绝 EW	灭绝 EX

分类地位 Taxonomic Status

动物界 Animalia	脊索动物门 Chordata	哺乳纲 Mammalia	啮齿目 Rodentia	鼹型鼠科 Spalacidae
学　　名 **Scientific Name**	*Eospalax rothschildi*			
命名人 **Species Authority**	Thomas, 1911			
英文名 **English Name(s)**	Rothschild's Zokor			
同物异名 **Synonym(s)**	*Myospalax rothschildi* (Thomas, 1911); *hubeinensis* (Li and Chen, 1989); *minor* (Lönnberg, 1926)			
种下单元评估 **Infra-specific Taxa Assessed**	无 / None			

评估信息 Assessment Information

评估年份 **Year Assessed**	2020
评定人 **Assessor(s)**	蒋志刚、刘少英 / Zhigang Jiang, Shaoying Liu
审定人 **Reviewer(s)**	李保国 / Baoguo Li
其他贡献人 **Other Contributor(s)**	李立立、丁晨晨 / Lili Li, Chenchen Ding

理由 **Justification:** 罗氏鼢鼠分布广，种群数量大。因此，列为无危等级 / Rothschild's Zokor is a widespread species with large populations. Thus, it is listed as Least Concern

地理分布 Geographical Distribution

国内分布 Domestic Distribution

河南、陕西、甘肃、四川、湖北、重庆 / Henan, Shaanxi, Gansu, Sichuan, Hubei, Chongqing

世界分布 World Distribution

中国 / China

分布标注 Distribution Note

特有种 / Endemic

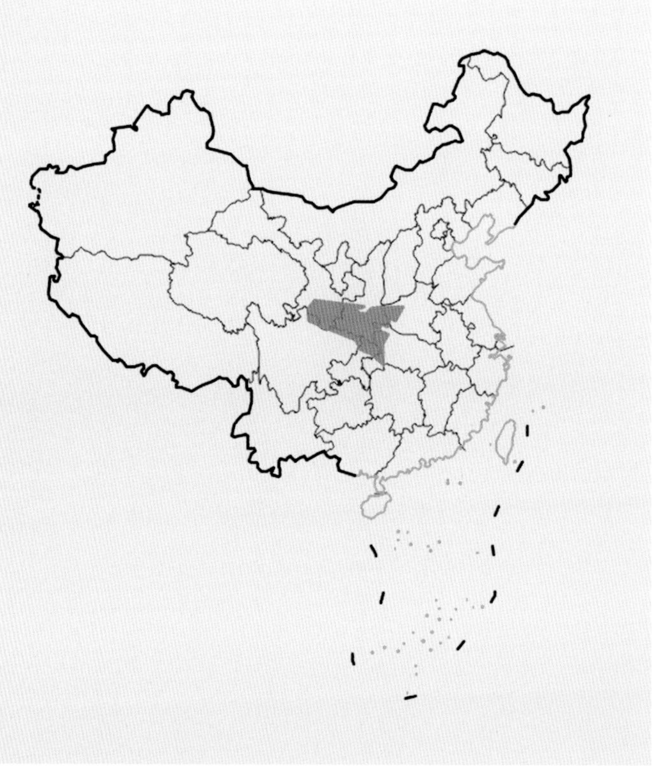

国内分布图
Map of Domestic Distribution

种群 Population

种群数量 Population Size	未知 / Unknown
种群趋势 Population Trend	未知 / Unknown

生境与生态系统 Habitat (s) and Ecosystem (s)

生　　境 Habitat(s)	森林、灌丛、草地、耕地 / Forest, Shrubland, Grassland, Arable Land
生态系统 Ecosystem(s)	森林生态系统、灌丛生态系统、草地生态系统、农田生态系统 / Forest Ecosystem, Shrubland Ecosystem, Grassland Ecosystem, Cropland Ecosystem

威胁 Threat (s)

主要威胁 Major Threat(s)	无 / None

保护级别与保护行动 Protection Category and Conservation Action (s)

国家重点保护野生动物等级 (2021) Category of National Key Protected Wild Animals (2021)	无 / NA
IUCN 红色名录 (2020-2) IUCN Red List (2020-2)	无危 / LC
CITES 附录 (2019) CITES Appendix (2019)	无 / NA
保护行动 Conservation Action(s)	无 / None

相关文献 Relevant References

Burgin *et al*., 2018; Jiang *et al.* (蒋志刚等), 2017; Wilson *et al*., 2016; Zheng *et al.* (郑智民等), 2012; Zhang *et al.* (张三亮等), 2008; Li (李瑛), 1997; Li and Chen (李保国和陈服官), 1989

草原鼢鼠

Myospalax aspalax

无危 LC

数据缺乏 DD	无危 LC	近危 NT	易危 VU	濒危 EN	极危 CR	区域灭绝 RE	野外灭绝 EW	灭绝 EX

分类地位 Taxonomic Status

动物界 Animalia	脊索动物门 Chordata	哺乳纲 Mammalia	啮齿目 Rodentia	鼹型鼠科 Spalacidae
学名 Scientific Name		*Myospalax aspalax*		
命名人 Species Authority		Pallas, 1776		
英文名 English Name(s)		False Zokor		
同物异名 Synonym(s)		Steppe Zokor; *Myospalax armandii* (Milne-Edwards, 1867); *dybowskii* (Tscherski, 1873); *talpinus* (Pallas, 1811); *zokor* (Desmarest, 1822); *angaicus* (Orlov and Baskevich, 1992)		
种下单元评估 Infra-specific Taxa Assessed		无 / None		

评估信息 Assessment Information

评估年份 Year Assessed	2020
评定人 Assessor(s)	蒋志刚、刘少英 / Zhigang Jiang, Shaoying Liu
审定人 Reviewer(s)	李保国 / Baoguo Li
其他贡献人 Other Contributor(s)	李立立、丁晨晨 / Lili Li, Chenchen Ding

理由 Justification: 草原鼢鼠分布较广，种群数量大。因此，列为无危等级 / False Zokor is a widespread species with large populations. Thus, it is listed as Least Concern

地理分布 Geographical Distribution

国内分布 Domestic Distribution

黑龙江、吉林、辽宁、内蒙古、河北、山西 / Heilongjiang, Jilin, Liaoning, Inner Mongolia (Nei Mongol), Hebei, Shanxi

世界分布 World Distribution

东亚 / East Asia

分布标注 Distribution Note

非特有种 / Non-endemic

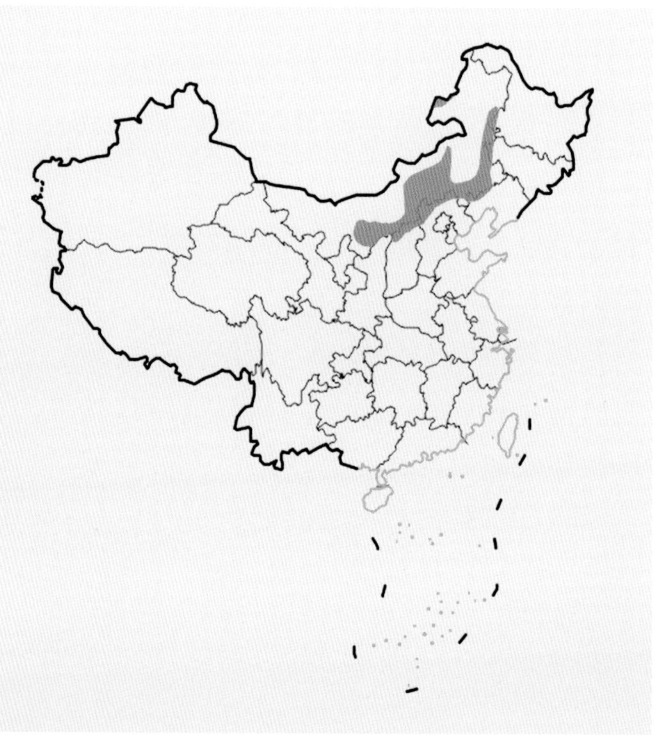

国内分布图
Map of Domestic Distribution

种群 Population

种群数量 Population Size	种群数量大 / Large populations
种群趋势 Population Trend	未知 / Unknown

生境与生态系统 Habitat (s) and Ecosystem (s)

生　境 Habitat(s)	草地、耕地 / Grassland, Arable Land
生态系统 Ecosystem(s)	草地生态系统、农田生态系统 / Grassland Ecosystem, Cropland Ecosystem

威胁 Threat (s)

主要威胁 Major Threat(s)	无 / None

保护级别与保护行动 Protection Category and Conservation Action (s)

国家重点保护野生动物等级 (2021) Category of National Key Protected Wild Animals (2021)	无 / NA
IUCN 红色名录 (2020-2) IUCN Red List (2020-2)	无危 / LC
CITES 附录 (2019) CITES Appendix (2019)	无 / NA
保护行动 Conservation Action(s)	无 / None

相关文献 Relevant References

Burgin *et al*., 2018; Jiang *et al.* (蒋志刚等), 2017; Wilson *et al*., 2016; Zheng *et al.* (郑智民等), 2012; Smith *et al.* (史密斯等), 2009; Wang (王应祥), 2003; Schauer, 1987

东北鼢鼠

Myospalax psilurus

无危 LC

数据缺乏 DD	无危 LC	近危 NT	易危 VU	濒危 EN	极危 CR	区域灭绝 RE	野外灭绝 EW	灭绝 EX

分类地位 Taxonomic Status

动物界 Animalia	脊索动物门 Chordata	哺乳纲 Mammalia	啮齿目 Rodentia	鼹型鼠科 Spalacidae

学名 Scientific Name	*Myospalax psilurus*
命名人 Species Authority	Milne-Edwards, 1874
英文名 English Name(s)	Transbaikal Zokor
同物异名 Synonym(s)	North China Zokor; *Myospalax epsilanus* (Thomas, 1912); *spilurus* (Trouessart, 1897)
种下单元评估 Infra-specific Taxa Assessed	无 / None

评估信息 Assessment Information

评估年份 Year Assessed	2020
评定人 Assessor(s)	蒋志刚、刘少英 / Zhigang Jiang, Shaoying Liu
审定人 Reviewer(s)	李保国 / Baoguo Li
其他贡献人 Other Contributor(s)	李立立、丁晨晨 / Lili Li, Chenchen Ding

理由 Justification: 东北鼢鼠分布广，种群数量大。因此，列为无危等级 / Transbaikal Zokor is a widespread species with large populations. Thus, it is listed as Least Concern

地理分布 Geographical Distribution

国内分布 Domestic Distribution

黑龙江、吉林、辽宁、内蒙古、河北、北京、天津、山东、河南、安徽 / Heilongjiang, Jilin, Liaoning, Inner Mongolia (Nei Mongol), Hebei, Beijing, Tianjin, Shandong, Henan, Anhui

世界分布 World Distribution

中国、蒙古、俄罗斯 / China, Mongolia, Russia

分布标注 Distribution Note

非特有种 / Non-endemic

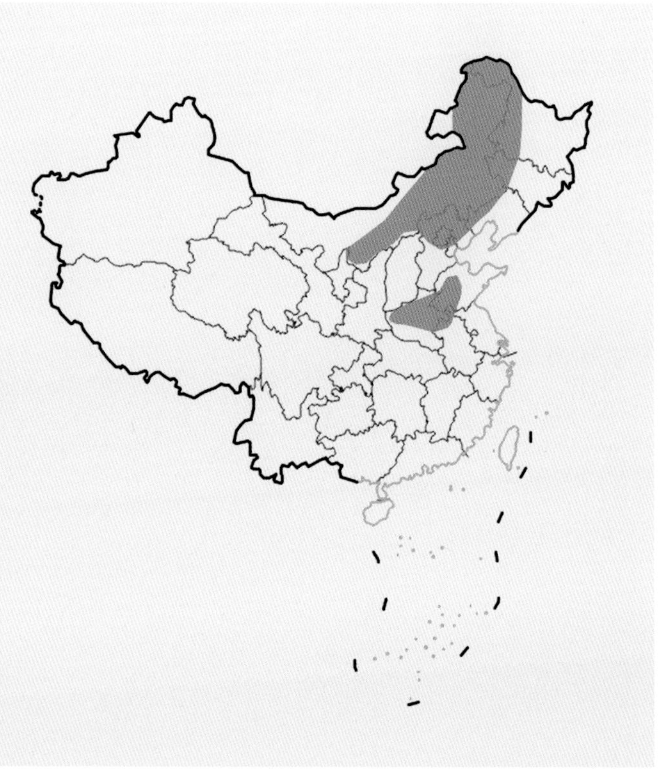

国内分布图
Map of Domestic Distribution

种群 Population

种群数量 **Population Size**	种群数量大 / Large populations
种群趋势 **Population Trend**	稳定 / Stable

生境与生态系统 Habitat (s) and Ecosystem (s)

生　　境 **Habitat(s)**	低地草地、耕地 / Lowland Grassland, Arable Land
生态系统 **Ecosystem(s)**	草地生态系统、农田生态系统 / Grassland Ecosystem, Cropland Ecosystem

威胁 Threat (s)

主要威胁 **Major Threat(s)**	无 / None

保护级别与保护行动 Protection Category and Conservation Action (s)

国家重点保护野生动物等级 **(2021) Category of National Key Protected Wild Animals (2021)**	无 / NA
IUCN 红色名录 **(2020-2) IUCN Red List (2020-2)**	无危 / LC
CITES 附录 **(2019) CITES Appendix (2019)**	无 / NA
保护行动 **Conservation Action(s)**	无 / None

相关文献 Relevant References

Burgin *et al.*, 2018; Jiang *et al.* (蒋志刚等), 2017; Wilson *et al.*, 2016; Zheng *et al.* (郑智民等), 2012; Smith *et al.* (史密斯等), 2009; Liu *et al.* (刘仁华等), 1989

分类地位 Taxonomic Status

动 物 界 **Animalia**	脊索动物门 **Chordata**	哺 乳 纲 **Mammalia**	啮 齿 目 **Rodentia**	跳 鼠 科 **Dipodidae**
学　　名 **Scientific Name**		*Sicista concolor*		
命 名 人 **Species Authority**		Büchner, 1892		
英 文 名 **English Name(s)**		Chinese Birch Mouse		
同物异名 **Synonym(s)**		*Sicista flavus* (True, 1894); *leathemi* (Thomas, 1893); *weigoldi* (Jacobi, 1923)		
种下单元评估 **Infra-specific Taxa Assessed**		无 / None		

评估信息 Assessment Information

评 估 年 份 **Year Assessed**	2020
评　定　人 **Assessor(s)**	蒋志刚、刘少英 / Zhigang Jiang, Shaoying Liu
审　定　人 **Reviewer(s)**	马勇、刘伟 / Yong Ma, Wei Liu
其他贡献人 **Other Contributor(s)**	李立立、丁晨晨 / Lili Li, Chenchen Ding

理由 **Justification:** 中国蹶鼠分布广，种群数量大。因此，列为无危等级 / Chinese Birch Mouse is a widespread species with large populations. Thus, it is listed as Least Concern

地理分布 Geographical Distribution

国内分布 Domestic Distribution

甘肃、青海、四川、陕西、云南 / Gansu, Qinghai, Sichuan, Shaanxi, Yunnan

世界分布 World Distribution

中国；南亚 / China; South Asia

分布标注 Distribution Note

非特有种 / Non-endemic

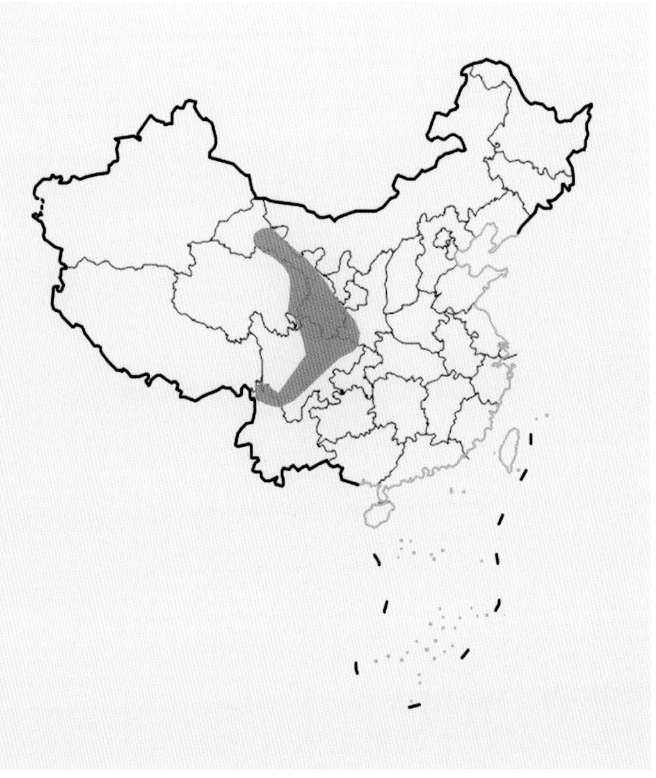

国内分布图
Map of Domestic Distribution

种群 Population

种群数量 **Population Size**	种群数量大 / Large populations
种群趋势 **Population Trend**	未知 / Unknown

生境与生态系统 Habitat (s) and Ecosystem (s)

生　　境 **Habitat(s)**	温带森林、灌木、草地 / Temperate Forest, Shrubland, Grassland
生态系统 **Ecosystem(s)**	森林生态系统、草地生态系统 / Forest Ecosystem, Grassland Ecosystem

威胁 Threat (s)

主要威胁 **Major Threat(s)**	无 / None

保护级别与保护行动 Protection Category and Conservation Action (s)

国家重点保护野生动物等级 **(2021) Category of National Key Protected Wild Animals (2021)**	无 / NA
IUCN 红色名录 **(2020-2) IUCN Red List (2020-2)**	无危 / LC
CITES 附录 **(2019) CITES Appendix (2019)**	无 / NA
保护行动 **Conservation Action(s)**	无 / None

相关文献 Relevant References

Burgin *et al*., 2018; Jiang *et al.* (蒋志刚等), 2017; Wilson *et al*., 2016; Zheng *et al.* (郑智民等), 2012; Smith *et al.* (史密斯等), 2009; Liu *et al.* (刘少英等), 2005

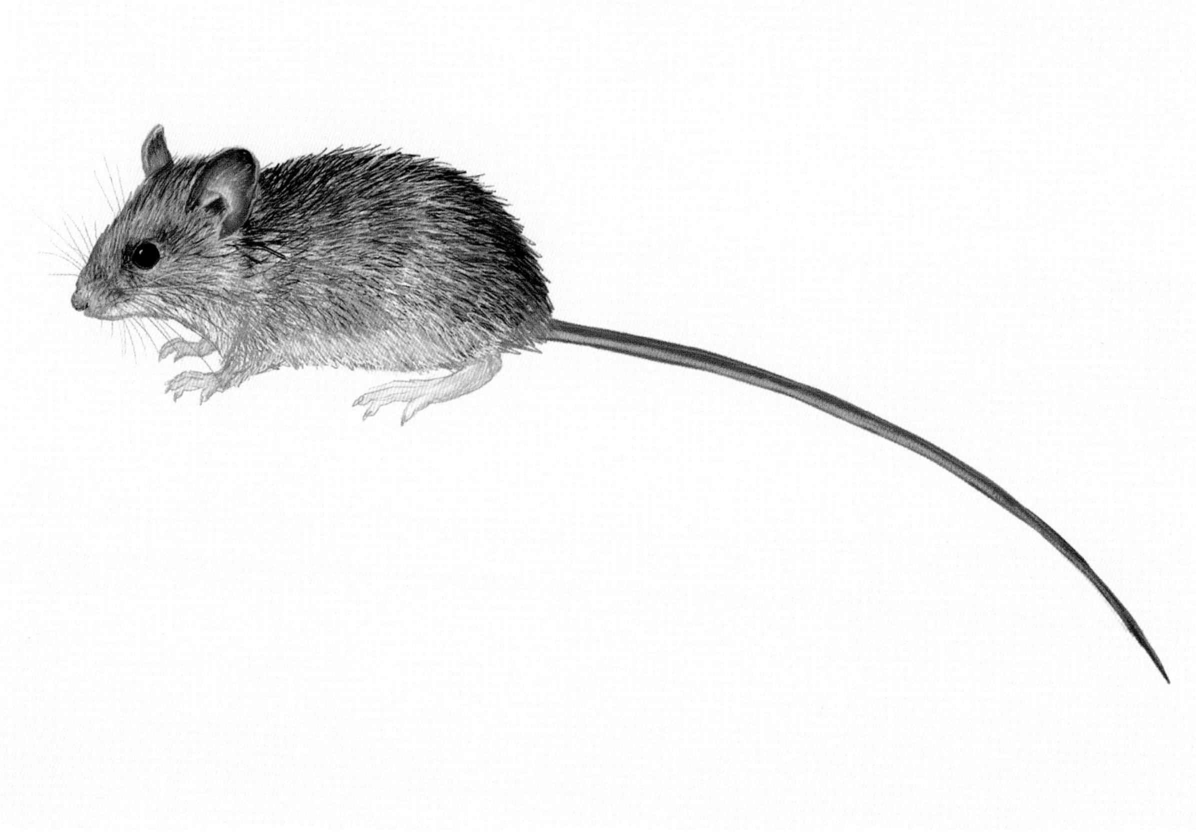

草原蹶鼠

Sicista subtilis

无危 LC

数据缺乏 DD	无危 LC	近危 NT	易危 VU	濒危 EN	极危 CR	区域灭绝 RE	野外灭绝 EW	灭绝 EX

分类地位 Taxonomic Status

动物界 Animalia	脊索动物门 Chordata	哺乳纲 Mammalia	啮齿目 Rodentia	跳鼠科 Dipodidae

学名 **Scientific Name**	*Sicista subtilis*
命名人 **Species Authority**	Pallas, 1773
英文名 **English Name(s)**	Southern Birch Mouse
同物异名 **Synonym(s)**	*Sicista interstriatus* (Petenyi, 1882); *interzonus* (Petenyi, 1882); *lineatus* (Lichtenstein, 1823); *loriger* (Nathusius, 1840); *nordmanni* (Keyserling and Blasius, 1840); (转下页)
种下单元评估 **Infra-specific Taxa Assessed**	无 / None

评估信息 Assessment Information

评估年份 **Year Assessed**	2020
评定人 **Assessor(s)**	蒋志刚、刘少英 / Zhigang Jiang, Shaoying Liu
审定人 **Reviewer(s)**	马勇、刘伟 / Yong Ma, Wei Liu
其他贡献人 **Other Contributor(s)**	李立立、丁晨晨 / Lili Li, Chenchen Ding

理由 **Justification:** 草原蹶鼠分布广，种群数量大。因此，列为无危等级 / Southern Birch Mouse is a widespread species with large populations. Thus, it is listed as Least Concern

地理分布 Geographical Distribution

国内分布 Domestic Distribution

新疆 / Xinjiang

世界分布 World Distribution

亚洲、欧洲 / Asia, Europe

分布标注 Distribution Note

非特有种 / Non-endemic

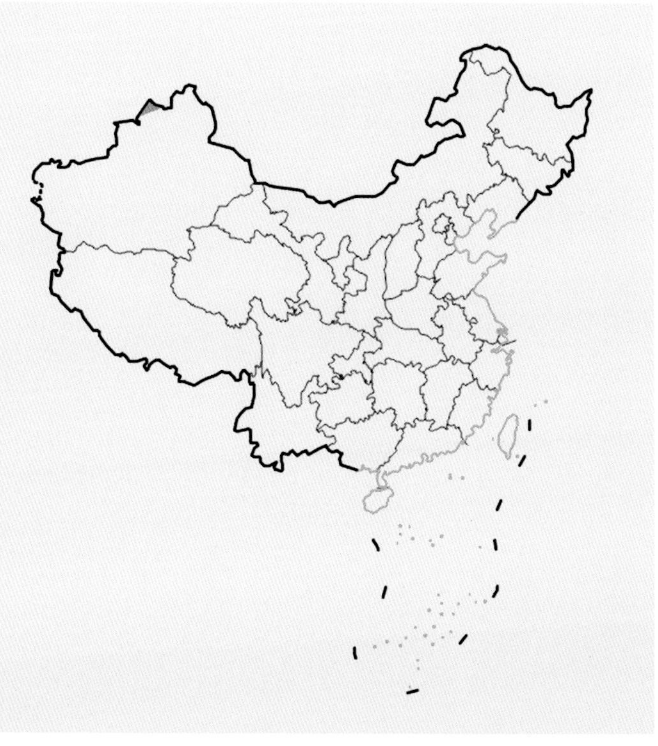

国内分布图
Map of Domestic Distribution

种群 Population

种群数量 **Population Size**	种群数量大 / Large populations
种群趋势 **Population Trend**	下降 / Decreasing

生境与生态系统 Habitat (s) and Ecosystem (s)

生　　境 **Habitat(s)**	草地、半荒漠、草甸 / Grassland, Semi-desert, Meadow
生态系统 **Ecosystem(s)**	草地生态系统、荒漠生态系统 / Grassland Ecosystem, Desert Ecosystem

威胁 Threat (s)

主要威胁 **Major Threat(s)**	无 / None

保护级别与保护行动 Protection Category and Conservation Action (s)

国家重点保护野生动物等级 **(2021) Category of National Key Protected Wild Animals (2021)**	无 / NA
IUCN 红色名录 (2020-2) IUCN Red List (2020-2)	无危 / LC
CITES 附录 (2019) CITES Appendix (2019)	无 / NA
保护行动 **Conservation Action(s)**	无 / None

相关文献 Relevant References

Burgin *et al.*, 2018; Jiang *et al.* (蒋志刚等), 2017; Wilson *et al.*, 2016; Zheng *et al.* (郑智民等), 2012; Smith *et al.* (史密斯等), 2009; Liu *et al.* (刘少英等), 2005

(接上页)
pallida (Kashkarov, 1926); *siberica* (Ognev, 1935); *tripartitus* (Petenyi, 1882); *tristriatus* (Petenyi, 1882); *trizona* (Petenyi, 1882); *vagus* (Pallas, 1779); *virgulosus* (Petenyi, 1882)

草原蹶鼠 *Sicista subtilis*

天山蹶鼠

Sicista tianshanica

无危 LC

数据缺乏 DD | 无危 LC | 近危 NT | 易危 VU | 濒危 EN | 极危 CR | 区域灭绝 RE | 野外灭绝 EW | 灭绝 EX

分类地位 Taxonomic Status

动物界 Animalia	脊索动物门 Chordata	哺乳纲 Mammalia	啮齿目 Rodentia	跳鼠科 Dipodidae
学　名 Scientific Name		*Sicista tianshanica*		
命名人 Species Authority		Salensky, 1903		
英文名 English Name(s)		Tien Shan Birch Mouse		
同物异名 Synonym(s)		*Sicista tianschanica* (Salensky, 1903)		
种下单元评估 Infra-specific Taxa Assessed		无 / None		

评估信息 Assessment Information

评估年份 Year Assessed	2020
评定人 Assessor(s)	蒋志刚、刘少英 / Zhigang Jiang, Shaoying Liu
审定人 Reviewer(s)	马勇、刘伟 / Yong Ma, Wei Liu
其他贡献人 Other Contributor(s)	李立立、丁晨晨 / Lili Li, Chenchen Ding

理由 Justification: 天山蹶鼠分布广，种群数量大。因此，列为无危等级 / Tien Shan Birch Mouse is a widespread species with large populations. Thus, it is listed as Least Concern

地理分布 Geographical Distribution

国内分布 Domestic Distribution

新疆 / Xinjiang

世界分布 World Distribution

中国、哈萨克斯坦、吉尔吉斯斯坦 / China, Kazakhstan, Kyrgyzstan

分布标注 Distribution Note

非特有种 / Non-endemic

国内分布图
Map of Domestic Distribution

种群 Population

种群数量 Population Size	种群数量大 / Large populations
种群趋势 Population Trend	下降 / Decreasing

生境与生态系统 Habitat (s) and Ecosystem (s)

生　　境 Habitat(s)	森林、草甸 / Forest, Meadow
生态系统 Ecosystem(s)	森林生态系统、草地生态系统 / Forest Ecosystem, Grassland Ecosystem

威胁 Threat (s)

主要威胁 Major Threat(s)	无 / None

保护级别与保护行动 Protection Category and Conservation Action (s)

国家重点保护野生动物等级 (2021) Category of National Key Protected Wild Animals (2021)	无 / NA
IUCN 红色名录 (2020-2) IUCN Red List (2020-2)	无危 / LC
CITES 附录 (2019) CITES Appendix (2019)	无 / NA
保护行动 Conservation Action(s)	无 / None

相关文献 Relevant References

Burgin *et al*., 2018; Jiang *et al.* (蒋志刚等), 2017; Wilson *et al*., 2016; Zheng *et al.* (郑智民等), 2012; Ye and Lei (叶生荣和雷刚), 2010; Smith *et al.* (史密斯等), 2009; Shenbrot *et al*., 1995

天山蹶鼠 *Sicista tianshanica*

中国生物多样性红色名录
China's Red List of Biodiversity

林跳鼠

Eozapus setchuanus

无危 LC

数据缺乏 DD	无危 LC	近危 NT	易危 VU	濒危 EN	极危 CR	区域灭绝 RE	野外灭绝 EW	灭绝 EX

分类地位 Taxonomic Status

动物界 Animalia	脊索动物门 Chordata	哺乳纲 Mammalia	啮齿目 Rodentia	跳鼠科 Dipodidae
学名 Scientific Name		*Eozapus setchuanus*		
命名人 Species Authority		Pousargues, 1896		
英文名 English Name(s)		Chinese Jumping Mouse		
同物异名 Synonym(s)		*Eozapus vicinus* (Thomas, 1912)		
种下单元评估 Infra-specific Taxa Assessed		无 / None		

评估信息 Assessment Information

评估年份 Year Assessed	2020
评定人 Assessor(s)	蒋志刚、刘少英 / Zhigang Jiang, Shaoying Liu
审定人 Reviewer(s)	马勇、刘伟 / Yong Ma, Wei Liu
其他贡献人 Other Contributor(s)	李立立、丁晨晨 / Lili Li, Chenchen Ding

理由 Justification: 林跳鼠分布广，种群数量大。因此，列为无危等级 / Chinese Jumping Mouse is a widespread species with large populations. Thus, it is listed as Least Concern

地理分布 Geographical Distribution

国内分布 Domestic Distribution

陕西、甘肃、青海、宁夏、四川、云南 / Shaanxi, Gansu, Qinghai, Ningxia, Sichuan, Yunnan

世界分布 World Distribution

中国 / China

分布标注 Distribution Note

特有种 / Endemic

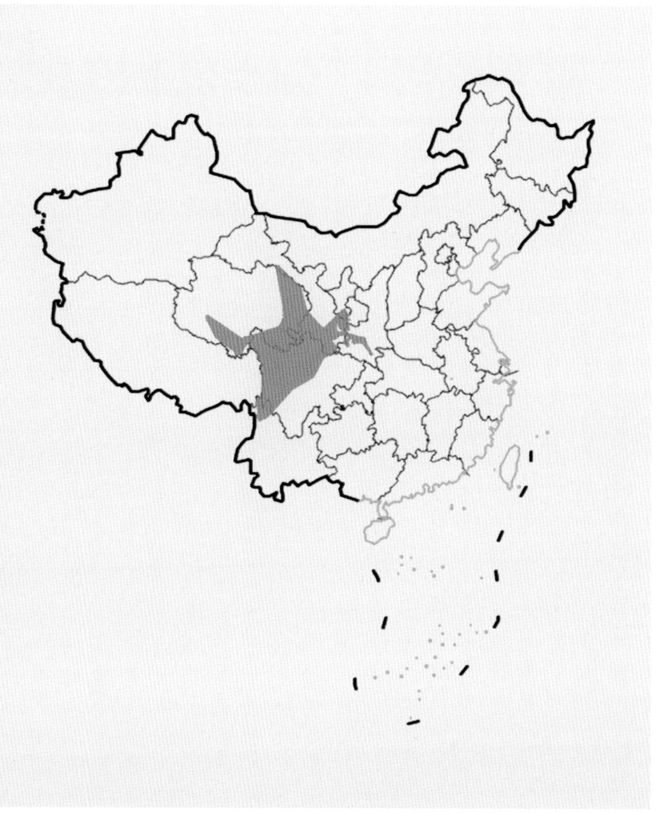

国内分布图
Map of Domestic Distribution

种群 Population

种群数量 Population Size	种群数量大 / Large populations
种群趋势 Population Trend	未知 / Unknown

生境与生态系统 Habitat (s) and Ecosystem (s)

生　　境 Habitat(s)	灌丛、草地、草甸 / Shrubland, Grassland, Meadow
生态系统 Ecosystem(s)	灌丛生态系统、草地生态系统 / Shrubland Ecosystem, Grassland Ecosystem

威胁 Threat (s)

主要威胁 Major Threat(s)	无 / None

保护级别与保护行动 Protection Category and Conservation Action (s)

国家重点保护野生动物等级 (2021) Category of National Key Protected Wild Animals (2021)	无 / NA
IUCN 红色名录 (2020-2) IUCN Red List (2020-2)	无危 / LC
CITES 附录 (2019) CITES Appendix (2019)	无 / NA
保护行动 Conservation Action(s)	无 / None

相关文献 Relevant References

Burgin *et al*., 2018; Zheng *et al.* (郑智民等), 2012; Wilson *et al*., 2016; Fan *et al.* (范振鑫等), 2009; Smith *et al.* (史密斯等), 2009; Liu *et al.* (刘少英等), 2005

巨泡五趾跳鼠
Allactaga bullata

无危 LC

数据缺乏 DD	无危 LC	近危 NT	易危 VU	濒危 EN	极危 CR	区域灭绝 RE	野外灭绝 EW	灭绝 EX

分类地位 Taxonomic Status

动物界 Animalia	脊索动物门 Chordata	哺乳纲 Mammalia	啮齿目 Rodentia	跳鼠科 Dipodidae

学名 Scientific Name	*Allactaga bullata*
命名人 Species Authority	Allen, 1925
英文名 English Name(s)	Gobi Jerboa
同物异名 Synonym(s)	无 / None
种下单元评估 Infra-specific Taxa Assessed	无 / None

评估信息 Assessment Information

评估年份 Year Assessed	2020
评定人 Assessor(s)	蒋志刚、刘少英 / Zhigang Jiang, Shaoying Liu
审定人 Reviewer(s)	马勇、刘伟 / Yong Ma, Wei Liu
其他贡献人 Other Contributor(s)	李立立、丁晨晨 / Lili Li, Chenchen Ding

理由 Justification: 巨泡五趾跳鼠分布广，种群数量大。因此，列为无危等级 / Gobi Jerboa is a widespread species with large populations. Thus, it is listed as Least Concern

地理分布 Geographical Distribution

国内分布 Domestic Distribution

内蒙古、新疆、甘肃 / Inner Mongolia (Nei Mongol), Xinjiang, Gansu

世界分布 World Distribution

中国、蒙古 / China, Mongolia

分布标注 Distribution Note

非特有种 / Non-endemic

国内分布图
Map of Domestic Distribution

种群 Population

种群数量 **Population Size**	种群数量大 / Large populations
种群趋势 **Population Trend**	未知 / Unknown

生境与生态系统 Habitat (s) and Ecosystem (s)

生　　境 **Habitat(s)**	灌丛、荒漠 / Shrubland, Desert
生态系统 **Ecosystem(s)**	荒漠生态系统 / Desert Ecosystem

威胁 Threat (s)

主要威胁 **Major Threat(s)**	无 / None

保护级别与保护行动 Protection Category and Conservation Action (s)

国家重点保护野生动物等级 **(2021) Category of National Key Protected Wild Animals (2021)**	无 / NA
IUCN 红色名录 **(2020-2) IUCN Red List (2020-2)**	无危 / LC
CITES 附录 **(2019) CITES Appendix (2019)**	无 / NA
保护行动 **Conservation Action(s)**	无 / None

相关文献 Relevant References

Burgin *et al*., 2018; Jiang *et al.* (蒋志刚等), 2017; Wilson *et al*., 2016; Zheng *et al.* (郑智民等), 2012; Fu *et al.* (付和平等), 2003; Wang (王应祥), 2003; Wang and Sun (王思博和孙玉珍), 1997

巨泡五趾跳鼠 *Allactaga bullata*　　蒋志刚 绘　Drawn by Zhigang Jiang

小五趾跳鼠

Allactaga elater

无危 LC

数据缺乏 DD	无危 LC	近危 NT	易危 VU	濒危 EN	极危 CR	区域灭绝 RE	野外灭绝 EW	灭绝 EX

分类地位 Taxonomic Status

动物界 Animalia	脊索动物门 Chordata	哺乳纲 Mammalia	啮齿目 Rodentia	跳鼠科 Dipodidae

学名 **Scientific Name**	*Allactaga elater*
命名人 **Species Authority**	Lichtenstein, 1825
英文名 **English Name(s)**	Small Five-toed Jerboa
同物异名 **Synonym(s)**	*Allactaga aralychensis* (Satunin, 1901); *bactriana* (Blyth, 1863); *caucasicus* (Nehring, 1900); *dzungariae* (Thomas, 1912); *heptneri* (Pavlenko and Denisov), 1976; (转下页)
种下单元评估 **Infra-specific Taxa Assessed**	无 / None

评估信息 Assessment Information

评估年份 **Year Assessed**	2020
评定人 **Assessor(s)**	蒋志刚、刘少英 / Zhigang Jiang, Shaoying Liu
审定人 **Reviewer(s)**	马勇、刘伟 / Yong Ma, Wei Liu
其他贡献人 **Other Contributor(s)**	李立立、丁晨晨 / Lili Li, Chenchen Ding

理由 **Justification:** 小五趾跳鼠分布广，种群数量大。因此，列为无危等级 / Small Five-toed Jerboa is a widespread species with large populations. Thus, it is listed as Least Concern

地理分布 Geographical Distribution

国内分布 Domestic Distribution

新疆 / Xinjiang

世界分布 World Distribution

东亚、中亚、南亚、西亚 / East Asia, Central Asia, South Asia, West Asia

分布标注 Distribution Note

非特有种 / Non-endemic

国内分布图
Map of Domestic Distribution

种群 Population

种群数量 **Population Size**	种群数量大 / Large populations
种群趋势 **Population Trend**	下降 / Decreasing

生境与生态系统 Habitat (s) and Ecosystem (s)

生　　境 **Habitat(s)**	沙漠、半荒漠 / Desert, Semi-desert
生态系统 **Ecosystem(s)**	荒漠生态系统 / Desert Ecosystem

威胁 Threat (s)

主要威胁 **Major Threat(s)**	无 / None

保护级别与保护行动 Protection Category and Conservation Action (s)

国家重点保护野生动物等级 **(2021) Category of National Key Protected Wild Animals (2021)**	无 / NA
IUCN 红色名录 **(2020-2) IUCN Red List (2020-2)**	无危 / LC
CITES 附录 **(2019) CITES Appendix (2019)**	无 / NA
保护行动 **Conservation Action(s)**	无 / None

相关文献 Relevant References

Burgin *et al*., 2018; Jiang *et al.* (蒋志刚等), 2017; Wilson *et al*., 2016; Bao *et al.* (包新康等), 2014; Zheng *et al.* (郑智民等), 2012; Smith *et al.* (史密斯等), 2009; Zhou *et al.* (周旭东等), 2005a

(接上页)
indica (Gray, 1842); *kizljaricus* (Satunin, 1907); *strandi* (Hepner, 1934); *turkmeni* (Goodwin, 1940); *zaisanicus* (Shenbrot, 1993)

小五趾跳鼠 *Allactaga elater*　　蒋志刚 绘　Drawn by Zhigang Jiang

数据缺乏 DD	无危 LC	近危 NT	易危 VU	濒危 EN	极危 CR	区域灭绝 RE	野外灭绝 EW	灭绝 EX

分类地位 Taxonomic Status

动物界 Animalia	脊索动物门 Chordata	哺乳纲 Mammalia	啮齿目 Rodentia	跳鼠科 Dipodidae
学　名 **Scientific Name**		*Allactaga major*		
命名人 **Species Authority**		Kerr, 1792		
英文名 **English Name(s)**		Great Jerboa		
同物异名 **Synonym(s)**		*Allactaga aulacotis* (Wagner, 1840); *brachyotis* (Brandt, 1844); *chachlovi* (Martino, 1921); *decumanus* (Lichtenstein, 1825); *djetysuensis* (Shenbrot, 1993); *flavescens* (转下页)		
种下单元评估 **Infra-specific Taxa Assessed**		无 / None		

评估信息 Assessment Information

评估年份 **Year Assessed**	2020
评定人 **Assessor(s)**	蒋志刚、刘少英 / Zhigang Jiang, Shaoying Liu
审定人 **Reviewer(s)**	马勇、刘伟 / Yong Ma, Wei Liu
其他贡献人 **Other Contributor(s)**	李立立、丁晨晨 / Lili Li, Chenchen Ding

理由 **Justification:** 大五趾跳鼠分布较广，种群数量大。因此，列为无危等级 / Great Jerboa is a widespread species with large populations. Thus, it is listed as Least Concern

地理分布 Geographical Distribution

国内分布 Domestic Distribution

新疆 / Xinjiang

世界分布 World Distribution

中国；中亚 / China; Central Asia

分布标注 Distribution Note

非特有种 / Non-endemic

国内分布图
Map of Domestic Distribution

种群 Population

种群数量 Population Size	种群数量大 / Large populations
种群趋势 Population Trend	下降 / Decreasing

生境与生态系统 Habitat (s) and Ecosystem (s)

生　　境 Habitat(s)	沙漠、草地 / Desert, Grassland
生态系统 Ecosystem(s)	草地生态系统、荒漠生态系统 / Grassland Ecosystem, Desert Ecosystem

威胁 Threat (s)

主要威胁 Major Threat(s)	无 / None

保护级别与保护行动 Protection Category and Conservation Action (s)

国家重点保护野生动物等级 (2021) Category of National Key Protected Wild Animals (2021)	无 / NA
IUCN 红色名录 (2020-2) IUCN Red List (2020-2)	无危 / LC
CITES 附录 (2019) CITES Appendix (2019)	无 / NA
保护行动 Conservation Action(s)	无 / None

相关文献 Relevant References

Burgin *et al.*, 2018; Jiang *et al.* (蒋志刚等), 2017; Wilson *et al.*, 2016; Smith *et al.* (史密斯等), 2009; Pan *et al.* (潘清华等), 2007; Wang (王应祥), 2003

(接上页)

(Brandt, 1844); *fuscus* (Ognev, 1924); *hochlovi* (Martino, 1922); *intermedius* (Ognev, 1948); *jaculus* (Pallas, 1779); *macrotis* (Brandt, 1844); *nigricans* (Brandt, 1844); *spiculum* (Lichtenstein, 1825); *vexillarius* (Eversmann, 1840)

大五趾跳鼠 *Allactaga major*　　　　胡慧建 摄　By Huijian Hu

五趾跳鼠

Allactaga sibirica

无危 LC

数据缺乏 DD	无危 LC	近危 NT	易危 VU	濒危 EN	极危 CR	区域灭绝 RE	野外灭绝 EW	灭绝 EX

分类地位 Taxonomic Status

动 物 界 Animalia	脊索动物门 Chordata	哺 乳 纲 Mammalia	啮 齿 目 Rodentia	跳 鼠 科 Dipodidae
学　　名 **Scientific Name**		*Allactaga sibirica*		
命 名 人 **Species Authority**		Forster, 1778		
英 文 名 **English Name(s)**		Siberian Jerboa		
同物异名 **Synonym(s)**		Mongolian Five-toed Jerboa; *Allactaga alactaga* (Olivier, 1800); *alpinus* (Shnitnikov, 1936); *altorum* (Ognev, 1946); *annulata* (Milne-Edwards, 1867); *brachyurus* (转下页)		
种下单元评估 **Infra-specific Taxa Assessed**		无 / None		

评估信息 Assessment Information

评 估 年 份 **Year Assessed**	2020
评　定　人 **Assessor(s)**	蒋志刚、刘少英 / Zhigang Jiang, Shaoying Liu
审　定　人 **Reviewer(s)**	马勇、刘伟 / Yong Ma, Wei Liu
其他贡献人 **Other Contributor(s)**	李立立、丁晨晨 / Lili Li, Chenchen Ding

理由 Justification: 五趾跳鼠分布广，种群数量大。因此，列为无危等级 / Siberian Jerboa is a widespread species with large populations. Thus, it is listed as Least Concern

地理分布 Geographical Distribution

国内分布 Domestic Distribution

黑龙江、吉林、辽宁、内蒙古、河北、山西、陕西、甘肃、宁夏、青海、新疆 / Heilongjiang, Jilin, Liaoning, Inner Mongolia (Nei Mongol), Hebei, Shanxi, Shaanxi, Gansu, Ningxia, Qinghai, Xinjiang

世界分布 World Distribution

中国、哈萨克斯坦、吉尔吉斯斯坦、蒙古、俄罗斯、土库曼斯坦、乌兹别克斯坦 / China, Kazakhstan, Kyrgyzstan, Mongolia, Russia, Turkmenistan, Uzbekistan

分布标注 Distribution Note

非特有种 / Non-endemic

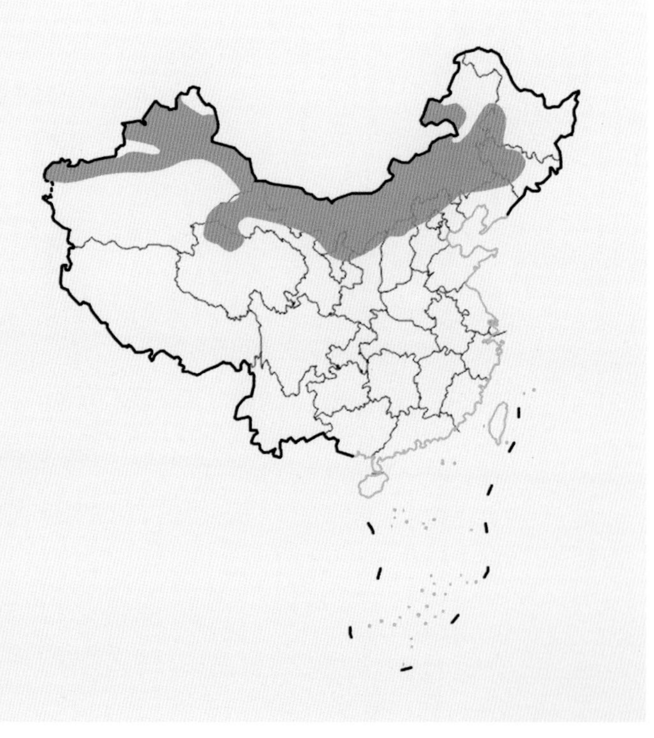

国内分布图
Map of Domestic Distribution

种群 Population

种群数量 **Population Size**	种群数量大 / Large populations
种群趋势 **Population Trend**	未知 / Unknown

生境与生态系统 Habitat (s) and Ecosystem (s)

生　　境 **Habitat(s)**	沙漠、半沙漠 / Desert, Semi-desert
生态系统 **Ecosystem(s)**	草地生态系统、荒漠生态系统 / Grassland Ecosystem, Desert Ecosystem

威胁 Threat (s)

主要威胁 **Major Threat(s)**	无 / None

保护级别与保护行动 Protection Category and Conservation Action (s)

国家重点保护野生动物等级 **(2021) Category of National Key Protected Wild Animals (2021)**	无 / NA
IUCN 红色名录 **(2020-2) IUCN Red List (2020-2)**	无危 / LC
CITES 附录 **(2019) CITES Appendix (2019)**	无 / NA
保护行动 **Conservation Action(s)**	无 / None

相关文献 Relevant References

Burgin *et al*., 2018; Jiang *et al.* (蒋志刚等), 2017; Wilson *et al*., 2016; Zheng *et al.* (郑智民等), 2012; Na *et al.* (娜日苏等), 2009; Dong *et al.* (董维惠等), 2006; Huang and Wu (黄英和武晓东), 2004

(接上页)
(Blainville, 1817); *bulganensis* (Shenbrot, 1993); *dementiewi* (Toktosunov, 1958); *grisescens* (Hollister, 1912); *halticus* (Illiger, 1825); *longior* (Miller, 1911); *media* (Pallas, 1779); *mongolica* (Rae, 1861); *ognevi* (Shenbrot, 1991); *ruckbeili* (Thomas, 1914); *salicus* (Ognev, 1924); *saliens* (Gmelin, 1760); *saliens* (Shaw, 1790); *saltator* (Eversmann, 1848); *semideserta* (Bannikov, 1947); *suschkini* (Satunin, 1900)

五趾跳鼠 *Allactaga sibirica*　　邢睿 摄　By Rui Xing

数据缺乏 DD	无危 LC	近危 NT	易危 VU	濒危 EN	极危 CR	区域灭绝 RE	野外灭绝 EW	灭绝 EX

分类地位 Taxonomic Status

动物界 Animalia	脊索动物门 Chordata	哺乳纲 Mammalia	啮齿目 Rodentia	跳鼠科 Dipodidae

学名 **Scientific Name**	*Pygeretmus pumilio*
命名人 **Species Authority**	Kerr, 1792
英文名 **English Name(s)**	Dwarf Fat-tailed Jerboa
同物异名 **Synonym(s)**	*Pygeretmus acontion* (Pallas, 1811); *aralensis* (Ognev, 1948); *brachyotis* (Ostrouchov, 1889); *dinniki* (Satunin, 1920); *minor* (Pallas, 1779); *minutus* (Blainville, 1817); (转下页)
种下单元评估 **Infra-specific Taxa Assessed**	无 / None

评估信息 Assessment Information

评估年份 **Year Assessed**	2020
评定人 **Assessor(s)**	蒋志刚、刘少英 / Zhigang Jiang, Shaoying Liu
审定人 **Reviewer(s)**	马勇、刘伟 / Yong Ma, Wei Liu
其他贡献人 **Other Contributor(s)**	李立立、丁晨晨 / Lili Li, Chenchen Ding

理由 **Justification:** 小地兔分布广，种群数量大。因此，列为无危等级 / Dwarf Fat-tailed Jerboa is a widespread species with large populations. Thus, it is listed as Least Concern

地理分布 Geographical Distribution

国内分布 Domestic Distribution

内蒙古、新疆、宁夏 / Inner Mongolia (Nei Mongol), Xinjiang, Ningxia

世界分布 World Distribution

中国、伊朗、哈萨克斯坦、蒙古、俄罗斯 / China, Iran, Kazakhstan, Mongolia, Russia

分布标注 Distribution Note

非特有种 / Non-endemic

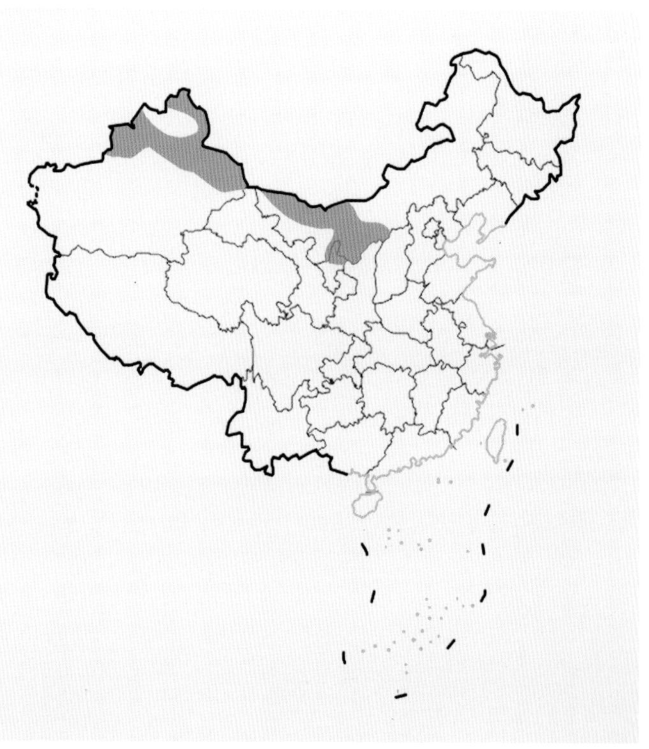

国内分布图
Map of Domestic Distribution

种群 Population

种群数量 **Population Size**	种群数量大 / Large populations
种群趋势 **Population Trend**	未知 / Unknown

生境与生态系统 Habitat (s) and Ecosystem (s)

生　　境 **Habitat(s)**	沙漠、半沙漠、草地 / Desert, Semi-desert, Grassland
生态系统 **Ecosystem(s)**	荒漠生态系统 / Desert Ecosystem

威胁 Threat (s)

主要威胁 **Major Threat(s)**	无 / None

保护级别与保护行动 Protection Category and Conservation Action (s)

国家重点保护野生动物等级 **(2021) Category of National Key Protected Wild Animals (2021)**	无 / NA
IUCN 红色名录 **(2020-2) IUCN Red List (2020-2)**	无危 / LC
CITES 附录 **(2019) CITES Appendix (2019)**	无 / NA
保护行动 **Conservation Action(s)**	无 / None

相关文献 Relevant References

Burgin *et al*., 2018; Jiang *et al.* (蒋志刚等), 2017; Wilson *et al*., 2016; Zheng *et al.* (郑智民等), 2012; Wang *et al.* (王开锋等), 2010; Wang (王应祥), 2003

(接上页)
pallidus (Vinogradov, 1933); *potanini* (Vinogradov, 1926); *pumilio* (Kerr, 1792); *pygmaea* (Pallas, 1779); *tanaiticus* (Ognev, 1948); *turcomanus* (Heptner and Samorodov, 1939)

小地兔 *Pygeretmus pumilio*

五趾心颅跳鼠

Cardiocranius paradoxus

无危 LC

数据缺乏 DD	无危 LC	近危 NT	易危 VU	濒危 EN	极危 CR	区域灭绝 RE	野外灭绝 EW	灭绝 EX

分类地位 Taxonomic Status

动物界 Animalia	脊索动物门 Chordata	哺乳纲 Mammalia	啮齿目 Rodentia	跳鼠科 Dipodidae

学　名 **Scientific Name**	*Cardiocranius paradoxus*
命名人 **Species Authority**	Satunin, 1903
英文名 **English Name(s)**	Five-toed Pygmy Jerboa
同物异名 **Synonym(s)**	无 / None
种下单元评估 **Infra-specific Taxa Assessed**	无 / None

评估信息 Assessment Information

评估年份 **Year Assessed**	2020
评定人 **Assessor(s)**	蒋志刚、刘少英 / Zhigang Jiang, Shaoying Liu
审定人 **Reviewer(s)**	马勇、刘伟 / Yong Ma, Wei Liu
其他贡献人 **Other Contributor(s)**	李立立、丁晨晨 / Lili Li, Chenchen Ding

理由 Justification: 五趾心颅跳鼠分布广，种群数量大。因此，列为无危等级 / Five-toed Pygmy Jerboa is a widespread species with large populations. Thus, it is listed as Least Concern

地理分布 Geographical Distribution

国内分布 Domestic Distribution

内蒙古、甘肃、宁夏、新疆 / Inner Mongolia (Nei Mongol), Gansu, Ningxia, Xinjiang

世界分布 World Distribution

东亚、中亚 / East Asia, Central Asia

分布标注 Distribution Note

非特有种 / Non-endemic

国内分布图
Map of Domestic Distribution

种群 Population

种群数量 Population Size	种群数量大 / Large populations
种群趋势 Population Trend	未知 / Unknown

生境与生态系统 Habitat (s) and Ecosystem (s)

生　　境 Habitat(s)	荒漠 / Desert
生态系统 Ecosystem(s)	荒漠生态系统 / Desert Ecosystem

威胁 Threat (s)

主要威胁 Major Threat(s)	无 / None

保护级别与保护行动 Protection Category and Conservation Action (s)

国家重点保护野生动物等级 (2021) Category of National Key Protected Wild Animals (2021)	无 / NA
IUCN 红色名录 (2020-2) IUCN Red List (2020-2)	无危 / LC
CITES 附录 (2019) CITES Appendix (2019)	无 / NA
保护行动 Conservation Action(s)	无 / None

相关文献 Relevant References

Burgin *et al*., 2018; Jiang *et al.* (蒋志刚等), 2017; Wilson *et al*., 2016; Zheng *et al.* (郑智民等), 2012; Hou and Ouyang (侯兰新和欧阳霞辉), 2010; Zhao (赵肯堂), 1977; Xia (夏武平), 1964

肥尾心颅跳鼠

Salpingotus crassicauda

无危 LC

数据缺乏 DD | 无危 LC | 近危 NT | 易危 VU | 濒危 EN | 极危 CR | 区域灭绝 RE | 野外灭绝 EW | 灭绝 EX

分类地位 Taxonomic Status

动物界 Animalia	脊索动物门 Chordata	哺乳纲 Mammalia	啮齿目 Rodentia	跳鼠科 Dipodidae

学　名 Scientific Name	*Salpingotus crassicauda*
命名人 Species Authority	Vinogradov, 1924
英文名 English Name(s)	Thick-tailed Pygmy Jerboa
同物异名 Synonym(s)	*Salpingotus crassicauda* (Sokolov and Shenbrot, 1988) subsp. *gobicus*
种下单元评估 Infra-specific Taxa Assessed	无 / None

评估信息 Assessment Information

评估年份 Year Assessed	2020
评定人 Assessor(s)	蒋志刚、刘少英 / Zhigang Jiang, Shaoying Liu
审定人 Reviewer(s)	马勇、刘伟 / Yong Ma, Wei Liu
其他贡献人 Other Contributor(s)	李立立、丁晨晨 / Lili Li, Chenchen Ding

理由 Justification: 肥尾心颅跳鼠分布广，种群数量大。因此，列为无危等级 / Thick-tailed Pygmy Jerboa is a widespread species with large populations. Thus, it is listed as Least Concern

地理分布 Geographical Distribution

国内分布 Domestic Distribution

新疆、甘肃、内蒙古 / Xinjiang, Gansu, Inner Mongolia (Nei Mongol)

世界分布 World Distribution

中国、哈萨克斯坦、蒙古 / China, Kazakhstan, Mongolia

分布标注 Distribution Note

非特有种 / Non-endemic

国内分布图
Map of Domestic Distribution

种群 Population

种群数量 Population Size	种群数量大 / Large populations
种群趋势 Population Trend	未知 / Unknown

生境与生态系统 Habitat (s) and Ecosystem (s)

生　　境 Habitat(s)	植被稳定的沙地 / Sandland with Stable Vegetation
生态系统 Ecosystem(s)	荒漠生态系统 / Desert Ecosystem

威胁 Threat (s)

主要威胁 Major Threat(s)	无 / None

保护级别与保护行动 Protection Category and Conservation Action (s)

国家重点保护野生动物等级 (2021) Category of National Key Protected Wild Animals (2021)	无 / NA
IUCN 红色名录 (2020-2) IUCN Red List (2020-2)	无危 / LC
CITES 附录 (2019) CITES Appendix (2019)	无 / NA
保护行动 Conservation Action(s)	无 / None

相关文献 Relevant References

Burgin *et al.*, 2018; Jiang *et al.* (蒋志刚等), 2017; Wilson *et al.*, 2016; Cha *et al.* (查木哈等), 2013; Zheng *et al.* (郑智民等), 2012; Hou and Ouyang (侯兰新和欧阳霞辉), 2010; Smith *et al.* (史密斯等), 2009

三趾心颅跳鼠

Salpingotus kozlovi

无危 LC

数据缺乏 DD	无危 LC	近危 NT	易危 VU	濒危 EN	极危 CR	区域灭绝 RE	野外灭绝 EW	灭绝 EX

分类地位 Taxonomic Status

动物界 Animalia	脊索动物门 Chordata	哺乳纲 Mammalia	啮齿目 Rodentia	跳鼠科 Dipodidae
学名 **Scientific Name**		*Salpingotus kozlovi*		
命名人 **Species Authority**		Vinogradov, 1922		
英文名 **English Name(s)**		Kozlov's Pygmy Jerboa		
同物异名 **Synonym(s)**		*Salpingotus xiangi* (Hou and Jiang, 1994)		
种下单元评估 **Infra-specific Taxa Assessed**		无 / None		

评估信息 Assessment Information

评估年份 **Year Assessed**	2020
评定人 **Assessor(s)**	蒋志刚、刘少英 / Zhigang Jiang, Shaoying Liu
审定人 **Reviewer(s)**	马勇、刘伟 / Yong Ma, Wei Liu
其他贡献人 **Other Contributor(s)**	李立立、丁晨晨 / Lili Li, Chenchen Ding

理由 **Justification:** 三趾心颅跳鼠分布广，种群数量大。因此，列为无危等级 / Kozlov's Pygmy Jerboa is a widespread species with large populations. Thus, it is listed as Least Concern

地理分布 Geographical Distribution

国内分布 Domestic Distribution

内蒙古、甘肃、陕西、宁夏、新疆 / Inner Mongolia (Nei Mongol), Gansu, Shaanxi, Ningxia, Xinjiang

世界分布 World Distribution

中国、蒙古 / China, Mongolia

分布标注 Distribution Note

非特有种 / Non-endemic

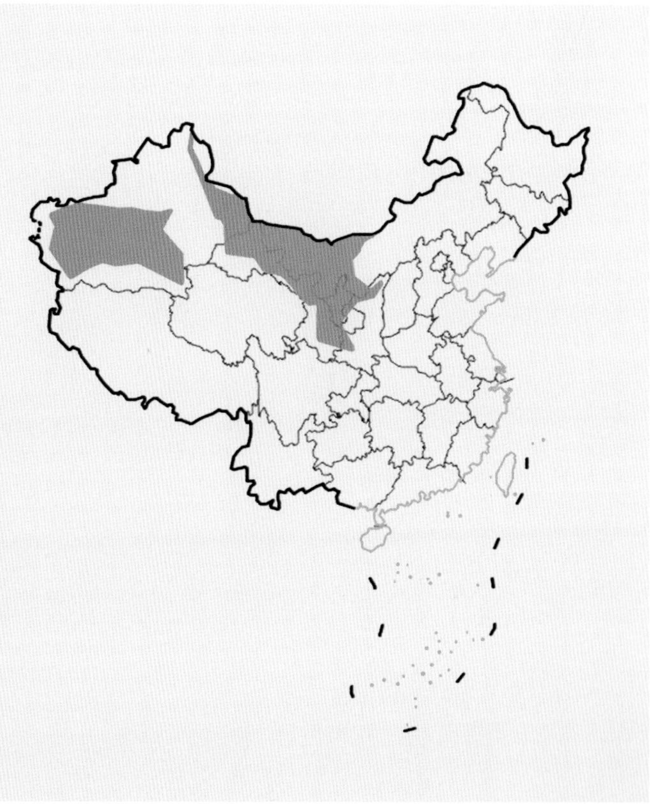

国内分布图
Map of Domestic Distribution

种群 Population

种群数量 **Population Size**	种群数量大 / Large populations
种群趋势 **Population Trend**	未知 / Unknown

生境与生态系统 Habitat (s) and Ecosystem (s)

生　　境 **Habitat(s)**	荒漠 / Desert
生态系统 **Ecosystem(s)**	荒漠生态系统 / Desert Ecosystem

威胁 Threat (s)

主要威胁 **Major Threat(s)**	无 / None

保护级别与保护行动 Protection Category and Conservation Action (s)

国家重点保护野生动物等级 **(2021) Category of National Key Protected Wild Animals (2021)**	无 / NA
IUCN 红色名录 **(2020-2) IUCN Red List (2020-2)**	无危 / LC
CITES 附录 **(2019) CITES Appendix (2019)**	无 / NA
保护行动 **Conservation Action(s)**	无 / None

相关文献 Relevant References

Burgin *et al*., 2018; Jiang *et al.* (蒋志刚等), 2017; Wilson *et al*., 2016; Zheng *et al.* (郑智民等), 2012; Hou and Ouyang (侯兰新和欧阳霞辉), 2010; Shuai *et al.* (帅凌鹰等), 2006; Hou (侯兰新), 1995

三趾跳鼠

Dipus sagitta

无危 LC

数据缺乏 DD	无危 LC	近危 NT	易危 VU	濒危 EN	极危 CR	区域灭绝 RE	野外灭绝 EW	灭绝 EX

分类地位 Taxonomic Status

动物界 Animalia	脊索动物门 Chordata	哺乳纲 Mammalia	啮齿目 Rodentia	跳鼠科 Dipodidae
学名 Scientific Name		*Dipus sagitta*		
命名人 Species Authority		Pallas, 1773		
英文名 English Name(s)		Hairy-footed Jerboa		
同物异名 Synonym(s)		Northern Three-toed Jerboa; *Dipus aksuensis* (Wang, 1964); *austrouralensis* (Shenbrot, 1991); *bulganensis* (Shenbrot, 1991); *deasyi* (Barrett-Hamilton, 1900); *fuscocanus* (转下页)		
种下单元评估 Infra-specific Taxa Assessed		无 / None		

评估信息 Assessment Information

评估年份 Year Assessed	2020
评定人 Assessor(s)	蒋志刚、刘少英 / Zhigang Jiang, Shaoying Liu
审定人 Reviewer(s)	马勇、刘伟 / Yong Ma, Wei Liu
其他贡献人 Other Contributor(s)	李立立、丁晨晨 / Lili Li, Chenchen Ding

理由 Justification: 三趾跳鼠分布广，种群数量大。因此，列为无危等级 / Hairy-footed Jerboa is a widespread species with large populations. Thus, it is listed as Least Concern

地理分布 Geographical Distribution

国内分布 Domestic Distribution

黑龙江、吉林、辽宁、内蒙古、陕西、宁夏、甘肃、青海、新疆 / Heilongjiang, Jilin, Liaoning, Inner Mongolia (Nei Mongol), Shaanxi, Ningxia, Gansu, Qinghai, Xinjiang

世界分布 World Distribution

中亚、东亚、西亚 / Central Asia, East Asia, West Asia

分布标注 Distribution Note

非特有种 / Non-endemic

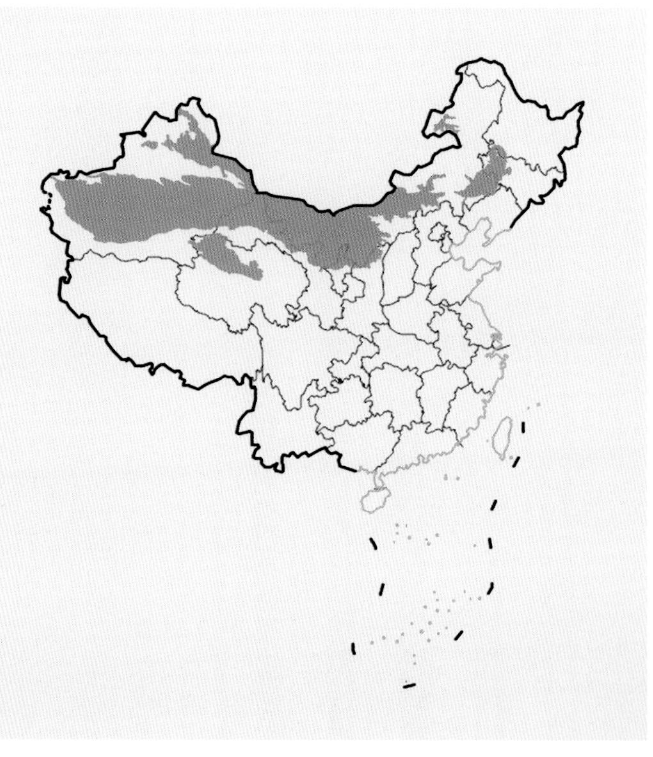

国内分布图
Map of Domestic Distribution

种群 Population

种群数量 Population Size	种群数量大 / Large populations
种群趋势 Population Trend	稳定 / Stable

生境与生态系统 Habitat (s) and Ecosystem (s)

生　　境 Habitat(s)	荒漠、半荒漠 / Desert, Semi-desert
生态系统 Ecosystem(s)	荒漠生态系统 / Desert Ecosystem

威胁 Threat (s)

主要威胁 Major Threat(s)	无 / None

保护级别与保护行动 Protection Category and Conservation Action (s)

国家重点保护野生动物等级 (2021) Category of National Key Protected Wild Animals (2021)	无 / NA
IUCN 红色名录 (2020-2) IUCN Red List (2020-2)	无危 / LC
CITES 附录 (2019) CITES Appendix (2019)	无 / NA
保护行动 Conservation Action(s)	无 / None

相关文献 Relevant References

Burgin *et al*., 2018; Jiang *et al.* (蒋志刚等), 2017; Wilson *et al*., 2016; Zheng *et al.* (郑智民等), 2012; Ji *et al.* (吉晟男等), 2009; Dong *et al.* (董维惠等), 2008; Shuai *et al.* (帅凌鹰等), 2006

(接上页)
(Wang, 1964); *halli* (Sowerby, 1920); *innae* (Ognev, 1930); *kalmikensis* (Kazantseva, 1940); *agopus* (Lichtenstein, 1823); *megacranius* (Shenbrot, 1991); *nogai* (Satunin, 1907); *sowerbyi* (Thomas, 1908); *turanicus* (Shenbrot, 1991); *ubsanensis* (Bannikov, 1947); *usuni* (Shenbrot, 1991); *zaissanensis* (Selevin, 1934)

三趾跳鼠 *Dipus sagitta*

数据缺乏 DD	无危 LC	近危 NT	易危 VU	濒危 EN	极危 CR	区域灭绝 RE	野外灭绝 EW	灭绝 EX

分类地位 Taxonomic Status

动物界 **Animalia**	脊索动物门 **Chordata**	哺乳纲 **Mammalia**	啮齿目 **Rodentia**	跳鼠科 **Dipodidae**

学　名 **Scientific Name**	*Stylodipus andrewsi*
命名人 **Species Authority**	Allen, 1925
英文名 **English Name(s)**	Andrews' Three-toed Jerboa
同物异名 **Synonym(s)**	*Scirtopoda andrewsi* (Allen, 1925)
种下单元评估 **Infra-specific Taxa Assessed**	无 / None

评估信息 Assessment Information

评估年份 **Year Assessed**	2020
评定人 **Assessor(s)**	蒋志刚、刘少英 / Zhigang Jiang, Shaoying Liu
审定人 **Reviewer(s)**	马勇、刘伟 / Yong Ma, Wei Liu
其他贡献人 **Other Contributor(s)**	李立立、丁晨晨 / Lili Li, Chenchen Ding

理由 **Justification:** 内蒙羽尾跳鼠分布广，种群数量大。因此，列为无危等级 / Andrews' Three-toed Jerboa is a widespread species with large populations. Thus, it is listed as Least Concern

地理分布 Geographical Distribution

国内分布 Domestic Distribution

甘肃、宁夏、内蒙古 / Gansu, Ningxia, Inner Mongolia (Nei Mongol)

世界分布 World Distribution

中国、蒙古 / China, Mongolia

分布标注 Distribution Note

非特有种 / Non-endemic

国内分布图
Map of Domestic Distribution

种群 Population

种群数量 Population Size	种群数量大 / Large populations
种群趋势 Population Trend	未知 / Unknown

生境与生态系统 Habitat (s) and Ecosystem (s)

生　境 Habitat(s)	半荒漠、草地、泰加林、灌丛 / Semi-desert, Grassland, Taiga Forest, Shrubland
生态系统 Ecosystem(s)	森林生态系统、灌丛生态系统、草地生态系统、荒漠荒原生态系统 / Forest Ecosystem, Shrubland Ecosystem, Grassland Ecosystem, Desert Ecosystem

威胁 Threat (s)

主要威胁 Major Threat(s)	无 / None

保护级别与保护行动 Protection Category and Conservation Action (s)

国家重点保护野生动物等级 (2021) Category of National Key Protected Wild Animals (2021)	无 / NA
IUCN 红色名录 (2020-2) IUCN Red List (2020-2)	无危 / LC
CITES 附录 (2019) CITES Appendix (2019)	无 / NA
保护行动 Conservation Action(s)	无 / None

相关文献 Relevant References

Burgin *et al.*, 2018; Jiang *et al.* (蒋志刚等), 2017; Wilson *et al.*, 2016; Zheng *et al.* (郑智民等), 2012; Yuan *et al.* (袁帅等), 2011; Smith *et al.* (史密斯等), 2009

分类地位 Taxonomic Status

动 物 界 Animalia	脊索动物门 Chordata	哺 乳 纲 Mammalia	啮 齿 目 Rodentia	跳 鼠 科 Dipodidae
学　名 Scientific Name		*Stylodipus telum*		
命 名 人 Species Authority		Lichtenstein, 1823		
英 文 名 English Name(s)		Thick-tailed Three-toed Jerboa		
同物异名 Synonym(s)		*Stylodipus amankaragai* (Selewin, 1934); *birulae* (Martino, 1922); *falzfeini* (Brauner, 1913); *halticus* (Brandt, 1844); *karelini* (Selewin, 1934); *nastjukovi* (Shenbrot, （转下页）		
种下单元评估 Infra-specific Taxa Assessed		无 / None		

评估信息 Assessment Information

评 估 年 份 Year Assessed	2020
评　定　人 Assessor(s)	蒋志刚、刘少英 / Zhigang Jiang, Shaoying Liu
审　定　人 Reviewer(s)	马勇、刘伟 / Yong Ma, Wei Liu
其他贡献人 Other Contributor(s)	李立立、丁晨晨 / Lili Li, Chenchen Ding

理由 Justification: 羽尾跳鼠分布广，种群数量大。因此，列为无危等级 / Thick-tailed Three-toed Jerboa is a widespread species with large populations. Thus, it is listed as Least Concern

地理分布 Geographical Distribution

国内分布 Domestic Distribution

新疆 / Xinjiang

世界分布 World Distribution

中国、哈萨克斯坦、俄罗斯、土库曼斯坦、乌克兰、乌兹别克斯坦 / China, Kazakhstan, Russia, Turkmenistan, Ukraine, Uzbekistan

分布标注 Distribution Note

非特有种 / Non-endemic

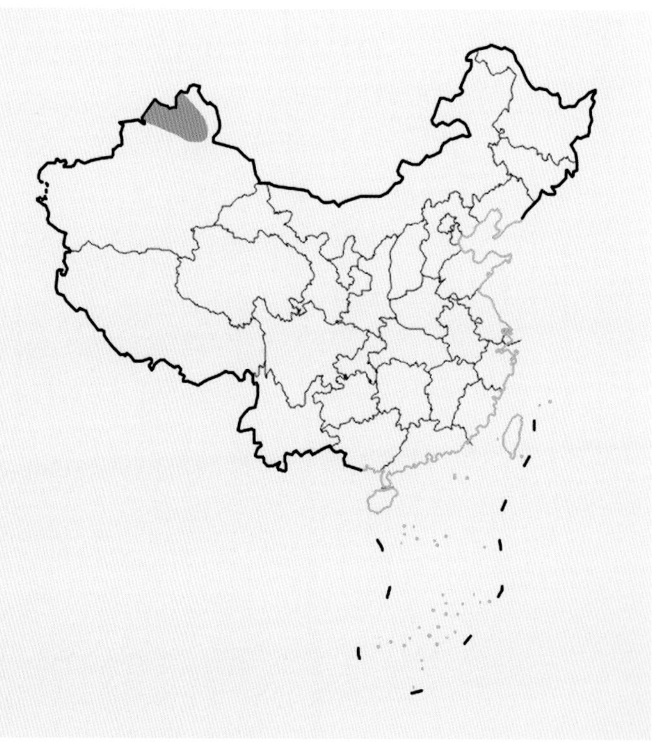

国内分布图
Map of Domestic Distribution

种群 Population

种群数量 Population Size	种群数量大 / Large populations
种群趋势 Population Trend	下降 / Decreasing

生境与生态系统 Habitat (s) and Ecosystem (s)

生　　境 Habitat(s)	沙漠、高海拔草地 / Desert, High Altitude Grassland
生态系统 Ecosystem(s)	草地生态系统、荒漠生态系统 / Grassland Ecosystem, Desert Ecosystem

威胁 Threat (s)

主要威胁 Major Threat(s)	无 / None

保护级别与保护行动 Protection Category and Conservation Action (s)

国家重点保护野生动物等级 (2021) Category of National Key Protected Wild Animals (2021)	无 / NA
IUCN 红色名录 (2020-2) IUCN Red List (2020-2)	无危 / LC
CITES 附录 (2019) CITES Appendix (2019)	无 / NA
保护行动 Conservation Action(s)	无 / None

相关文献 Relevant References

Burgin *et al*., 2018; Jiang *et al.* (蒋志刚等), 2017; Wilson *et al*., 2016; Zheng *et al.* (郑智民等), 2012; Wang (王应祥), 2003; Li and Han (李枝林和韩建芳), 1988; Zhao (赵肯堂), 1981

(接上页)
1991); *proximus* (Fairmaire, 1853); *turovi* (Heptner, 1934)

数据缺乏 DD	无危 LC	近危 NT	易危 VU	濒危 EN	极危 CR	区域灭绝 RE	野外灭绝 EW	灭绝 EX

分类地位 Taxonomic Status

动物界 Animalia	脊索动物门 Chordata	哺乳纲 Mammalia	啮齿目 Rodentia	跳鼠科 Dipodidae

学　名 Scientific Name	*Euchoreutes naso*
命名人 Species Authority	Sclater, 1891
英文名 English Name(s)	Long-eared Jerboa
同物异名 Synonym(s)	*Euchoreutes alashanicus* (Howell, 1928); *yiwuensis* (Ma and Li, 1979)
种下单元评估 Infra-specific Taxa Assessed	无 / None

评估信息 Assessment Information

评估年份 Year Assessed	2020
评定人 Assessor(s)	蒋志刚、刘少英 / Zhigang Jiang, Shaoying Liu
审定人 Reviewer(s)	马勇、刘伟 / Yong Ma, Wei Liu
其他贡献人 Other Contributor(s)	李立立、丁晨晨 / Lili Li, Chenchen Ding

理由 Justification: 长耳跳鼠分布广，种群数量大。因此，列为无危等级 / Long-eared Jerboa is a widespread species with large populations. Thus, it is listed as Least Concern

地理分布 Geographical Distribution

国内分布 Domestic Distribution

内蒙古、新疆、甘肃、宁夏、青海 / Inner Mongolia (Nei Mongol), Xinjiang, Gansu, Ningxia, Qinghai

世界分布 World Distribution

中国、蒙古 / China, Mongolia

分布标注 Distribution Note

非特有种 / Non-endemic

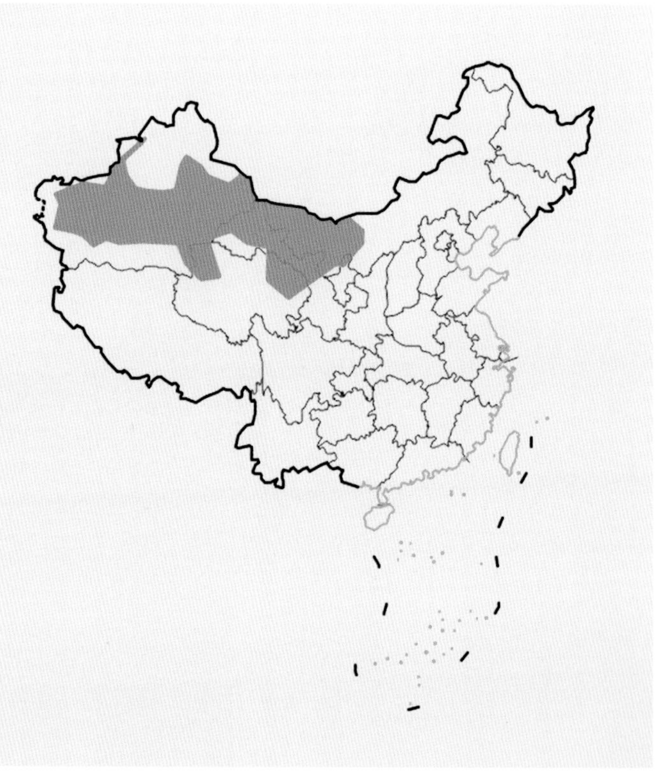

国内分布图
Map of Domestic Distribution

种群 Population

种群数量 Population Size	种群数量大 / Large populations
种群趋势 Population Trend	未知 / Unknown

生境与生态系统 Habitat (s) and Ecosystem (s)

生　　境 Habitat(s)	荒漠、绿洲 / Desert, Oasis
生态系统 Ecosystem(s)	荒漠生态系统 / Desert Ecosystem

威胁 Threat (s)

主要威胁 Major Threat(s)	无 / None

保护级别与保护行动 Protection Category and Conservation Action (s)

国家重点保护野生动物等级 (2021) Category of National Key Protected Wild Animals (2021)	无 / NA
IUCN 红色名录 (2020-2) IUCN Red List (2020-2)	无危 / LC
CITES 附录 (2019) CITES Appendix (2019)	无 / NA
保护行动 Conservation Action(s)	无 / None

相关文献 Relevant References

Burgin *et al*., 2018; Jiang *et al.* (蒋志刚等), 2017; Wilson *et al*., 2016; Zheng *et al.* (郑智民等), 2012; Smith *et al.* (史密斯等), 2009; Wu *et al.* (武晓东等), 2003; Ma and Li (马勇和李思华), 1979

数据缺乏 DD	无危 LC	近危 NT	易危 VU	濒危 EN	极危 CR	区域灭绝 RE	野外灭绝 EW	灭绝 EX

分类地位 Taxonomic Status

动物界 Animalia	脊索动物门 Chordata	哺乳纲 Mammalia	啮齿目 Rodentia	豪猪科 Hystricidae
学名 Scientific Name		*Atherurus macrourus*		
命名人 Species Authority		Linnaeus, 1758		
英文名 English Name(s)		Asiatic Brush-tailed Porcupine		
同物异名 Synonym(s)		*Atherurus angustiramus* (Mohr, 1964); *assamensis* (Thomas, 1921); *hainanus* (Allen, 1906); *pemangilis* (Robinson, 1912); *retardatus* (Mohr, 1964); *stevensi* (Thomas, 1925); (转下页)		
种下单元评估 Infra-specific Taxa Assessed		无 / None		

评估信息 Assessment Information

评估年份 Year Assessed	2020
评定人 Assessor(s)	蒋志刚、刘少英 / Zhigang Jiang, Shaoying Liu
审定人 Reviewer(s)	蒋学龙 / Xuelong Jiang
其他贡献人 Other Contributor(s)	李立立、丁晨晨 / Lili Li, Chenchen Ding

理由 Justification: 帚尾豪猪分布较广，种群数量大。因此，列为无危等级 / Asiatic Brush-tailed Porcupine is a widespread species with large populations. Thus, it is listed as Least Concern

地理分布 Geographical Distribution

国内分布 Domestic Distribution

广西、海南、湖北、四川、贵州、云南、重庆 / Guangxi, Hainan, Hubei, Sichuan, Guizhou, Yunnan, Chongqing

世界分布 World Distribution

中国；南亚、东南亚 / China; South Asia, Southeast Asia

分布标注 Distribution Note

非特有种 / Non-endemic

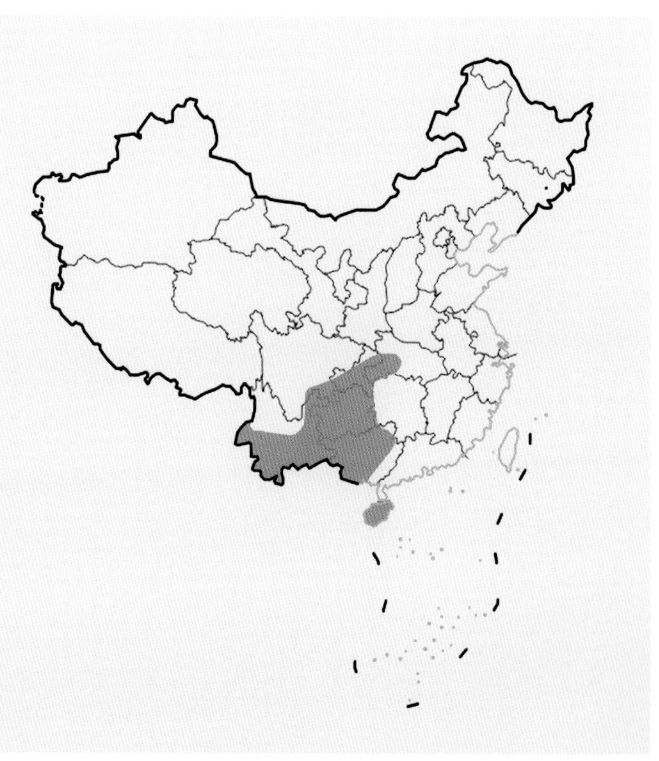

国内分布图
Map of Domestic Distribution

种群 Population

种群数量 Population Size	种群数量大 / Large populations
种群趋势 Population Trend	下降 / Decreasing

生境与生态系统 Habitat (s) and Ecosystem (s)

生　　境 Habitat(s)	森林 / Forest
生态系统 Ecosystem(s)	森林生态系统 / Forest Ecosystem

威胁 Threat (s)

主要威胁 Major Threat(s)	无 / None

保护级别与保护行动 Protection Category and Conservation Action (s)

国家重点保护野生动物等级 (2021) Category of National Key Protected Wild Animals (2021)	无 / NA
IUCN 红色名录 (2020-2) IUCN Red List (2020-2)	无危 / LC
CITES 附录 (2019) CITES Appendix (2019)	无 / NA
保护行动 Conservation Action(s)	无 / None

相关文献 Relevant References

Burgin *et al*., 2018; Jiang *et al.* (蒋志刚等), 2017; Wilson *et al*., 2016; Zheng *et al.* (郑智民等), 2012; Smith *et al.* (史密斯等), 2009; Pan *et al.* (潘清华等), 2007; Wang (王应祥), 2003

(接上页)
terutaus (Lyon, 1907); *tionis* (Thomas, 1908); *zygomatica* (Miller, 1903)

中国生物多样性红色名录
China's Red List of Biodiversity

马来豪猪
Hystrix brachyura

无危 LC

数据缺乏 DD	无危 LC	近危 NT	易危 VU	濒危 EN	极危 CR	区域灭绝 RE	野外灭绝 EW	灭绝 EX

分类地位 Taxonomic Status

动 物 界 Animalia	脊索动物门 Chordata	哺 乳 纲 Mammalia	啮 齿 目 Rodentia	豪 猪 科 Hystricidae
学 名 Scientific Name		*Hystrix brachyura*		
命 名 人 Species Authority		Linnaeus, 1758		
英 文 名 English Name(s)		Malayan Porcupine		
同物异名 Synonym(s)		无 / None		
种下单元评估 Infra-specific Taxa Assessed		无 / None		

评估信息 Assessment Information

评 估 年 份 Year Assessed	2020
评 定 人 Assessor(s)	蒋志刚、刘少英 / Zhigang Jiang, Shaoying Liu
审 定 人 Reviewer(s)	蒋学龙 / Xuelong Jiang
其他贡献人 Other Contributor(s)	李立立、丁晨晨 / Lili Li, Chenchen Ding

理由 Justification: 马来豪猪分布广，种群数量大。因此，列为无危等级 / Malayan Porcupine is a widespread species with large populations. Thus, it is listed as Least Concern

地理分布 Geographical Distribution

国内分布 Domestic Distribution

云南 / Yunnan

世界分布 World Distribution

中国；南亚、东南亚 / China; South Asia, Southeast Asia

分布标注 Distribution Note

非特有种 / Non-endemic

国内分布图
Map of Domestic Distribution

种群 Population

种群数量 Population Size	种群数量大 / Large populations
种群趋势 Population Trend	下降 / Decreasing

生境与生态系统 Habitat (s) and Ecosystem (s)

生　　境 Habitat(s)	森林 / Forest
生态系统 Ecosystem(s)	森林生态系统 / Forest Ecosystem

威胁 Threat (s)

主要威胁 Major Threat(s)	无 / None

保护级别与保护行动 Protection Category and Conservation Action (s)

国家重点保护野生动物等级 (2021) Category of National Key Protected Wild Animals (2021)	无 / NA
IUCN 红色名录 (2020-2) IUCN Red List (2020-2)	无危 / LC
CITES 附录 (2019) CITES Appendix (2019)	无 / NA
保护行动 Conservation Action(s)	无 / None

相关文献 Relevant References

Burgin *et al*., 2018; Jiang *et al.* (蒋志刚等), 2017; Wilson *et al*., 2016; Xu *et al.* (徐爱春等), 2014; Zheng *et al.* (郑智民等), 2012; Chung and Corlett, 2006

马来豪猪 *Hystrix brachyura*

中国豪猪

Hystrix hodgsoni

无危 LC

数据缺乏 DD	无危 LC	近危 NT	易危 VU	濒危 EN	极危 CR	区域灭绝 RE	野外灭绝 EW	灭绝 EX

分类地位 Taxonomic Status

动物界 Animalia	脊索动物门 Chordata	哺乳纲 Mammalia	啮齿目 Rodentia	豪猪科 Hystricidae

学　名 Scientific Name	*Hystrix hodgsoni*
命名人 Species Authority	Gray, 1847
英文名 English Name(s)	Chinese Porcupine
同物异名 Synonym(s)	Himalayan Porcupine
种下单元评估 Infra-specific Taxa Assessed	无 / None

评估信息 Assessment Information

评估年份 Year Assessed	2020
评定人 Assessor(s)	蒋志刚、刘少英 / Zhigang Jiang, Shaoying Liu
审定人 Reviewer(s)	蒋学龙 / Xuelong Jiang
其他贡献人 Other Contributor(s)	李立立、丁晨晨 / Lili Li, Chenchen Ding

理由 Justification: 中国豪猪分布广，种群数量大。因此，列为无危等级 / Chinese Porcupine is a widespread species with large populations. Thus, it is listed as Least Concern

地理分布 Geographical Distribution

国内分布 Domestic Distribution

陕西、西藏、四川、重庆、湖北、安徽、江苏、上海、浙江、福建、江西、湖南、贵州、云南、广西、广东、海南、甘肃、河南 / Shaanxi, Tibet (Xizang), Sichuan, Chongqing, Hubei, Anhui, Jiangsu, Shanghai, Zhejiang, Fujian, Jiangxi, Hunan, Guizhou, Yunnan, Guangxi, Guangdong, Hainan, Gansu, Henan

世界分布 World Distribution

中国；不详 / China; Unknown

分布标注 Distribution Note

非特有种 / Non-endemic

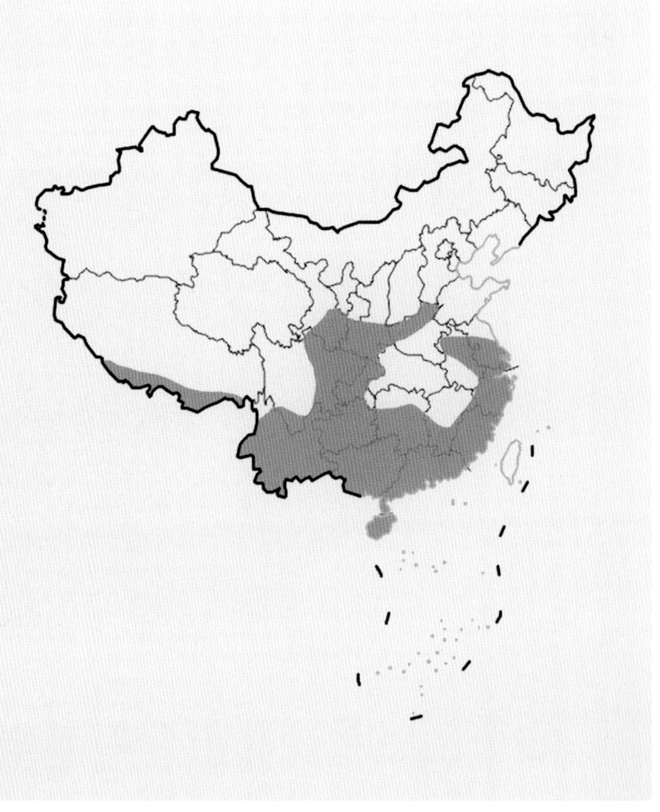

国内分布图
Map of Domestic Distribution

种群 Population

种群数量 Population Size	种群数量大 / Large populations
种群趋势 Population Trend	下降 / Decreasing

生境与生态系统 Habitat (s) and Ecosystem (s)

生　　境 Habitat(s)	森林 / Forest
生态系统 Ecosystem(s)	森林生态系统 / Forest Ecosystem

威胁 Threat (s)

主要威胁 Major Threat(s)	无 / None

保护级别与保护行动 Protection Category and Conservation Action (s)

国家重点保护野生动物等级 (2021) Category of National Key Protected Wild Animals (2021)	无 / NA
IUCN 红色名录 (2020-2) IUCN Red List (2020-2)	无危 / LC
CITES 附录 (2019) CITES Appendix (2019)	无 / NA
保护行动 Conservation Action(s)	无 / None

相关文献 Relevant References

Burgin *et al*., 2018; Jiang *et al*. (蒋志刚等), 2017; Wilson *et al*., 2016; Zheng *et al*. (郑智民等), 2012; Pan *et al*. (潘清华等), 2007

中国豪猪 *Hystrix hodgsoni*　　蒋志刚 摄　By Zhigang Jiang

高山鼠兔

Ochotona alpina

无危 LC

数据缺乏 DD	无危 LC	近危 NT	易危 VU	濒危 EN	极危 CR	区域灭绝 RE	野外灭绝 EW	灭绝 EX

分类地位 Taxonomic Status

动物界 Animalia	脊索动物门 Chordata	哺乳纲 Mammalia	兔形目 Lagomorpha	鼠兔科 Ochotonidae
学名 Scientific Name		*Ochotona alpina*		
命名人 Species Authority		Pallas, 1773		
英文名 English Name(s)		Alpine Pika		
同物异名 Synonym(s)		无 / None		
种下单元评估 Infra-specific Taxa Assessed		无 / None		

评估信息 Assessment Information

评估年份 Year Assessed	2020
评定人 Assessor(s)	刘少英、蒋志刚 / Shaoying Liu, Zhigang Jiang
审定人 Reviewer(s)	苏建平、冯祚建、宗浩、廖继承 / Jianping Su, Zuojian Feng, Hao Zong, Jicheng Liao
其他贡献人 Other Contributor(s)	李立立、丁晨晨 / Lili Li, Chenchen Ding

理由 Justification: 高山鼠兔分布广，种群数量大。因此，列为无危等级 / Alpine Pika is a widespread species with large populations. Thus, it is listed as Least Concern

地理分布 Geographical Distribution

国内分布 Domestic Distribution

新疆、黑龙江 / Xinjiang, Heilongjiang

世界分布 World Distribution

中亚、东亚 / Central Asia, East Asia

分布标注 Distribution Note

非特有种 / Non-endemic

国内分布图
Map of Domestic Distribution

种群 Population

种群数量 **Population Size**	种群数量大 / Large populations
种群趋势 **Population Trend**	下降 / Decreasing

生境与生态系统 Habitat (s) and Ecosystem (s)

生　　境 **Habitat(s)**	森林、岩石区域 / Forest, Rocky Area
生态系统 **Ecosystem(s)**	森林生态系统、草地生态系统 / Forest Ecosystem, Grassland Ecosystem

威胁 Threat (s)

主要威胁 **Major Threat(s)**	无 / None

保护级别与保护行动 Protection Category and Conservation Action (s)

国家重点保护野生动物等级 **(2021) Category of National Key Protected Wild Animals (2021)**	无 / NA
IUCN 红色名录 **(2020-2) IUCN Red List (2020-2)**	无危 / LC
CITES 附录 **(2019) CITES Appendix (2019)**	无 / NA
保护行动 **Conservation Action(s)**	无 / None

相关文献 Relevant References

Burgin *et al.*, 2018; Jiang *et al.* (蒋志刚等), 2017; Wilson *et al.*, 2016; Zheng *et al.* (郑智民等), 2012; Lissovsky *et al.*, 2007; Ma and Jiang (马瑞俊和蒋志刚), 2006; Zhou *et al.* (周立志等), 2002

中国生物多样性红色名录
China's Red List of Biodiversity

间颅鼠兔

Ochotona cansus

无危 LC

数据缺乏 DD	无危 LC	近危 NT	易危 VU	濒危 EN	极危 CR	区域灭绝 RE	野外灭绝 EW	灭绝 EX

分类地位 Taxonomic Status

动物界 Animalia	脊索动物门 Chordata	哺乳纲 Mammalia	兔形目 Lagomorpha	鼠兔科 Ochotonidae
学名 Scientific Name		*Ochotona cansus*		
命名人 Species Authority		Lyon, 1907		
英文名 English Name(s)		Gansu Pika		
同物异名 Synonym(s)		无 / None		
种下单元评估 Infra-specific Taxa Assessed		无 / None		

评估信息 Assessment Information

评估年份 Year Assessed	2020
评定人 Assessor(s)	刘少英、蒋志刚 / Shaoying Liu, Zhigang Jiang
审定人 Reviewer(s)	苏建平、宗浩、廖继承 / Jianping Su, Hao Zong, Jicheng Liao
其他贡献人 Other Contributor(s)	李立立、丁晨晨 / Lili Li, Chenchen Ding

理由 Justification: 间颅鼠兔分布广，种群数量大。因此，列为无危等级 / Gansu Pika is a widespread species with large populations. Thus, it is listed as Least Concern

地理分布 Geographical Distribution

国内分布 Domestic Distribution

甘肃、青海、四川、陕西、山西 / Gansu, Qinghai, Sichuan, Shaanxi, Shanxi

世界分布 World Distribution

中国 / China

分布标注 Distribution Note

特有种 / Endemic

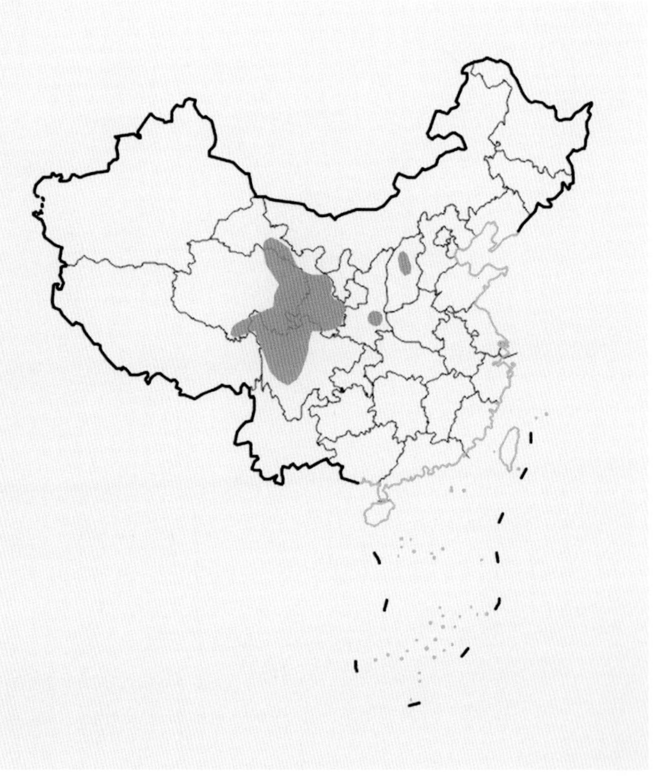

国内分布图
Map of Domestic Distribution

种群 Population

种群数量 Population Size	种群数量大 / Large populations
种群趋势 Population Trend	未知 / Unknown

生境与生态系统 Habitat (s) and Ecosystem (s)

生　　境 Habitat(s)	草甸、灌丛 / Meadow, Shrubland
生态系统 Ecosystem(s)	灌丛生态系统、草地生态系统 / Shrubland Ecosystem, Grassland Ecosystem

威胁 Threat (s)

主要威胁 Major Threat(s)	无 / None

保护级别与保护行动 Protection Category and Conservation Action (s)

国家重点保护野生动物等级 (2021) Category of National Key Protected Wild Animals (2021)	无 / NA
IUCN 红色名录 (2020-2) IUCN Red List (2020-2)	无危 / LC
CITES 附录 (2019) CITES Appendix (2019)	无 / NA
保护行动 Conservation Action(s)	无 / None

相关文献 Relevant References

Burgin *et al*., 2018; Jiang *et al.* (蒋志刚等), 2017; Wilson *et al*., 2016; Zheng *et al.* (郑智民等), 2012; Smith *et al.* (史密斯等), 2009; Wu *et al.* (武晓东等), 2002; Zhou *et al.* (周立志等), 2002

高原鼠兔

Ochotona curzoniae

无危 LC

数据缺乏 DD	无危 LC	近危 NT	易危 VU	濒危 EN	极危 CR	区域灭绝 RE	野外灭绝 EW	灭绝 EX

分类地位 Taxonomic Status

动物界 Animalia	脊索动物门 Chordata	哺乳纲 Mammalia	兔形目 Lagomorpha	鼠兔科 Ochotonidae

学名 Scientific Name	*Ochotona curzoniae*
命名人 Species Authority	Hodgson, 1858
英文名 English Name(s)	Plateau Pika
同物异名 Synonym(s)	Black-lipped Pika; *Ochotona melanostoma* (Büchner, 1890)
种下单元评估 Infra-specific Taxa Assessed	无 / None

评估信息 Assessment Information

评估年份 Year Assessed	2020
评定人 Assessor(s)	刘少英、蒋志刚 / Shaoying Liu, Zhigang Jiang
审定人 Reviewer(s)	苏建平、宗浩、廖继承 / Jianping Su, Hao Zong, Jicheng Liao
其他贡献人 Other Contributor(s)	李立立、丁晨晨 / Lili Li, Chenchen Ding

理由 **Justification:** 高原鼠兔在青藏高原分布广，种群数量大。因此，列为无危等级 / Plateau Pika is a widespread species with large populations on the Qinghai-Tibet (Xizang) Plateau. Thus, it is listed as Least Concern

地理分布 Geographical Distribution

国内分布 Domestic Distribution

青海、新疆、四川、甘肃、西藏 / Qinghai, Xinjiang, Sichuan, Gansu, Tibet (Xizang)

世界分布 World Distribution

中国、印度、尼泊尔 / China, India, Nepal

分布标注 Distribution Note

非特有种 / Non-endemic

国内分布图
Map of Domestic Distribution

种群 Population

种群数量 **Population Size**	种群数量大 / Large populations
种群趋势 **Population Trend**	下降 / Decreasing

生境与生态系统 Habitat (s) and Ecosystem (s)

生　　境 **Habitat(s)**	草甸、草地、荒漠 / Meadow, Grassland, Desert
生态系统 **Ecosystem(s)**	草地生态系统、荒漠生态系统 / Grassland Ecosystem, Desert Ecosystem

威胁 Threat (s)

主要威胁 **Major Threat(s)**	毒杀 / Poisoning

保护级别与保护行动 Protection Category and Conservation Action (s)

国家重点保护野生动物等级 **(2021) Category of National Key Protected Wild Animals (2021)**	无 / NA
IUCN 红色名录 **(2020-2) IUCN Red List (2020-2)**	无危 / LC
CITES 附录 **(2019) CITES Appendix (2019)**	无 / NA
保护行动 **Conservation Action(s)**	无 / None

相关文献 Relevant References

Burgin *et al*., 2018; Jiang *et al.* (蒋志刚等), 2017; Wilson *et al*., 2016; Zheng *et al.* (郑智民等), 2012; Pech *et al*., 2007; Dai *et al.* (戴强等), 2006; Lai and Smith, 2003

高原鼠兔 *Ochotona curzoniae*　　蒋志刚 摄　By Zhigang Jiang

中国生物多样性红色名录
China's Red List of Biodiversity

达乌尔鼠兔

Ochotona dauurica

无危 LC

数据缺乏 DD	无危 LC	近危 NT	易危 VU	濒危 EN	极危 CR	区域灭绝 RE	野外灭绝 EW	灭绝 EX

分类地位 Taxonomic Status

动物界 Animalia	脊索动物门 Chordata	哺乳纲 Mammalia	兔形目 Lagomorpha	鼠兔科 Ochotonidae

学　名 **Scientific Name**	*Ochotona dauurica*
命名人 **Species Authority**	Pallas, 1776
英文名 **English Name(s)**	Daurian Pika
同物异名 **Synonym(s)**	无 / None
种下单元评估 **Infra-specific Taxa Assessed**	无 / None

评估信息 Assessment Information

评估年份 **Year Assessed**	2020
评定人 **Assessor(s)**	刘少英、蒋志刚 / Shaoying Liu, Zhigang Jiang
审定人 **Reviewer(s)**	苏建平、宗浩、廖继承 / Jianping Su, Hao Zong, Jicheng Liao
其他贡献人 **Other Contributor(s)**	李立立、丁晨晨 / Lili Li, Chenchen Ding

理由 **Justification:** 达乌尔鼠兔分布广，种群数量大。因此，列为无危等级 / Daurian Pika is a widespread species with large populations. Thus, it is listed as Least Concern

地理分布 Geographical Distribution

国内分布 Domestic Distribution

陕西、山西、河北、内蒙古、河南、甘肃、青海、宁夏 / Shaanxi, Shanxi, Hebei, Inner Mongolia (Nei Mongol), Henan, Gansu, Qinghai, Ningxia

世界分布 World Distribution

中国、蒙古、俄罗斯 / China, Mongolia, Russia

分布标注 Distribution Note

非特有种 / Non-endemic

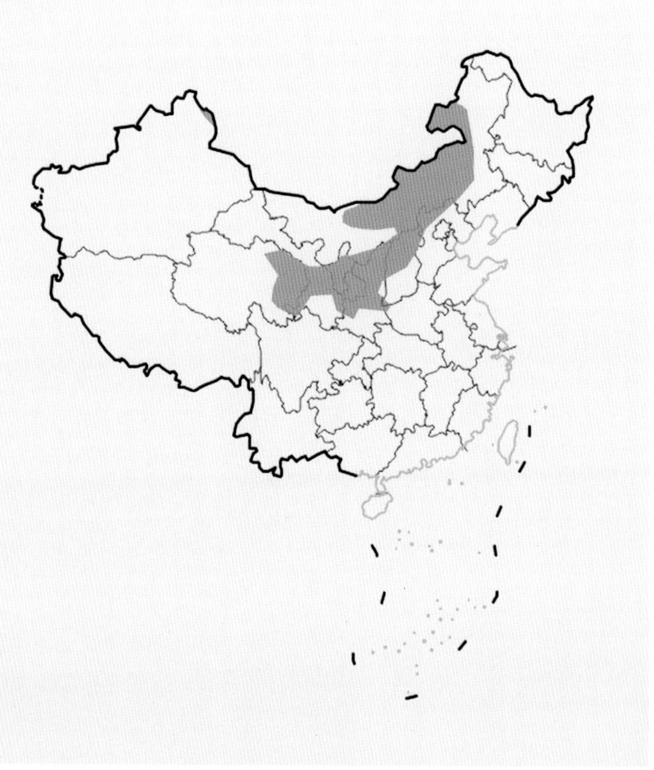

国内分布图
Map of Domestic Distribution

种群 Population

种群数量 Population Size	种群数量大 / Large populations
种群趋势 Population Trend	未知 / Unknown

生境与生态系统 Habitat (s) and Ecosystem (s)

生　　境 Habitat(s)	荒漠、草地 / Desert, Grassland
生态系统 Ecosystem(s)	草地生态系统、荒漠生态系统 / Grassland Ecosystem, Desert Ecosystem

威胁 Threat (s)

主要威胁 Major Threat(s)	无 / None

保护级别与保护行动 Protection Category and Conservation Action (s)

国家重点保护野生动物等级 (2021) Category of National Key Protected Wild Animals (2021)	无 / NA
IUCN 红色名录 (2020-2) IUCN Red List (2020-2)	无危 / LC
CITES 附录 (2019) CITES Appendix (2019)	无 / NA
保护行动 Conservation Action(s)	无 / None

相关文献 Relevant References

Burgin *et al*., 2018; Jiang *et al.* (蒋志刚等), 2017; Wilson *et al*., 2016; Chen *et al.* (陈立军等), 2014; Zheng *et al.* (郑智民等), 2012; Xing *et al.* (邢雅俊等), 2008; Zhong *et al*., 2008

红耳鼠兔
Ochotona erythrotis

无危 LC

数据缺乏 DD	无危 LC	近危 NT	易危 VU	濒危 EN	极危 CR	区域灭绝 RE	野外灭绝 EW	灭绝 EX

分类地位 Taxonomic Status

动物界 Animalia	脊索动物门 Chordata	哺乳纲 Mammalia	兔形目 Lagomorpha	鼠兔科 Ochotonidae

学　名 **Scientific Name**	*Ochotona erythrotis*
命名人 **Species Authority**	Büchner, 1890
英文名 **English Name(s)**	Chinese Red Pika
同物异名 **Synonym(s)**	无 / None
种下单元评估 **Infra-specific Taxa Assessed**	无 / None

评估信息 Assessment Information

评估年份 **Year Assessed**	2020
评定人 **Assessor(s)**	刘少英、蒋志刚 / Shaoying Liu, Zhigang Jiang
审定人 **Reviewer(s)**	苏建平、宗浩、廖继承 / Jianping Su, Hao Zong, Jicheng Liao
其他贡献人 **Other Contributor(s)**	李立立、丁晨晨 / Lili Li, Chenchen Ding

理由 **Justification:** 红耳鼠兔分布广，种群数量大。因此，列为无危等级 / Chinese Red Pika is a widespread species with large populations. Thus, it is listed as Least Concern

地理分布 Geographical Distribution

国内分布 Domestic Distribution

甘肃、青海、四川、云南、西藏 / Gansu, Qinghai, Sichuan, Yunnan, Tibet (Xizang)

世界分布 World Distribution

中国 / China

分布标注 Distribution Note

特有种 / Endemic

国内分布图
Map of Domestic Distribution

种群 Population

种群数量 **Population Size**	种群数量大 / Large populations
种群趋势 **Population Trend**	未知 / Unknown

生境与生态系统 Habitat (s) and Ecosystem (s)

生　境 **Habitat(s)**	岩石区域、江河附近 / Rocky Area, Near River
生态系统 **Ecosystem(s)**	湖泊河流生态系统 / Lake and River Ecosystem

威胁 Threat (s)

主要威胁 **Major Threat(s)**	无 / None

保护级别与保护行动 Protection Category and Conservation Action (s)

国家重点保护野生动物等级 **(2021) Category of National Key Protected Wild Animals (2021)**	无 / NA
IUCN 红色名录 **(2020-2) IUCN Red List (2020-2)**	无危 / LC
CITES 附录 **(2019) CITES Appendix (2019)**	无 / NA
保护行动 **Conservation Action(s)**	无 / None

相关文献 Relevant References

Burgin *et al*., 2018; Jiang *et al.* (蒋志刚等), 2017; Wilson *et al*., 2016; Zheng *et al.* (郑智民等), 2012; Li *et al*., 2003; Wang (王应祥), 2003; Song *et al.* (宋延龄等), 2002

红耳鼠兔 *Ochotona erythrotis* 　　　　巫嘉伟 摄 By Jiawei Wu

川西鼠兔

Ochotona gloveri

无危 LC

数据缺乏 DD	无危 LC	近危 NT	易危 VU	濒危 EN	极危 CR	区域灭绝 RE	野外灭绝 EW	灭绝 EX

分类地位 Taxonomic Status

动物界 Animalia	脊索动物门 Chordata	哺乳纲 Mammalia	兔形目 Lagomorpha	鼠兔科 Ochotonidae
学名 Scientific Name		*Ochotona gloveri*		
命名人 Species Authority		Thomas, 1922		
英文名 English Name(s)		Glover's Pika		
同物异名 Synonym(s)		无 / None		
种下单元评估 Infra-specific Taxa Assessed		无 / None		

评估信息 Assessment Information

评估年份 Year Assessed	2020
评定人 Assessor(s)	刘少英、蒋志刚 / Shaoying Liu, Zhigang Jiang
审定人 Reviewer(s)	苏建平、宗浩、廖继承 / Jianping Su, Hao Zong, Jicheng Liao
其他贡献人 Other Contributor(s)	李立立、丁晨晨 / Lili Li, Chenchen Ding

理由 Justification: 川西鼠兔有一定分布区，种群数量大。因此，列为无危等级 / Glover's Pika has certain ranges and large populations. Thus, it is listed as Least Concern

地理分布 Geographical Distribution

国内分布 Domestic Distribution

四川、青海、西藏 / Sichuan, Qinghai, Tibet (Xizang)

世界分布 World Distribution

中国 / China

分布标注 Distribution Note

特有种 / Endemic

国内分布图
Map of Domestic Distribution

种群 Population

种群数量 Population Size	种群数量大 / Large populations
种群趋势 Population Trend	未知 / Unknown

生境与生态系统 Habitat (s) and Ecosystem (s)

生　境 Habitat(s)	针阔混交林、灌丛、草甸的岩石区域 / Rocky Area in Coniferous and Broad-leaved Mixed Forest, Shrubland, Meadow
生态系统 Ecosystem(s)	森林生态系统、灌丛生态系统、草甸生态系统 / Forest Ecosystem, Shrubland Ecosystem, Meadow Ecosystem

威胁 Threat (s)

主要威胁 Major Threat(s)	无 / None

保护级别与保护行动 Protection Category and Conservation Action (s)

国家重点保护野生动物等级 (2021) Category of National Key Protected Wild Animals (2021)	无 / NA
IUCN 红色名录 (2020-2) IUCN Red List (2020-2)	无危 / LC
CITES 附录 (2019) CITES Appendix (2019)	无 / NA
保护行动 Conservation Action(s)	无 / None

相关文献 Relevant References

Burgin *et al.*, 2018; Jiang *et al.* (蒋志刚等), 2017; Liu *et al.* (刘少英等), 2017; Wilson *et al.*, 2016

东北鼠兔

Ochotona hyperborea

无危 LC

数据缺乏 DD	无危 LC	近危 NT	易危 VU	濒危 EN	极危 CR	区域灭绝 RE	野外灭绝 EW	灭绝 EX

分类地位 Taxonomic Status

动物界 Animalia	脊索动物门 Chordata	哺乳纲 Mammalia	兔形目 Lagomorpha	鼠兔科 Ochotonidae

学　名 **Scientific Name**	*Ochotona hyperborea*
命名人 **Species Authority**	Pallas, 1811
英文名 **English Name(s)**	Northeast Pika
同物异名 **Synonym(s)**	无 / None
种下单元评估 **Infra-specific Taxa Assessed**	无 / None

评估信息 Assessment Information

评估年份 **Year Assessed**	2020
评定人 **Assessor(s)**	刘少英、蒋志刚 / Shaoying Liu, Zhigang Jiang
审定人 **Reviewer(s)**	苏建平、宗浩、廖继承 / Jianping Su, Hao Zong, Jicheng Liao
其他贡献人 **Other Contributor(s)**	李立立、丁晨晨 / Lili Li, Chenchen Ding

理由 **Justification:** 东北鼠兔分布较广，种群数量大。因此，列为无危等级 / Northeast Pika is a widespread species with large populations. Thus, it is listed as Least Concern

地理分布 Geographical Distribution

国内分布 Domestic Distribution

吉林、辽宁、黑龙江、内蒙古 / Jilin, Liaoning, Heilongjiang, Inner Mongolia (Nei Mongol)

世界分布 World Distribution

东亚 / East Asia

分布标注 Distribution Note

非特有种 / Non-endemic

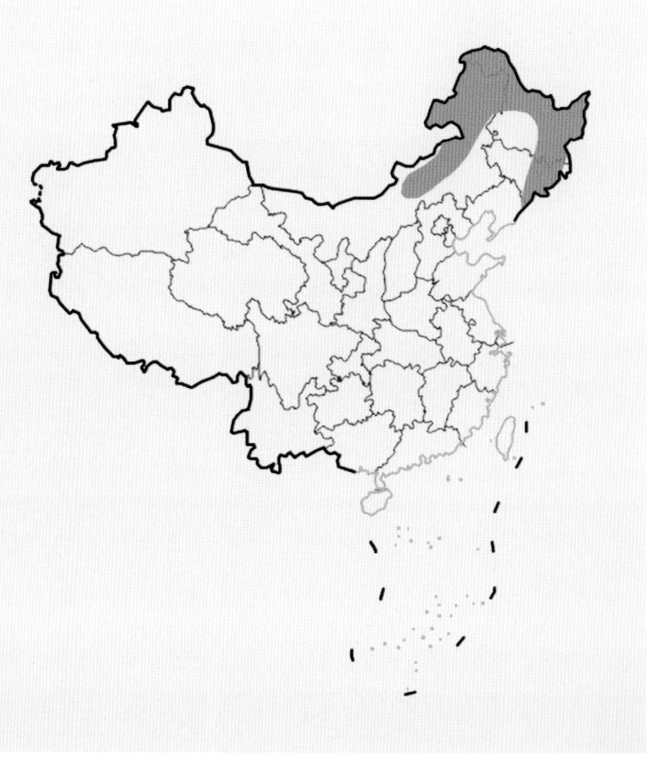

国内分布图
Map of Domestic Distribution

种群 Population

种群数量 **Population Size**	种群数量大 / Large populations
种群趋势 **Population Trend**	下降 / Decreasing

生境与生态系统 Habitat (s) and Ecosystem (s)

生　　境 **Habitat(s)**	森林、内陆岩石区域 / Forest, Inland Rocky Area
生态系统 **Ecosystem(s)**	森林生态系统 / Forest Ecosystem

威胁 Threat (s)

主要威胁 **Major Threat(s)**	无 / None

保护级别与保护行动 Protection Category and Conservation Action (s)

国家重点保护野生动物等级 **(2021) Category of National Key Protected Wild Animals (2021)**	无 / NA
IUCN 红色名录 **(2020-2) IUCN Red List (2020-2)**	无危 / LC
CITES 附录 **(2019) CITES Appendix (2019)**	无 / NA
保护行动 **Conservation Action(s)**	无 / None

相关文献 Relevant References

Burgin *et al.*, 2018; Jiang *et al.* (蒋志刚等), 2017; Wilson *et al.*, 2016; Zheng *et al.* (郑智民等), 2012; Smith *et al.* (史密斯等), 2009; Lissovsky *et al.*, 2007; Wang (王应祥), 2003

拉达克鼠兔
Ochotona ladacensis

无危 LC

数据缺乏 DD	无危 LC	近危 NT	易危 VU	濒危 EN	极危 CR	区域灭绝 RE	野外灭绝 EW	灭绝 EX

分类地位 Taxonomic Status

动物界 Animalia	脊索动物门 Chordata	哺乳纲 Mammalia	兔形目 Lagomorpha	鼠兔科 Ochotonidae
学　名 Scientific Name		*Ochotona ladacensis*		
命名人 Species Authority		Günther, 1875		
英文名 English Name(s)		Ladak Pika		
同物异名 Synonym(s)		无 / None		
种下单元评估 Infra-specific Taxa Assessed		无 / None		

评估信息 Assessment Information

评估年份 Year Assessed	2020
评　定　人 Assessor(s)	刘少英、蒋志刚 / Shaoying Liu, Zhigang Jiang
审　定　人 Reviewer(s)	苏建平、宗浩、廖继承 / Jianping Su, Hao Zong, Jicheng Liao
其他贡献人 Other Contributor(s)	李立立、丁晨晨 / Lili Li, Chenchen Ding

理由 Justification: 拉达克鼠兔分布广，种群数量大。因此，列为无危等级 / Ladak Pika is a widespread species with large populations. Thus, it is listed as Least Concern

地理分布 Geographical Distribution

国内分布 Domestic Distribution

新疆、青海、西藏 / Xinjiang, Qinghai, Tibet (Xizang)

世界分布 World Distribution

中国、印度、巴基斯坦 / China, India, Pakistan

分布标注 Distribution Note

非特有种 / Non-endemic

国内分布图
Map of Domestic Distribution

种群 Population

种群数量 Population Size	种群数量大 / Large populations
种群趋势 Population Trend	未知 / Unknown

生境与生态系统 Habitat (s) and Ecosystem (s)

生　　境 Habitat(s)	灌丛、草甸 / Shrubland, Meadow
生态系统 Ecosystem(s)	灌丛生态系统、草地生态系统 / Shrubland Ecosystem, Grassland Ecosystem

威胁 Threat (s)

主要威胁 Major Threat(s)	无 / None

保护级别与保护行动 Protection Category and Conservation Action (s)

国家重点保护野生动物等级 (2021) Category of National Key Protected Wild Animals (2021)	无 / NA
IUCN 红色名录 (2020-2) IUCN Red List (2020-2)	无危 / LC
CITES 附录 (2019) CITES Appendix (2019)	无 / NA
保护行动 Conservation Action(s)	无 / None

相关文献 Relevant References

Burgin *et al*., 2018; Jiang *et al.* (蒋志刚等), 2017; Wilson *et al*., 2016; Zheng *et al.* (郑智民等), 2012; Huang *et al.* (黄薇等), 2008, 2007; Zhou *et al.* (周立志等), 2002

大耳鼠兔

Ochotona macrotis

无危 LC

数据缺乏 DD	无危 LC	近危 NT	易危 VU	濒危 EN	极危 CR	区域灭绝 RE	野外灭绝 EW	灭绝 EX

分类地位 Taxonomic Status

动 物 界 Animalia	脊索动物门 Chordata	哺 乳 纲 Mammalia	兔 形 目 Lagomorpha	鼠 兔 科 Ochotonidae
学　名 Scientific Name		*Ochotona macrotis*		
命 名 人 Species Authority		Günther, 1875		
英 文 名 English Name(s)		Large-eared Pika		
同物异名 Synonym(s)		无 / None		
种下单元评估 Infra-specific Taxa Assessed		无 / None		

评估信息 Assessment Information

评 估 年 份 Year Assessed	2020
评　定　人 Assessor(s)	刘少英、蒋志刚 / Shaoying Liu, Zhigang Jiang
审　定　人 Reviewer(s)	苏建平、宗浩、廖继承 / Jianping Su, Hao Zong, Jicheng Liao
其他贡献人 Other Contributor(s)	李立立、丁晨晨 / Lili Li, Chenchen Ding

理由 Justification: 大耳鼠兔分布广，种群数量大。因此，列为无危等级 / Large-eared Pika is a widespread species with large populations. Thus, it is listed as Least Concern

地理分布 Geographical Distribution

国内分布 Domestic Distribution

西藏、新疆、甘肃、青海、四川、云南 / Tibet (Xizang), Xinjiang, Gansu, Qinghai, Sichuan, Yunnan

世界分布 World Distribution

中国；中亚、南亚 / China; Central Asia, South Asia

分布标注 Distribution Note

非特有种 / Non-endemic

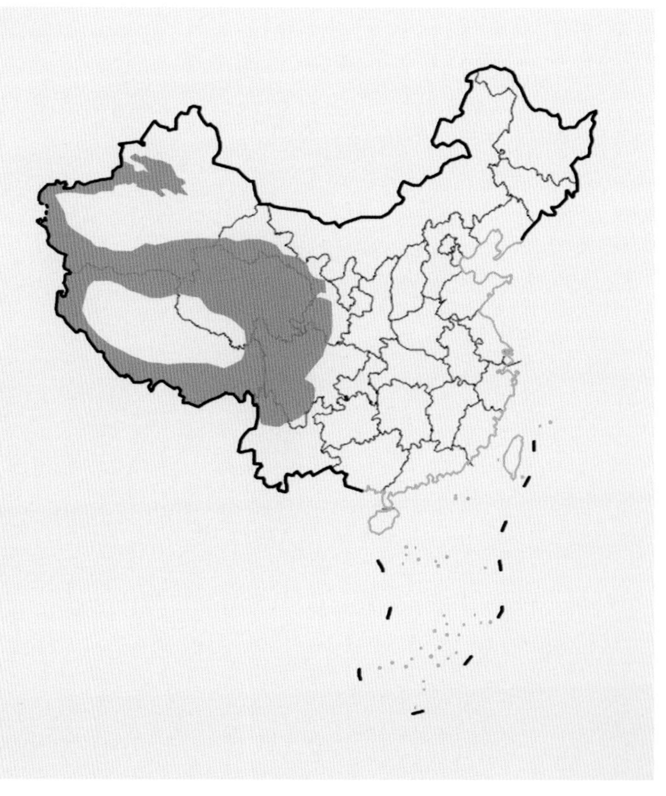

国内分布图
Map of Domestic Distribution

种群 Population

种群数量 Population Size	种群数量大 / Large populations
种群趋势 Population Trend	未知 / Unknown

生境与生态系统 Habitat (s) and Ecosystem (s)

生　　境 Habitat(s)	森林、内陆岩石区域 / Forest, Inland Rocky Area
生态系统 Ecosystem(s)	森林生态系统、草原生态系统 / Forest Ecosystem, Grassland Ecosystem

威胁 Threat (s)

主要威胁 Major Threat(s)	无 / None

保护级别与保护行动 Protection Category and Conservation Action (s)

国家重点保护野生动物等级 (2021) Category of National Key Protected Wild Animals (2021)	无 / NA
IUCN 红色名录 (2020-2) IUCN Red List (2020-2)	无危 / LC
CITES 附录 (2019) CITES Appendix (2019)	无 / NA
保护行动 Conservation Action(s)	无 / None

相关文献 Relevant References

Burgin *et al*., 2018; Hu *et al.* (胡一鸣等), 2014; Wilson *et al*., 2016; Zheng *et al.* (郑智民等), 2012; Huang *et al.* (黄薇等), 2008, 2007; Xing *et al.* (邢雅俊等), 2008; Zhou *et al.* (周立志等), 2002

满洲里鼠兔

Ochotona mantchurica

无危 LC

数据缺乏 DD	无危 LC	近危 NT	易危 VU	濒危 EN	极危 CR	区域灭绝 RE	野外灭绝 EW	灭绝 EX

分类地位 Taxonomic Status

动 物 界 Animalia	脊索动物门 Chordata	哺 乳 纲 Mammalia	兔 形 目 Lagomorpha	鼠 兔 科 Ochotonidae

学　名 Scientific Name	*Ochotona mantchurica*
命 名 人 Species Authority	Thomas, 1909
英 文 名 English Name(s)	Manchurian Pika
同物异名 Synonym(s)	无 / None
种下单元评估 Infra-specific Taxa Assessed	无 / None

评估信息 Assessment Information

评 估 年 份 Year Assessed	2020
评　定　人 Assessor(s)	刘少英、蒋志刚 / Shaoying Liu, Zhigang Jiang
审　定　人 Reviewer(s)	苏建平、冯祚建、宗浩、廖继承 / Jianping Su, Zuojian Feng, Hao Zong, Jicheng Liao
其他贡献人 Other Contributor(s)	李立立、丁晨晨 / Lili Li, Chenchen Ding

理由 Justification: 满洲里鼠兔分布广，种群数量大。因此，列为无危等级 / Manchurian Pika is a widespread species with large populations. Thus, it is listed as Least Concern

地理分布 Geographical Distribution

国内分布 Domestic Distribution

黑龙江、内蒙古 / Heilongjiang, Inner Mongolia (Nei Mongol)

世界分布 World Distribution

中国、俄罗斯 / China, Russia

分布标注 Distribution Note

非特有种 / Non-endemic

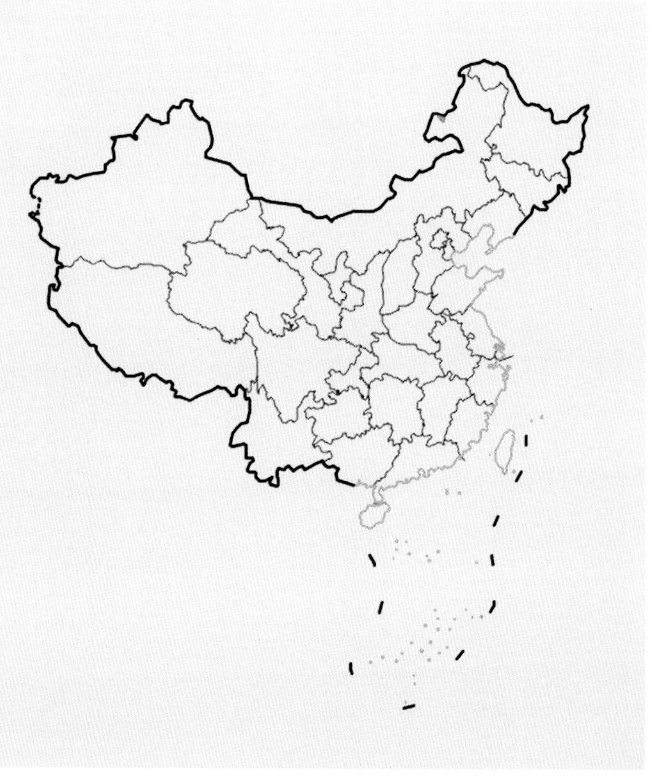

国内分布图
Map of Domestic Distribution

种群 Population

种群数量 Population Size	种群数量大 / Large populations
种群趋势 Population Trend	下降 / Decreasing

生境与生态系统 Habitat (s) and Ecosystem (s)

生　　境 Habitat(s)	草原岩石区域 / Rocky Areas of Grassland
生态系统 Ecosystem(s)	草原生态系统 / Grassland Ecosystem

威胁 Threat (s)

主要威胁 Major Threat(s)	无 / None

保护级别与保护行动 Protection Category and Conservation Action (s)

国家重点保护野生动物等级 (2021) Category of National Key Protected Wild Animals (2021)	无 / NA
IUCN 红色名录 (2020-2) IUCN Red List (2020-2)	无危 / LC
CITES 附录 (2019) CITES Appendix (2019)	无 / NA
保护行动 Conservation Action(s)	无 / None

相关文献 Relevant References

Burgin *et al*., 2018; Jiang *et al*. (蒋志刚等), 2017; Liu *et al*. (刘少英等), 2017; Wilson *et al*., 2016

中国生物多样性红色名录
China's Red List of Biodiversity

奴布拉鼠兔

Ochotona nubrica

无危 LC

数据缺乏 DD	无危 LC	近危 NT	易危 VU	濒危 EN	极危 CR	区域灭绝 RE	野外灭绝 EW	灭绝 EX

分类地位 Taxonomic Status

动物界 Animalia	脊索动物门 Chordata	哺乳纲 Mammalia	兔形目 Lagomorpha	鼠兔科 Ochotonidae

学名 Scientific Name	*Ochotona nubrica*
命名人 Species Authority	Thomas, 1922
英文名 English Name(s)	Nubra Pika
同物异名 Synonym(s)	无 / None
种下单元评估 Infra-specific Taxa Assessed	无 / None

评估信息 Assessment Information

评估年份 Year Assessed	2020
评定人 Assessor(s)	刘少英、蒋志刚 / Shaoying Liu, Zhigang Jiang
审定人 Reviewer(s)	苏建平、宗浩、廖继承 / Jianping Su, Hao Zong, Jicheng Liao
其他贡献人 Other Contributor(s)	李立立、丁晨晨 / Lili Li, Chenchen Ding

理由 **Justification:** 奴布拉鼠兔分布广，种群数量大。因此，列为无危等级 / Nubra Pika is a widespread species with large populations. Thus, it is listed as Least Concern

地理分布 Geographical Distribution

国内分布 Domestic Distribution	西藏 / Tibet (Xizang)
世界分布 World Distribution	中国、印度 / China, India
分布标注 Distribution Note	非特有种 / Non-endemic

国内分布图
Map of Domestic Distribution

种群 Population

种群数量 Population Size	种群数量大 / Large populations
种群趋势 Population Trend	未知 / Unknown

生境与生态系统 Habitat (s) and Ecosystem (s)

生　　境 Habitat(s)	荒漠、灌丛 / Desert, Shrubland
生态系统 Ecosystem(s)	灌丛生态系统、荒漠生态系统 / Shrubland Ecosystem, Desert Ecosystem

威胁 Threat (s)

主要威胁 Major Threat(s)	无 / None

保护级别与保护行动 Protection Category and Conservation Action (s)

国家重点保护野生动物等级 (2021) Category of National Key Protected Wild Animals (2021)	无 / NA
IUCN 红色名录 (2020-2) IUCN Red List (2020-2)	无危 / LC
CITES 附录 (2019) CITES Appendix (2019)	无 / NA
保护行动 Conservation Action(s)	无 / None

相关文献 Relevant References

Burgin *et al*., 2018; Jiang *et al*. (蒋志刚等), 2017; Liu *et al*. (刘少英等), 2017; Wilson *et al*., 2016

奴布拉鼠兔 *Ochotona nubrica*

蒙古鼠兔

Ochotona pallasi

无危 LC

数据缺乏 DD	无危 LC	近危 NT	易危 VU	濒危 EN	极危 CR	区域灭绝 RE	野外灭绝 EW	灭绝 EX

分类地位 Taxonomic Status

动物界 Animalia	脊索动物门 Chordata	哺乳纲 Mammalia	兔形目 Lagomorpha	鼠兔科 Ochotonidae
学名 Scientific Name		*Ochotona pallasi*		
命名人 Species Authority		Gray, 1867		
英文名 English Name(s)		Pallas's Pika		
同物异名 Synonym(s)		Mongolian Pika		
种下单元评估 Infra-specific Taxa Assessed		无 / None		

评估信息 Assessment Information

评估年份 Year Assessed	2020
评定人 Assessor(s)	刘少英、蒋志刚 / Shaoying Liu, Zhigang Jiang
审定人 Reviewer(s)	苏建平、宗浩、廖继承 / Jianping Su, Hao Zong, Jicheng Liao
其他贡献人 Other Contributor(s)	李立立、丁晨晨 / Lili Li, Chenchen Ding

理由 Justification: 蒙古鼠兔分布广，种群数量大。因此，列为无危等级 / Pallas's Pika is a widespread species with large populations. Thus, it is listed as Least Concern

地理分布 Geographical Distribution

国内分布 Domestic Distribution

新疆、内蒙古 / Xinjiang, Inner Mongolia (Nei Mongol)

世界分布 World Distribution

中亚、东亚 / Central Asia, East Asia

分布标注 Distribution Note

非特有种 / Non-endemic

国内分布图
Map of Domestic Distribution

种群 Population

种群数量 Population Size	种群数量大 / Large populations
种群趋势 Population Trend	下降 / Decreasing

生境与生态系统 Habitat (s) and Ecosystem (s)

生　　境 Habitat(s)	内陆岩石区域、荒漠、草地 / Inland Rocky Area, Desert, Grassland
生态系统 Ecosystem(s)	草地生态系统、荒漠生态系统 / Grassland Ecosystem, Desert Ecosystem

威胁 Threat (s)

主要威胁 Major Threat(s)	无 / None

保护级别与保护行动 Protection Category and Conservation Action (s)

国家重点保护野生动物等级 (2021) Category of National Key Protected Wild Animals (2021)	无 / NA
IUCN 红色名录 (2020-2) IUCN Red List (2020-2)	无危 / LC
CITES 附录 (2019) CITES Appendix (2019)	无 / NA
保护行动 Conservation Action(s)	无 / None

相关文献 Relevant References

Burgin *et al*., 2018; Jiang *et al*. (蒋志刚等), 2017; Liu *et al*. (刘少英等), 2017; Wilson *et al*., 2016

蒙古鼠兔 *Ochotona pallasi*　　方红霞 绘　Drawn By Hongxia Fang

邛崃鼠兔

Ochotona qionglaiensis

无危 LC

数据缺乏 DD	无危 LC	近危 NT	易危 VU	濒危 EN	极危 CR	区域灭绝 RE	野外灭绝 EW	灭绝 EX

分类地位 Taxonomic Status

动物界 Animalia	脊索动物门 Chordata	哺乳纲 Mammalia	兔形目 Lagomorpha	鼠兔科 Ochotonidae

学　名 **Scientific Name**	*Ochotona qionglaiensis*
命名人 **Species Authority**	Liu *et al.*, 2017
英文名 **English Name(s)**	Qionglai Pika
同物异名 **Synonym(s)**	无 / None
种下单元评估 **Infra-specific Taxa Assessed**	无 / None

评估信息 Assessment Information

评估年份 **Year Assessed**	2020
评定人 **Assessor(s)**	刘少英、蒋志刚 / Shaoying Liu, Zhigang Jiang
审定人 **Reviewer(s)**	苏建平、宗浩、廖继承 / Jianping Su, Hao Zong, Jicheng Liao
其他贡献人 **Other Contributor(s)**	李立立、丁晨晨 / Lili Li, Chenchen Ding

理由 **Justification:** 邛崃鼠兔分布较广，种群数量大。因此，列为无危等级 / Qionglai Pika is a widespread species with large populations. Thus, it is listed as Least Concern

地理分布 Geographical Distribution

国内分布 Domestic Distribution

四川 / Sichuan

世界分布 World Distribution

中国 / China

分布标注 Distribution Note

特有种 / Endemic

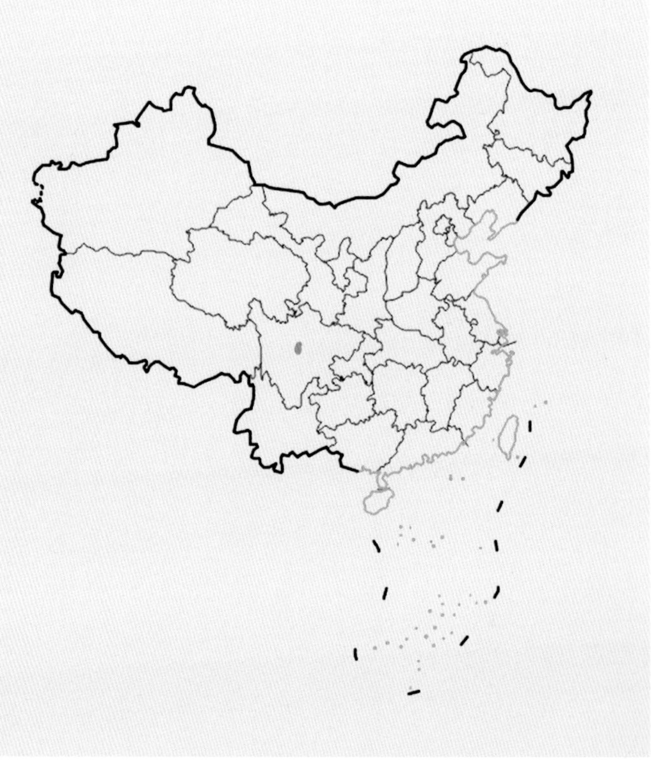

国内分布图
Map of Domestic Distribution

种群 Population

种群数量 Population Size	种群数量大 / Large populations
种群趋势 Population Trend	未知 / Unknown

生境与生态系统 Habitat (s) and Ecosystem (s)

生　　境 Habitat(s)	草原岩石区域 / Rocky Area of Grassland
生态系统 Ecosystem(s)	草原生态系统 / Grassland Ecosystem

威胁 Threat (s)

主要威胁 Major Threat(s)	无 / None

保护级别与保护行动 Protection Category and Conservation Action (s)

国家重点保护野生动物等级 (2021) Category of National Key Protected Wild Animals (2021)	无 / NA
IUCN 红色名录 (2020-2) IUCN Red List (2020-2)	无危 / LC
CITES 附录 (2019) CITES Appendix (2019)	无 / NA
保护行动 Conservation Action(s)	无 / None

相关文献 Relevant References

Burgin *et al*., 2018; Jiang *et al*. (蒋志刚等), 2017; Liu *et al*. (刘少英等), 2017

邛崃鼠兔 *Ochotona qionglaiensis*　　方红霞 绘　Drawn By Hongxia Fang

锡金鼠兔

Ochotona sikimaria

无危 LC

数据缺乏 DD	无危 LC	近危 NT	易危 VU	濒危 EN	极危 CR	区域灭绝 RE	野外灭绝 EW	灭绝 EX

分类地位 Taxonomic Status

动物界 Animalia	脊索动物门 Chordata	哺乳纲 Mammalia	兔形目 Lagomorpha	鼠兔科 Ochotonidae
学名 **Scientific Name**		*Ochotona sikimaria*		
命名人 **Species Authority**		Liu *et al.*, 2017		
英文名 **English Name(s)**		Sikim Pika		
同物异名 **Synonym(s)**		无 / None		
种下单元评估 **Infra-specific Taxa Assessed**		无 / None		

评估信息 Assessment Information

评估年份 **Year Assessed**	2020
评定人 **Assessor(s)**	刘少英、蒋志刚 / Shaoying Liu, Zhigang Jiang
审定人 **Reviewer(s)**	苏建平、宗浩、廖继承 / Jianping Su, Hao Zong, Jicheng Liao
其他贡献人 **Other Contributor(s)**	李立立、丁晨晨 / Lili Li, Chenchen Ding

理由 **Justification:** 锡金鼠兔在其分布区有相当种群数量。因此，列为无危等级 / There is a considerable population of Sikim Pika in its area of occurences. Thus, it is listed as Least Concern

地理分布 Geographical Distribution

国内分布 Domestic Distribution

西藏 / Tibet (Xizang)

世界分布 World Distribution

中国、印度 / China, India

分布标注 Distribution Note

非特有种 / Non-endemic

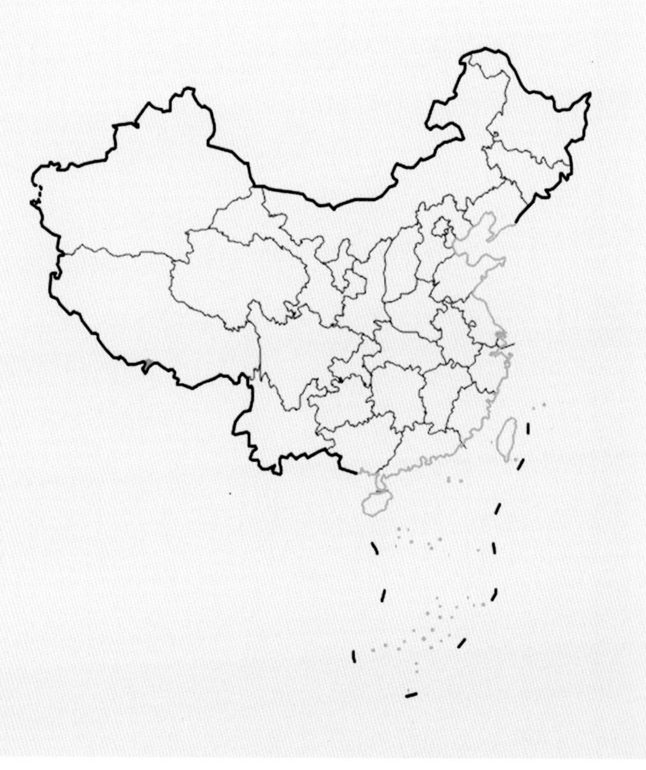

国内分布图
Map of Domestic Distribution

种群 Population

种群数量 **Population Size**	未知 / Unknown
种群趋势 **Population Trend**	未知 / Unknown

生境与生态系统 Habitat (s) and Ecosystem (s)

生　　境 **Habitat(s)**	草原岩石区域 / Rocky Area of Grassland
生态系统 **Ecosystem(s)**	草原生态系统 / Grassland Ecosystem

威胁 Threat (s)

主要威胁 **Major Threat(s)**	无 / None

保护级别与保护行动 Protection Category and Conservation Action (s)

国家重点保护野生动物等级 **(2021) Category of National Key Protected Wild Animals (2021)**	无 / NA
IUCN 红色名录 **(2020-2) IUCN Red List (2020-2)**	未列入 / NA
CITES 附录 **(2019) CITES Appendix (2019)**	无 / NA
保护行动 **Conservation Action(s)**	无 / None

相关文献 Relevant References

Burgin *et al.*, 2018; Jiang *et al.* (蒋志刚等), 2017; Liu *et al.* (刘少英等), 2017

锡金鼠兔 *Ochotona sikimaria*　　　方红霞 绘　Drawn By Hongxia Fang

藏鼠兔

Ochotona thibetana

无危 LC

数据缺乏 DD	无危 LC	近危 NT	易危 VU	濒危 EN	极危 CR	区域灭绝 RE	野外灭绝 EW	灭绝 EX

分类地位 Taxonomic Status

动物界 Animalia	脊索动物门 Chordata	哺乳纲 Mammalia	兔形目 Lagomorpha	鼠兔科 Ochotonidae

学　名 Scientific Name	*Ochotona thibetana*
命名人 Species Authority	Milne-Edwards, 1871
英文名 English Name(s)	Moupin Pika
同物异名 Synonym(s)	无 / None
种下单元评估 Infra-specific Taxa Assessed	无 / None

评估信息 Assessment Information

评估年份 Year Assessed	2020
评　定　人 Assessor(s)	刘少英、蒋志刚 / Shaoying Liu, Zhigang Jiang
审　定　人 Reviewer(s)	苏建平、冯祚建、宗浩、廖继承 / Jianping Su, Zuojian Feng, Hao Zong, Jicheng Liao
其他贡献人 Other Contributor(s)	李立立、丁晨晨 / Lili Li, Chenchen Ding

理由 Justification: 藏鼠兔分布广，种群数量大。因此，列为无危等级 / Moupin Pika is a widespread species with large populations. Thus, it is listed as Least Concern

地理分布 Geographical Distribution

国内分布 Domestic Distribution

甘肃、四川、云南、西藏、青海 / Gansu, Sichuan, Yunnan, Tibet (Xizang), Qinghai

世界分布 World Distribution

中国；南亚 / China; South Asia

分布标注 Distribution Note

非特有种 / Non-endemic

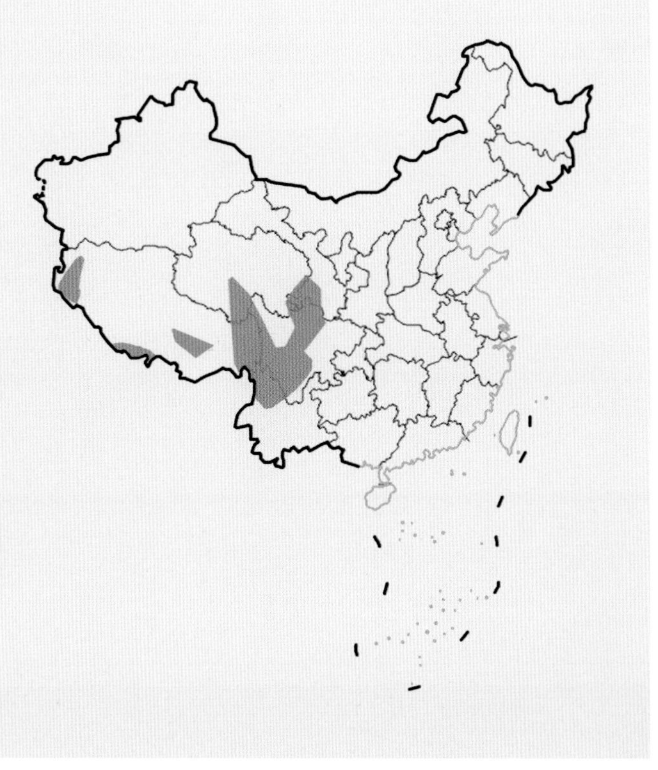

国内分布图
Map of Domestic Distribution

种群 Population

种群数量 Population Size	种群数量大 / Large populations
种群趋势 Population Trend	未知 / Unknown

生境与生态系统 Habitat (s) and Ecosystem (s)

生　　境 Habitat(s)	草原岩石区域 / Rocky Area of Grassland
生态系统 Ecosystem(s)	草原生态系统 / Grassland Ecosystem

威胁 Threat (s)

主要威胁 Major Threat(s)	无 / None

保护级别与保护行动 Protection Category and Conservation Action (s)

国家重点保护野生动物等级 (2021) Category of National Key Protected Wild Animals (2021)	无 / NA
IUCN 红色名录 (2020-2) IUCN Red List (2020-2)	无危 / LC
CITES 附录 (2019) CITES Appendix (2019)	无 / NA
保护行动 Conservation Action(s)	无 / None

相关文献 Relevant References

Burgin *et al*., 2018; Jiang *et al.* (蒋志刚等), 2017; Zheng *et al.* (郑智民等), 2012; Huang *et al.* (黄薇等), 2007; Li *et al.*, 2003; Giraudoux *et al.*, 1998; Yu *et al.*, 1997

藏鼠兔 *Ochotona thibetana*

循化鼠兔
Ochotona xunhuaensis

无危 LC

数据缺乏 DD	无危 LC	近危 NT	易危 VU	濒危 EN	极危 CR	区域灭绝 RE	野外灭绝 EW	灭绝 EX

分类地位 Taxonomic Status

动物界 Animalia	脊索动物门 Chordata	哺乳纲 Mammalia	兔形目 Lagomorpha	鼠兔科 Ochotonidae

学　名 Scientific Name	*Ochotona xunhuaensis*
命名人 Species Authority	Liu *et al*., 2017
英文名 English Name(s)	Xunhua Pika
同物异名 Synonym(s)	无 / None
种下单元评估 Infra-specific Taxa Assessed	无 / None

评估信息 Assessment Information

评估年份 Year Assessed	2020
评定人 Assessor(s)	刘少英、蒋志刚 / Shaoying Liu, Zhigang Jiang
审定人 Reviewer(s)	苏建平、冯祚建、宗浩、廖继承 / Jianping Su, Zuojian Feng, Hao Zong, Jicheng Liao
其他贡献人 Other Contributor(s)	李立立、丁晨晨 / Lili Li, Chenchen Ding

理由 Justification: 循化鼠兔仅分布在青海循化县，但其种群数量大，在其分布区常见。因此，列为无危等级 / Xunhua Pika is restricted in Xunhua County but has large populations, and is common in its range. Thus, it is listed as Least Concern

地理分布 Geographical Distribution

国内分布 Domestic Distribution

青海 / Qinghai

世界分布 World Distribution

中国 / China

分布标注 Distribution Note

特有种 / Endemic

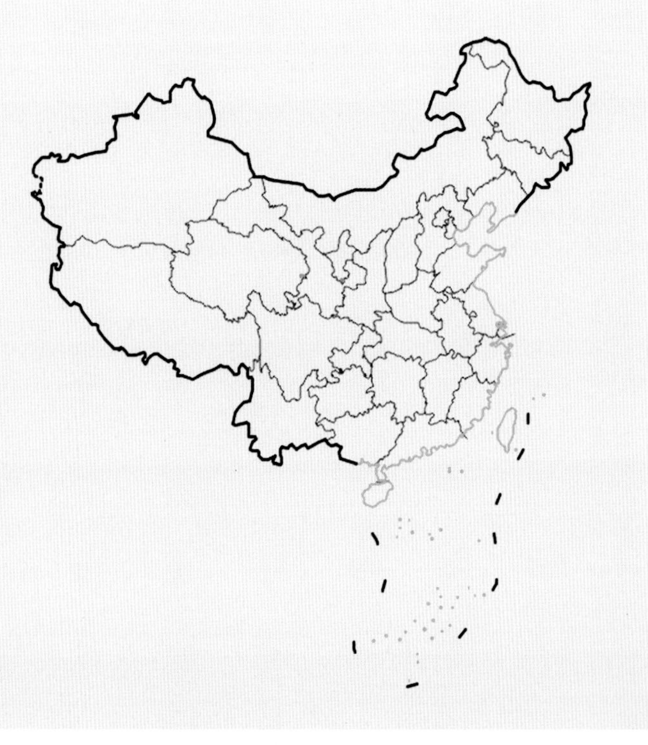

国内分布图
Map of Domestic Distribution

种群 Population

种群数量 Population Size	种群数量大 / Large populations
种群趋势 Population Trend	未知 / Unknown

生境与生态系统 Habitat (s) and Ecosystem (s)

生　　境 Habitat(s)	灌丛 / Shrubland
生态系统 Ecosystem(s)	灌丛生态系统 / Shrubland Ecosystem

威胁 Threat (s)

主要威胁 Major Threat(s)	无 / None

保护级别与保护行动 Protection Category and Conservation Action (s)

国家重点保护野生动物等级 (2021) Category of National Key Protected Wild Animals (2021)	无 / NA
IUCN 红色名录 (2020-2) IUCN Red List (2020-2)	无危 / LC
CITES 附录 (2019) CITES Appendix (2019)	无 / NA
保护行动 Conservation Action(s)	无 / None

相关文献 Relevant References

Burgin *et al.*, 2018; Jiang *et al.* (蒋志刚等), 2017; Liu *et al.* (刘少英等), 2017

循化鼠兔 *Ochotona xunhuaensis*　　方红霞 绘　Drawn By Hongxia Fang

数据缺乏 DD	无危 LC	近危 NT	易危 VU	濒危 EN	极危 CR	区域灭绝 RE	野外灭绝 EW	灭绝 EX

分类地位 Taxonomic Status

动物界 **Animalia**	脊索动物门 **Chordata**	哺乳纲 **Mammalia**	兔形目 **Lagomorpha**	鼠兔科 **Ochotonidae**
学 名 **Scientific Name**		*Ochotona yarlungensis*		
命名人 **Species Authority**		Liu *et al.*, 2017		
英文名 **English Name(s)**		Yarlung Pika		
同物异名 **Synonym(s)**		无 / None		
种下单元评估 **Infra-specific Taxa Assessed**		无 / None		

评估信息 Assessment Information

评估年份 **Year Assessed**	2020
评定人 **Assessor(s)**	刘少英、蒋志刚 / Shaoying Liu, Zhigang Jiang
审定人 **Reviewer(s)**	苏建平、冯祚建、宗浩、廖继承 / Jianping Su, Zuojian Feng, Hao Zong, Jicheng Liao
其他贡献人 **Other Contributor(s)**	李立立、丁晨晨 / Lili Li, Chenchen Ding

理由 **Justification:** 雅鲁藏布鼠兔种群数量大，在其分布区常见。因此，列为无危等级 / Yarlung Pika has large populations, and is common in its range. Thus, it is listed as Least Concern

地理分布 Geographical Distribution

国内分布 Domestic Distribution

西藏 / Tibet (Xizang)

世界分布 World Distribution

中国 / China

分布标注 Distribution Note

特有种 / Endemic

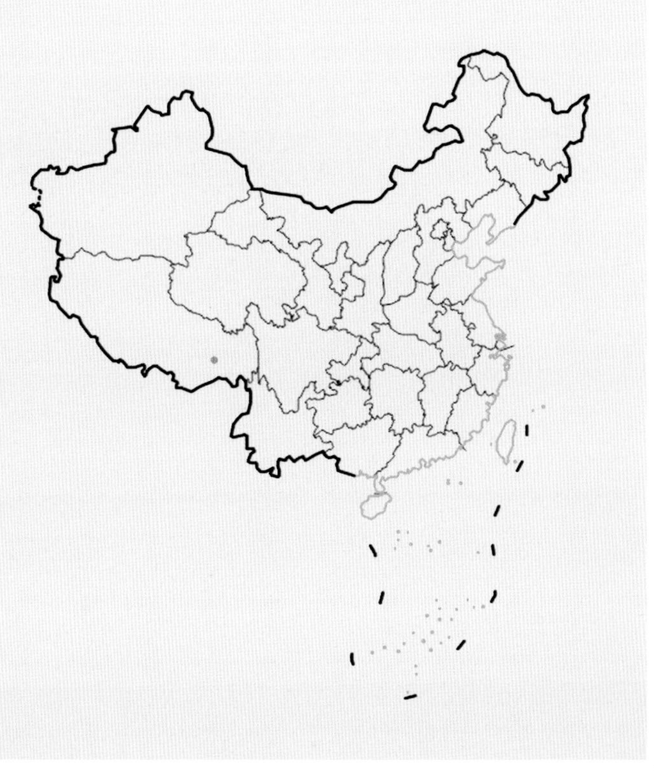

国内分布图
Map of Domestic Distribution

种群 Population

种群数量 Population Size	种群数量大 / Large populations
种群趋势 Population Trend	未知 / Unknown

生境与生态系统 Habitat (s) and Ecosystem (s)

生　　境 Habitat(s)	灌丛 / Shrubland
生态系统 Ecosystem(s)	灌丛生态系统 / Shrubland Ecosystem

威胁 Threat (s)

主要威胁 Major Threat(s)	无 / None

保护级别与保护行动 Protection Category and Conservation Action (s)

国家重点保护野生动物等级 (2021) Category of National Key Protected Wild Animals (2021)	无 / NA
IUCN 红色名录 (2020-2) IUCN Red List (2020-2)	无危 / LC
CITES 附录 (2019) CITES Appendix (2019)	无 / NA
保护行动 Conservation Action(s)	无 / None

相关文献 Relevant References

Burgin *et al*., 2018; Jiang *et al.* (蒋志刚等), 2017; Liu *et al*. (刘少英等), 2017

雅鲁藏布鼠兔 *Ochotona yarlungensis*　　方红霞 绘　Drawn By Hongxia Fang

数据缺乏 DD	无危 LC	近危 NT	易危 VU	濒危 EN	极危 CR	区域灭绝 RE	野外灭绝 EW	灭绝 EX

分类地位 Taxonomic Status

动 物 界 **Animalia**	脊索动物门 **Chordata**	哺 乳 纲 **Mammalia**	兔 形 目 **Lagomorpha**	兔 科 **Leporidae**
学 名 **Scientific Name**		*Lepus coreanu*		
命 名 人 **Species Authority**		Thomas, 1892		
英 文 名 **English Name(s)**		Korean Hare		
同物异名 **Synonym(s)**		无 / None		
种下单元评估 **Infra-specific Taxa Assessed**		无 / None		

评估信息 Assessment Information

评 估 年 份 **Year Assessed**	2020
评 定 人 **Assessor(s)**	蒋志刚 / Zhigang Jiang
审 定 人 **Reviewer(s)**	夏霖、杨奇森 / Lin Xia, Qisen Yang
其他贡献人 **Other Contributor(s)**	李立立、丁晨晨 / Lili Li, Chenchen Ding

理由 **Justification:** 高丽兔分布区大，在分布区常见。因此，列为无危等级 / Korean Hare has a large distribution range, and is common in its range. Thus, it is listed as Least Concern

地理分布 Geographical Distribution

国内分布 Domestic Distribution

吉林 / Jilin

世界分布 World Distribution

中国、朝鲜、韩国 / China, Korea (the Democratic People's Republic of), Korea (the Republic of)

分布标注 Distribution Note

非特有种 / Non-endemic

国内分布图
Map of Domestic Distribution

种群 Population

种群数量 **Population Size**	尚有一定种群数量 / It still has a certain population size
种群趋势 **Population Trend**	未知 / Unknown

生境与生态系统 Habitat (s) and Ecosystem (s)

生　　境 **Habitat(s)**	草甸、灌丛、森林 / Meadow, Shrubland, Forest
生态系统 **Ecosystem(s)**	森林生态系统、灌丛生态系统、草地生态系统 / Forest Ecosystem, Shrubland Ecosystem, Grassland Ecosystem

威胁 Threat (s)

主要威胁 **Major Threat(s)**	无 / None

保护级别与保护行动 Protection Category and Conservation Action (s)

国家重点保护野生动物等级 **(2021) Category of National Key Protected Wild Animals (2021)**	无 / NA
IUCN 红色名录 **(2020-2) IUCN Red List (2020-2)**	无危 / LC
CITES 附录 **(2019) CITES Appendix (2019)**	无 / NA
保护行动 **Conservation Action(s)**	无 / None

相关文献 Relevant References

Burgin *et al*., 2018; Jiang *et al.* (蒋志刚等), 2017; Wilson *et al*., 2016; Yu *et al*., 2013; Smith *et al.* (史密斯等), 2009; Pan *et al.* (潘清华等), 2007

东北兔
Lepus mandshuricus

无危 LC

数据缺乏 DD	无危 LC	近危 NT	易危 VU	濒危 EN	极危 CR	区域灭绝 RE	野外灭绝 EW	灭绝 EX

分类地位 Taxonomic Status

动物界 Animalia	脊索动物门 Chordata	哺乳纲 Mammalia	兔形目 Lagomorpha	兔科 Leporidae

学　名 Scientific Name	*Lepus mandshuricus*
命名人 Species Authority	Radde, 1861
英文名 English Name(s)	Manchurian Hare
同物异名 Synonym(s)	*Lepus melainus* (Li and Luo, 1979)
种下单元评估 Infra-specific Taxa Assessed	无 / None

评估信息 Assessment Information

评估年份 Year Assessed	2020
评定人 Assessor(s)	蒋志刚 / Zhigang Jiang
审定人 Reviewer(s)	夏霖、杨奇森 / Xia Lin, Qisen Yang
其他贡献人 Other Contributor(s)	李立立、丁晨晨 / Lili Li, Chenchen Ding

理由 Justification: 东北兔分布较广，种群数量大。因此，列为无危等级 / Manchurian Hare is a widespread species with large populations. Thus, it is listed as Least Concern

地理分布 Geographical Distribution

国内分布 Domestic Distribution

内蒙古、辽宁、吉林、黑龙江 / Inner Mongolia (Nei Mongol), Liaoning, Jilin, Heilongjiang

世界分布 World Distribution

中国、俄罗斯 / China, Russia

分布标注 Distribution Note

非特有种 / Non-endemic

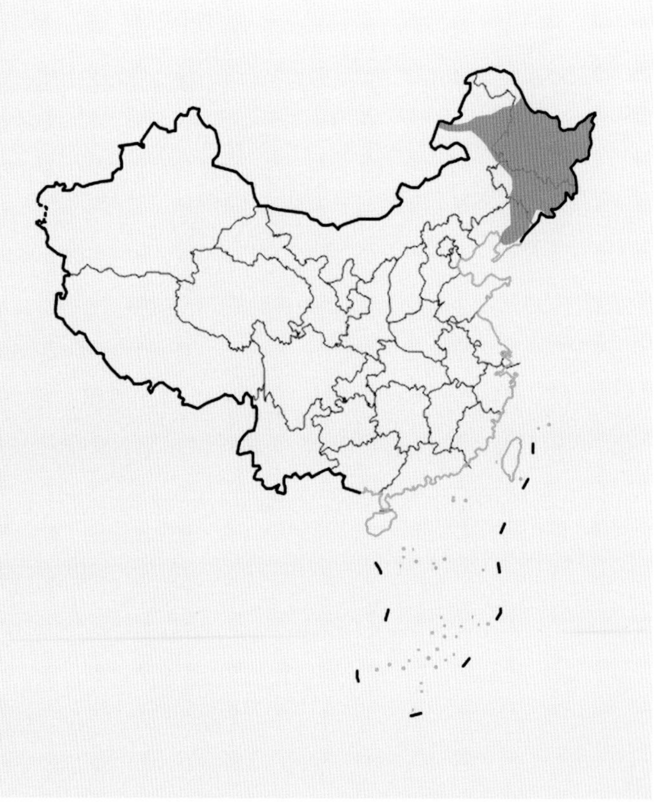

国内分布图
Map of Domestic Distribution

种群 Population

种群数量 Population Size	种群数量大 / Large populations
种群趋势 Population Trend	未知 / Unknown

生境与生态系统 Habitat (s) and Ecosystem (s)

生　　境 Habitat(s)	泰加林、森林 / Taiga Forest, Forest
生态系统 Ecosystem(s)	森林生态系统 / Forest Ecosystem

威胁 Threat (s)

主要威胁 Major Threat(s)	无 / None

保护级别与保护行动 Protection Category and Conservation Action (s)

国家重点保护野生动物等级 (2021) Category of National Key Protected Wild Animals (2021)	无 / NA
IUCN 红色名录 (2020-2) IUCN Red List (2020-2)	无危 / LC
CITES 附录 (2019) CITES Appendix (2019)	无 / NA
保护行动 Conservation Action(s)	无 / None

相关文献 Relevant References

Burgin *et al*., 2018; Jiang *et al.* (蒋志刚等), 2017; Wilson *et al*., 2016; Zheng *et al.* (郑智民等), 2012; Ren and Huang (任梦非和黄海娇), 2009; Smith *et al.* (史密斯等), 2009; Wang (王应祥), 2003

东北兔 *Lepus mandshuricus*　　　冯利民 摄　By Limin Feng

数据缺乏 DD	无危 LC	近危 NT	易危 VU	濒危 EN	极危 CR	区域灭绝 RE	野外灭绝 EW	灭绝 EX

分类地位 Taxonomic Status

动 物 界 Animalia	脊索动物门 Chordata	哺 乳 纲 Mammalia	兔 形 目 Lagomorpha	兔 科 Leporidae
学 名 Scientific Name		*Lepus nigricollis*		
命 名 人 Species Authority		F. Cuvier, 1823		
英 文 名 English Name(s)		Indian Hare		
同物异名 Synonym(s)		Black-naped Hare		
种下单元评估 Infra-specific Taxa Assessed		无 / None		

评估信息 Assessment Information

评 估 年 份 Year Assessed	2020
评 定 人 Assessor(s)	蒋志刚 / Zhigang Jiang
审 定 人 Reviewer(s)	胡慧建、胡一鸣 / Huijian Hu, Yiming Hu
其他贡献人 Other Contributor(s)	李立立、丁晨晨 / Lili Li, Chenchen Ding

理由 Justification: 尼泊尔黑兔为广布常见种。因此，列为无危等级 / Indian Hare is a widespread species and it is common in the field. Thus, it is listed as Least Concern

地理分布 Geographical Distribution

国内分布 Domestic Distribution

西藏 / Tibet (Xizang)

世界分布 World Distribution

中国；南亚 / China; South Asia

分布标注 Distribution Note

非特有种 / Non-endemic

国内分布图
Map of Domestic Distribution

种群 Population

种群数量 Population Size	未知 / Unknown
种群趋势 Population Trend	未知 / Unknown

生境与生态系统 Habitat (s) and Ecosystem (s)

生　　境 Habitat(s)	除红树林和高草原栖息地外，在矮草原、裸地、农田和森林道路等生境也有分布 / In addition to Mangrove and Tall Grassland, it also can be seen in Low Grassland, Barren Agricultural Field, Cropland and along Forest Road
生态系统 Ecosystem(s)	草原、农田和森林生态系统 / Grassland Ecosystem, Farmland Ecosystem, Forest Ecosystem

威胁 Threat (s)

主要威胁 Major Threat(s)	为了获取肉食或防止农作物受损而猎杀的比例高达 40% / Up to 40% of the species is Hunted for meat or Removed to prevent crop damage

保护级别与保护行动 Protection Category and Conservation Action (s)

国家重点保护野生动物等级 (2021) Category of National Key Protected Wild Animals (2021)	无 / NA
IUCN 红色名录 (2020-2) IUCN Red List (2020-2)	无危 / LC
CITES 附录 (2019) CITES Appendix (2019)	无 / NA
保护行动 Conservation Action(s)	无 / None

相关文献 Relevant References

Burgin *et al*., 2018; Jiang *et al.* (蒋志刚等), 2017; Choudhury, 2003

尼泊尔黑兔 *Lepus nigricollis*

数据缺乏 DD	无危 LC	近危 NT	易危 VU	濒危 EN	极危 CR	区域灭绝 RE	野外灭绝 EW	灭绝 EX

分类地位 Taxonomic Status

动物界 Animalia	脊索动物门 Chordata	哺乳纲 Mammalia	兔形目 Lagomorpha	兔科 Leporidae
学名 Scientific Name		*Lepus oiostolus*		
命名人 Species Authority		Hodgson, 1840		
英文名 English Name(s)		Woolly Hare		
同物异名 Synonym(s)		无 / None		
种下单元评估 Infra-specific Taxa Assessed		无 / None		

评估信息 Assessment Information

评估年份 Year Assessed	2020
评定人 Assessor(s)	蒋志刚 / Zhigang Jiang
审定人 Reviewer(s)	夏霖、杨奇森 / Lin Xia, Qisen Yang
其他贡献人 Other Contributor(s)	李立立、丁晨晨 / Lili Li, Chenchen Ding

理由 Justification: 灰尾兔分布较广，种群数量大。因此，列为无危等级 / Woolly Hare is a widespread species with large populations. Thus, it is listed as Least Concern

地理分布 Geographical Distribution

国内分布 Domestic Distribution

新疆、青海、甘肃、四川、云南、西藏 / Xinjiang, Qinghai, Gansu, Sichuan, Yunnan, Tibet (Xizang)

世界分布 World Distribution

中国、印度、尼泊尔 / China, India, Nepal

分布标注 Distribution Note

非特有种 / Non-endemic

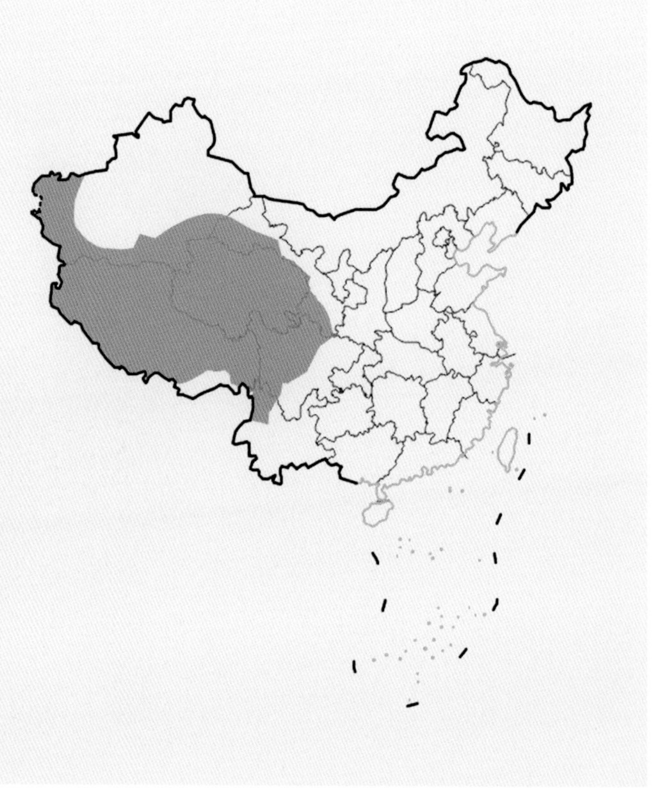

国内分布图
Map of Domestic Distribution

种群 Population

种群数量 Population Size	种群数量大 / Large populations
种群趋势 Population Trend	未知 / Unknown

生境与生态系统 Habitat (s) and Ecosystem (s)

生　　境 Habitat(s)	高海拔草地、草甸、灌丛、荒漠 / High Altitude Grassland, Meadow, Shrubland, Desert
生态系统 Ecosystem(s)	高寒灌丛生态系统、高寒草地生态系统、荒漠生态系统 / Alpine Shrubland Ecosystem, Alpine Grassland Ecosystem, Desert Ecosystem

威胁 Threat (s)

主要威胁 Major Threat(s)	无 / None

保护级别与保护行动 Protection Category and Conservation Action (s)

国家重点保护野生动物等级 (2021) Category of National Key Protected Wild Animals (2021)	无 / NA
IUCN 红色名录 (2020-2) IUCN Red List (2020-2)	无危 / LC
CITES 附录 (2019) CITES Appendix (2019)	无 / NA
保护行动 Conservation Action(s)	无 / None

相关文献 Relevant References

Burgin *et al*., 2018; Jiang *et al.* (蒋志刚等), 2017; Wilson *et al*., 2016; Zheng *et al.* (郑智民等), 2012; Lu, 2011; Smith *et al.* (史密斯等), 2009; Gao and Feng (高耀亭和冯祚建), 1964

灰尾兔 *Lepus oiostolus*

数据缺乏 DD	无危 LC	近危 NT	易危 VU	濒危 EN	极危 CR	区域灭绝 RE	野外灭绝 EW	灭绝 EX

分类地位 Taxonomic Status

动 物 界 Animalia	脊索动物门 Chordata	哺 乳 纲 Mammalia	兔 形 目 Lagomorpha	兔 科 Leporidae
学 名 Scientific Name	*Lepus sinensis*			
命 名 人 Species Authority	Gray, 1832			
英 文 名 English Name(s)	Chinese Hare			
同物异名 Synonym(s)	无 / None			
种下单元评估 Infra-specific Taxa Assessed	无 / None			

评估信息 Assessment Information

评 估 年 份 Year Assessed	2020
评 定 人 Assessor(s)	蒋志刚 / Zhigang Jiang
审 定 人 Reviewer(s)	夏霖、杨奇森 / Lin Xia, Qisen Yang
其他贡献人 Other Contributor(s)	李立立、丁晨晨 / Lili Li, Chenchen Ding

理由 Justification: 华南兔分布较广，种群数量大。因此，列为无危等级 / Chinese Hare is a widespread species with large populations. Thus, it is listed as Least Concern

地理分布 Geographical Distribution

国内分布 Domestic Distribution

广西、湖南、上海、江苏、浙江、安徽、江西、广东、贵州、台湾、福建 / Guangxi, Hunan, Shanghai, Jiangsu, Zhejiang, Anhui, Jiangxi, Guangdong, Guizhou, Taiwan, Fujian

世界分布 World Distribution

中国、越南 / China, Viet Nam

分布标注 Distribution Note

非特有种 / Non-endemic

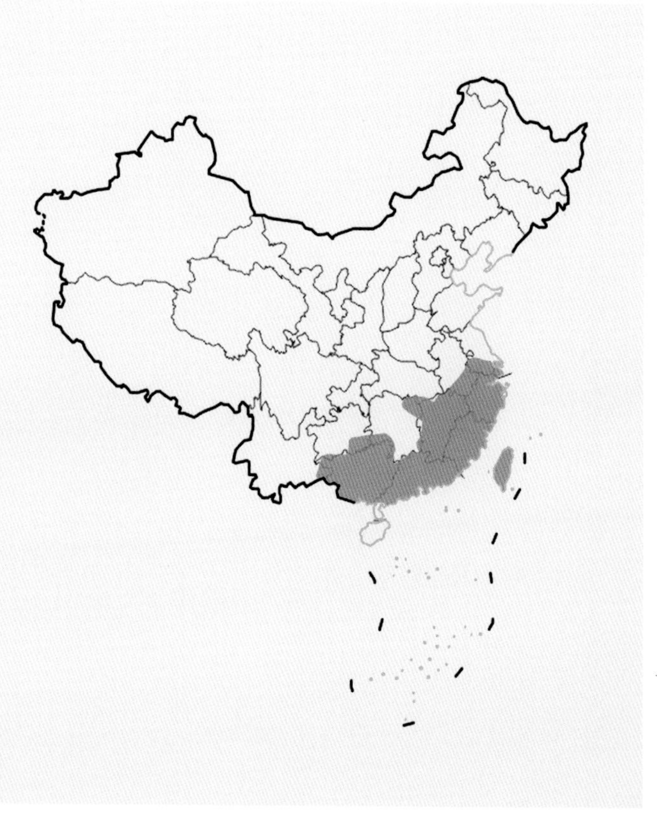

国内分布图
Map of Domestic Distribution

种群 Population

种群数量 **Population Size**	种群数量大 / Large populations
种群趋势 **Population Trend**	未知 / Unknown

生境与生态系统 Habitat (s) and Ecosystem (s)

生　　境 **Habitat(s)**	草地、林地 / Grassland, Woodland
生态系统 **Ecosystem(s)**	森林生态系统、草地生态系统 / Forest Ecosystem, Grassland Ecosystem

威胁 Threat (s)

主要威胁 **Major Threat(s)**	无 / None

保护级别与保护行动 Protection Category and Conservation Action (s)

国家重点保护野生动物等级 **(2021) Category of National Key Protected Wild Animals (2021)**	无 / NA
IUCN 红色名录 **(2020-2) IUCN Red List (2020-2)**	无危 / LC
CITES 附录 **(2019) CITES Appendix (2019)**	无 / NA
保护行动 **Conservation Action(s)**	无 / None

相关文献 Relevant References

Burgin *et al.*, 2018; Jiang *et al.* (蒋志刚等), 2017; Wilson *et al.*, 2016; Ding *et al.*, 2014; Zheng *et al.* (郑智民等), 2012; Smith *et al.* (史密斯等), 2009; Wang (王应祥), 2003

华南兔 *Lepus sinensis*

数据缺乏 DD	无危 LC	近危 NT	易危 VU	濒危 EN	极危 CR	区域灭绝 RE	野外灭绝 EW	灭绝 EX

分类地位 Taxonomic Status

动物界 Animalia	脊索动物门 Chordata	哺乳纲 Mammalia	兔形目 Lagomorpha	兔科 Leporidae
学　名 Scientific Name		*Lepus tibetanus*		
命名人 Species Authority		Waterhouse, 1841		
英文名 English Name(s)		Desert Hare		
同物异名 Synonym(s)		无 / None		
种下单元评估 Infra-specific Taxa Assessed		无 / None		

评估信息 Assessment Information

评估年份 Year Assessed	2020
评　定　人 Assessor(s)	蒋志刚 / Zhigang Jiang
审　定　人 Reviewer(s)	夏霖、杨奇森 / Lin Xia, Qisen Yang
其他贡献人 Other Contributor(s)	李立立、丁晨晨 / Lili Li, Chenchen Ding

理由 Justification: 中亚兔分布广，种群数量大。因此，列为无危等级 / Desert Hare is a widespread species with large populations. Thus, it is listed as Least Concern

地理分布 Geographical Distribution

国内分布 Domestic Distribution

新疆、甘肃、内蒙古 / Xinjiang, Gansu, Inner Mongolia (Nei Mongol)

世界分布 World Distribution

中国；中亚、南亚 / China; Central Asia, South Asia

分布标注 Distribution Note

非特有种 / Non-endemic

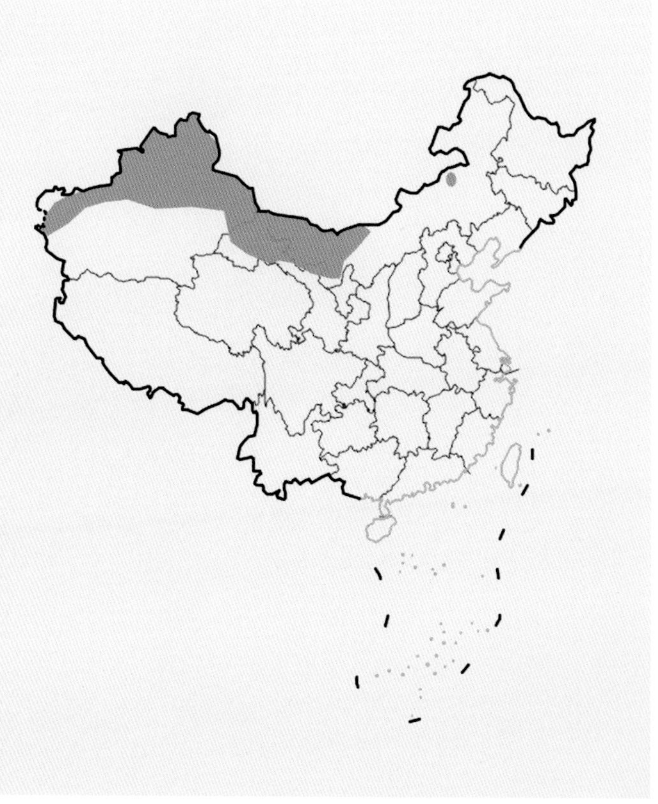

国内分布图
Map of Domestic Distribution

种群 Population

种群数量 Population Size	未知 / Unknown
种群趋势 Population Trend	未知 / Unknown

生境与生态系统 Habitat (s) and Ecosystem (s)

生　　境 Habitat(s)	荒漠、半荒漠、干旱草地、灌丛 / Desert, Semi-desert, Arid Grassland, Shrubland
生态系统 Ecosystem(s)	灌丛生态系统、草地生态系统、荒漠生态系统 / Shrubland Ecosystem, Grassland Ecosystem, Desert Ecosystem

威胁 Threat (s)

主要威胁 Major Threat(s)	无 / None

保护级别与保护行动 Protection Category and Conservation Action (s)

国家重点保护野生动物等级 (2021) Category of National Key Protected Wild Animals (2021)	无 / NA
IUCN 红色名录 (2020-2) IUCN Red List (2020-2)	无危 / LC
CITES 附录 (2019) CITES Appendix (2019)	无 / NA
保护行动 Conservation Action(s)	无 / None

相关文献 Relevant References

Burgin *et al.*, 2018; Jiang *et al.* (蒋志刚等), 2017; Wilson *et al.*, 2016; Zheng *et al.* (郑智民等), 2012; Smith *et al.* (史密斯等), 2009; Pan *et al.* (潘清华等), 2007

中亚兔 *Lepus tibetanus*

雪兔
Lepus timidus

无危 LC

数据缺乏 DD	无危 LC	近危 NT	易危 VU	濒危 EN	极危 CR	区域灭绝 RE	野外灭绝 EW	灭绝 EX

分类地位 Taxonomic Status

动物界 Animalia	脊索动物门 Chordata	哺乳纲 Mammalia	兔形目 Lagomorpha	兔科 Leporidae
学名 Scientific Name		*Lepus timidus*		
命名人 Species Authority		Linnaeus, 1758		
英文名 English Name(s)		Mountain Hare		
同物异名 Synonym(s)		无 / None		
种下单元评估 Infra-specific Taxa Assessed		无 / None		

评估信息 Assessment Information

评估年份 Year Assessed	2020
评定人 Assessor(s)	蒋志刚 / Zhigang Jiang
审定人 Reviewer(s)	夏霖、杨奇森 / Lin Xia, Qisen Yang
其他贡献人 Other Contributor(s)	李立立、丁晨晨 / Lili Li, Chenchen Ding

理由 Justification: 雪兔分布较广，种群数量大。因此，列为无危等级 / Mountain Hare is a widespread species with large populations. Thus, it is listed as Least Concern

地理分布 Geographical Distribution

国内分布 Domestic Distribution

黑龙江、新疆、内蒙古 / Heilongjiang, Xinjiang, Inner Mongolia (Nei Mongol)

世界分布 World Distribution

欧亚大陆 / Eurasian Continent

分布标注 Distribution Note

非特有种 / Non-endemic

国内分布图
Map of Domestic Distribution

种群 Population

种群数量 Population Size	种群数量大 / Large populations
种群趋势 Population Trend	未知 / Unknown

生境与生态系统 Habitat (s) and Ecosystem (s)

生　境 Habitat(s)	泰加林、草地 / Taiga Forest, Grassland
生态系统 Ecosystem(s)	森林生态系统、草地生态系统 / Forest Ecosystem, Grassland Ecosystem

威胁 Threat (s)

主要威胁 Major Threat(s)	无 / None

保护级别与保护行动 Protection Category and Conservation Action (s)

国家重点保护野生动物等级 (2021) Category of National Key Protected Wild Animals (2021)	二级 / Category II
IUCN 红色名录 (2020-2) IUCN Red List (2020-2)	无危 / LC
CITES 附录 (2019) CITES Appendix (2019)	无 / NA
保护行动 Conservation Action(s)	无 / None

相关文献 Relevant References

Burgin *et al*., 2018; Jiang *et al.* (蒋志刚等), 2017; Zheng *et al.* (郑智民等), 2012; Smith *et al.* (史密斯等), 2009; Luo and Li (罗泽珣和李振营), 1982

雪兔 *Lepus timidus*

蒙古兔

Lepus tolai

无危 LC

数据缺乏 DD	无危 LC	近危 NT	易危 VU	濒危 EN	极危 CR	区域灭绝 RE	野外灭绝 EW	灭绝 EX

分类地位 Taxonomic Status

动物界 Animalia	脊索动物门 Chordata	哺乳纲 Mammalia	兔形目 Lagomorpha	兔科 Leporidae
学名 Scientific Name		*Lepus tolai*		
命名人 Species Authority		Pallas, 1778		
英文名 English Name(s)		Tolai Hare		
同物异名 Synonym(s)		无 / None		
种下单元评估 Infra-specific Taxa Assessed		无 / None		

评估信息 Assessment Information

评估年份 Year Assessed	2020
评定人 Assessor(s)	蒋志刚 / Zhigang Jiang
审定人 Reviewer(s)	夏霖、杨奇森 / Xia Lin, Qisen Yang
其他贡献人 Other Contributor(s)	李立立、丁晨晨 / Lili Li, Chenchen Ding

理由 Justification: 蒙古兔分布广，种群数量大。因此，列为无危等级 / Tolai Hare is a widespread species with large populations. Thus, it is listed as Least Concern

地理分布 Geographical Distribution

国内分布 Domestic Distribution

新疆、内蒙古、河北、河南、山东、辽宁、吉林、黑龙江、山西、陕西、四川、云南、重庆、湖北、湖南、江西、江苏、北京、天津、宁夏、甘肃、青海、安徽、贵州 / Xinjiang, Inner Mongolia (Nei Mongol), Hebei, Henan, Shandong, Liaoning, Jilin, Heilongjiang, Shanxi, Shaanxi, Sichuan, Yunnan, Chongqing, Hubei, Hunan, Jiangxi, Jiangsu, Beijing, Tianjin, Ningxia, Gansu, Qinghai, Anhui, Guizhou

世界分布 World Distribution

中国；中亚 / China; Central Asia

分布标注 Distribution Note

非特有种 / Non-endemic

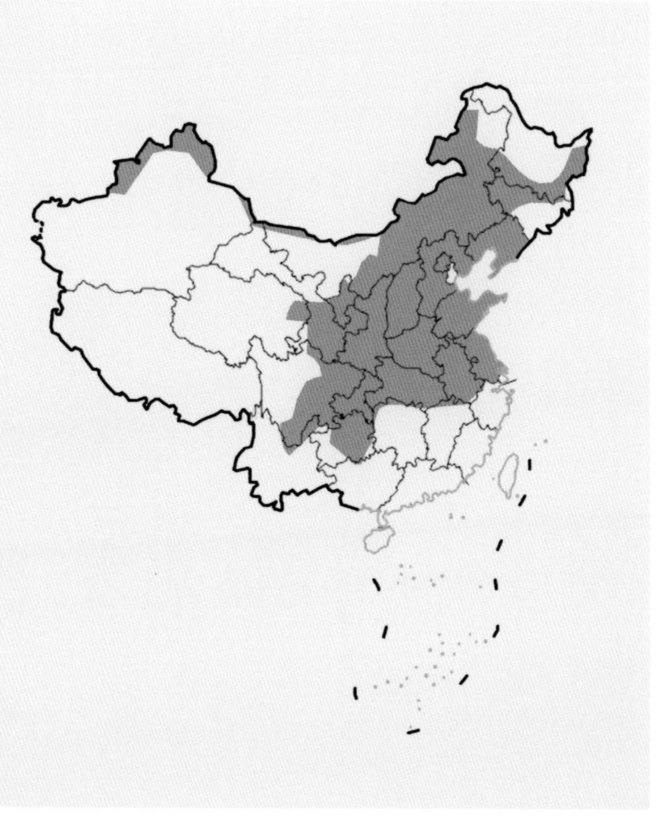

国内分布图
Map of Domestic Distribution

种群 Population

种群数量 Population Size	未知 / Unknown
种群趋势 Population Trend	未知 / Unknown

生境与生态系统 Habitat (s) and Ecosystem (s)

生　境 Habitat(s)	灌丛、草地 / Shrubland, Grassland
生态系统 Ecosystem(s)	灌丛生态系统、草地生态系统 / Shrubland Ecosystem, Grassland Ecosystem

威胁 Threat (s)

主要威胁 Major Threat(s)	无 / None

保护级别与保护行动 Protection Category and Conservation Action (s)

国家重点保护野生动物等级 (2021) Category of National Key Protected Wild Animals (2021)	无 / NA
IUCN 红色名录 (2020-2) IUCN Red List (2020-2)	无危 / LC
CITES 附录 (2019) CITES Appendix (2019)	无 / NA
保护行动 Conservation Action(s)	无 / None

相关文献 Relevant References

Burgin *et al.*, 2018; Jiang *et al.* (蒋志刚等), 2017; Zheng *et al.* (郑智民等), 2012; Smith *et al.* (史密斯等), 2009; Pan *et al.* (潘清华等), 2007

蒙古兔 *Lepus tolai*

双角犀

Dicerorhinus sumatrensis

区域灭绝 RE

数据缺乏 DD	无危 LC	近危 NT	易危 VU	濒危 EN	极危 CR	区域灭绝 RE	野外灭绝 EW	灭绝 EX

分类地位 Taxonomic Status

动物界 Animalia	脊索动物门 Chordata	哺乳纲 Mammalia	奇蹄目 Perissodactyla	犀科 Rhinocerotidae

学名 Scientific Name	*Dicerorhinus sumatrensis*
命名人 Species Authority	Fischer, 1814
英文名 English Name(s)	Sumatran Rhinoceros
同物异名 Synonym(s)	无 / None
种下单元评估 Infra-specific Taxa Assessed	无 / None

评估信息 Assessment Information

评估年份 Year Assessed	2020
评定人 Assessor(s)	蒋志刚 / Zhigang Jiang
审定人 Reviewer(s)	胡慧建 / Huijian Hu
其他贡献人 Other Contributor(s)	李立立、丁晨晨 / Lili Li, Chenchen Ding

理由 Justification: 20 世纪 50 年代初，双角犀在云南消失。此后，中国境内再没有发现双角犀。因此，双角犀列为区域灭绝等级 / In the early 1950s, Sumatran Rhinoceros disappeared in Yunnan and none has been found in China since then. Thus, it is listed as Regionally Extinct

地理分布 Geographical Distribution

国内分布 Domestic Distribution

云南（历史上）/ Yunnan (Historically)

世界分布 World Distribution

中国；南亚、东南亚 / China; South Asia, Southeast Asia

分布标注 Distribution Note

非特有种 / Non-endemic

国内分布图
Map of Domestic Distribution

种群 Population

种群数量 **Population Size**	0
种群趋势 **Population Trend**	未知 / Unknown

生境与生态系统 Habitat (s) and Ecosystem (s)

生　　境 **Habitat(s)**	热带和亚热带低山河谷 / Tropical and Subtropical Low Mountain Valley
生态系统 **Ecosystem(s)**	森林生态系统 / Forest Ecosystem

威胁 Threat (s)

主要威胁 **Major Threat(s)**	栖息地丧失、猎捕 / Habitat Loss, Hunting

保护级别与保护行动 Protection Category and Conservation Action (s)

国家重点保护野生动物等级 **(2021) Category of National Key Protected Wild Animals (2021)**	无 / NA
IUCN 红色名录 **(2020-2) IUCN Red List (2020-2)**	极危 / CR
CITES 附录 **(2019) CITES Appendix (2019)**	无 / NA
保护行动 **Conservation Action(s)**	未知 / Unknown

相关文献 Relevant References

Burgin *et al*., 2018; Jiang *et al.* (蒋志刚等), 2017; Wilson and Mittermeier, 2012; Groves and Grubb, 2011; Xu (许再富), 2000; He (何业恒), 1993; Rookmaaker, 1980

数据缺乏 DD	无危 LC	近危 NT	易危 VU	濒危 EN	极危 CR	**区域灭绝 RE**	野外灭绝 EW	灭绝 EX

分类地位 Taxonomic Status

动 物 界 Animalia	脊索动物门 Chordata	哺 乳 纲 Mammalia	奇 蹄 目 Perissodactyla	犀 科 Rhinocerotidae

学 名 **Scientific Name**	*Rhinoceros sondaicus*
命 名 人 **Species Authority**	Desmarest, 1822
英 文 名 **English Name(s)**	Javan Rhinoceros
同物异名 **Synonym(s)**	无 / None
种下单元评估 **Infra-specific Taxa Assessed**	无 / None

评估信息 Assessment Information

评 估 年 份 **Year Assessed**	2020
评 定 人 **Assessor(s)**	蒋志刚 / Zhigang Jiang
审 定 人 **Reviewer(s)**	胡慧建 / Huijian Hu
其他贡献人 **Other Contributor(s)**	李立立、丁晨晨 / Lili Li, Chenchen Ding

理由 Justification: 20 世纪 50 年代初，爪哇犀在中国云南消失。此后，中国境内再没有发现爪哇犀。因此，列为区域灭绝等级 / In the early 1950s, Javan Rhinoceros disappeared in Yunnan and none has been found in China since then. Thus, it is listed as Regionally Extinct

地理分布 Geographical Distribution

国内分布 Domestic Distribution

云南 (历史上) / Yunnan (Historically)

世界分布 World Distribution

中国、印度尼西亚、越南 / China, Indonesia, Viet Nam

分布标注 Distribution Note

非特有种 / Non-endemic

国内分布图
Map of Domestic Distribution

种群 Population

种群数量 Population Size	0
种群趋势 Population Trend	未知 / Unknown

生境与生态系统 Habitat (s) and Ecosystem (s)

生　　境 Habitat(s)	热带湿润低地森林 / Tropical Moist Lowland Forest
生态系统 Ecosystem(s)	森林生态系统 / Forest Ecosystem

威胁 Threat (s)

主要威胁 Major Threat(s)	栖息地丧失、猎捕 / Habitat Loss, Hunting

保护级别与保护行动 Protection Category and Conservation Action (s)

国家重点保护野生动物等级 (2021) Category of National Key Protected Wild Animals (2021)	无 / NA
IUCN 红色名录 (2020-2) IUCN Red List (2020-2)	极危 / CR
CITES 附录 (2019) CITES Appendix (2019)	I
保护行动 Conservation Action(s)	未知 / Unknown

相关文献 Relevant References

Burgin *et al*., 2018; Jiang *et al.* (蒋志刚等), 2017; Wilson and Mittermeier, 2012; Groves and Grubb, 2011; Smith *et al.* (史密斯等), 2009; Xu (许再富), 2000; He (何业恒), 1993

爪哇犀 *Rhinoceros sondaicus*

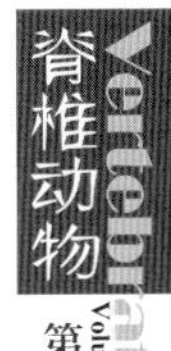

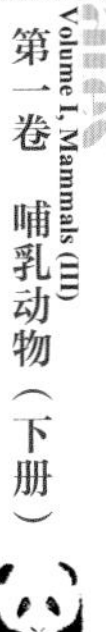

大独角犀

Rhinoceros unicornis

区域灭绝 RE

数据缺乏 DD	无危 LC	近危 NT	易危 VU	濒危 EN	极危 CR	区域灭绝 RE	野外灭绝 EW	灭绝 EX

分类地位 Taxonomic Status

动物界 Animalia	脊索动物门 Chordata	哺乳纲 Mammalia	奇蹄目 Perissodactyla	犀科 Rhinocerotidae

学　名 **Scientific Name**	*Rhinoceros unicornis*
命名人 **Species Authority**	Linnaeus, 1758
英文名 **English Name(s)**	Indian Rhinoceros
同物异名 **Synonym(s)**	无 / None
种下单元评估 **Infra-specific Taxa Assessed**	无 / None

评估信息 Assessment Information

评估年份 **Year Assessed**	2020
评定人 **Assessor(s)**	蒋志刚 / Zhigang Jiang
审定人 **Reviewer(s)**	胡慧建 / Huijian Hu
其他贡献人 **Other Contributor(s)**	李立立、丁晨晨 / Lili Li, Chenchen Ding

理由 Justification: 20世纪50年代初，大独角犀在云南、西藏消失。此后，中国境内再没有发现大独角犀。因此，列为区域灭绝等级 / In the early 1950s, Indian Rhinoceros has disappeared in Yunnan and Tibet (Xizang), and none has been found in China since then. Thus, it is listed as Regionally Extinct

地理分布 Geographical Distribution

国内分布 Domestic Distribution

西藏和云南(历史上) / Tibet (Xizang) and Yunnan (Historically)

世界分布 World Distribution

中国；南亚、东南亚 / China; South Asia, Southeast Asia

分布标注 Distribution Note

非特有种 / Non-endemic

国内分布图
Map of Domestic Distribution

种群 Population

种群数量 **Population Size**	0
种群趋势 **Population Trend**	未知 / Unknown

生境与生态系统 Habitat (s) and Ecosystem (s)

生　　境 **Habitat(s)**	草地、森林、沼泽、农耕地 / Grassland, Forest, Swamp, Arable Land
生态系统 **Ecosystem(s)**	森林生态系统、草地生态系统、农田生态系统、湿地生态系统 / Forest Ecosystem, Grassland Ecosystem, Cropland Ecosystem, Wetland Ecosystem

威胁 Threat (s)

主要威胁 **Major Threat(s)**	栖息地丧失、猎捕 / Habitat Loss, Hunting

保护级别与保护行动 Protection Category and Conservation Action (s)

国家重点保护野生动物等级 **(2021) Category of National Key Protected Wild Animals (2021)**	无 / NA
IUCN 红色名录 **(2020-2) IUCN Red List (2020-2)**	易危 / VU
CITES 附录 **(2019) CITES Appendix (2019)**	I
保护行动 **Conservation Action(s)**	有重引入的计划 / Re-introduction planned

相关文献 Relevant References

Burgin *et al.*, 2018; Jiang *et al.* (蒋志刚等), 2017, 2016a; Wilson and Mittermeier, 2012; Groves and Grubb, 2011; Smith *et al.* (史密斯等), 2009; Xu (许再富), 2000; He (何业恒), 1993

大独角犀 *Rhinoceros unicornis*

爪哇野牛

Bos javanicus

区域灭绝 RE

数据缺乏 DD	无危 LC	近危 NT	易危 VU	濒危 EN	极危 CR	区域灭绝 RE	野外灭绝 EW	灭绝 EX

分类地位 Taxonomic Status

动物界 Animalia	脊索动物门 Chordata	哺乳纲 Mammalia	偶蹄目 Artiodactyla	牛科 Bovidae
学名 Scientific Name		*Bos javanicus*		
命名人 Species Authority		d'Alton, 1823		
英文名 English Name(s)		Banteng		
同物异名 Synonym(s)		*Bos birmanicus* (Lydekker, 1898); *lowi* (Lydekker, 1912)		
种下单元评估 Infra-specific Taxa Assessed		无 / None		

评估信息 Assessment Information

评估年份 Year Assessed	2020
评定人 Assessor(s)	蒋志刚 / Zhigang Jiang
审定人 Reviewer(s)	蒋学龙 / Xuelong Jiang
其他贡献人 Other Contributor(s)	李立立、丁晨晨 / Lili Li, Chenchen Ding

理由 Justification: 爪哇野牛曾被报道分布于云南普洱市，但人们很长一段时间以来都未在野外发现爪哇野牛个体。因此，列为区域灭绝等级 / Banteng was once reported that distributed in Pu'er City, Yunnan. However, it has not been found in the wild during the field surveys for a very long time. Thus, it is listed as Regionally Extinct

地理分布 Geographical Distribution

国内分布 Domestic Distribution

云南 (历史上) / Yunnan (Historically)

世界分布 World Distribution

中国；南亚、东南亚 / China; South Asia, Southeast Asia

分布标注 Distribution Note

非特有种 / Non-endemic

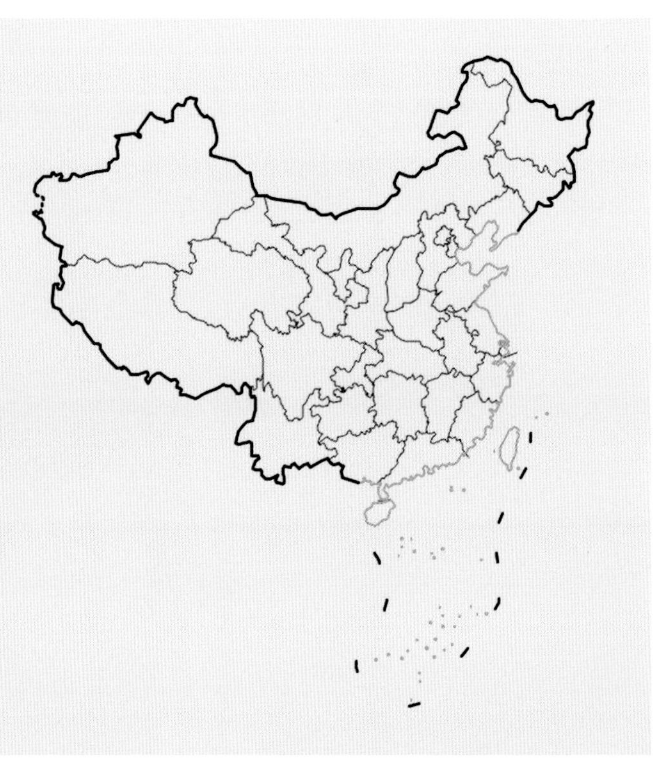

国内分布图
Map of Domestic Distribution

种群 Population

种群数量 Population Size	0
种群趋势 Population Trend	未知 / Unknown

生境与生态系统 Habitat (s) and Ecosystem (s)

生　　境 Habitat(s)	热带湿润森林、次生林 / Tropical Moist Forest, Secondary Forest
生态系统 Ecosystem(s)	森林生态系统 / Forest Ecosystem

威胁 Threat (s)

主要威胁 Major Threat(s)	狩猎、耕种 / Hunting, Plantation

保护级别与保护行动 Protection Category and Conservation Action (s)

国家重点保护野生动物等级 (2021) Category of National Key Protected Wild Animals (2021)	一级 / Category I
IUCN 红色名录 (2020-2) IUCN Red List (2020-2)	濒危 / EN
CITES 附录 (2019) CITES Appendix (2019)	无 / NA
保护行动 Conservation Action(s)	无 / None

相关文献 Relevant References

Burgin *et al*., 2018; Jiang *et al.* (蒋志刚等), 2017; Castelló, 2016; Groves and Grubb, 2011; Wilson and Mittermeier, 2011; Smith *et al.* (史密斯等), 2009; Pan *et al.* (潘清华等), 2007; Wang (王应祥), 2003

爪哇野牛 *Bos javanicus*　　蒋志刚 摄　By Zhigang Jiang

野水牛
Bubalus arnee

区域灭绝 RE

数据缺乏 DD	无危 LC	近危 NT	易危 VU	濒危 EN	极危 CR	区域灭绝 RE	野外灭绝 EW	灭绝 EX

分类地位 Taxonomic Status

动物界 Animalia	脊索动物门 Chordata	哺乳纲 Mammalia	偶蹄目 Artiodactyla	牛科 Bovidae

学 名 **Scientific Name**	*Bubalus arnee*
命名人 **Species Authority**	Kerr, 1792
英文名 **English Name(s)**	Asian Buffalo
同物异名 **Synonym(s)**	*Bos arni* (Hamilton Smith, 1827); *B. bubalus* var. *fulvus* (Blanford, 1891); *Bubalis bubalis* subsp. *migona* (Deraniyagala, 1953); *Bubalus arna* (Hodgson, 1841); *B. arna* (转下页)
种下单元评估 **Infra-specific Taxa Assessed**	无 / None

评估信息 Assessment Information

评估年份 **Year Assessed**	2020
评定人 **Assessor(s)**	蒋志刚 / Zhigang Jiang
审定人 **Reviewer(s)**	蒋学龙 / Xuelong Jiang
其他贡献人 **Other Contributor(s)**	李立立、丁晨晨 / Lili Li, Chenchen Ding

理由 Justification: 野水牛在西藏东南部米什米山曾有分布记录(王应祥，2003)。近年来，野水牛由于近交、猎杀、栖息地丧失而种群数量急剧下降。目前，评估人未查到米什米山发现野水牛的资料。按照 Don E. Wilson 的意见，家养水牛是一个独立的种 (*Bulalus arnee bubalis*)。因此，列为区域灭绝等级 / Asian Buffalo was recorded in Mishimi Shan in the southeast of Tibet (Xizang) (Wang, 2003). In recent years, due to inbreeding, hunting, habitat loss, the populations of Asian Buffalo suffered a sharp decline. We did not find any record of Asian Buffalo in Mishimi Shan. According to Don E. Wilson, the domestic buffalo in China is another species of *Bulalus arnee bubalis*. Thus, it is listed as Regionally Extinct

地理分布 Geographical Distribution

国内分布 Domestic Distribution

西藏(历史上) / Tibet (Xizang) (Historically)

世界分布 World Distribution

中国；南亚 / China; South Asia

分布标注 Distribution Note

非特有种 / Non-endemic

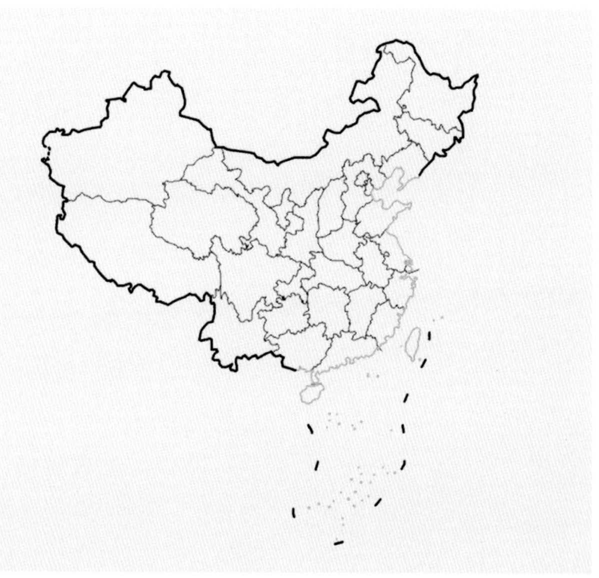

国内分布图
Map of Domestic Distribution

种群 Population

种群数量 Population Size	0
种群趋势 Population Trend	未知 / Unknown

生境与生态系统 Habitat (s) and Ecosystem (s)

生　　境 Habitat(s)	森林、溪流边、草地 / Forest, Near Stream, Grassland
生态系统 Ecosystem(s)	森林生态系统、草地生态系统、湖泊河流生态系统 / Forest Ecosystem, Grassland Ecosystem, Lake and River Ecosystem

威胁 Threat (s)

主要威胁 Major Threat(s)	杂交、耕种、堤坝与水道改变、疾病 / Hybridization, Plantation, Dam and Water Management Use, Disease

保护级别与保护行动 Protection Category and Conservation Action (s)

国家重点保护野生动物等级 (2021) Category of National Key Protected Wild Animals (2021)	无 / NA
IUCN 红色名录 (2020-2) IUCN Red List (2020-2)	濒危 / EN
CITES 附录 (2019) CITES Appendix (2019)	III
保护行动 Conservation Action(s)	无 / None

相关文献 Relevant References

Burgin *et al*., 2018; Jiang *et al.* (蒋志刚等), 2017; Castelló, 2016; Groves and Grubb, 2011; Wilson and Mittermeier, 2011; Pan *et al.* (潘清华等), 2007; Wang (王应祥), 2003

(接上页)
var. *macrocerus* (Hodgson, 1842); *B. bubalus* subsp. *septentrionalis* (Matschie, 1912)

野水牛 *Bubalus arnee*　　By Raju Kasambe

数据缺乏 DD	无危 LC	近危 NT	易危 VU	濒危 EN	极危 CR	区域灭绝 RE	野外灭绝 EW	灭绝 EX

分类地位 Taxonomic Status

动 物 界 **Animalia**	脊索动物门 **Chordata**	哺 乳 纲 **Mammalia**	偶 蹄 目 **Artiodactyla**	鹿 科 **Cervidae**
学 名 **Scientific Name**		*Axis porcinus*		
命 名 人 **Species Authority**		Zimmermann, 1780		
英 文 名 **English Name(s)**		Hog Deer		
同物异名 **Synonym(s)**		无 / None		
种下单元评估 **Infra-specific Taxa Assessed**		无 / None		

评估信息 Assessment Information

评 估 年 份 **Year Assessed**	2020
评 定 人 **Assessor(s)**	蒋志刚 / Zhigang Jiang
审 定 人 **Reviewer(s)**	蒋学龙、陈辈乐、李飞 / Xuelong Jiang, Bosco P. L. Chan, Fei Li
其他贡献人 **Other Contributor(s)**	李立立、丁晨晨 / Lili Li, Chenchen Ding

理由 **Justification:** 中国是豚鹿的边缘分布区。1962 年，彭鸿授等在云南耿马县和西盟县发现了豚鹿角和皮，估计当地有 10 余只豚鹿。1965 年，杨德华等调查时仅发现 4 只豚鹿。20 世纪 70 年代中后期，云南孟定南丁河地区开办农场，毁坏了豚鹿栖息地，加之猎捕，到 80 年代末期，豚鹿在耿马地区绝迹。2018 ～ 2019 年，蒋志刚研究组在野外架设红外相机监测未发现豚鹿。因此，列为区域灭绝等级 / China is the marginal range of the Hog Deer. Peng *et al.* found the antlers and skins of Hog Deer in Gengma County and Ximeng County in 1962 in Yunnan, they estimated that there were more than 10 Hog Deer individuals in the area. Yang *et al.* found only 4 Hog Deer individuals in the field in 1965. From the middle to late 1970s, the riverbed of the Nanding River of Mengding was opened as farm lands, thus destroyed the habitat of Hog Deer. Besides land reclamation, the Hog Deer was also hunted at that time. In the late 1980s, Hog Deer was extinct in the Gengma area. From 2018 to 2019, Zhigang Jiang *et al.* set up infrared camera traps to survey the Hog Deer in its former range, but found no one in the wild. Thus, it is listed as Regionally Extinct

地理分布 Geographical Distribution

国内分布 Domestic Distribution

云南 (历史上) / Yunnan (Historically)

世界分布 World Distribution

中国；南亚、东南亚 / China; South Asia, Southeast Asia

分布标注 Distribution Note

非特有种 / Non-endemic

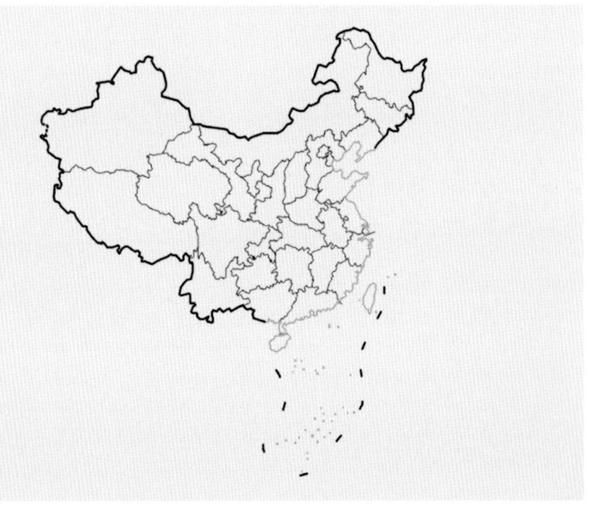

国内分布示意图
Schematic Map of Domestic Distribution

种群 Population

种群数量 **Population Size**	未知 / Unknown
种群趋势 **Population Trend**	下降 / Decreasing

生境与生态系统 Habitat (s) and Ecosystem (s)

生　　境 **Habitat(s)**	沼泽、永久性内陆三角洲 / Swamp, Permanent Inland Delta
生态系统 **Ecosystem(s)**	湿地生态系统 / Wetland Ecosystem

威胁 Threat (s)

主要威胁 **Major Threat(s)**	狩猎、耕种、家畜放牧、洪水 / Hunting, Plantation, Livestock Ranching, Flood

保护级别与保护行动 Protection Category and Conservation Action (s)

国家重点保护野生动物等级 **(2021) Category of National Key Protected Wild Animals (2021)**	一级 / Category I
IUCN 红色名录 **(2020-2) IUCN Red List (2020-2)**	濒危 / EN
CITES 附录 **(2019) CITES Appendix (2019)**	印支亚种被列入附录 I / Indo-Chinese subspecies is listed in appendix I
保护行动 **Conservation Action(s)**	未知 / Unknown

相关文献 Relevant References

Burgin *et al*., 2018; Jiang *et al*. (蒋志刚等), 2017; Wilson and Mittermeier, 2012; Groves and Grubb, 2011; Smith *et al*. (史密斯等), 2009; Pan *et al*. (潘清华等), 2007; Wang (王应祥), 2003

豚鹿 *Axis porcinus*　　　　© Lynx Edicions

中国生物多样性红色名录
China's Red List of Biodiversity

野马

Equus ferus

野外灭绝 EW

数据缺乏 DD	无危 LC	近危 NT	易危 VU	濒危 EN	极危 CR	区域灭绝 RE	野外灭绝 EW	灭绝 EX

分类地位 Taxonomic Status

动物界 Animalia	脊索动物门 Chordata	哺乳纲 Mammalia	奇蹄目 Perissodactyla	马科 Equidae

学名 Scientific Name	*Equus ferus*
命名人 Species Authority	Poliakov, 1881
英文名 English Name(s)	Przewalski's Horse
同物异名 Synonym(s)	普氏野马
种下单元评估 Infra-specific Taxa Assessed	无 / None

评估信息 Assessment Information

评估年份 Year Assessed	2020
评定人 Assessor(s)	蒋志刚 / Zhigang Jiang
审定人 Reviewer(s)	初红军、杨维康 / Hongjun Chu,Weikang Yang
其他贡献人 Other Contributor(s)	李立立、丁晨晨 / Lili Li, Chenchen Ding

理由 Justification: 野马在野外已没有踪迹，但目前正在新疆卡拉麦里山自然保护区、甘肃安西极旱荒漠自然保护区和敦煌西湖自然保护区进行实验野放。尚未建立真正的野生种群。因此，列为野外灭绝等级 / Przewalski's Horse disappeared in the wild. So far, some re-introduced individuals are released in the wild of Kalamaili Shan Nature Reserve, Xinjiang and Anxi Extreme Arid Nature Reserve, Dunhuang West Lake Nature Reserve, Gansu on trial. No ture wild population has been established. Thus, it is listed as Extinct in the Wild

地理分布 Geographical Distribution

国内分布 Domestic Distribution

新疆和甘肃（野放）/ Xinjiang and Gansu (Re-wild)

世界分布 World Distribution

欧亚大陆 / Eurasian Continent

分布标注 Distribution Note

非特有种 / Non-endemic

国内分布图
Map of Domestic Distribution

种群 Population

种群数量 **Population Size**	306 头 / 306 individuals
种群趋势 **Population Trend**	上升 / Increasing

生境与生态系统 Habitat (s) and Ecosystem (s)

生　　境 **Habitat(s)**	半荒漠地区 / Semi-desert Area
生态系统 **Ecosystem(s)**	荒漠生态系统 / Desert Ecosystem

威胁 Threat (s)

主要威胁 **Major Threat(s)**	人类活动干扰、气候变化、家畜放牧、家马遗传物质引入 / Human Disturbance, Climate Change, Livestock Ranching, Introduced Genetic Material of Domestic Horse

保护级别与保护行动 Protection Category and Conservation Action (s)

国家重点保护野生动物等级 **(2021) Category of National Key Protected Wild Animals (2021)**	一级 / Category I
IUCN 红色名录 **(2020-2) IUCN Red List (2020-2)**	濒危 / EN
CITES 附录 **(2019) CITES Appendix (2019)**	I
保护行动 **Conservation Action(s)**	正在开展重野化计划 / Undergoing re-wilding plan

相关文献 Relevant References

Jiang and Zong, 2019; Burgin *et al*., 2018; Jiang *et al.* (蒋志刚等), 2017; Meng *et al.* (孟玉萍等), 2009; Zhang *et al.* (张峰等), 2009; Jiang (蒋志刚), 2004

野马 *Equus ferus*　　蒋志刚 摄　By Zhigang Jiang

驯鹿

Rangifer tarandus

野外灭绝 EW

数据缺乏 DD | 无危 LC | 近危 NT | 易危 VU | 濒危 EN | 极危 CR | 区域灭绝 RE | **野外灭绝 EW** | 灭绝 EX

分类地位 Taxonomic Status

动物界 Animalia	脊索动物门 Chordata	哺乳纲 Mammalia	偶蹄目 Artiodactyla	鹿科 Cervidae

学名 Scientific Name	*Rangifer tarandus*
命名人 Species Authority	Linnaeus, 1758
英文名 English Name(s)	Reindeer
同物异名 Synonym(s)	无 / None
种下单元评估 Infra-specific Taxa Assessed	无 / None

评估信息 Assessment Information

评估年份 Year Assessed	2020
评定人 Assessor(s)	蒋志刚 / Zhigang Jiang
审定人 Reviewer(s)	孟秀祥 / Xiuxiang Meng
其他贡献人 Other Contributor(s)	李立立、丁晨晨 / Lili Li, Chenchen Ding

理由 Justification: 驯鹿种群小，仅分布于大兴安岭西北麓根河地区，仅 700 余头，存在于 12 个隔离种群，处于半驯化状态。驯鹿的生存受气候变化影响极大，在可预测的未来，情况难以好转。因此，列为野外灭绝等级 / The Reindeer's population in China is small. The species is distributed only in the Genhe area that is in the northwest of the Da Hinggan Ling, where about 700 individuals live in 12 isolated sub-populations. The Reindeer in China is not fully domesticated. The survival of Reindeer is greatly impacted by climate change and it is difficult to foresee any improvement in the future. Thus, it is listed as Extinct in the Wild

地理分布 Geographical Distribution

国内分布 Domestic Distribution

内蒙古 (驯养) / Inner Mongolia (Nei Mongol) (Domesticated)

世界分布 World Distribution

中国；环北极圈 / China; Circum-Arctic

分布标注 Distribution Note

非特有种 / Non-endemic

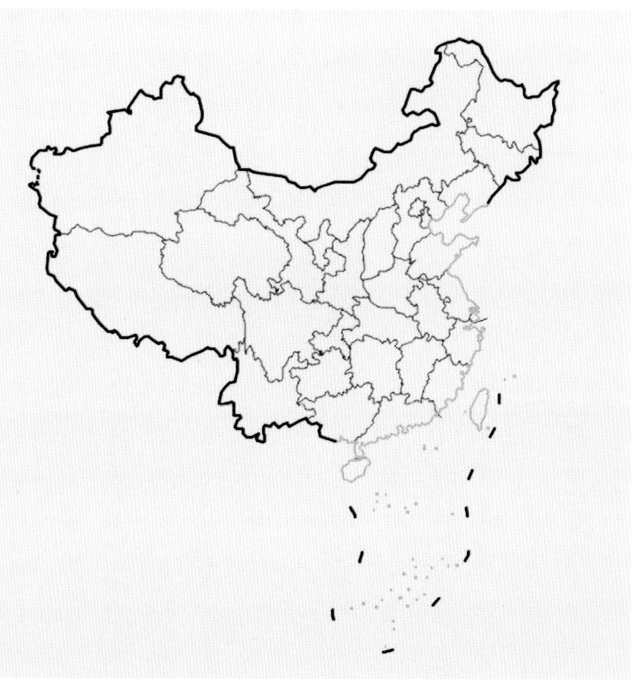

国内分布图
Map of Domestic Distribution

种群 Population

种群数量 **Population Size**	无野生个体 / No wild individuals
种群趋势 **Population Trend**	稳定 / Stable

生境与生态系统 Habitat (s) and Ecosystem (s)

生　　境 **Habitat(s)**	泰加林、苔原 / Taiga Forest, Tundra
生态系统 **Ecosystem(s)**	森林生态系统、苔原生态系统 / Forest Ecosystem, Tundra Ecosystem

威胁 Threat (s)

主要威胁 **Major Threat(s)**	气候变化、疾病 / Climate Change, Disease

保护级别与保护行动 Protection Category and Conservation Action (s)

国家重点保护野生动物等级 **(2021) Category of National Key Protected Wild Animals (2021)**	无 / NA
IUCN 红色名录 **(2020-2) IUCN Red List (2020-2)**	无危 / LC
CITES 附录 **(2019) CITES Appendix (2019)**	无 / NA
保护行动 **Conservation Action(s)**	无 / None

相关文献 Relevant References

Burgin *et al*., 2018; Jiang *et al.* (蒋志刚等), 2017; Ge *et al.* (葛小芳等), 2015; Smith *et al.* (史密斯等), 2009; Wang (王应祥), 2003

驯鹿 *Rangifer tarandus*

蒋志刚 摄　By Zhigang Jiang

大额牛
Bos frontalis

野外灭绝 EW

数据缺乏 DD	无危 LC	近危 NT	易危 VU	濒危 EN	极危 CR	区域灭绝 RE	**野外灭绝 EW**	灭绝 EX

分类地位 Taxonomic Status

动物界 Animalia	脊索动物门 Chordata	哺乳纲 Mammalia	偶蹄目 Artiodactyla	牛科 Bovidae

学　名 **Scientific Name**	*Bos frontalis*
命名人 **Species Authority**	Lambert, 1804
英文名 **English Name(s)**	Gayal
同物异名 **Synonym(s)**	*Bos gaurus* (Hamilton Smith, 1827)
种下单元评估 **Infra-specific Taxa Assessed**	无 / None

评估信息 Assessment Information

评估年份 **Year Assessed**	2020
评定人 **Assessor(s)**	蒋志刚 / Zhigang Jiang
审定人 **Reviewer(s)**	蒋学龙 / Xuelong Jiang
其他贡献人 **Other Contributor(s)**	李立立、丁晨晨 / Lili Li, Chenchen Ding

理由 Justification: 2018 年，蒋志刚等在考察中没有发现野生大额牛，大额牛已经野外灭绝。现在高黎贡山的大额牛已经被人类驯化，被放牧于高山牧场。20 世纪 80 年代初期，怒江傈僳族自治州（怒江州）养殖存栏 77 头大额牛。1986 年以来，怒江州农业部门将原仅存于独龙江的大额牛引种到怒江流域的贡山、福贡及泸水境内。截至 2012 年，怒江州全州共有大额牛养殖户 37 户，大额牛存栏达 3,726 头。因此，列为野外灭绝等级 / Zhigang Jiang *et al.* did not find the species in field survey in 2018. *Bos frontalis* were extinct in the wild in the country. Now *Bos frontalis* of Gaoligong Shan are domesticated and are herded on alpine pastures. In the early 1980s, there were 77 *Bos frontalis* in the cattle farms in the Nujiang River area. Since 1986, the domesticated *Bos frontalis* were introduced to Gongshan, Fugong and Lushui counties. By 2012, there were 37 large cattle farms and 3,726 domesticated *Bos frontalis* in the area. Thus, it is listed as Extinct in the wild

地理分布 Geographical Distribution

国内分布 Domestic Distribution

云南（驯养）/ Yunnan (Domesticated)

世界分布 World Distribution

中国；南亚、东南亚 / China; South Asia, Southeast Asia

分布标注 Distribution Note

非特有种 / Non-endemic

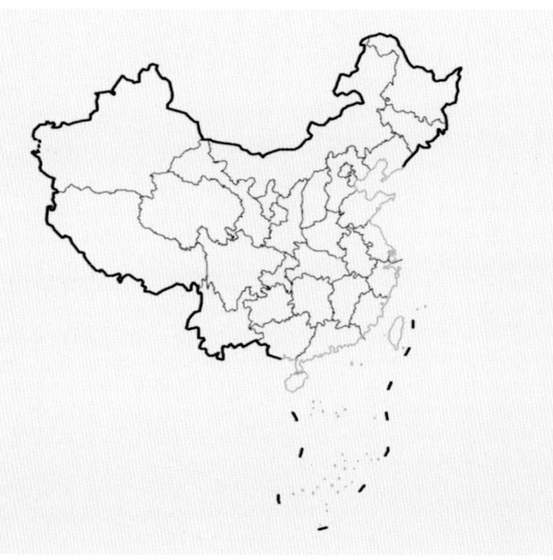

国内分布图
Map of Domestic Distribution

种群 Population

种群数量 **Population Size**	0
种群趋势 **Population Trend**	未知 / Unknown

生境与生态系统 Habitat (s) and Ecosystem (s)

生　　境 **Habitat(s)**	森林 / Forest
生态系统 **Ecosystem(s)**	森林生态系统 / Forest Ecosystem

威胁 Threat (s)

主要威胁 **Major Threat(s)**	人类驯化 / Domestication

保护级别与保护行动 Protection Category and Conservation Action (s)

国家重点保护野生动物等级 **(2021) Category of National Key Protected Wild Animals (2021)**	无 / NA
IUCN 红色名录 **(2020-2) IUCN Red List (2020-2)**	未列入 / NA
CITES 附录 **(2019) CITES Appendix (2019)**	无 / NA
保护行动 **Conservation Action(s)**	无 / None

相关文献 Relevant References

Burgin *et al.*, 2018; Jiang *et al.* (蒋志刚等), 2017; Smith *et al.* (史密斯等), 2009; Wang *et al.* (王兰萍等), 2009; Wang (王应祥), 2003

大额牛 *Bos frontalis*　　蒋志刚 摄　By Zhigang Jiang

高鼻羚羊

Saiga tatarica

野外灭绝 EW

数据缺乏 DD	无危 LC	近危 NT	易危 VU	濒危 EN	极危 CR	区域灭绝 RE	野外灭绝 EW	灭绝 EX

分类地位 Taxonomic Status

动物界 Animalia	脊索动物门 Chordata	哺乳纲 Mammalia	偶蹄目 Artiodactyla	牛科 Bovidae

学名 Scientific Name	*Saiga tatarica*
命名人 Species Authority	Linnaeus, 1766
英文名 English Name(s)	Saiga
同物异名 Synonym(s)	Saiga Antelope, Mongolian Antelope
种下单元评估 Infra-specific Taxa Assessed	无 / None

评估信息 Assessment Information

评估年份 Year Assessed	2020
评定人 Assessor(s)	蒋志刚 / Zhigang Jiang
审定人 Reviewer(s)	胡慧建 / Huijian Hu
其他贡献人 Other Contributor(s)	李立立、丁晨晨 / Lili Li, Chenchen Ding

理由 Justification: 高鼻羚羊 20 世纪 50 年代在中国灭绝。尽管于 80 年代从国外重新引入，在甘肃建立了圈养种群，但高鼻羚羊仍生存于人工圈养状态下，尚未在野外形成可生存种群。因此，列为野外灭绝等级 / In 1950s, Saiga is extincted in china. In 1980s, Saiga was reintroduced into China and established a captive breeding population. However, the population of Saiga still live in enclosures, and no wild population has been established. Thus, it is listed as Extinct in the Wild

地理分布 Geographical Distribution

国内分布 Domestic Distribution

甘肃 (保护与繁育中心圈养) / Gansu (Captive Breeding in Conservation and Breeding Center)

世界分布 World Distribution

中国; 中亚、东亚 / China; Central Asia, East Asia

分布标注 Distribution Note

非特有种 / Non-endemic

国内分布图
Map of Domestic Distribution

种群 Population

种群数量 **Population Size**	圈养繁育 170 只 / 170 captive breeding individuals
种群趋势 **Population Trend**	未知 / Unknown

生境与生态系统 Habitat (s) and Ecosystem (s)

生　　境 **Habitat(s)**	温带草地 / Temperate Steppe
生态系统 **Ecosystem(s)**	草地生态系统 / Grassland Ecosystem

威胁 Threat (s)

主要威胁 **Major Threat(s)**	狩猎、疾病、家畜放牧、疾病 / Hunting, Disease, Livestock Ranching, Disease

保护级别与保护行动 Protection Category and Conservation Action (s)

国家重点保护野生动物等级 **(2021) Category of National Key Protected Wild Animals (2021)**	一级 / Category I
IUCN 红色名录 **(2020-2) IUCN Red List (2020-2)**	极危 / CR
CITES 附录 **(2019) CITES Appendix (2019)**	II
保护行动 **Conservation Action(s)**	保护性繁育 / Conservational breeding

相关文献 Relevant References

Burgin *et al*., 2018; Jiang *et al.* (蒋志刚等), 2018; Cui *et al.*, 2017; Castelló, 2016; Groves and Grubb, 2011; Wilson and Mittermeier, 2011; Zhao *et al.*, 2013a

高鼻羚羊 *Saiga tatarica*　　蒋志刚 摄　By Zhigang Jiang

小齿猬
Mesechinus miodon

数据缺乏 DD

数据缺乏 DD	无危 LC	近危 NT	易危 VU	濒危 EN	极危 CR	区域灭绝 RE	野外灭绝 EW	灭绝 EX

分类地位 Taxonomic Status

动物界 Animalia	脊索动物门 Chordata	哺乳纲 Mammalia	劳亚食虫目 Eulipotyphla	猬科 Erinaceidae

学　　名 Scientific Name	*Mesechinus miodon*
命 名 人 Species Authority	Thomas, 1908
英 文 名 English Name(s)	Small-toothed Hedgehog
同物异名 Synonym(s)	无 / None
种下单元评估 Infra-specific Taxa Assessed	无 / None

评估信息 Assessment Information

评 估 年 份 Year Assessed	2020
评　定　人 Assessor(s)	蒋志刚 / Zhigang Jiang
审　定　人 Reviewer(s)	蒋学龙、冯祚建 / Xuelong Jiang, Zuojian Feng
其他贡献人 Other Contributor(s)	李立立、丁晨晨 / Lili Li, Chenchen Ding

理由 Justification: 人们对小齿猬所知甚少，目前仅存少数几号小齿猬标本，须对其进一步研究。因此，列为数据缺乏等级 / Few information is known about Small-toothed Hedgehog, and only a few specimens have been collected. Further study on this species is needed. Thus, it is listed as Data Deficient

地理分布 Geographical Distribution

国内分布 Domestic Distribution

内蒙古、山西、陕西、甘肃 / Inner Mongolia (Nei Mongol), Shanxi, Shaanxi, Gansu

世界分布 World Distribution

中国 / China

分布标注 Distribution Note

特有种 / Endemic

国内分布图
Map of Domestic Distribution

种群 Population

种群数量 Population Size	未知 / Unknown
种群趋势 Population Trend	未知 / Unknown

生境与生态系统 Habitat (s) and Ecosystem (s)

生　　境 Habitat(s)	森林、草地、耕地 / Forest, Grassland, Farm Land
生态系统 Ecosystem(s)	森林生态系统、草地生态系统、农田生态系统 / Forest Ecosystem, Grassland Ecosystem, Cropland Ecosystem

威胁 Threat (s)

主要威胁 Major Threat(s)	未知 / Unknown

保护级别与保护行动 Protection Category and Conservation Action (s)

国家重点保护野生动物等级 (2021) Category of National Key Protected Wild Animals (2021)	无 / NA
IUCN 红色名录 (2020-2) IUCN Red List (2020-2)	未列入 / NA
CITES 附录 (2019) CITES Appendix (2019)	无 / NA
保护行动 Conservation Action(s)	无 / None

相关文献 Relevant References

Burgin *et al.*, 2018; Jiang *et al.* (蒋志刚等), 2017; Pan *et al.* (潘清华等), 2007; Wang (王应祥), 2003

中国生物多样性红色名录
China's Red List of Biodiversity

高黎贡林猬

Mesechinus wangi

数据缺乏 DD

数据缺乏 DD	无危 LC	近危 NT	易危 VU	濒危 EN	极危 CR	区域灭绝 RE	野外灭绝 EW	灭绝 EX

分类地位 Taxonomic Status

动物界 Animalia	脊索动物门 Chordata	哺乳纲 Mammalia	劳亚食虫目 Eulipotyphla	猬科 Erinaceidae

学名 Scientific Name	*Mesechinus wangi*
命名人 Species Authority	Ai *et al.*, 2018
英文名 English Name(s)	Wang's Hedgehog
同物异名 Synonym(s)	Gaoligong Forest Hedgehog, Wang's Forest Hedgehog
种下单元评估 Infra-specific Taxa Assessed	无 / None

评估信息 Assessment Information

评估年份 Year Assessed	2020
评定人 Assessor(s)	蒋学龙 / Xuelong Jiang
审定人 Reviewer(s)	蒋志刚 / Zhigang Jiang
其他贡献人 Other Contributor(s)	丁晨晨 / Chenchen Ding

理由 Justification: 高黎贡林猬是2018年新近发现的种，为纪念兽类学家王应祥而命名。目前发现它仅分布在云南高黎贡山自然保护区，其种群与分布区现状不明。因此，列为数据缺乏 / Named after orcologists Yinxiang Wang, Wang's Hedgehog is a new species found in 2018, and distributed only in Gaoligong Shan Nature Reserve in Yunnan as we know. Its population and habitat status are unknown. Thus, it is listed as Data Deficient

地理分布 Geographical Distribution

国内分布 Domestic Distribution

云南 / Yunnan

世界分布 World Distribution

中国 / China

分布标注 Distribution Note

特有种 / Endemic

国内分布图
Map of Domestic Distribution

种群 Population

种群数量 Population Size	未知 / Unknown
种群趋势 Population Trend	未知 / Unknown

生境与生态系统 Habitat (s) and Ecosystem (s)

生　　境 Habitat(s)	落叶阔叶林 / Deciduous Broad-leaved Forest
生态系统 Ecosystem(s)	森林生态系统 / Forest Ecosystem

威胁 Threat (s)

主要威胁 Major Threat(s)	未知 / Unknown

保护级别与保护行动 Protection Category and Conservation Action (s)

国家重点保护野生动物等级 (2021) Category of National Key Protected Wild Animals (2021)	无 / NA
IUCN 红色名录 (2020-2) IUCN Red List (2020-2)	未列入 / NA
CITES 附录 (2019) CITES Appendix (2019)	无 / NA
保护行动 Conservation Action(s)	无 / None

相关文献 Relevant References

Ai *et al.*, 2018; Burgin *et al.*, 2018

高黎贡林猬 *Mesechinus wangi*

蒋学龙 摄　By Xuelong Jiang

中国生物多样性红色名录
China's Red List of Biodiversity

台湾缺齿鼹

Mogera kanoana

数据缺乏 DD

数据缺乏 DD	无危 LC	近危 NT	易危 VU	濒危 EN	极危 CR	区域灭绝 RE	野外灭绝 EW	灭绝 EX

分类地位 Taxonomic Status

动物界 Animalia	脊索动物门 Chordata	哺乳纲 Mammalia	劳亚食虫目 Eulipotyphla	鼹科 Talpidae

学名 Scientific Name	*Mogera kanoana*
命名人 Species Authority	Kawada, 2007
英文名 English Name(s)	Taiwan Mole
同物异名 Synonym(s)	Kano Mole
种下单元评估 Infra-specific Taxa Assessed	无 / None

评估信息 Assessment Information

评估年份 Year Assessed	2020
评定人 Assessor(s)	蒋志刚 / Zhigang Jiang
审定人 Reviewer(s)	蒋学龙、冯祚建 / Xuelong Jiang, Zuojian Feng
其他贡献人 Other Contributor(s)	李立立、丁晨晨 / Lili Li, Chenchen Ding

理由 Justification: 台湾缺齿鼹在中国仅发现于台湾东南部的中央山脉，为 Kawada *et al.* (2007) 基于采集的 11 号标本描述和记录的物种，未有其他更多濒危等级的报道。因此，列为数据缺乏等级 / Taiwan Mole is only found in the Central Mountains in the southeast of Taiwan, China. Based on the 11 specimens collected by Kawada *et al.* (2007), the species was named. However, more reports about the endangered category of this species is deficient. Thus, it is listed as Data Deficient

地理分布 Geographical Distribution

国内分布 Domestic Distribution

台湾 / Taiwan

世界分布 World Distribution

中国 / China

分布标注 Distribution Note

特有种 / Endemic

国内分布图
Map of Domestic Distribution

种群 Population

种群数量 **Population Size**	未知 / Unknown
种群趋势 **Population Trend**	未知 / Unknown

生境与生态系统 Habitat (s) and Ecosystem (s)

生　　境 **Habitat(s)**	森林 / Forest
生态系统 **Ecosystem(s)**	森林生态系统 / Forest Ecosystem

威胁 Threat (s)

主要威胁 **Major Threat(s)**	无 / None

保护级别与保护行动 Protection Category and Conservation Action (s)

国家重点保护野生动物等级 **(2021) Category of National Key Protected Wild Animals (2021)**	无 / NA
IUCN 红色名录 **(2020-2) IUCN Red List (2020-2)**	数据缺乏 / DD
CITES 附录 **(2019) CITES Appendix (2019)**	无 / NA
保护行动 **Conservation Action(s)**	无 / None

相关文献 Relevant References

Burgin *et al*., 2018; Wilson and Mittermeier, 2018; Jiang *et al.* (蒋志刚等), 2017; Kawada *et al.*, 2007

© Lynx Edicions

数据缺乏 DD	无危 LC	近危 NT	易危 VU	濒危 EN	极危 CR	区域灭绝 RE	野外灭绝 EW	灭绝 EX

分类地位 Taxonomic Status

动物界 Animalia	脊索动物门 Chordata	哺乳纲 Mammalia	劳亚食虫目 Eulipotyphla	鼹科 Talpidae

学名 Scientific Name	*Mogera uchidai*
命名人 Species Authority	Abe, Shiraishi and Arai, 1991
英文名 English Name(s)	Senkaku Mole
同物异名 Synonym(s)	Ryukyu Mole; *Nesoscaptor uchidai* (Abe, Shiraishi and Arai, 1991)
种下单元评估 Infra-specific Taxa Assessed	无 / None

评估信息 Assessment Information

评估年份 Year Assessed	2020
评定人 Assessor(s)	蒋志刚 / Zhigang Jiang
审定人 Reviewer(s)	蒋学龙、冯祚建 / Xuelong Jiang, Zuojian Feng
其他贡献人 Other Contributor(s)	李立立、丁晨晨 / Lili Li, Chenchen Ding

理由 Justification: 钓鱼岛鼹分布在钓鱼岛，分布区面积小，针对该物种的信息较少。因此，列为数据缺乏 / Senkaku Mole is found in the Diaoyu Islands, with small distributed area. Information on this species is deficient. Thus, it is listed as Data Deficient

地理分布 Geographical Distribution

国内分布 Domestic Distribution

钓鱼岛 / Diaoyu Islands

世界分布 World Distribution

中国 / China

分布标注 Distribution Note

特有种 / Endemic

国内分布图
Map of Domestic Distribution

种群 Population

种群数量 Population Size	未知 / Unknown
种群趋势 Population Trend	未知 / Unknown

生境与生态系统 Habitat (s) and Ecosystem (s)

生　　境 Habitat(s)	草地 / Grassland
生态系统 Ecosystem(s)	草地生态系统 / Grassland Ecosystem

威胁 Threat (s)

主要威胁 Major Threat(s)	外来入侵物种 / Alien Invasive Species

保护级别与保护行动 Protection Category and Conservation Action (s)

国家重点保护野生动物等级 (2021) Category of National Key Protected Wild Animals (2021)	无 / NA
IUCN 红色名录 (2020-2) IUCN Red List (2020-2)	易危 / VU
CITES 附录 (2019) CITES Appendix (2019)	无 / NA
保护行动 Conservation Action(s)	无 / None

相关文献 Relevant References

Burgin *et al*., 2018; Wilson and Mittermeier, 2018; Jiang *et al.* (蒋志刚等), 2017; Pan *et al.* (潘清华等), 2007; Wang (王应祥), 2003

柯氏鼩鼱
Sorex kozlovi

数据缺乏 DD

数据缺乏 DD	无危 LC	近危 NT	易危 VU	濒危 EN	极危 CR	区域灭绝 RE	野外灭绝 EW	灭绝 EX

分类地位 Taxonomic Status

动物界 Animalia	脊索动物门 Chordata	哺乳纲 Mammalia	劳亚食虫目 Eulipotyphla	鼩鼱科 Soricidae

学名 Scientific Name	*Sorex kozlovi*
命名人 Species Authority	Stroganov, 1952
英文名 English Name(s)	Kozlov's Shrew
同物异名 Synonym(s)	无 / None
种下单元评估 Infra-specific Taxa Assessed	无 / None

评估信息 Assessment Information

评估年份 Year Assessed	2020
评定人 Assessor(s)	蒋志刚 / Zhigang Jiang
审定人 Reviewer(s)	蒋学龙、冯祚建 / Xuelong Jiang, Zuojian Feng
其他贡献人 Other Contributor(s)	李立立、丁晨晨 / Lili Li, Chenchen Ding

理由 Justification: 柯氏鼩鼱仅分布在澜沧江上游，对其研究较少，有关信息极少。因此，列为数据缺乏等级 / Kozlov's Shrew is an endemic species in upper river of Lancang River. Little research has been conducted on this species, information on the species is deficient. Thus, it is listed as Data Deficient

地理分布 Geographical Distribution

国内分布 Domestic Distribution

西藏、青海 / Tibet (Xizang), Qinghai

世界分布 World Distribution

中国 / China

分布标注 Distribution Note

特有种 / Endemic

国内分布图
Map of Domestic Distribution

种群 Population

种群数量 **Population Size**	未知 / Unknown
种群趋势 **Population Trend**	未知 / Unknown

生境与生态系统 Habitat (s) and Ecosystem (s)

生　　境 **Habitat(s)**	森林、灌丛 / Forest, Shrubland
生态系统 **Ecosystem(s)**	森林生态系统、灌丛生态系统 / Forest Ecosystem, Shrubland Ecosystem

威胁 Threat (s)

主要威胁 **Major Threat(s)**	未知 / Unknown

保护级别与保护行动 Protection Category and Conservation Action (s)

国家重点保护野生动物等级 **(2021) Category of National Key Protected Wild Animals (2021)**	无 / NA
IUCN 红色名录 **(2020-2) IUCN Red List (2020-2)**	数据缺乏 / DD
CITES 附录 **(2019) CITES Appendix (2019)**	无 / NA
保护行动 **Conservation Action(s)**	无 / None

相关文献 Relevant References

Burgin *et al*., 2018; Jiang *et al.* (蒋志刚等), 2017; Pan *et al.* (潘清华等), 2007; Hutterer, 2005

柯氏鼩鼱 *Sorex kozlovi*　　蒋志刚 绘　Drawn by Zhigang Jiang

米什米长尾鼩鼱

Episoriculus baileyi

数据缺乏 DD

数据缺乏 DD	无危 LC	近危 NT	易危 VU	濒危 EN	极危 CR	区域灭绝 RE	野外灭绝 EW	灭绝 EX

分类地位 Taxonomic Status

动物界 Animalia	脊索动物门 Chordata	哺乳纲 Mammalia	劳亚食虫目 Eulipotyphla	鼩鼱科 Soricidae

学名 Scientific Name	*Episoriculus baileyi*
命名人 Species Authority	Thomas, 1914
英文名 English Name(s)	Mishmi Brown-toothed Shrew
同物异名 Synonym(s)	无 / None
种下单元评估 Infra-specific Taxa Assessed	无 / None

评估信息 Assessment Information

评估年份 Year Assessed	2020
评定人 Assessor(s)	蒋志刚 / Zhigang Jiang
审定人 Reviewer(s)	蒋学龙、冯祚建 / Xuelong Jiang, Zuojian Feng
其他贡献人 Other Contributor(s)	李立立、丁晨晨 / Lili Li, Chenchen Ding

理由 Justification: 米什米长尾鼩鼱仅分布于西藏东南部，目前缺乏该种的有关信息。因此，列为数据缺乏等级 / Mishmi Brown-toothed Shrew is only distributed in southeast Tibet (Xizang). The information on the species is deficient. Thus, it is listed as Data Deficient

地理分布 Geographical Distribution

国内分布 Domestic Distribution

西藏 / Tibet (Xizang)

世界分布 World Distribution

中国、尼泊尔 / China, Nepal

分布标注 Distribution Note

非特有种 / Non-endemic

国内分布图
Map of Domestic Distribution

种群 Population

种群数量 Population Size	未知 / Unknown
种群趋势 Population Trend	未知 / Unknown

生境与生态系统 Habitat (s) and Ecosystem (s)

生　　境 Habitat(s)	针叶林、落叶阔叶林 / Coniferous Forest, Deciduous Broad-leaved Forest
生态系统 Ecosystem(s)	森林生态系统 / Forest Ecosystem

威胁 Threat (s)

主要威胁 Major Threat(s)	未知 / Unknown

保护级别与保护行动 Protection Category and Conservation Action (s)

国家重点保护野生动物等级 (2021) Category of National Key Protected Wild Animals (2021)	无 / NA
IUCN 红色名录 (2020-2) IUCN Red List (2020-2)	未列入 / NA
CITES 附录 (2019) CITES Appendix (2019)	无 / NA
保护行动 Conservation Action(s)	无 / None

相关文献 Relevant References

Burgin *et al*., 2018; Wilson and Mittermeier, 2018; Jiang *et al.* (蒋志刚等), 2017; Corbet and Hill, 1986

中国生物多样性红色名录
China's Red List of Biodiversity

灰腹长尾鼩鼱

Episoriculus sacratus

数据缺乏 DD

数据缺乏 DD	无危 LC	近危 NT	易危 VU	濒危 EN	极危 CR	区域灭绝 RE	野外灭绝 EW	灭绝 EX

分类地位 Taxonomic Status

动物界 Animalia	脊索动物门 Chordata	哺乳纲 Mammalia	劳亚食虫目 Eulipotyphla	鼩鼱科 Soricidae

学　　名 Scientific Name	*Episoriculus sacratus*
命名人 Species Authority	Thomas, 1911
英文名 English Name(s)	Gray-bellied Shrew
同物异名 Synonym(s)	无 / None
种下单元评估 Infra-specific Taxa Assessed	无 / None

评估信息 Assessment Information

评估年份 Year Assessed	2020
评　定　人 Assessor(s)	蒋志刚 / Zhigang Jiang
审　定　人 Reviewer(s)	蒋学龙、冯祚建 / Xuelong Jiang, Zuojian Feng
其他贡献人 Other Contributor(s)	李立立、丁晨晨 / Lili Li, Chenchen Ding

理由 Justification: 灰腹长尾鼩鼱分布区狭窄，记录较少，对其的研究缺乏。因此，列为数据缺乏 / Gray-bellied Shrew is narrowly distributed with few records, and study on this species is deficient. Thus, it is listed as Data Deficient

地理分布 Geographical Distribution

国内分布 Domestic Distribution

四川、云南 / Sichuan, Yunnan

世界分布 World Distribution

中国 / China

分布标注 Distribution Note

特有种 / Endemic

国内分布图
Map of Domestic Distribution

种群 Population

种群数量 **Population Size**	未知 / Unknown
种群趋势 **Population Trend**	未知 / Unknown

生境与生态系统 Habitat (s) and Ecosystem (s)

生　　境 **Habitat(s)**	针叶林 / Coniferous Forest
生态系统 **Ecosystem(s)**	森林生态系统 / Forest Ecosystem

威胁 Threat (s)

主要威胁 **Major Threat(s)**	未知 / Unknown

保护级别与保护行动 Protection Category and Conservation Action (s)

国家重点保护野生动物等级 **(2021) Category of National Key Protected Wild Animals (2021)**	无 / NA
IUCN 红色名录 **(2020-2) IUCN Red List (2020-2)**	未列入 / NA
CITES 附录 **(2019) CITES Appendix (2019)**	无 / NA
保护行动 **Conservation Action(s)**	无 / None

相关文献 Relevant References

Burgin *et al.*, 2018; Wilson and Mittermeier, 2018; Jiang *et al.* (蒋志刚等), 2017; Pan *et al.* (潘清华等), 2007; Wang(王应祥), 2003

烟黑缺齿鼩

Chodsigoa furva

数据缺乏 DD

数据缺乏 DD	无危 LC	近危 NT	易危 VU	濒危 EN	极危 CR	区域灭绝 RE	野外灭绝 EW	灭绝 EX

分类地位 Taxonomic Status

动物界 Animalia	脊索动物门 Chordata	哺乳纲 Mammalia	劳亚食虫目 Eulipotyphla	鼩鼱科 Soricidae
学名 Scientific Name		*Chodsigoa furva*		
命名人 Species Authority		Anthony, 1941		
英文名 English Name(s)		Dusky Long-tailed Shrew		
同物异名 Synonym(s)		*Chodsigoa smithii furva* (Anthony, 1941)		
种下单元评估 Infra-specific Taxa Assessed		无 / None		

评估信息 Assessment Information

评估年份 Year Assessed	2020
评定人 Assessor(s)	蒋志刚 / Zhigang Jiang
审定人 Reviewer(s)	蒋学龙、冯祚建 / Xuelong Jiang, Zuojian Feng
其他贡献人 Other Contributor(s)	李立立、丁晨晨 / Lili Li, Chenchen Ding

理由 **Justification:** 烟黑缺齿鼩的种群与生境的研究缺乏。因此，列为数据缺乏等级 / Information about the population and habitat status of Dusky Long-tailed Shrew are deficient. Thus, it is listed as Data Deficient

地理分布 Geographical Distribution

国内分布 Domestic Distribution

云南 / Yunnan

世界分布 World Distribution

中国、缅甸 / China, Myanmar

分布标注 Distribution Note

非特有种 / Non-endemic

国内分布图

Map of Domestic Distribution

种群 Population

种群数量 **Population Size**	未知 / Unknown
种群趋势 **Population Trend**	未知 / Unknown

生境与生态系统 Habitat (s) and Ecosystem (s)

生　　境 **Habitat(s)**	海拔 2,000m 以上的山区 / Mountain region at elevation over 2,000m
生态系统 **Ecosystem(s)**	山地生态系统 / Mountain Ecosystem

威胁 Threat (s)

主要威胁 **Major Threat(s)**	未知 / Unknown

保护级别与保护行动 Protection Category and Conservation Action (s)

国家重点保护野生动物等级 **(2021) Category of National Key Protected Wild Animals (2021)**	无 / NA
IUCN 红色名录 **(2020-2) IUCN Red List (2020-2)**	未列入 / NA
CITES 附录 **(2019) CITES Appendix (2019)**	无 / NA
保护行动 **Conservation Action(s)**	无 / None

相关文献 Relevant References

Burgin *et al.*, 2018; Wilson and Mittermeier, 2018; Chen *et al.*, 2017b; Jiang *et al.* (蒋志刚等), 2017

小缺齿鼩鼱
Chodsigoa lamula

数据缺乏 DD

数据缺乏 DD	无危 LC	近危 NT	易危 VU	濒危 EN	极危 CR	区域灭绝 RE	野外灭绝 EW	灭绝 EX

分类地位 Taxonomic Status

动物界 Animalia	脊索动物门 Chordata	哺乳纲 Mammalia	劳亚食虫目 Eulipotyphla	鼩鼱科 Soricidae
学名 Scientific Name		*Chodsigoa lamula*		
命名人 Species Authority		Thomas, 1912		
英文名 English Name(s)		Lamulate Shrew		
同物异名 Synonym(s)		*Soriculus lamula* (Thomas, 1912)		
种下单元评估 Infra-specific Taxa Assessed		无 / None		

评估信息 Assessment Information

评估年份 Year Assessed	2020
评定人 Assessor(s)	蒋志刚 / Zhigang Jiang
审定人 Reviewer(s)	蒋学龙、冯祚建 / Xuelong Jiang, Zuojian Feng
其他贡献人 Other Contributor(s)	李立立、丁晨晨 / Lili Li, Chenchen Ding

理由 Justification: 小缺齿鼩鼱的分布区较狭窄，且其物种地位尚待厘清。因此，列为数据缺乏等级 / Lamulate Shrew is narrowly distributed in China and the status of the species needs further clarifying. Thus, it is listed as Data Deficient

地理分布 Geographical Distribution

国内分布 Domestic Distribution

甘肃、四川、云南 / Gansu, Sichuan, Yunnan

世界分布 World Distribution

中国 / China

分布标注 Distribution Note

特有种 / Endemic

国内分布图
Map of Domestic Distribution

种群 Population

种群数量 Population Size	未知 / Unknown
种群趋势 Population Trend	未知 / Unknown

生境与生态系统 Habitat (s) and Ecosystem (s)

生　　境 Habitat(s)	亚热带湿润山地森林 / Subtropical Moist Montane Forest
生态系统 Ecosystem(s)	森林生态系统 / Forest Ecosystem

威胁 Threat (s)

主要威胁 Major Threat(s)	未知 / Unknown

保护级别与保护行动 Protection Category and Conservation Action (s)

国家重点保护野生动物等级 (2021) Category of National Key Protected Wild Animals (2021)	无 / NA
IUCN 红色名录 (2020-2) IUCN Red List (2020-2)	无危 / LC
CITES 附录 (2019) CITES Appendix (2019)	无 / NA
保护行动 Conservation Action(s)	无 / None

相关文献 Relevant References

Burgin *et al.*, 2018; Shi *et al.* (师蕾等), 2013; Smith *et al.* (史密斯等), 2009; Liu *et al.* (刘少英等), 2005; Thomas, 1912a, 1912b

小缺齿鼩鼱 *Chodsigoa lamula*　　蒋志刚 绘　Drawn by Zhigang Jiang

中国生物多样性红色名录
China's Red List of Biodiversity

大缺齿鼩鼱
Chodsigoa salenskii

数据缺乏 DD

数据缺乏 DD	无危 LC	近危 NT	易危 VU	濒危 EN	极危 CR	区域灭绝 RE	野外灭绝 EW	灭绝 EX

分类地位 Taxonomic Status

动物界 Animalia	脊索动物门 Chordata	哺乳纲 Mammalia	劳亚食虫目 Eulipotyphla	鼩鼱科 Soricidae

学　名 **Scientific Name**	*Chodsigoa salenskii*
命名人 **Species Authority**	Kastschenko, 1907
英文名 **English Name(s)**	Salenski's Shrew
同物异名 **Synonym(s)**	*Soriculus salenskii* (Kastschenko, 1907)
种下单元评估 **Infra-specific Taxa Assessed**	无 / None

评估信息 Assessment Information

评估年份 **Year Assessed**	2020
评定人 **Assessor(s)**	蒋志刚 / Zhigang Jiang
审定人 **Reviewer(s)**	蒋学龙 / Xuelong Jiang
其他贡献人 **Other Contributor(s)**	李立立、丁晨晨 / Lili Li, Chenchen Ding

理由 **Justification:** 大缺齿鼩鼱为中国特有种，其分布区狭窄，目前仅有模式标本，无更多信息。因此，列为数据缺乏等级 / Salenski's Shrew is an endemic species with a narrow distributed region in China, knowledge about the species is only based on the type specimen. Thus, it is listed as Data Deficient

地理分布 Geographical Distribution

国内分布 Domestic Distribution

四川、贵州 / Sichuan, Guizhou

世界分布 World Distribution

中国 / China

分布标注 Distribution Note

特有种 / Endemic

国内分布图
Map of Domestic Distribution

种群 Population

种群数量 **Population Size**	未知 / Unknown
种群趋势 **Population Trend**	未知 / Unknown

生境与生态系统 Habitat (s) and Ecosystem (s)

生　　境 **Habitat(s)**	温带森林 / Temperate Forest
生态系统 **Ecosystem(s)**	森林生态系统 / Forest Ecosystem

威胁 Threat (s)

主要威胁 **Major Threat(s)**	未知 / Unknown

保护级别与保护行动 Protection Category and Conservation Action (s)

国家重点保护野生动物等级 **(2021) Category of National Key Protected Wild Animals (2021)**	无 / NA
IUCN 红色名录 **(2020-2) IUCN Red List (2020-2)**	数据缺乏 / DD
CITES 附录 **(2019) CITES Appendix (2019)**	无 / NA
保护行动 **Conservation Action(s)**	无 / None

相关文献 Relevant References

Burgin *et al.*, 2018; Wilson and Mittermeier, 2018; Chen *et al.*, 2017b; Jiang *et al.* (蒋志刚等), 2017; Sun *et al.* (孙治宇等), 2013; Tu *et al.*, 2012; Qin *et al.* (秦岭等), 2007

细尾缺齿鼩鼱

Chodsigoa sodalis

数据缺乏 DD

数据缺乏 DD	无危 LC	近危 NT	易危 VU	濒危 EN	极危 CR	区域灭绝 RE	野外灭绝 EW	灭绝 EX

分类地位 Taxonomic Status

动物界 Animalia	脊索动物门 Chordata	哺乳纲 Mammalia	劳亚食虫目 Eulipotyphla	鼩鼱科 Soricidae

学名 Scientific Name	*Chodsigoa sodalis*
命名人 Species Authority	Thomas, 1913
英文名 English Name(s)	Lesser Taiwanese Shrew
同物异名 Synonym(s)	无 / None
种下单元评估 Infra-specific Taxa Assessed	无 / None

评估信息 Assessment Information

评估年份 Year Assessed	2020
评定人 Assessor(s)	蒋志刚 / Zhigang Jiang
审定人 Reviewer(s)	蒋学龙 / Xuelong Jiang
其他贡献人 Other Contributor(s)	李立立、丁晨晨 / Lili Li, Chenchen Ding

理由 Justification: 细尾缺齿鼩鼱为中国特有种，仅分布在台湾，有关该种的信息缺乏。因此，列为数据缺乏等级 / Lesser Taiwanese Shrew is an endemic species to China, only found in Taiwan. The information on this species is scarce. Thus, it is listed as Data Deficient

地理分布 Geographical Distribution

国内分布 Domestic Distribution
台湾 / Taiwan
世界分布 World Distribution
中国 / China
分布标注 Distribution Note
特有种 / Endemic

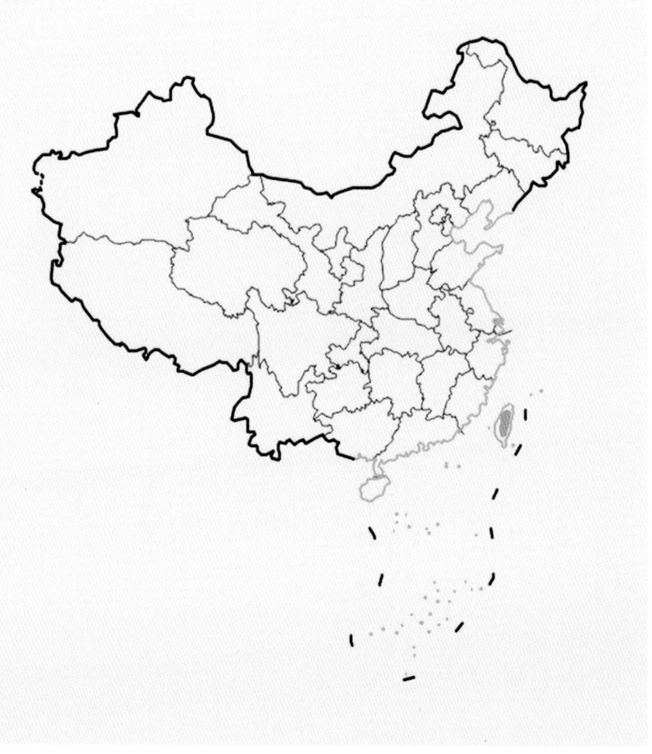

国内分布图
Map of Domestic Distribution

种群 Population

种群数量 **Population Size**	未知 / Unknown
种群趋势 **Population Trend**	未知 / Unknown

生境与生态系统 Habitat (s) and Ecosystem (s)

生　　境 **Habitat(s)**	亚热带高海拔灌丛 / Subtropical High Altitude Shrubland
生态系统 **Ecosystem(s)**	灌丛生态系统 / Shrubland Ecosystem

威胁 Threat (s)

主要威胁 **Major Threat(s)**	伐木 / Logging

保护级别与保护行动 Protection Category and Conservation Action (s)

国家重点保护野生动物等级 **(2021) Category of National Key Protected Wild Animals (2021)**	无 / NA
IUCN 红色名录 **(2020-2) IUCN Red List (2020-2)**	数据缺乏 / DD
CITES 附录 **(2019) CITES Appendix (2019)**	无 / NA
保护行动 **Conservation Action(s)**	无 / None

相关文献 Relevant References

Burgin *et al*., 2018; Wilson and Mittermeier, 2018; Jiang *et al*. (蒋志刚等), 2017; Motokawa *et al*., 1998; Yu, 1993

台湾短尾鼩

Anourosorex yamashinai

数据缺乏 DD

数据缺乏 DD	无危 LC	近危 NT	易危 VU	濒危 EN	极危 CR	区域灭绝 RE	野外灭绝 EW	灭绝 EX

分类地位 Taxonomic Status

动物界 Animalia	脊索动物门 Chordata	哺乳纲 Mammalia	劳亚食虫目 Eulipotyphla	鼩鼱科 Soricidae

学　名 **Scientific Name**	*Anourosorex yamashinai*
命名人 **Species Authority**	Kuroda, 1935
英文名 **English Name(s)**	Taiwanese Mole Shrew
同物异名 **Synonym(s)**	无 / None
种下单元评估 **Infra-specific Taxa Assessed**	无 / None

评估信息 Assessment Information

评估年份 **Year Assessed**	2020
评　定　人 **Assessor(s)**	蒋志刚 / Zhigang Jiang
审　定　人 **Reviewer(s)**	蒋学龙 / Xuelong Jiang
其他贡献人 **Other Contributor(s)**	李立立、丁晨晨 / Lili Li, Chenchen Ding

理由 Justification: 台湾短尾鼩为中国特有种，仅分布在台湾，目前对其的研究缺乏。因此，列为数据缺乏等级 / Taiwanese Mole Shrew is an endemic species in China, distributed only in Taiwan. To date, studies on this species are deficient. Thus, it is listed as Data Deficient

地理分布 Geographical Distribution

国内分布 Domestic Distribution

台湾 / Taiwan

世界分布 World Distribution

中国 / China

分布标注 Distribution Note

特有种 / Endemic

国内分布图
Map of Domestic Distribution

种群 Population

种群数量 **Population Size**	未知 / Unknown
种群趋势 **Population Trend**	未知 / Unknown

生境与生态系统 Habitat (s) and Ecosystem (s)

生　　境 **Habitat(s)**	亚热带湿润山地森林、农田 / Subtropical Moist Montane Forest, Cropland
生态系统 **Ecosystem(s)**	森林生态系统、农田生态系统 / Forest Ecosystem, Cropland Ecosystem

威胁 Threat (s)

主要威胁 **Major Threat(s)**	未知 / Unknown

保护级别与保护行动 Protection Category and Conservation Action (s)

国家重点保护野生动物等级 **(2021) Category of National Key Protected Wild Animals (2021)**	无 / NA
IUCN 红色名录 **(2020-2) IUCN Red List (2020-2)**	无危 / LC
CITES 附录 **(2019) CITES Appendix (2019)**	无 / NA
保护行动 **Conservation Action(s)**	无 / None

相关文献 Relevant References

Burgin *et al.*, 2018; Wilson and Mittermeier, 2018; Jiang *et al.* (蒋志刚等), 2017; Hutterer, 2005; Motokawa and Lin, 2002; Jameson and Jones, 1977

利安得水麝鼩

Chimarrogale leander

数据缺乏 DD

数据缺乏 DD	无危 LC	近危 NT	易危 VU	濒危 EN	极危 CR	区域灭绝 RE	野外灭绝 EW	灭绝 EX

分类地位 Taxonomic Status

动 物 界 Animalia	脊索动物门 Chordata	哺 乳 纲 Mammalia	劳亚食虫目 Eulipotyphla	鼩 鼱 科 Soricidae

学　名 Scientific Name	*Chimarrogale leander*
命 名 人 Species Authority	Thomas, 1902
英 文 名 English Name(s)	Leander Water Shrew
同物异名 Synonym(s)	无 / None
种下单元评估 Infra-specific Taxa Assessed	无 / None

评估信息 Assessment Information

评 估 年 份 Year Assessed	2020
评　定　人 Assessor(s)	蒋志刚 / Zhigang Jiang
审　定　人 Reviewer(s)	蒋学龙 / Xuelong Jiang
其他贡献人 Other Contributor(s)	李立立、丁晨晨 / Lili Li, Chenchen Ding

理由 Justification: 利安得水麝鼩在2013年被提升为种，对其种群与分布区的研究尚缺乏。因此，列为数据缺乏等级 / Leander Water Shrew was elevated from subspecies to full species status in 2013. However, information on its populations and habitats are remain deficient. Thus, it is listed as Data Deficient

地理分布 Geographical Distribution

国内分布 Domestic Distribution

浙江、福建、广东、广西、江苏、江西、台湾 / Zhejiang, Fujian, Guangdong, Guangxi, Jiangsu, Jiangxi, Taiwan

世界分布 World Distribution

中国；南亚、东南亚 / China; South Asia, Southeast Asia

分布标注 Distribution Note

非特有种 / Non-endemic

国内分布图
Map of Domestic Distribution

种群 Population

种群数量 Population Size	未知 / Unknown
种群趋势 Population Trend	未知 / Unknown

生境与生态系统 Habitat (s) and Ecosystem (s)

生　境 Habitat(s)	森林、湖泊、河流 / Forest, Lake, River
生态系统 Ecosystem(s)	森林生态系统、湖泊河流生态系统 / Forest Ecosystem, Lake and River Ecosystem

威胁 Threat (s)

主要威胁 Major Threat(s)	未知 / Unknown

保护级别与保护行动 Protection Category and Conservation Action (s)

国家重点保护野生动物等级 (2021) Category of National Key Protected Wild Animals (2021)	无 / NA
IUCN 红色名录 (2020-2) IUCN Red List (2020-2)	未列入 / NA
CITES 附录 (2019) CITES Appendix (2019)	无 / NA
保护行动 Conservation Action(s)	无 / None

相关文献 Relevant References

Burgin *et al.*, 2018; Wilson and Mittermeier, 2018; Jiang *et al.* (蒋志刚等), 2017; Yuan *et al.*, 2013

印度大狐蝠

Pteropus giganteus

数据缺乏 DD

数据缺乏 DD	无危 LC	近危 NT	易危 VU	濒危 EN	极危 CR	区域灭绝 RE	野外灭绝 EW	灭绝 EX

分类地位 Taxonomic Status

动物界 Animalia	脊索动物门 Chordata	哺乳纲 Mammalia	翼手目 Chiroptera	狐蝠科 Pteropodidae

学名 Scientific Name	*Pteropus giganteus*
命名人 Species Authority	Brünnich, 1782
英文名 English Name(s)	Indian Flying Fox
同物异名 Synonym(s)	Greater Indian Fruit Bat; *Pteropus ariel* (Allen, 1908); *P. assamensis* (McClelland, 1839); *P. edwardsi* (I. Geoffroy, 1828); *P. kelaarti* (Gray, 1871); *P. leucocephalus* (转下页)
种下单元评估 Infra-specific Taxa Assessed	无 / None

评估信息 Assessment Information

评估年份 Year Assessed	2020
评定人 Assessor(s)	吴毅、蒋志刚 / Yi Wu, Zhigang Jiang
审定人 Reviewer(s)	张礼标、毛秀光、张树义 / Libiao Zhang, Xiuguang Mao, Shuyi Zhang
其他贡献人 Other Contributor(s)	李立立、丁晨晨 / Lili Li, Chenchen Ding

理由 Justification: 印度大狐蝠记录较少，中国仅采集到 1 号标本，目前对其的研究缺乏。因此，列为数据缺乏等级 / Studies on Indian Flying Fox are deficient, and only one specimen was collected in China. Little information is known about this species. Thus, it is listed as Data Deficient

地理分布 Geographical Distribution

国内分布 Domestic Distribution

青海 / Qinghai

世界分布 World Distribution

中国；印度洋岛屿；南亚 / China; Indian Ocean Islands; South Asia

分布标注 Distribution Note

非特有种 / Non-endemic

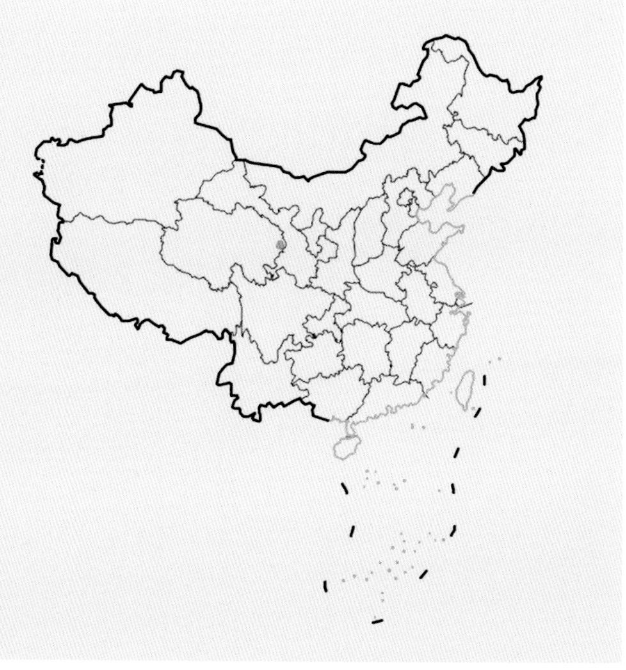

国内分布图
Map of Domestic Distribution

种群 Population

种群数量 Population Size	未知 / Unknown
种群趋势 Population Trend	未知 / Unknown

生境与生态系统 Habitat (s) and Ecosystem (s)

生　　境 Habitat(s)	未知 / Unknown
生态系统 Ecosystem(s)	未知 / Unknown

威胁 Threat (s)

主要威胁 Major Threat(s)	未知 / Unknown

保护级别与保护行动 Protection Category and Conservation Action (s)

国家重点保护野生动物等级 (2021) Category of National Key Protected Wild Animals (2021)	无 / NA
IUCN 红色名录 (2020-2) IUCN Red List (2020-2)	无危 / LC
CITES 附录 (2019) CITES Appendix (2019)	II
保护行动 Conservation Action(s)	无 / None

相关文献 Relevant References

Burgin *et al.*, 2018; Wilson and Mittermeier, 2018; Jiang *et al.* (蒋志刚等), 2017; Smith *et al.* (史密斯等), 2009; Wang (王应祥), 2003; Wang and Wang, 1962

(接上页)
(Hodgson, 1835); *P. medius* (Temminck, 1825); *P. ruvicollis* (Ogilby, 1840); *Vespertilio gigantea* (Brunnich, 1782)

印度大狐蝠 *Pteropus giganteus*　　By Charles J Sharp

无尾果蝠
Megaerops ecaudatus

数据缺乏 DD

数据缺乏 DD | 无危 LC | 近危 NT | 易危 VU | 濒危 EN | 极危 CR | 区域灭绝 RE | 野外灭绝 EW | 灭绝 EX

分类地位 Taxonomic Status

动物界 Animalia	脊索动物门 Chordata	哺乳纲 Mammalia	翼手目 Chiroptera	狐蝠科 Pteropodidae

学　名 Scientific Name	*Megaerops ecaudatus*
命名人 Species Authority	Temminck, 1837
英文名 English Name(s)	Temminck's Tailless Fruit Bat
同物异名 Synonym(s)	*Pachysoma ecaudata* (Temminck, 1837); *P. ecaudatum* (Temminck, 1837)
种下单元评估 Infra-specific Taxa Assessed	无 / None

评估信息 Assessment Information

评估年份 Year Assessed	2020
评定人 Assessor(s)	吴毅、蒋志刚 / Yi Wu, Zhigang Jiang
审定人 Reviewer(s)	张礼标、毛秀光、张树义 / Libiao Zhang, Xiuguang Mao, Shuyi Zhang
其他贡献人 Other Contributor(s)	李立立、丁晨晨 / Lili Li, Chenchen Ding

理由 Justification: 无尾果蝠 2006 年在中国第一次被记录，目前对其的研究缺乏。因此，列为数据缺乏等级 / Temminck's Tailless Fruit Bat was first recorded in China in 2006, but research on this species remains scarce. Thus, it is listed as Data Deficient

地理分布 Geographical Distribution

国内分布 Domestic Distribution

云南 / Yunnan

世界分布 World Distribution

中国；东南亚 / China; Southeast Asia

分布标注 Distribution Note

非特有种 / Non-endemic

国内分布图
Map of Domestic Distribution

种群 Population

种群数量 **Population Size**	稀少 / Rare
种群趋势 **Population Trend**	未知 / Unknown

生境与生态系统 Habitat (s) and Ecosystem (s)

生 境 **Habitat(s)**	热带和亚热带湿润低地森林、次生林 / Tropical and Subtropical Moist Lowland Forest, Secondary Forest
生态系统 **Ecosystem(s)**	森林生态系统 / Forest Ecosystem

威胁 Threat (s)

主要威胁 **Major Threat(s)**	未知 / Unknown

保护级别与保护行动 Protection Category and Conservation Action (s)

国家重点保护野生动物等级 **(2021) Category of National Key Protected Wild Animals (2021)**	无 / NA
IUCN 红色名录 **(2020-2) IUCN Red List (2020-2)**	无危 / LC
CITES 附录 **(2019) CITES Appendix (2019)**	无 / NA
保护行动 **Conservation Action(s)**	无 / None

相关文献 Relevant References

Wilson and Mittermeier, 2019; Burgin *et al.*, 2018; Jiang *et al.* (蒋志刚等), 2017; Pan *et al.* (潘清华等), 2007; Feng *et al.* (冯庆等), 2006; Van Peenen *et al.*, 1969

无尾果蝠 *Megaerops ecaudatus* 张礼标 摄 By Libiao Zhang

泰国无尾果蝠

Megaerops niphanae

数据缺乏 DD

数据缺乏 DD	无危 LC	近危 NT	易危 VU	濒危 EN	极危 CR	区域灭绝 RE	野外灭绝 EW	灭绝 EX

分类地位 Taxonomic Status

动物界 Animalia	脊索动物门 Chordata	哺乳纲 Mammalia	翼手目 Chiroptera	狐蝠科 Pteropodidae
学名 **Scientific Name**		*Megaerops niphanae*		
命名人 **Species Authority**		Yenbutra and Felten, 1983		
英文名 **English Name(s)**		Ratanaworabhan's Fruit Bat		
同物异名 **Synonym(s)**		无 / None		
种下单元评估 **Infra-specific Taxa Assessed**		无 / None		

评估信息 Assessment Information

评估年份 **Year Assessed**	2020
评定人 **Assessor(s)**	吴毅、蒋志刚 / Yi Wu, Zhigang Jiang
审定人 **Reviewer(s)**	张礼标、毛秀光、张树义 / Libiao Zhang, Xiuguang Mao, Shuyi Zhang
其他贡献人 **Other Contributor(s)**	李立立、丁晨晨 / Lili Li, Chenchen Ding

理由 **Justification:** 2006 年，泰国无尾果蝠在中国第一次被记录，目前对其的研究尚缺乏。因此，列为数据缺乏等级 / Ratanaworabhan's Fruit Bat was first recorded in China in 2006, however, research on this species remains scarce. Thus, it is listed as Data Deficient

地理分布 Geographical Distribution

国内分布 Domestic Distribution

云南 / Yunnan

世界分布 World Distribution

中国；南亚、东南亚 / China; South Asia, Southeast Asia

分布标注 Distribution Note

非特有种 / Non-endemic

国内分布图
Map of Domestic Distribution

种群 Population

种群数量 **Population Size**	稀少 / Rare
种群趋势 **Population Trend**	未知 / Unknown

生境与生态系统 Habitat (s) and Ecosystem (s)

生　　境 **Habitat(s)**	在喜马拉雅山脚的落叶林、针叶林、竹林和亚热带混交林中都有发现 / It is found in Deciduous Forest, Coniferous Forest, Bamboo Forests and Subtropical Mixed Forest across the Himalayan Foothill
生态系统 **Ecosystem(s)**	森林生态系统 / Forest Ecosystem

威胁 Threat (s)

主要威胁 **Major Threat(s)**	未知 / Unknown

保护级别与保护行动 Protection Category and Conservation Action (s)

国家重点保护野生动物等级 **(2021) Category of National Key Protected Wild Animals (2021)**	无 / NA
IUCN 红色名录 **(2020-2) IUCN Red List (2020-2)**	无危 / LC
CITES 附录 **(2019) CITES Appendix (2019)**	无 / NA
保护行动 **Conservation Action(s)**	无 / None

相关文献 Relevant References

Wilson and Mittermeier, 2019; Burgin *et al*., 2018; Jiang *et al.* (蒋志刚等), 2017; Feng *et al.* (冯庆等), 2006; Simmons, 2005

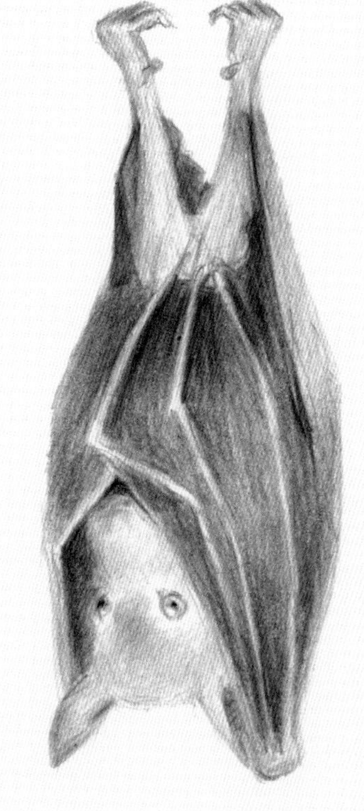

泰国无尾果蝠 *Megaerops niphanae*　　蒋志刚 绘　Drawn by Zhigang Jiang

数据缺乏 DD	无危 LC	近危 NT	易危 VU	濒危 EN	极危 CR	区域灭绝 RE	野外灭绝 EW	灭绝 EX

分类地位 Taxonomic Status

动 物 界 Animalia	脊索动物门 Chordata	哺 乳 纲 Mammalia	翼 手 目 Chiroptera	假吸血蝠科 Megadermatidae
学　　名 Scientific Name		*Megaderma spasma*		
命 名 人 Species Authority		Linnaeus, 1758		
英 文 名 English Name(s)		Lesser False Vampire		
同物异名 Synonym(s)		无 / None		
种下单元评估 Infra-specific Taxa Assessed		无 / None		

评估信息 Assessment Information

评 估 年 份 Year Assessed	2020
评　定　人 Assessor(s)	吴毅、蒋志刚 / Yi Wu, Zhigang Jiang
审　定　人 Reviewer(s)	张礼标、毛秀光、张树义 / Libiao Zhang, Xiuguang Mao, Shuyi Zhang
其他贡献人 Other Contributor(s)	李立立、丁晨晨 / Lili Li, Chenchen Ding

理由 Justification: 马来假吸血蝠在 2010 年被记录，目前对其的研究缺乏。因此，列为数据缺乏等级 / Lesser False Vampire was first recorded in China in 2010, but research on this species remains scarce. Thus, it is listed as Data Deficient

地理分布 Geographical Distribution

国内分布 Domestic Distribution

云南 / Yunnan

世界分布 World Distribution

中国；南亚、东南亚 / China; South Asia, Southeast Asia

分布标注 Distribution Note

非特有种 / Non-endemic

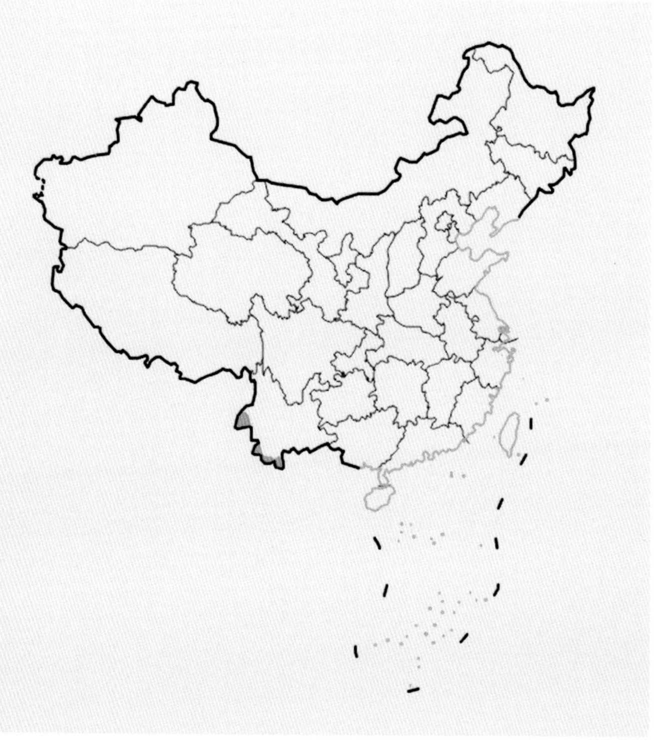

国内分布图
Map of Domestic Distribution

种群 Population

种群数量 **Population Size**	未知 / Unknown
种群趋势 **Population Trend**	未知 / Unknown

生境与生态系统 Habitat (s) and Ecosystem (s)

生　　境 **Habitat(s)**	森林 / Forest
生态系统 **Ecosystem(s)**	森林生态系统 / Forest Ecosystem

威胁 Threat (s)

主要威胁 **Major Threat(s)**	旅游、采石场 / Tourism, Quarrying Field

保护级别与保护行动 Protection Category and Conservation Action (s)

国家重点保护野生动物等级 **(2021) Category of National Key Protected Wild Animals (2021)**	无 / NA
IUCN 红色名录 **(2020-2) IUCN Red List (2020-2)**	无危 / LC
CITES 附录 **(2019) CITES Appendix (2019)**	无 / NA
保护行动 **Conservation Action(s)**	无 / None

相关文献 Relevant References

Wilson and Mittermeier, 2019; Burgin *et al.*, 2018; Jiang *et al.* (蒋志刚等), 2017; Zhang *et al.* (张礼标等), 2010

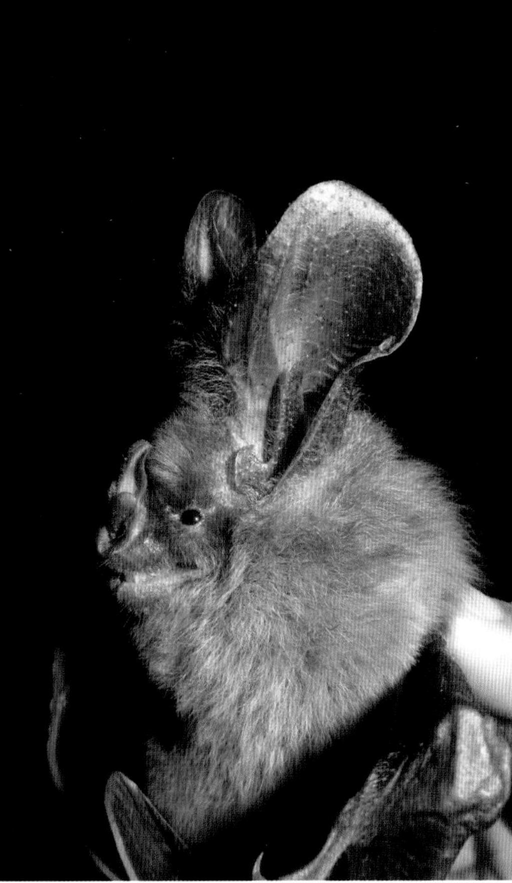

马来假吸血蝠 *Megaderma spasma*

By Alice Hughes

大墓蝠

Taphozous theobaldi

数据缺乏 DD

数据缺乏 DD	无危 LC	近危 NT	易危 VU	濒危 EN	极危 CR	区域灭绝 RE	野外灭绝 EW	灭绝 EX

分类地位 Taxonomic Status

动物界 Animalia	脊索动物门 Chordata	哺乳纲 Mammalia	翼手目 Chiroptera	鞘尾蝠科 Emballonuridae
学名 Scientific Name		*Taphozous theobaldi*		
命名人 Species Authority		Dobson, 1872		
英文名 English Name(s)		Theobold's Bat		
同物异名 Synonym(s)		Theobald's Tomb Bat; *Taphozous secatus* Thomas, 1915		
种下单元评估 Infra-specific Taxa Assessed		无 / None		

评估信息 Assessment Information

评估年份 Year Assessed	2020
评定人 Assessor(s)	吴毅、蒋志刚 / Yi Wu, Zhigang Jiang
审定人 Reviewer(s)	石红艳、江廷磊、余文华 / Hongyan Shi, Tinglei Jiang, Wenhua Yu
其他贡献人 Other Contributor(s)	李立立、丁晨晨 / Lili Li, Chenchen Ding

理由 Justification: 大墓蝠在中国的种群数量较少， 且其占有区面积减小。目前，关于该物种的信息较少。因此， 列为数据缺乏等级 / Theobold's Bat has small populations in China, and its areas of occupancy are declining. Currently, information about this species is scarce. Thus, it is listed as Data Deficient

地理分布 Geographical Distribution

国内分布 Domestic Distribution

广东、云南 / Guangdong, Yunnan

世界分布 World Distribution

中国；东南亚 / China; Southeast Asia

分布标注 Distribution Note

非特有种 / Non-endemic

国内分布图
Map of Domestic Distribution

种群 Population

种群数量 **Population Size**	稀少 / Rare
种群趋势 **Population Trend**	未知 / Unknown

生境与生态系统 Habitat (s) and Ecosystem (s)

生　　境 **Habitat(s)**	森林 (洞穴) / Forest (Cave)
生态系统 **Ecosystem(s)**	森林生态系统 / Forest Ecosystem

威胁 Threat (s)

主要威胁 **Major Threat(s)**	采集 / Collection

保护级别与保护行动 Protection Category and Conservation Action (s)

国家重点保护野生动物等级 **(2021) Category of National Key Protected Wild Animals (2021)**	无 / NA
IUCN 红色名录 **(2020-2) IUCN Red List (2020-2)**	无危 / LC
CITES 附录 **(2019) CITES Appendix (2019)**	无 / NA
保护行动 **Conservation Action(s)**	无 / None

相关文献 Relevant References

Wilson and Mittermeier, 2019; Burgin *et al*., 2018; Jiang *et al*. (蒋志刚等), 2017; Zhou *et al*. (周全等), 2012; Wang (王应祥), 2003

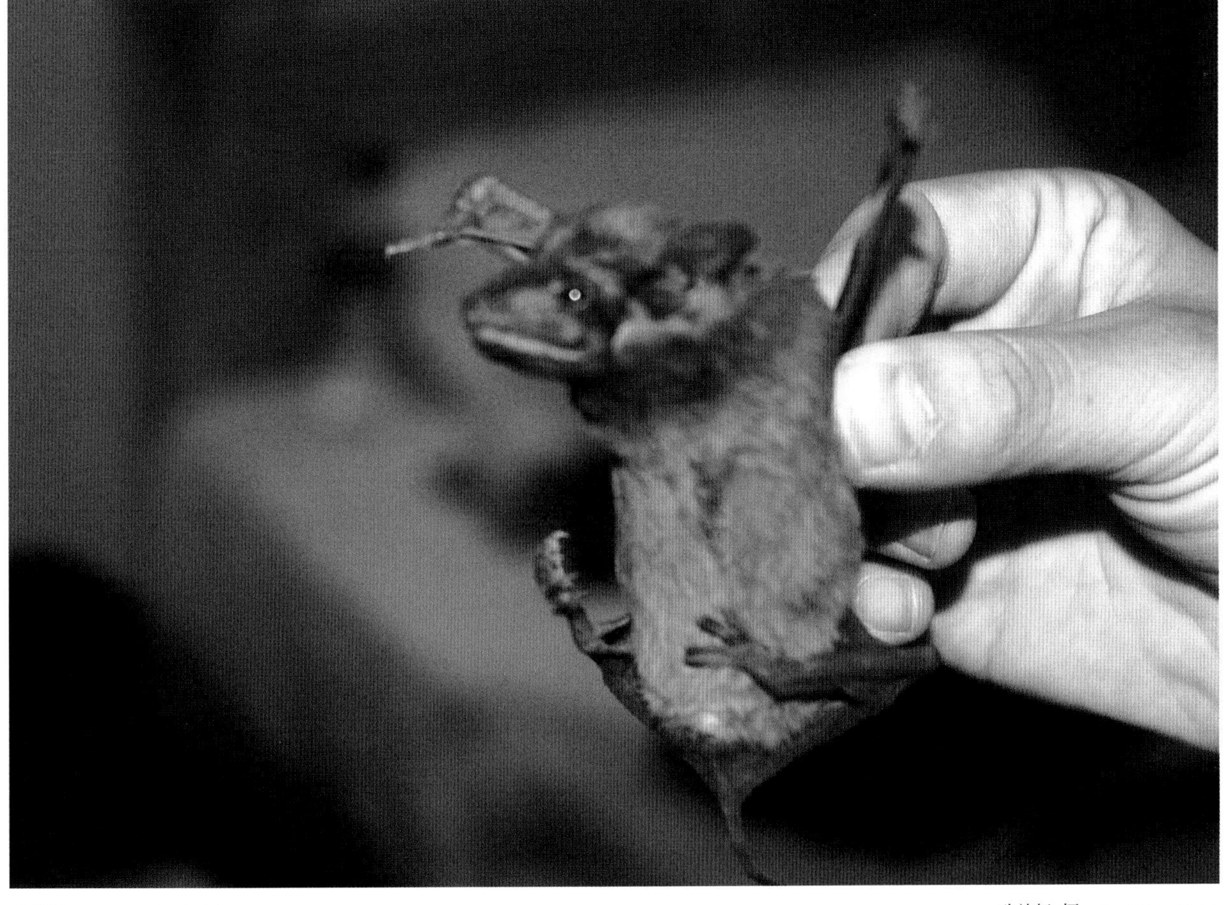

大墓蝠 *Taphozous theobaldi*　　张礼标 摄 By Libiao Zhang

数据缺乏 DD	无危 LC	近危 NT	易危 VU	濒危 EN	极危 CR	区域灭绝 RE	野外灭绝 EW	灭绝 EX

分类地位 Taxonomic Status

动物界 Animalia	脊索动物门 Chordata	哺乳纲 Mammalia	翼手目 Chiroptera	菊头蝠科 Rhinolophidae
学名 Scientific Name		*Rhinolophus lepidus*		
命名人 Species Authority		Blyth, 1844		
英文名 English Name(s)		Blyth's Horseshoe Bat		
同物异名 Synonym(s)		*Rhinolophus shortridgei* Anderson, 1918		
种下单元评估 Infra-specific Taxa Assessed		无 / None		

评估信息 Assessment Information

评估年份 Year Assessed	2020
评定人 Assessor(s)	吴毅、蒋志刚 / Yi Wu, Zhigang Jiang
审定人 Reviewer(s)	张礼标、毛秀光、张树义 / Libiao Zhang, Xiuguang Mao, Shuyi Zhang
其他贡献人 Other Contributor(s)	李立立、丁晨晨 / Lili Li, Chenchen Ding

理由 Justification: 短翼菊头蝠在历史记录中在中国分布较广，但目前其分布及标本信息十分有限。因此，列为数据缺乏等级 / Blyth's Horseshoe Bat is widespread in China in historical records, however, currently the information about its distribution and specimens are very limited. Thus, it is listed as Data Deficient

地理分布 Geographical Distribution

国内分布 Domestic Distribution

江苏、浙江、安徽、江西、四川、云南、广西、贵州、湖南、福建、重庆、广东 / Jiangsu, Zhejiang, Anhui, Jiangxi, Sichuan, Yunnan, Guangxi, Guizhou, Hunan, Fujian, Chongqing, Guangdong

世界分布 World Distribution

阿富汗、孟加拉国、柬埔寨、中国、印度、印度尼西亚、马来西亚、缅甸、尼泊尔、巴基斯坦、泰国、越南 / Afghanistan, Bangladesh, Cambodia, China, India, Indonesia, Malaysia, Myanmar, Nepal, Pakistan, Thailand, Viet Nam

分布标注 Distribution Note

非特有种 / Non-endemic

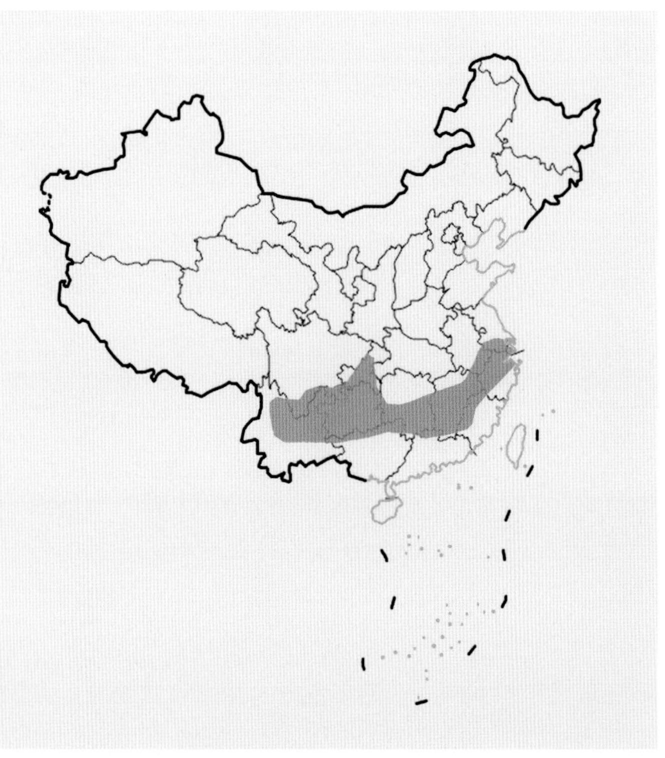

国内分布图
Map of Domestic Distribution

种群 Population

种群数量 Population Size	未知 / Unknown
种群趋势 Population Trend	未知 / Unknown

生境与生态系统 Habitat (s) and Ecosystem (s)

生　　境 Habitat(s)	洞穴、人造建筑、热带湿润低地森林 / Cave, Building, Tropical Moist Lowland Forest
生态系统 Ecosystem(s)	森林生态系统、人类聚落生态系统 / Forest Ecosystem, Human Settlement Ecosystem

威胁 Threat (s)

主要威胁 Major Threat(s)	旅游休闲区建设 / Construction of Tourism and Recreation Area

保护级别与保护行动 Protection Category and Conservation Action (s)

国家重点保护野生动物等级 (2021) Category of National Key Protected Wild Animals (2021)	无 / NA
IUCN 红色名录 (2020-2) IUCN Red List (2020-2)	无危 / LC
CITES 附录 (2019) CITES Appendix (2019)	无 / NA
保护行动 Conservation Action(s)	无 / None

相关文献 Relevant References

Wilson and Mittermeier, 2019; Burgin *et al.*, 2018; Jiang *et al.* (蒋志刚等), 2017; Huang *et al.* (黄薇等), 2008; Wang (王应祥), 2003; Luo and Gao (罗键和高红英), 2002

短翼菊头蝠 *Rhinolophus lepidus*

丽江菊头蝠

Rhinolophus osgoodi

数据缺乏 DD

数据缺乏 DD	无危 LC	近危 NT	易危 VU	濒危 EN	极危 CR	区域灭绝 RE	野外灭绝 EW	灭绝 EX

分类地位 Taxonomic Status

动物界 Animalia	脊索动物门 Chordata	哺乳纲 Mammalia	翼手目 Chiroptera	菊头蝠科 Rhinolophidae
学 名 Scientific Name		*Rhinolophus osgoodi*		
命名人 Species Authority		Sanborn, 1939		
英文名 English Name(s)		Osgood's Horseshoe Bat		
同物异名 Synonym(s)		无 / None		
种下单元评估 Infra-specific Taxa Assessed		无 / None		

评估信息 Assessment Information

评估年份 Year Assessed	2020
评定人 Assessor(s)	吴毅、蒋志刚 / Yi Wu, Zhigang Jiang
审定人 Reviewer(s)	张礼标、毛秀光、张树义 / Libiao Zhang, Xiuguang Mao, Shuyi Zhang
其他贡献人 Other Contributor(s)	李立立、丁晨晨 / Lili Li, Chenchen Ding

理由 Justification: 丽江菊头蝠除了模式标本外，后来没有其他标本记录，对其的研究缺乏。因此，列为数据缺乏等级 / Apart from the original type specimens, no other specimen of Osgood's Horseshoe Bat has been collected. Studies on this species are scarce. Thus, it is listed as Data Deficient

地理分布 Geographical Distribution

国内分布 Domestic Distribution

云南 / Yunnan

世界分布 World Distribution

中国 / China

分布标注 Distribution Note

特有种 / Endemic

国内分布图
Map of Domestic Distribution

种群 Population

种群数量 Population Size	未知 / Unknown
种群趋势 Population Trend	未知 / Unknown

生境与生态系统 Habitat (s) and Ecosystem (s)

生　　境 Habitat(s)	未知 / Unknown
生态系统 Ecosystem(s)	喀斯特生态系统 / Karst Ecosystem

威胁 Threat (s)

主要威胁 Major Threat(s)	未知 / Unknown

保护级别与保护行动 Protection Category and Conservation Action (s)

国家重点保护野生动物等级 (2021) Category of National Key Protected Wild Animals (2021)	无 / NA
IUCN 红色名录 (2020-2) IUCN Red List (2020-2)	数据缺乏 / DD
CITES 附录 (2019) CITES Appendix (2019)	无 / NA
保护行动 Conservation Action(s)	无 / None

相关文献 Relevant References

Wilson and Mittermeier, 2019; Burgin *et al*., 2018; Jiang *et al.* (蒋志刚等), 2017; Zhang *et al*., 2009a, 2009b, 2009c; Sanborn, 1939

施氏菊头蝠

Rhinolophus schnitzleri

数据缺乏 DD

数据缺乏 DD	无危 LC	近危 NT	易危 VU	濒危 EN	极危 CR	区域灭绝 RE	野外灭绝 EW	灭绝 EX

分类地位 Taxonomic Status

动物界 Animalia	脊索动物门 Chordata	哺乳纲 Mammalia	翼手目 Chiroptera	菊头蝠科 Rhinolophidae

学名 Scientific Name	*Rhinolophus schnitzleri*
命名人 Species Authority	Wu and Thong, 2011
英文名 English Name(s)	Schnitzler's Horseshoe Bat
同物异名 Synonym(s)	无 / None
种下单元评估 Infra-specific Taxa Assessed	无 / None

评估信息 Assessment Information

评估年份 Year Assessed	2020
评定人 Assessor(s)	吴毅、蒋志刚 / Yi Wu, Zhigang Jiang
审定人 Reviewer(s)	张礼标、毛秀光、张树义 / Libiao Zhang, Xiuguang Mao, Shuyi Zhang
其他贡献人 Other Contributor(s)	李立立、丁晨晨 / Lili Li, Chenchen Ding

理由 Justification: 2010 年，施氏菊头蝠在云南被发现，为中国特有种，但目前对其的研究缺乏。因此，列为数据缺乏等级 / Schnitzler's Horseshoe Bat was firstly found in Yunnan in 2010. It is endemic to China. However, studies on this species are deficient. Thus, it is listed as Data Deficient

地理分布 Geographical Distribution

国内分布 Domestic Distribution

云南 / Yunnan

世界分布 World Distribution

中国 / China

分布标注 Distribution Note

特有种 / Endemic

国内分布图
Map of Domestic Distribution

种群 Population

种群数量 **Population Size**	罕见 / Very rare
种群趋势 **Population Trend**	未知 / Unknown

生境与生态系统 Habitat (s) and Ecosystem (s)

生　　境 **Habitat(s)**	洞穴 / Cave
生态系统 **Ecosystem(s)**	洞穴生态系统 / Cave Ecosystem

威胁 Threat (s)

主要威胁 **Major Threat(s)**	未知 / Unknown

保护级别与保护行动 Protection Category and Conservation Action (s)

国家重点保护野生动物等级 **(2021) Category of National Key Protected Wild Animals (2021)**	无 / NA
IUCN 红色名录 **(2020-2) IUCN Red List (2020-2)**	未列入 / NA
CITES 附录 **(2019) CITES Appendix (2019)**	无 / NA
保护行动 **Conservation Action(s)**	无 / None

相关文献 Relevant References

Burgin *et al*., 2018; Jiang *et al.* (蒋志刚等), 2017; Wu and Thong, 2011

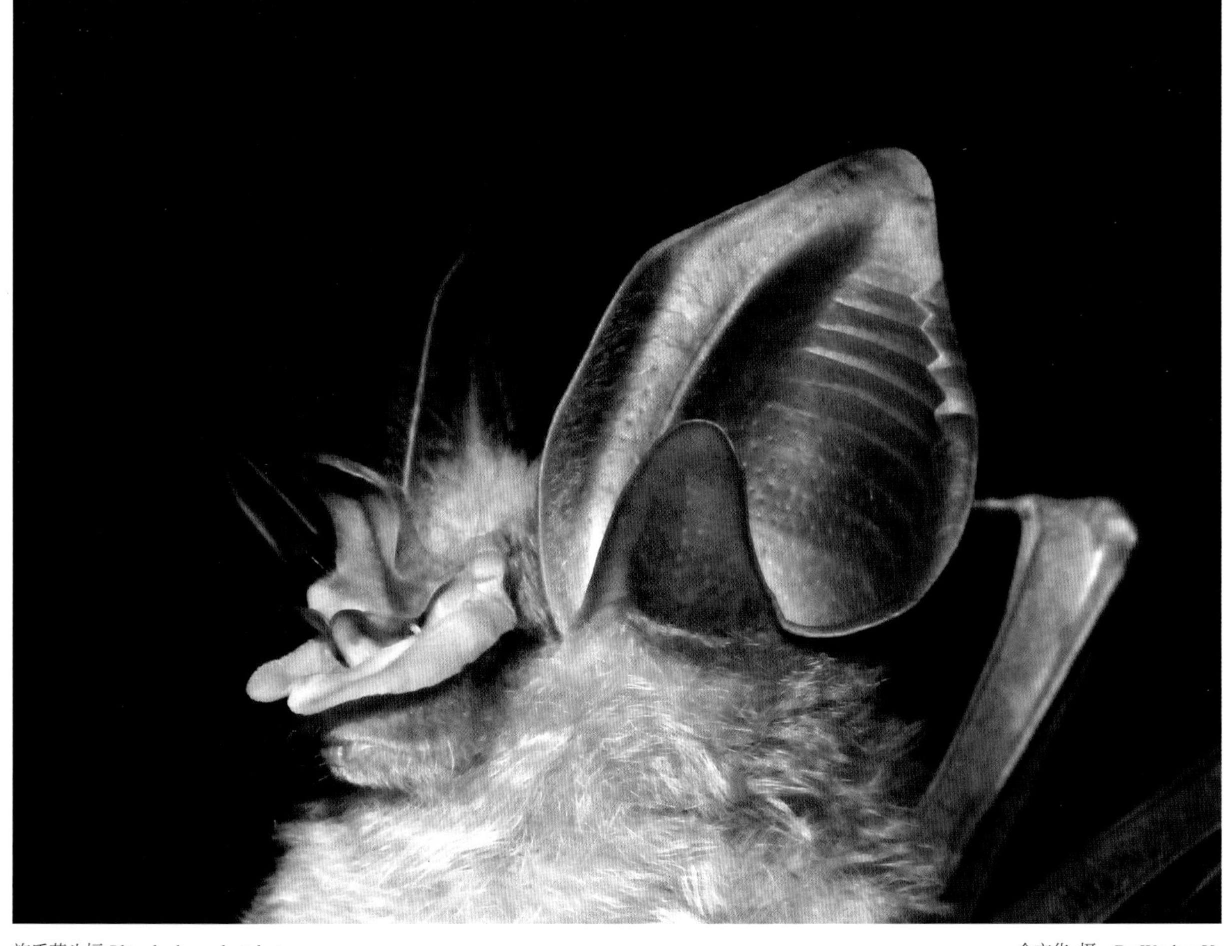

施氏菊头蝠 *Rhinolophus schnitzleri*　　余文华 摄　By Wenhua Yu

数据缺乏 DD	无危 LC	近危 NT	易危 VU	濒危 EN	极危 CR	区域灭绝 RE	野外灭绝 EW	灭绝 EX

分类地位 Taxonomic Status

动 物 界 Animalia	脊索动物门 Chordata	哺 乳 纲 Mammalia	翼 手 目 Chiroptera	菊头蝠科 Rhinolophidae

学　　名 **Scientific Name**	*Rhinolophus subbadius*
命 名 人 **Species Authority**	Blyth, 1844
英 文 名 **English Name(s)**	Little Nepalese Horseshoe Bat
同物异名 **Synonym(s)**	*Rhinolophus garoensis* (Dobson, 1872); *subbadius* (Hodgson, 1841)
种下单元评估 **Infra-specific Taxa Assessed**	无 / None

评估信息 Assessment Information

评 估 年 份 **Year Assessed**	2020
评　定　人 **Assessor(s)**	吴毅、蒋志刚 / Yi Wu, Zhigang Jiang
审　定　人 **Reviewer(s)**	张礼标、毛秀光、张树义 / Libiao Zhang, Xiuguang Mao, Shuyi Zhang
其他贡献人 **Other Contributor(s)**	李立立、丁晨晨 / Lili Li, Chenchen Ding

理由 **Justification:** 葛氏菊头蝠的种群与栖息地研究现状还不清楚。因此，列为数据缺乏等级 / There are limited researches on its population and habitat. Thus, it is listed as Data Deficient

地理分布 Geographical Distribution

国内分布 **Domestic Distribution**

西藏 / Tibet (Xizang)

世界分布 **World Distribution**

中国、孟加拉国、印度、缅甸、尼泊尔 / China, Bangladesh, India, Myanmar, Nepal

分布标注 **Distribution Note**

非特有种 / Non-endemic

国内分布图
Map of Domestic Distribution

种群 Population

种群数量 **Population Size**	罕见 / Very rare
种群趋势 **Population Trend**	未知 / Unknown

生境与生态系统 Habitat (s) and Ecosystem (s)

生　　境 **Habitat(s)**	有竹丛的密林地带 / Dense Forest with Clumpy Bamboo
生态系统 **Ecosystem(s)**	森林生态系统 / Forest Ecosystem

威胁 Threat (s)

主要威胁 **Major Threat(s)**	未知 / Unknown

保护级别与保护行动 Protection Category and Conservation Action (s)

国家重点保护野生动物等级 **(2021) Category of National Key Protected Wild Animals (2021)**	无 / NA
IUCN 红色名录 **(2020-2) IUCN Red List (2020-2)**	无危 / LC
CITES 附录 **(2019) CITES Appendix (2019)**	无 / NA
保护行动 **Conservation Action(s)**	无 / None

相关文献 Relevant References

Wilson and Mittermeier, 2019; Burgin *et al*., 2018; Jiang *et al*. (蒋志刚等), 2017

葛氏菊头蝠 *Rhinolophus subbadius*

数据缺乏 DD	无危 LC	近危 NT	易危 VU	濒危 EN	极危 CR	区域灭绝 RE	野外灭绝 EW	灭绝 EX

分类地位 Taxonomic Status

动物界 Animalia	脊索动物门 Chordata	哺乳纲 Mammalia	翼手目 Chiroptera	菊头蝠科 Rhinolophidae

学　名 **Scientific Name**	*Rhinolophus xinanzhongguoensis*
命名人 **Species Authority**	Zhou, 2009
英文名 **English Name(s)**	Xinan Horseshoe Bat
同物异名 **Synonym(s)**	无 / None
种下单元评估 **Infra-specific Taxa Assessed**	无 / None

评估信息 Assessment Information

评估年份 **Year Assessed**	2020
评定人 **Assessor(s)**	吴毅、蒋志刚 / Yi Wu, Zhigang Jiang
审定人 **Reviewer(s)**	张礼标、毛秀光、张树义 / Libiao Zhang, Xiuguang Mao, Shuyi Zhang
其他贡献人 **Other Contributor(s)**	李立立、丁晨晨 / Lili Li, Chenchen Ding

理由 **Justification:** 锲鞍菊头蝠在 2010 年被发现，但目前对其的研究缺乏。因此，列为数据缺乏 / Xinan Horseshoe Bat was first discovered in 2010 in China. However, studies on this species are deficient. Thus, it is listed as Data Deficient

地理分布 Geographical Distribution

国内分布 Domestic Distribution

云南、贵州 / Yunnan, Guizhou

世界分布 World Distribution

中国 / China

分布标注 Distribution Note

特有种 / Endemic

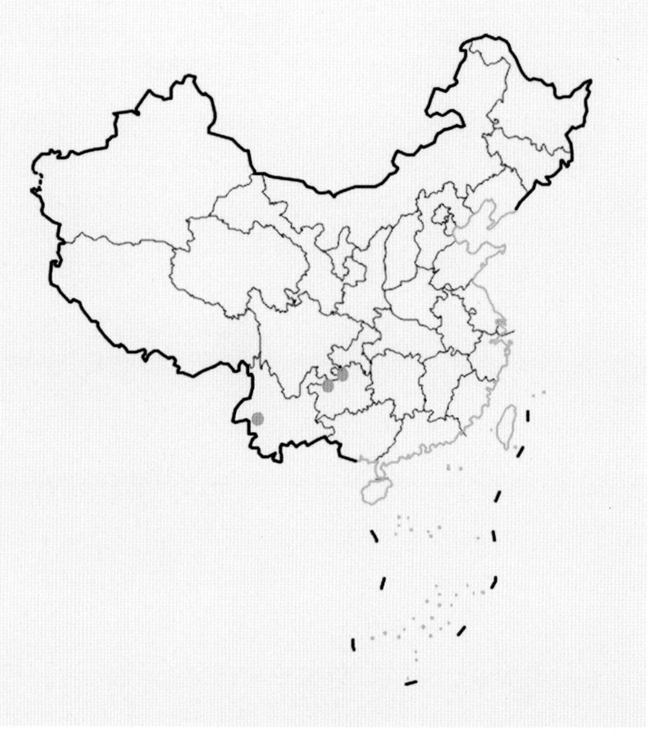

国内分布图
Map of Domestic Distribution

种群 Population

种群数量 Population Size	极罕见 / Very rare
种群趋势 Population Trend	未知 / Unknown

生境与生态系统 Habitat (s) and Ecosystem (s)

生　　境 Habitat(s)	洞穴 / Cave
生态系统 Ecosystem(s)	洞穴生态系统 / Cave Ecosystem

威胁 Threat (s)

主要威胁 Major Threat(s)	伐木、耕种、火灾、采矿、采石场、旅游 / Logging, Plantation, Fire, Mining, Quarrying Field, Tourism

保护级别与保护行动 Protection Category and Conservation Action (s)

国家重点保护野生动物等级 (2021) Category of National Key Protected Wild Animals (2021)	无 / NA
IUCN 红色名录 (2020-2) IUCN Red List (2020-2)	未列入 / NA
CITES 附录 (2019) CITES Appendix (2019)	无 / NA
保护行动 Conservation Action(s)	无 / None

相关文献 Relevant References

Burgin *et al*., 2018; Jiang *et al.* (蒋志刚等), 2017; Zhou *et al*., 2009

锲鞍菊头蝠 *Rhinolophus xinanzhongguoensis*

数据缺乏 DD	无危 LC	近危 NT	易危 VU	濒危 EN	极危 CR	区域灭绝 RE	野外灭绝 EW	灭绝 EX

分类地位 Taxonomic Status

动物界 **Animalia**	脊索动物门 **Chordata**	哺乳纲 **Mammalia**	翼手目 **Chiroptera**	蹄蝠科 **Hipposideridae**

学名 **Scientific Name**	*Hipposideros fulvus*
命名人 **Species Authority**	Gray, 1838
英文名 **English Name(s)**	Fulvus Leaf-nosed Bat
同物异名 **Synonym(s)**	Fulvus Roundleaf Bat; *Hipposideros bicolor* (Gray, 1838) subsp. *fulvus*; *H. bicolor* (Andersen, 1918) subsp. *pallidus*; *H. fulvus* (Andersen, 1918) subsp. *pallidus*; *H. murinus* （转下页）
种下单元评估 **Infra-specific Taxa Assessed**	无 / None

评估信息 Assessment Information

评估年份 **Year Assessed**	2020
评定人 **Assessor(s)**	吴毅、蒋志刚 / Yi Wu, Zhigang Jiang
审定人 **Reviewer(s)**	张礼标、毛秀光、张树义 / Libiao Zhang, Xiuguang Mao, Shuyi Zhang
其他贡献人 **Other Contributor(s)**	李立立、丁晨晨 / Lili Li, Chenchen Ding

理由 **Justification:** 大耳小蹄蝠在中国境内仅分布在云南，对其的研究缺乏。因此，列为数据缺乏等级 / In China, Fulvus Leaf-nosed Bat is distributed only in Yunnan. Studies on it are deficient. Thus, it is listed as Data Deficient

地理分布 Geographical Distribution

国内分布 Domestic Distribution

云南 / Yunnan

世界分布 World Distribution

中国；中亚、南亚 / China; Central Asia, South Asia

分布标注 Distribution Note

非特有种 / Non-endemic

国内分布图
Map of Domestic Distribution

种群 Population

种群数量 Population Size	稀少 / Rare
种群趋势 Population Trend	未知 / Unknown

生境与生态系统 Habitat (s) and Ecosystem (s)

生　境 Habitat(s)	洞穴、人造建筑、森林、草地 / Cave, Building, Forest, Grassland
生态系统 Ecosystem(s)	喀斯特生态系统、人类聚落生态系统 / Karst Ecosystem, Human Settlement Ecosystem

威胁 Threat (s)

主要威胁 Major Threat(s)	未知 / Unknown

保护级别与保护行动 Protection Category and Conservation Action (s)

国家重点保护野生动物等级 (2021) Category of National Key Protected Wild Animals (2021)	无 / NA
IUCN 红色名录 (2020-2) IUCN Red List (2020-2)	无危 / LC
CITES 附录 (2019) CITES Appendix (2019)	无 / NA
保护行动 Conservation Action(s)	无 / None

相关文献 Relevant References

Burgin *et al*., 2018; Jiang *et al.* (蒋志刚等), 2017; Wang *et al*., 2013; Smith *et al.* (史密斯等), 2009; Wang (王应祥), 2003; Koopman, 1993

(接上页)
(Gray, 1838); *Phyllorhina atra* (Fitzinger, 1870); *P. aurita* (Tomes, 1859); *Rhinolophus fulgens* (Elliot, 1839)

大耳小蹄蝠 *Hipposideros fulvus* By Rohit. chakravarty

栗鼠耳蝠
Myotis badius

数据缺乏 DD

数据缺乏 DD	无危 LC	近危 NT	易危 VU	濒危 EN	极危 CR	区域灭绝 RE	野外灭绝 EW	灭绝 EX

分类地位 Taxonomic Status

动 物 界 Animalia	脊索动物门 Chordata	哺 乳 纲 Mammalia	翼 手 目 Chiroptera	蝙 蝠 科 Vespertilionidae

学 名 **Scientific Name**	*Myotis badius*
命 名 人 **Species Authority**	Tiunov, 2011
英 文 名 **English Name(s)**	Bay Mouse-eared Bat
同物异名 **Synonym(s)**	Chestnut Myotis
种下单元评估 **Infra-specific Taxa Assessed**	无 / None

评估信息 Assessment Information

评 估 年 份 **Year Assessed**	2020
评 定 人 **Assessor(s)**	吴毅、蒋志刚 / Yi Wu, Zhigang Jiang
审 定 人 **Reviewer(s)**	江廷磊、余文华、石红艳 / Tinglei Jiang, Wenhua Yu, Hongyan Shi
其他贡献人 **Other Contributor(s)**	李立立、丁晨晨 / Lili Li, Chenchen Ding

理由 **Justification:** 栗鼠耳蝠是中国特有种，2011 年该种在云南被发现并记录，目前对其的研究缺乏。因此，列为数据缺乏等级 / Bay Mouse-eared Bat is an endemic species, and was firstly discovered and recorded in Yunnan in 2011. So far, studies on this species are deficient. Thus, it is listed as Data Deficient

地理分布 Geographical Distribution

国内分布 Domestic Distribution

云南 / Yunnan

世界分布 World Distribution

中国 / China

分布标注 Distribution Note

特有种 / Endemic

国内分布图
Map of Domestic Distribution

种群 Population

种群数量 **Population Size**	未知 / Unknown
种群趋势 **Population Trend**	下降 / Decreasing

生境与生态系统 Habitat (s) and Ecosystem (s)

生　　境 **Habitat(s)**	洞穴 / Cave
生态系统 **Ecosystem(s)**	喀斯特生态系统、森林生态系统 / Karst Ecosystem, Forest Ecosystem

威胁 Threat (s)

主要威胁 **Major Threat(s)**	未知 / Unknown

保护级别与保护行动 Protection Category and Conservation Action (s)

国家重点保护野生动物等级 **(2021) Category of National Key Protected Wild Animals (2021)**	无 / NA
IUCN 红色名录 **(2020-2) IUCN Red List (2020-2)**	未列入 / NA
CITES 附录 **(2019) CITES Appendix (2019)**	无 / NA
保护行动 **Conservation Action(s)**	无 / None

相关文献 Relevant References

Burgin *et al*., 2018; Jiang *et al.* (蒋志刚等), 2017; Tiunov *et al.*, 2011

栗鼠耳蝠 *Myotis badius*　　By Gabor Csorba

远东鼠耳蝠

Myotis bombinus

数据缺乏 DD

数据缺乏 DD	无危 LC	近危 NT	易危 VU	濒危 EN	极危 CR	区域灭绝 RE	野外灭绝 EW	灭绝 EX

分类地位 Taxonomic Status

动物界 Animalia	脊索动物门 Chordata	哺乳纲 Mammalia	翼手目 Chiroptera	蝙蝠科 Vespertilionidae

学名 **Scientific Name**	*Myotis bombinus*
命名人 **Species Authority**	Thomas, 1906
英文名 **English Name(s)**	Far Eastern Mouse-eared Bat
同物异名 **Synonym(s)**	Far Eastern Myotis; Bombinus Bat; *Myotis amurensis* (Ognev, 1927)
种下单元评估 **Infra-specific Taxa Assessed**	无 / None

评估信息 Assessment Information

评估年份 **Year Assessed**	2020
评定人 **Assessor(s)**	吴毅、蒋志刚 / Yi Wu, Zhigang Jiang
审定人 **Reviewer(s)**	江廷磊、余文华、石红艳 / Tinglei Jiang, Wenhua Yu, Hongyan Shi
其他贡献人 **Other Contributor(s)**	李立立、丁晨晨 / Lili Li, Chenchen Ding

理由 **Justification:** 远东鼠耳蝠在中国的分布区小，种群数量少，且其信息十分有限。因此，列为数据缺乏等级 / Far Eastern Mouse-eared Bat is narrowly distributed with small populations in China. Furthermore, information about this species is very limited. Thus, it is listed as Data Deficient

地理分布 Geographical Distribution

国内分布 Domestic Distribution

吉林、黑龙江 / Jilin, Heilongjiang

世界分布 World Distribution

中国；东北亚 / China; Northeast Asia

分布标注 Distribution Note

非特有种 / Non-endemic

国内分布图
Map of Domestic Distribution

种群 Population

种群数量 Population Size	稀少 / Rare
种群趋势 Population Trend	下降 / Decreasing

生境与生态系统 Habitat (s) and Ecosystem (s)

生　　境 Habitat(s)	森林、洞穴、人造建筑 / Forest, Cave, Building
生态系统 Ecosystem(s)	森林生态系统、喀斯特生态系统、人类聚落生态系统 / Forest Ecosystem, Karst Ecosystem, Human Settlement Ecosystem

威胁 Threat (s)

主要威胁 Major Threat(s)	人类活动干扰 / Human Disturbance

保护级别与保护行动 Protection Category and Conservation Action (s)

国家重点保护野生动物等级 (2021) Category of National Key Protected Wild Animals (2021)	无 / NA
IUCN 红色名录 (2020-2) IUCN Red List (2020-2)	近危 / NT
CITES 附录 (2019) CITES Appendix (2019)	无 / NA
保护行动 Conservation Action(s)	无 / None

相关文献 Relevant References

Wilson and Mittermeier, 2019; Burgin *et al.*, 2018; Jiang *et al.* (蒋志刚等), 2017; Smith *et al.* (史密斯等), 2009; Kawai *et al.*, 2003; Wang (王应祥), 2003; Horácek *et al.*, 2000; Yoon, 1990; Horácek and Hanák, 1984

长尾鼠耳蝠

Myotis frater

数据缺乏 DD

数据缺乏 DD	无危 LC	近危 NT	易危 VU	濒危 EN	极危 CR	区域灭绝 RE	野外灭绝 EW	灭绝 EX

分类地位 Taxonomic Status

动物界 Animalia	脊索动物门 Chordata	哺乳纲 Mammalia	翼手目 Chiroptera	蝙蝠科 Vespertilionidae

学名 Scientific Name	*Myotis frater*
命名人 Species Authority	G. M. Allen, 1923
英文名 English Name(s)	Fraternal Mouse-eared Bat
同物异名 Synonym(s)	Fraternal Myotis; Long-tailed Whiskered Bat; *Myotis frater* (Tsytsulina and Strelkov, 2001) subsp. *eniseensis*; *M. frater* (Imaizumi, 1956) subsp. *kguyae*; *longicaudatus* (Ognev, 1927)
种下单元评估 Infra-specific Taxa Assessed	无 / None

评估信息 Assessment Information

评估年份 Year Assessed	2020
评定人 Assessor(s)	吴毅、蒋志刚 / Yi Wu, Zhigang Jiang
审定人 Reviewer(s)	江廷磊、余文华、石红艳 / Tinglei Jiang, Wenhua Yu, Hongyan Shi
其他贡献人 Other Contributor(s)	李立立、丁晨晨 / Lili Li, Chenchen Ding

理由 **Justification:** 长尾鼠耳蝠分布广，但对其的研究缺乏。因此，列为数据缺乏等级 / Fraternal Mouse-eared Bat has a wide range of distribution, but studies on it are deficient. Thus, it is listed as Data Deficient

地理分布 Geographical Distribution

国内分布 Domestic Distribution

黑龙江、安徽、江西、福建、四川 / Heilongjiang, Anhui, Jiangxi, Fujian, Sichuan

世界分布 World Distribution

中国；东北亚 / China; Northeast Asia

分布标注 Distribution Note

非特有种 / Non-endemic

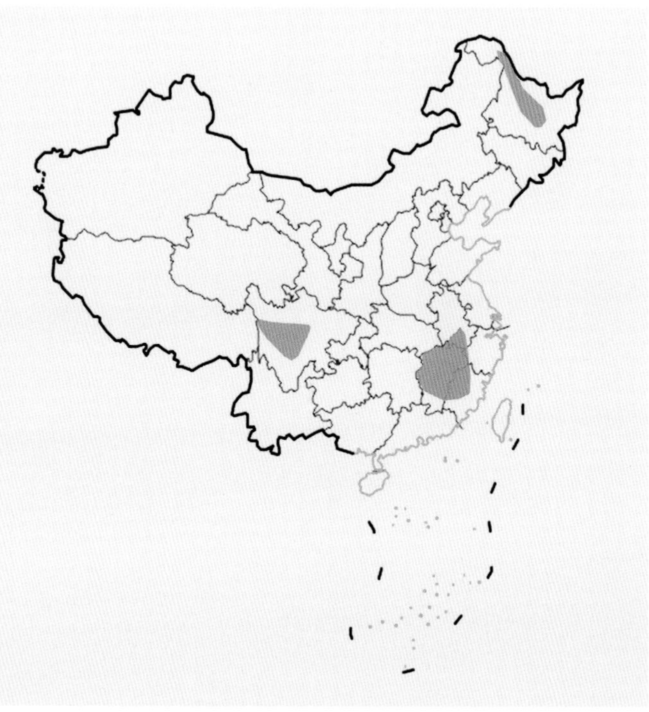

国内分布图
Map of Domestic Distribution

种群 Population

种群数量 **Population Size**	未知 / Unknown
种群趋势 **Population Trend**	未知 / Unknown

生境与生态系统 Habitat (s) and Ecosystem (s)

生　　境 **Habitat(s)**	洞穴、森林 / Cave, Forest
生态系统 **Ecosystem(s)**	喀斯特生态系统、森林生态系统 / Karst Ecosystem, Forest Ecosystem

威胁 Threat (s)

主要威胁 **Major Threat(s)**	未知 / Unknown

保护级别与保护行动 Protection Category and Conservation Action (s)

国家重点保护野生动物等级 **(2021) Category of National Key Protected Wild Animals (2021)**	无 / NA
IUCN 红色名录 **(2020-2) IUCN Red List (2020-2)**	数据缺乏 / DD
CITES 附录 **(2019) CITES Appendix (2019)**	无 / NA
保护行动 **Conservation Action(s)**	无 / None

相关文献 Relevant References

Burgin *et al*., 2018; Jiang *et al.* (蒋志刚等), 2017; Dang *et al.* (党飞红等), 2017; Liu (刘伟), 2012; Zhang *et al.* (张桢珍等), 2008; Zhang *et al.* (张树义等), 2000

印支鼠耳蝠
Myotis indochinensis

数据缺乏 DD

数据缺乏 DD	无危 LC	近危 NT	易危 VU	濒危 EN	极危 CR	区域灭绝 RE	野外灭绝 EW	灭绝 EX

分类地位 Taxonomic Status

动物界 Animalia	脊索动物门 Chordata	哺乳纲 Mammalia	翼手目 Chiroptera	蝙蝠科 Vespertilionidae

学名 Scientific Name	*Myotis indochinensis*
命名人 Species Authority	Son *et al*., 2013
英文名 English Name(s)	Indochinese Mouse-eared Bat
同物异名 Synonym(s)	无 / None
种下单元评估 Infra-specific Taxa Assessed	无 / None

评估信息 Assessment Information

评估年份 Year Assessed	2020
评定人 Assessor(s)	吴毅、蒋志刚 / Yi Wu, Zhigang Jiang
审定人 Reviewer(s)	江廷磊、余文华、石红艳 / Tinglei Jiang, Wenhua Yu, Hongyan Shi
其他贡献人 Other Contributor(s)	李立立、丁晨晨 / Lili Li, Chenchen Ding

理由 Justification: 印支鼠耳蝠是2017年发表的中国新记录，但其信息有限。因此，列为数据缺乏等级 / Indochinese Mouse-eared Bat is a newly record published in 2017 in China , but information about this species is limited. Thus, it is listed as Data Deficient

地理分布 Geographical Distribution

国内分布 Domestic Distribution

广东、江西 / Guangdong, Jiangxi

世界分布 World Distribution

中国、越南 / China, Viet Nam

分布标注 Distribution Note

非特有种 / Non-endemic

国内分布图
Map of Domestic Distribution

种群 Population

种群数量 Population Size	未知 / Unknown
种群趋势 Population Trend	未知 / Unknown

生境与生态系统 Habitat (s) and Ecosystem (s)

生　境 Habitat(s)	热带和亚热带湿润山地森林 / Tropical and Subtropical Moist Montane Forest
生态系统 Ecosystem(s)	喀斯特生态系统、森林生态系统 / Karst Ecosystem, Forest Ecosystem

威胁 Threat (s)

主要威胁 Major Threat(s)	无 / None

保护级别与保护行动 Protection Category and Conservation Action (s)

国家重点保护野生动物等级 (2021) Category of National Key Protected Wild Animals (2021)	无 / NA
IUCN 红色名录 (2020-2) IUCN Red List (2020-2)	无危 / LC
CITES 附录 (2019) CITES Appendix (2019)	无 / NA
保护行动 Conservation Action(s)	无 / None

相关文献 Relevant References

He *et al*. (何向阳等), 2019; Burgin *et al*., 2018; Dang *et al*. (党飞红等), 2017; Son *et al*., 2013

尼泊尔鼠耳蝠
Myotis nipalensis

数据缺乏 DD

数据缺乏 DD	无危 LC	近危 NT	易危 VU	濒危 EN	极危 CR	区域灭绝 RE	野外灭绝 EW	灭绝 EX

分类地位 Taxonomic Status

动物界 Animalia	脊索动物门 Chordata	哺乳纲 Mammalia	翼手目 Chiroptera	蝙蝠科 Vespertilionidae
学名 Scientific Name		*Myotis nipalensis*		
命名人 Species Authority		Dobson, 1871		
英文名 English Name(s)		Nepal Mouse-eared Bat		
同物异名 Synonym(s)		Nepal Myotis; *Myotis kukunoriensis* (Bobrinskii, 1929); *M. meinertzhageni* (Thomas, 1926); *M. mystacinus* (Dobson, 1871) subsp. *nipalensis*; *M. mystacinus* (Kuhl, 1817) (转下页)		
种下单元评估 Infra-specific Taxa Assessed		无 / None		

评估信息 Assessment Information

评估年份 Year Assessed	2020
评定人 Assessor(s)	吴毅、蒋志刚 / Yi Wu, Zhigang Jiang
审定人 Reviewer(s)	江廷磊、余文华、石红艳 / Tinglei Jiang, Wenhua Yu, Hongyan Shi
其他贡献人 Other Contributor(s)	李立立、丁晨晨 / Lili Li, Chenchen Ding

理由 Justification: 尼泊尔鼠耳蝠分布广，但对其的研究缺乏。因此，列为数据缺乏等级 / Nepal Mouse-eared Bat is widely distributed, but studies on it are deficient. Thus, it is listed as Data Deficient

地理分布 Geographical Distribution

国内分布 Domestic Distribution

青海、江苏、陕西、甘肃、新疆、西藏 / Qinghai, Jiangsu, Shaanxi, Gansu, Xinjiang, Tibet (Xizang)

世界分布 World Distribution

中国；中亚、西亚 / China; Central Asia, West Asia

分布标注 Distribution Note

非特有种 / Non-endemic

国内分布图
Map of Domestic Distribution

种群 Population

种群数量 **Population Size**	未知 / Unknown
种群趋势 **Population Trend**	未知 / Unknown

生境与生态系统 Habitat (s) and Ecosystem (s)

生　境 **Habitat(s)**	森林、灌丛、草地、荒漠、人造建筑、洞穴 / Forest, Shrubland, Grassland, Desert, Building, Cave
生态系统 **Ecosystem(s)**	森林生态系统、灌丛生态系统、农田生态系统、人类聚落生态系统、湖泊河流生态系统、荒漠生态系统 / Forest Ecosystem, Shrubland Ecosystem, Cropland Ecosystem, Human Settlement Ecosystem, Lake and River Ecosystem, Desert Ecosystem

威胁 Threat (s)

主要威胁 **Major Threat(s)**	无 / None

保护级别与保护行动 Protection Category and Conservation Action (s)

国家重点保护野生动物等级 **(2021) Category of National Key Protected Wild Animals (2021)**	无 / NA
IUCN 红色名录 **(2020-2) IUCN Red List (2020-2)**	无危 / LC
CITES 附录 **(2019) CITES Appendix (2019)**	无 / NA
保护行动 **Conservation Action(s)**	无 / None

相关文献 Relevant References

Wilson and Mittermeier, 2019; Burgin *et al*., 2018; Jiang *et al*. (蒋志刚等), 2017; Liu *et al.* (刘奇等), 2014; Wu and Pei (吴家炎和裴俊峰), 2011; Smith *et al.* (史密斯等), 2009

(接上页)
subsp. *partim*; *M. przewalskii* (Bobrinskoj, 1926); *M. sogdianus* (Kuzyakin, 1934); *M. transcaspicus* (Ognev and Heptner, 1928); *Vespertilio nipalensis* (Dobson, 1871); *V. pallidiventris* (Hodgson, 1844)

中国生物多样性红色名录
China's Red List of Biodiversity

北京鼠耳蝠
Myotis pequinius

数据缺乏 DD

数据缺乏 DD	无危 LC	近危 NT	易危 VU	濒危 EN	极危 CR	区域灭绝 RE	野外灭绝 EW	灭绝 EX

分类地位 Taxonomic Status

动物界 Animalia	脊索动物门 Chordata	哺乳纲 Mammalia	翼手目 Chiroptera	蝙蝠科 Vespertilionidae

学名 Scientific Name	*Myotis pequinius*
命名人 Species Authority	Thomas, 1908
英文名 English Name(s)	Peking Mouse-eared Bat
同物异名 Synonym(s)	Beijing Mouse-eared Bat, Peking Myotis
种下单元评估 Infra-specific Taxa Assessed	无 / None

评估信息 Assessment Information

评估年份 Year Assessed	2020
评定人 Assessor(s)	吴毅、蒋志刚 / Yi Wu, Zhigang Jiang
审定人 Reviewer(s)	江廷磊、余文华、石红艳 / Tinglei Jiang, Wenhua Yu, Hongyan Shi
其他贡献人 Other Contributor(s)	李立立、丁晨晨 / Lili Li, Chenchen Ding

理由 Justification: 北京鼠耳蝠为中国特有种，历史记录该种分布较广，但目前该种信息有限。因此，列为数据缺乏等级 / Peking Mouse-eared Bat is an endemic species, and has widely distributed historic records in China, but currently information about this species is limited. Thus, it is listed as Data Deficient

地理分布 Geographical Distribution

国内分布 Domestic Distribution

北京、河北、四川、山东、河南、安徽、江苏 / Beijing, Hebei, Sichuan, Shandong, Henan, Anhui, Jiangsu

世界分布 World Distribution

中国 / China

分布标注 Distribution Note

特有种 / Endemic

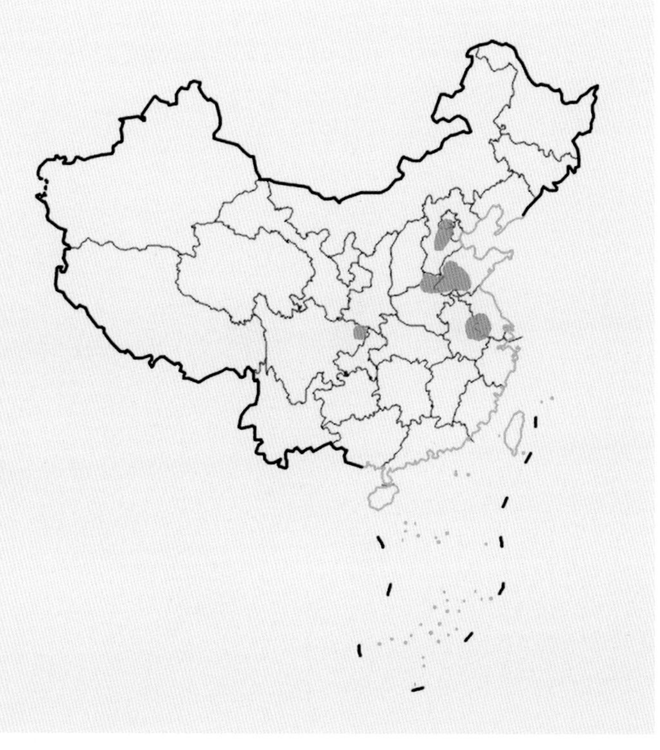

国内分布图
Map of Domestic Distribution

种群 Population

种群数量 **Population Size**	未知 / Unknown
种群趋势 **Population Trend**	未知 / Unknown

生境与生态系统 Habitat (s) and Ecosystem (s)

生　　境 **Habitat(s)**	洞穴、森林 / Cave, Forest
生态系统 **Ecosystem(s)**	喀斯特生态系统、森林生态系统 / Karst Ecosystem, Forest Ecosystem

威胁 Threat (s)

主要威胁 **Major Threat(s)**	未知 / Unknown

保护级别与保护行动 Protection Category and Conservation Action (s)

国家重点保护野生动物等级 **(2021) Category of National Key Protected Wild Animals (2021)**	无 / NA
IUCN 红色名录 **(2020-2) IUCN Red List (2020-2)**	无危 / LC
CITES 附录 **(2019) CITES Appendix (2019)**	无 / NA
保护行动 **Conservation Action(s)**	无 / None

相关文献 Relevant References

Wilson and Mittermeier, 2019; Burgin *et al*., 2018; Jiang *et al*. (蒋志刚等), 2017; Jones *et al*., 2006; Woodman, 1993

北京鼠耳蝠 *Myotis pequinius*

© Lynx Edicions

东亚水鼠耳蝠

Myotis petax

数据缺乏 DD

数据缺乏 DD	无危 LC	近危 NT	易危 VU	濒危 EN	极危 CR	区域灭绝 RE	野外灭绝 EW	灭绝 EX

分类地位 Taxonomic Status

动物界 **Animalia**	脊索动物门 **Chordata**	哺乳纲 **Mammalia**	翼手目 **Chiroptera**	蝙蝠科 **Vespertilionidae**

学名 **Scientific Name**	*Myotis petax*
命名人 **Species Authority**	Hollister, 1912
英文名 **English Name(s)**	Eastern Water Bat
同物异名 **Synonym(s)**	Sakhalin Bat
种下单元评估 **Infra-specific Taxa Assessed**	无 / None

评估信息 Assessment Information

评估年份 **Year Assessed**	2020
评定人 **Assessor(s)**	吴毅、蒋志刚 / Yi Wu, Zhigang Jiang
审定人 **Reviewer(s)**	江廷磊、余文华、石红艳 / Tinglei Jiang, Wenhua Yu, Hongyan Shi
其他贡献人 **Other Contributor(s)**	李立立、丁晨晨 / Lili Li, Chenchen Ding

理由 **Justification:** 东亚水鼠耳蝠分布广，但对其的研究缺乏。因此，列为数据缺乏等级 / Eastern Water Bat is widely distributed, but studies on it are scarce. Thus, it is listed as Data Deficient

地理分布 Geographical Distribution

国内分布 Domestic Distribution

黑龙江、吉林、内蒙古、新疆 / Heilongjiang, Jilin, Inner Mongolia (Nei Mongol), Xinjiang

世界分布 World Distribution

亚洲、欧洲 / Asia, Europe

分布标注 Distribution Note

非特有种 / Non-endemic

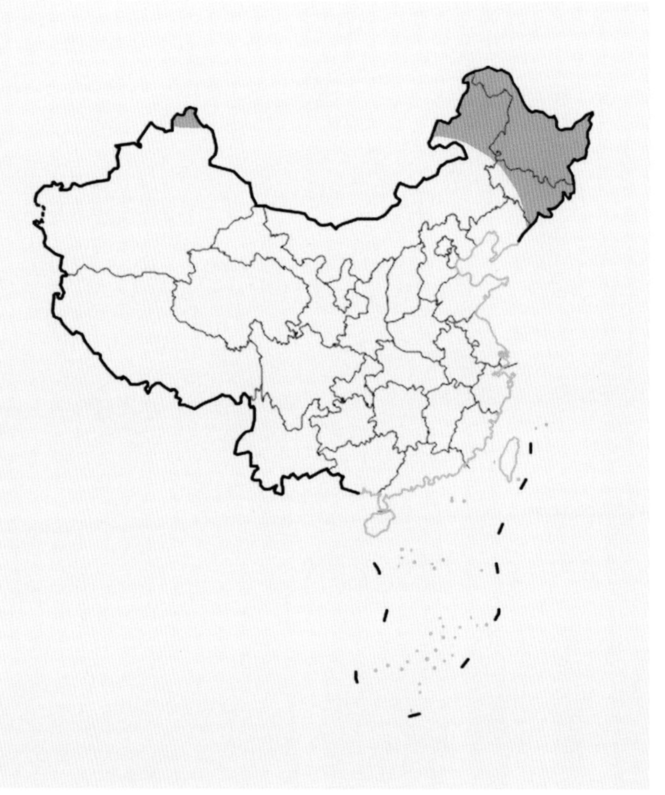

国内分布图
Map of Domestic Distribution

种群 Population

种群数量 **Population Size**	未知 / Unknown
种群趋势 **Population Trend**	未知 / Unknown

生境与生态系统 Habitat (s) and Ecosystem (s)

生　　境 **Habitat(s)**	洞穴、淡水湖、人工水域、人造建筑、森林、灌丛 / Caves, Freshwater Lakes, Artificial Water, Building, Forest, Shrubland
生态系统 **Ecosystem(s)**	喀斯特生态系统、湖泊生态系统、森林生态系统、草地生态系统 / Karst Ecosystem, Lake Ecosystem, Forest Ecosystem, Grassland Ecosystem

威胁 Threat (s)

主要威胁 **Major Threat(s)**	未知 / Unknown

保护级别与保护行动 Protection Category and Conservation Action (s)

国家重点保护野生动物等级 **(2021) Category of National Key Protected Wild Animals (2021)**	无 / NA
IUCN 红色名录 **(2020-2) IUCN Red List (2020-2)**	未列入 / NA
CITES 附录 **(2019) CITES Appendix (2019)**	无 / NA
保护行动 **Conservation Action(s)**	无 / None

相关文献 Relevant References

Burgin *et al*., 2018; Jiang *et al.* (蒋志刚等), 2017; Wang *et al.* (王磊等), 2010; Smith *et al.* (史密斯等), 2009

东亚水鼠耳蝠 *Myotis petax*　　郭东革 摄　By Dongge Guo

古氏伏翼

Pipistrellus kuhlii

数据缺乏 DD

数据缺乏 DD	无危 LC	近危 NT	易危 VU	濒危 EN	极危 CR	区域灭绝 RE	野外灭绝 EW	灭绝 EX

分类地位 Taxonomic Status

动物界 Animalia	脊索动物门 Chordata	哺乳纲 Mammalia	翼手目 Chiroptera	蝙蝠科 Vespertilionidae
学　名 Scientific Name		*Pipistrellus kuhlii*		
命名人 Species Authority		Kuhl, 1817		
英文名 English Name(s)		Kuhl's Pipistrelle		
同物异名 Synonym(s)		无 / None		
种下单元评估 Infra-specific Taxa Assessed		无 / None		

评估信息 Assessment Information

评估年份 Year Assessed	2020
评定人 Assessor(s)	吴毅、蒋志刚 / Yi Wu, Zhigang Jiang
审定人 Reviewer(s)	余文华、石红艳、江廷磊 / Wenhua Yu, Hongyan Shi, Tinglei Jiang
其他贡献人 Other Contributor(s)	李立立、丁晨晨 / Lili Li, Chenchen Ding

理由 Justification: 古氏伏翼仅分布于云南，其信息十分有限。因此，列为数据缺乏等级 / Kuhl's Pipistrelle distributed only in Yunnan, and information about this species is very limited. Thus, it is listed as Data Deficient

地理分布 Geographical Distribution

国内分布 Domestic Distribution

云南 / Yunnan

世界分布 World Distribution

亚洲、欧洲、北非 / Asia, Europe, North Africa

分布标注 Distribution Note

非特有种 / Non-endemic

国内分布图 Map of Domestic Distribution

种群 Population

种群数量 **Population Size**	未知 / Unknown
种群趋势 **Population Trend**	未知 / Unknown

生境与生态系统 Habitat (s) and Ecosystem (s)

生　　境 **Habitat(s)**	耕地、城市 / Arable Land, Urban Area
生态系统 **Ecosystem(s)**	农田生态系统、人类聚落生态系统 / Cropland Ecosystem, Human Settlement Ecosystem

威胁 Threat (s)

主要威胁 **Major Threat(s)**	未知 / Unknown

保护级别与保护行动 Protection Category and Conservation Action (s)

国家重点保护野生动物等级 (2021) **Category of National Key Protected Wild Animals (2021)**	无 / NA
IUCN 红色名录 **(2020-2) IUCN Red List (2020-2)**	无危 / LC
CITES 附录 **(2019) CITES Appendix (2019)**	无 / NA
保护行动 **Conservation Action(s)**	无 / None

相关文献 Relevant References

Wilson and Mittermeier, 2019; Burgin *et al.*, 2018; Jiang *et al.* (蒋志刚等), 2017; Smith *et al.* (史密斯等), 2009; Wang (王应祥), 2003; Zhang (张荣祖), 1997

古氏伏翼 *Pipistrellus kuhlii*

蒋志刚 绘 Drawn by Zhigang Jiang

大灰伏翼

Falsistrellus mordax

数据缺乏 DD

数据缺乏 DD	无危 LC	近危 NT	易危 VU	濒危 EN	极危 CR	区域灭绝 RE	野外灭绝 EW	灭绝 EX

分类地位 Taxonomic Status

动物界 Animalia	脊索动物门 Chordata	哺乳纲 Mammalia	翼手目 Chiroptera	蝙蝠科 Vespertilionidae

学名 Scientific Name	*Falsistrellus mordax*
命名人 Species Authority	Peters, 1866
英文名 English Name(s)	Pungent Pipistrelle
同物异名 Synonym(s)	*Pipistrellus mordax* (Peters, 1866)
种下单元评估 Infra-specific Taxa Assessed	无 / None

评估信息 Assessment Information

评估年份 Year Assessed	2020
评定人 Assessor(s)	吴毅、蒋志刚 / Yi Wu, Zhigang Jiang
审定人 Reviewer(s)	余文华、石红艳、江廷磊 / Wenhua Yu, Hongyan Shi, Tinglei Jiang
其他贡献人 Other Contributor(s)	李立立、丁晨晨 / Lili Li, Chenchen Ding

理由 Justification: 大灰伏翼在中国的分布区极为狭窄，种群数量较少，且信息十分有限。因此，列为数据缺乏等级 / Pungent Pipistrelle is extremely narrowly distributed in China, and its population size is small. Furthermore, information about this species is very limited. Thus, it is listed as Data Deficient

地理分布 Geographical Distribution

国内分布 Domestic Distribution

云南 / Yunnan

世界分布 World Distribution

中国、印度尼西亚 / China, Indonesia

分布标注 Distribution Note

非特有种 / Non-endemic

国内分布图
Map of Domestic Distribution

种群 Population

种群数量 **Population Size**	未知 / Unknown
种群趋势 **Population Trend**	未知 / Unknown

生境与生态系统 Habitat (s) and Ecosystem (s)

生　境 **Habitat(s)**	未知 / Unknown
生态系统 **Ecosystem(s)**	未知 / Unknown

威胁 Threat (s)

主要威胁 **Major Threat(s)**	未知 / Unknown

保护级别与保护行动 Protection Category and Conservation Action (s)

国家重点保护野生动物等级 **(2021) Category of National Key Protected Wild Animals (2021)**	无 / NA
IUCN 红色名录 **(2020-2) IUCN Red List (2020-2)**	数据缺乏 / DD
CITES 附录 **(2019) CITES Appendix (2019)**	无 / NA
保护行动 **Conservation Action(s)**	无 / None

相关文献 Relevant References

Wilson and Mittermeier, 2019; Burgin *et al.*, 2018; Jiang *et al.* (蒋志刚等), 2017; Wang (王应祥), 2003; Corbet and Hill, 1992

环颈伏翼

Thainycteris aureocollaris

数据缺乏 DD

数据缺乏 DD	无危 LC	近危 NT	易危 VU	濒危 EN	极危 CR	区域灭绝 RE	野外灭绝 EW	灭绝 EX

分类地位 Taxonomic Status

动物界 Animalia	脊索动物门 Chordata	哺乳纲 Mammalia	翼手目 Chiroptera	蝙蝠科 Vespertilionidae

学 名 **Scientific Name**	*Thainycteris aureocollaris*
命名人 **Species Authority**	Kock and Storch, 1996
英文名 **English Name(s)**	Collared Sprite
同物异名 **Synonym(s)**	*Arielulus aureocollaris*
种下单元评估 **Infra-specific Taxa Assessed**	无 / None

评估信息 Assessment Information

评估年份 **Year Assessed**	2020
评定人 **Assessor(s)**	吴毅、蒋志刚 / Yi Wu, Zhigang Jiang
审定人 **Reviewer(s)**	张礼标、毛秀光、张树义 / Libiao Zhang, Xiuguang Mao, Shuyi Zhang
其他贡献人 **Other Contributor(s)**	李立立、丁晨晨 / Lili Li, Chenchen Ding

理由 **Justification:** 环颈伏翼是2017年发表的新种，目前其信息有限。因此，列为数据缺乏等级 / Collared Sprite is a newly published species in 2017, information about it is very limited currently. Thus, it is listed as Data Deficient

地理分布 Geographical Distribution

国内分布 Domestic Distribution

贵州 / Guizhou

世界分布 World Distribution

中国；东南亚 / China; Southeast Asia

分布标注 Distribution Note

非特有种 / Non-endemic

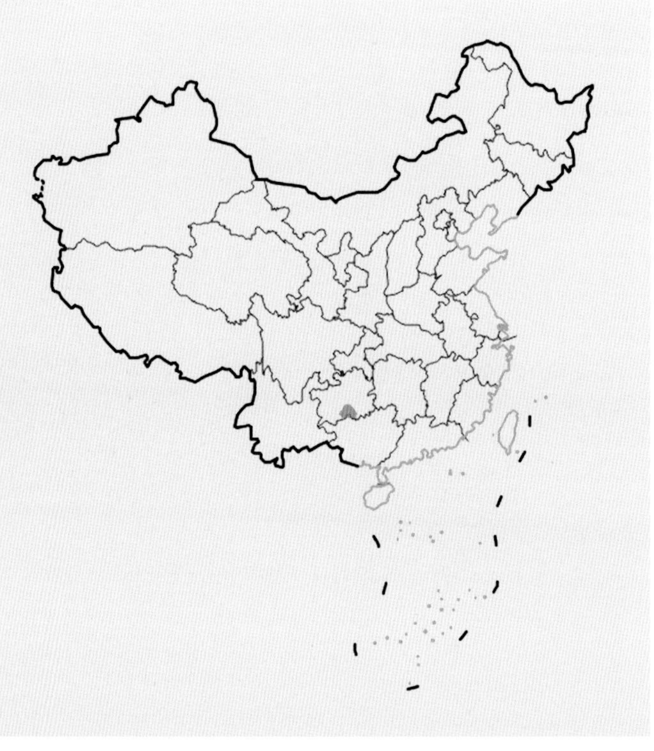

国内分布图
Map of Domestic Distribution

种群 Population

种群数量 Population Size	未知 / Unknown
种群趋势 Population Trend	未知 / Unknown

生境与生态系统 Habitat (s) and Ecosystem (s)

生　　境 Habitat(s)	森林 / Forest
生态系统 Ecosystem(s)	森林生态系统 / Forest Ecosystem

威胁 Threat (s)

主要威胁 Major Threat(s)	未知 / Unknown

保护级别与保护行动 Protection Category and Conservation Action (s)

国家重点保护野生动物等级 (2021) Category of National Key Protected Wild Animals (2021)	无 / NA
IUCN 红色名录 (2020-2) IUCN Red List (2020-2)	无危 / LC
CITES 附录 (2019) CITES Appendix (2019)	无 / NA
保护行动 Conservation Action(s)	无 / None

相关文献 Relevant References

Liu and Wu (刘少英和吴毅), 2019; Burgin *et al*., 2018; Guo *et al*., 2017

环颈伏翼 *Thainycteris aureocollaris*

中国生物多样性红色名录
China's Red List of Biodiversity

小扁颅蝠

Tylonycteris pygmaeus

数据缺乏 DD

数据缺乏 DD	无危 LC	近危 NT	易危 VU	濒危 EN	极危 CR	区域灭绝 RE	野外灭绝 EW	灭绝 EX

分类地位 Taxonomic Status

动物界 Animalia	脊索动物门 Chordata	哺乳纲 Mammalia	翼手目 Chiroptera	蝙蝠科 Vespertilionidae

学　名 **Scientific Name**	*Tylonycteris pygmaeus*
命名人 **Species Authority**	Feng, 2008
英文名 **English Name(s)**	Pygmy Bamboo Bat
同物异名 **Synonym(s)**	无 / None
种下单元评估 **Infra-specific Taxa Assessed**	无 / None

评估信息 Assessment Information

评估年份 **Year Assessed**	2020
评定人 **Assessor(s)**	吴毅、蒋志刚 / Yi Wu, Zhigang Jiang
审定人 **Reviewer(s)**	石红艳、江廷磊、余文华 / Hongyan Shi, Tinglei Jiang, Wenhua Yu
其他贡献人 **Other Contributor(s)**	李立立、丁晨晨 / Lili Li, Chenchen Ding

理由 **Justification:** 小扁颅蝠为中国特有种，其记录极少，对其的研究缺乏。因此，列为数据缺乏等级 / Pygmy Bamboo Bat is an endemic species in China. Little information about this species has been recorded and studies on this species are scarce. Thus, it is listed as Data Deficient

地理分布 Geographical Distribution

国内分布 Domestic Distribution

云南 / Yunnan

世界分布 World Distribution

中国 / China

分布标注 Distribution Note

特有种 / Endemic

国内分布图
Map of Domestic Distribution

种群 Population

种群数量 **Population Size**	未知 / Unknown
种群趋势 **Population Trend**	未知 / Unknown

生境与生态系统 Habitat (s) and Ecosystem (s)

生　　境 **Habitat(s)**	竹林 / Bamboo Forest
生态系统 **Ecosystem(s)**	竹林生态系统 / Bamboo Ecosystem

威胁 Threat (s)

主要威胁 **Major Threat(s)**	伐竹、耕种 / Bamboo Logging, Plantation

保护级别与保护行动 Protection Category and Conservation Action (s)

国家重点保护野生动物等级 **(2021) Category of National Key Protected Wild Animals (2021)**	无 / NA
IUCN 红色名录 **(2020-2) IUCN Red List (2020-2)**	未列入 / NA
CITES 附录 **(2019) CITES Appendix (2019)**	无 / NA
保护行动 **Conservation Action(s)**	无 / None

相关文献 Relevant References

Burgin *et al*., 2018; Jiang *et al.* (蒋志刚等), 2017; Feng *et al.*, 2008b

小扁颅蝠 *Tylonycteris pygmaeus*　　　蒋志刚 绘　Drawn by Zhigang Jiang

北京宽耳蝠

Barbastella beijingensis

数据缺乏 DD

数据缺乏 DD	无危 LC	近危 NT	易危 VU	濒危 EN	极危 CR	区域灭绝 RE	野外灭绝 EW	灭绝 EX

分类地位 Taxonomic Status

动物界 Animalia	脊索动物门 Chordata	哺乳纲 Mammalia	翼手目 Chiroptera	蝙蝠科 Vespertilionidae

学名 Scientific Name	*Barbastella beijingensis*
命名人 Species Authority	Zhang, 2007
英文名 English Name(s)	Beijing Barbastelle
同物异名 Synonym(s)	Beijing Wide-eared Bat
种下单元评估 Infra-specific Taxa Assessed	无 / None

评估信息 Assessment Information

评估年份 Year Assessed	2020
评定人 Assessor(s)	吴毅、蒋志刚 / Yi Wu, Zhigang Jiang
审定人 Reviewer(s)	石红艳、江廷磊、余文华 / Hongyan Shi, Tinglei Jiang, Wenhua Yu
其他贡献人 Other Contributor(s)	李立立、丁晨晨 / Lili Li, Chenchen Ding

理由 Justification: 北京宽耳蝠为中国特有种，其记录极少，对其的研究缺乏。因此，列为数据缺乏等级 / Beijing Barbastelle is an endemic species in China. Very little information about this species has been recorded, and studies on it are deficient. Thus, it is listed as Data Deficient

地理分布 Geographical Distribution

国内分布 Domestic Distribution

北京 / Beijing

世界分布 World Distribution

中国 / China

分布标注 Distribution Note

特有种 / Endemic

国内分布图
Map of Domestic Distribution

种群 Population

种群数量 Population Size	未知 / Unknown
种群趋势 Population Trend	未知 / Unknown

生境与生态系统 Habitat (s) and Ecosystem (s)

生　境 Habitat(s)	洞穴 / Cave
生态系统 Ecosystem(s)	洞穴生态系统 / Cave Ecosystem

威胁 Threat (s)

主要威胁 Major Threat(s)	未知 / Unknown

保护级别与保护行动 Protection Category and Conservation Action (s)

国家重点保护野生动物等级 (2021) Category of National Key Protected Wild Animals (2021)	无 / NA
IUCN 红色名录 (2020-2) IUCN Red List (2020-2)	未列入 / NA
CITES 附录 (2019) CITES Appendix (2019)	无 / NA
保护行动 Conservation Action(s)	无 / None

相关文献 Relevant References

Burgin *et al*., 2018; Jiang *et al.* (蒋志刚等), 2017; Zhang *et al.*, 2007

中国生物多样性红色名录
China's Red List of Biodiversity

金毛管鼻蝠

Murina chrysochaetes

数据缺乏 DD

数据缺乏 DD	无危 LC	近危 NT	易危 VU	濒危 EN	极危 CR	区域灭绝 RE	野外灭绝 EW	灭绝 EX

分类地位 Taxonomic Status

动物界 Animalia	脊索动物门 Chordata	哺乳纲 Mammalia	翼手目 Chiroptera	蝙蝠科 Vespertilionidae

学 名 Scientific Name	*Murina chrysochaetes*
命名人 Species Authority	Eger and Lim, 2011
英文名 English Name(s)	Golden-haired Tube-nosed Bat
同物异名 Synonym(s)	无 / None
种下单元评估 Infra-specific Taxa Assessed	无 / None

评估信息 Assessment Information

评估年份 Year Assessed	2020
评定人 Assessor(s)	吴毅、蒋志刚 / Yi Wu, Zhigang Jiang
审定人 Reviewer(s)	余文华、石红艳、江廷磊 / Wenhua Yu, Hongyan Shi, Tinglei Jiang
其他贡献人 Other Contributor(s)	李立立、丁晨晨 / Lili Li, Chenchen Ding

理由 Justification: 金毛管鼻蝠为中国特有种，仅在广西有发现。对其记录少，缺乏研究。因此，列为数据缺乏等级 / Golden-haired Tube-nosed Bat is an endemic species found only in Guangxi, China. Little information on this species has been recorded and studies on it are deficient. Thus, it is listed as Data Deficient

地理分布 Geographical Distribution

国内分布 Domestic Distribution

广西 / Guangxi

世界分布 World Distribution

中国 / China

分布标注 Distribution Note

特有种 / Endemic

国内分布图
Map of Domestic Distribution

种群 Population

种群数量 **Population Size**	未知 / Unknown
种群趋势 **Population Trend**	未知 / Unknown

生境与生态系统 Habitat (s) and Ecosystem (s)

生　境 **Habitat(s)**	森林、喀斯特地貌 / Forest, Karst Landscape
生态系统 **Ecosystem(s)**	森林生态系统、喀斯特生态系统 / Forest Ecosystem, Karst Ecosystem

威胁 Threat (s)

主要威胁 **Major Threat(s)**	未知 / Unknown

保护级别与保护行动 Protection Category and Conservation Action (s)

国家重点保护野生动物等级 **(2021) Category of National Key Protected Wild Animals (2021)**	无 / NA
IUCN 红色名录 **(2020-2) IUCN Red List (2020-2)**	未列入 / NA
CITES 附录 **(2019) CITES Appendix (2019)**	无 / NA
保护行动 **Conservation Action(s)**	无 / None

相关文献 Relevant References

Liu and Wu (刘少英和吴毅), 2019; Wilson and Mittermeier, 2019; Burgin *et al*., 2018; Jiang *et al.* (蒋志刚等), 2017; Eger and Lim, 2011

脊椎动物 Vertebrates 第一卷 哺乳动物（下册） Volume I, Mammals (III)

中国生物多样性红色名录
China's Red List of Biodiversity

梵净山管鼻蝠

Murina fanjingshanensis

数据缺乏 DD

数据缺乏 DD	无危 LC	近危 NT	易危 VU	濒危 EN	极危 CR	区域灭绝 RE	野外灭绝 EW	灭绝 EX

分类地位 Taxonomic Status

动 物 界 Animalia	脊索动物门 Chordata	哺 乳 纲 Mammalia	翼 手 目 Chiroptera	蝙 蝠 科 Vespertilionidae

学　名 **Scientific Name**	*Murina fanjingshanensis*
命 名 人 **Species Authority**	He *et al*., 2015
英 文 名 **English Name(s)**	Fanjingshan Tube-nosed Bat
同物异名 **Synonym(s)**	无 / None
种下单元评估 **Infra-specific Taxa Assessed**	无 / None

评估信息 Assessment Information

评 估 年 份 **Year Assessed**	2020
评　定　人 **Assessor(s)**	吴毅、蒋志刚 / Yi Wu, Zhigang Jiang
审　定　人 **Reviewer(s)**	余文华、石红艳、江廷磊 / Wenhua Yu, Hongyan Shi, Tinglei Jiang
其他贡献人 **Other Contributor(s)**	丁晨晨 / Chenchen Ding

理由 **Justification:** 梵净山管鼻蝠为中国特有种，现仅在贵州被发现。其本底资料和相关研究缺乏。故将梵净山管鼻蝠列为数据缺乏等级 / Fanjingshan Tube-nosed Bat is endemic to China, which is only recorded in Guizhou. Little information is known about this species as lacking of related studies. Thus, it is listed as Data Deficient

地理分布 Geographical Distribution

国内分布 Domestic Distribution

贵州 / Guizhou

世界分布 World Distribution

中国 / China

分布标注 Distribution Note

特有种 / Endemic

国内分布图
Map of Domestic Distribution

种群 Population

种群数量 **Population Size**	未知 / Unknown
种群趋势 **Population Trend**	未知 / Unknown

生境与生态系统 Habitat (s) and Ecosystem (s)

生　　境 **Habitat(s)**	森林、喀斯特地貌 / Forest, Karst Landscape
生态系统 **Ecosystem(s)**	森林生态系统、喀斯特生态系统 / Forest Ecosystem, Karst Ecosystem

威胁 Threat (s)

主要威胁 **Major Threat(s)**	未知 / Unknown

保护级别与保护行动 Protection Category and Conservation Action (s)

国家重点保护野生动物等级 **(2021) Category of National Key Protected Wild Animals (2021)**	无 / NA
IUCN 红色名录 **(2020-2) IUCN Red List (2020-2)**	未列入 / NA
CITES 附录 **(2019) CITES Appendix (2019)**	无 / NA
保护行动 **Conservation Action(s)**	无 / None

相关文献 Relevant References

Liu and Wu (刘少英和吴毅), 2019; Wilson and Mittermeier, 2019; Burgin *et al*., 2018; Jiang *et al.* (蒋志刚等), 2017; He *et al*., 2015

姬管鼻蝠

Murina gracilis

数据缺乏 DD

数据缺乏 DD	无危 LC	近危 NT	易危 VU	濒危 EN	极危 CR	区域灭绝 RE	野外灭绝 EW	灭绝 EX

分类地位 Taxonomic Status

动物界 Animalia	脊索动物门 Chordata	哺乳纲 Mammalia	翼手目 Chiroptera	蝙蝠科 Vespertilionidae
学名 Scientific Name		*Murina gracilis*		
命名人 Species Authority		Kuo, 2009		
英文名 English Name(s)		Taiwanese Little Tube-nosed Bat		
同物异名 Synonym(s)		Slender Tube-nosed Bat		
种下单元评估 Infra-specific Taxa Assessed		无 / None		

评估信息 Assessment Information

评估年份 Year Assessed	2020
评定人 Assessor(s)	吴毅、蒋志刚 / Yi Wu, Zhigang Jiang
审定人 Reviewer(s)	余文华、石红艳、江廷磊 / Wenhua Yu, Hongyan Shi, Tinglei Jiang
其他贡献人 Other Contributor(s)	李立立、丁晨晨 / Lili Li, Chenchen Ding

理由 Justification: 姬管鼻蝠为中国特有种，仅分布于台湾。对其的研究缺乏。因此，列为数据缺乏 / Taiwanese Little Tube-nosed Bat is an endemic species distributed only in Taiwan, China. Studies on this species are deficient. Thus, it is listed as Data Deficient

地理分布 Geographical Distribution

国内分布 Domestic Distribution
台湾 / Taiwan
世界分布 World Distribution
中国 / China
分布标注 Distribution Note
特有种 / Endemic

国内分布图
Map of Domestic Distribution

种群 Population

种群数量 Population Size	未知 / Unknown
种群趋势 Population Trend	未知 / Unknown

生境与生态系统 Habitat (s) and Ecosystem (s)

生　　境 **Habitat(s)**	森林、喀斯特地貌 / Forest, Karst Landscape
生态系统 **Ecosystem(s)**	森林生态系统 / Forest Ecosystem

威胁 Threat (s)

主要威胁 **Major Threat(s)**	未知 / Unknown

保护级别与保护行动 Protection Category and Conservation Action (s)

国家重点保护野生动物等级 **(2021) Category of National Key Protected Wild Animals (2021)**	无 / NA
IUCN 红色名录 **(2020-2) IUCN Red List (2020-2)**	未列入 / NA
CITES 附录 **(2019) CITES Appendix (2019)**	无 / NA
保护行动 **Conservation Action(s)**	无 / None

相关文献 Relevant References

Burgin *et al.*, 2018; Jiang *et al.* (蒋志刚等), 2017; Kuo *et al.*, 2015, 2009

哈氏管鼻蝠

Murina harrisoni

数据缺乏 DD

数据缺乏 DD	无危 LC	近危 NT	易危 VU	濒危 EN	极危 CR	区域灭绝 RE	野外灭绝 EW	灭绝 EX

分类地位 Taxonomic Status

动物界 Animalia	脊索动物门 Chordata	哺乳纲 Mammalia	翼手目 Chiroptera	蝙蝠科 Vespertilionidae

学　名 **Scientific Name**	*Murina harrisoni*
命名人 **Species Authority**	Csorba and Bates, 2005
英文名 **English Name(s)**	Harrison's Tube-nosed Bat
同物异名 **Synonym(s)**	无 / None
种下单元评估 **Infra-specific Taxa Assessed**	无 / None

评估信息 Assessment Information

评估年份 **Year Assessed**	2020
评定人 **Assessor(s)**	吴毅、蒋志刚 / Yi Wu, Zhigang Jiang
审定人 **Reviewer(s)**	余文华、石红艳、江廷磊 / Wenhua Yu, Hongyan Shi, Tinglei Jiang
其他贡献人 **Other Contributor(s)**	李立立、丁晨晨 / Lili Li, Chenchen Ding

理由 **Justification:** 哈氏管鼻蝠的记录少，对其的研究缺乏。因此，列为数据缺乏等级 / Little information regarding to Harrison's Tube-nosed Bat has been recorded, and studies on the species are deficient. Thus, it is listed as Data Deficient

地理分布 Geographical Distribution

国内分布 Domestic Distribution

海南 / Hainan

世界分布 World Distribution

中国、柬埔寨、越南 / China, Cambodia, Viet Nam

分布标注 Distribution Note

非特有种 / Non-endemic

国内分布图
Map of Domestic Distribution

种群 Population

种群数量 Population Size	未知 / Unknown
种群趋势 Population Trend	未知 / Unknown

生境与生态系统 Habitat (s) and Ecosystem (s)

生　　境 Habitat(s)	森林、次生林 / Forest, Secondary Forest
生态系统 Ecosystem(s)	森林生态系统 / Forest Ecosystem

威胁 Threat (s)

主要威胁 Major Threat(s)	砍伐 / Logging

保护级别与保护行动 Protection Category and Conservation Action (s)

国家重点保护野生动物等级 (2021) Category of National Key Protected Wild Animals (2021)	无 / NA
IUCN 红色名录 (2020-2) IUCN Red List (2020-2)	数据缺乏 / DD
CITES 附录 (2019) CITES Appendix (2019)	无 / NA
保护行动 Conservation Action(s)	无 / None

相关文献 Relevant References

Liu and Wu (刘少英和吴毅), 2019; Wilson and Mittermeier, 2019; Burgin *et al.*, 2018; Jiang *et al.* (蒋志刚等), 2017; Wu *et al.*, 2010; Csorba and Bates, 2005

哈氏管鼻蝠 *Murina harrisoni* 余文华 摄 By Wenhua Yu

锦矗管鼻蝠

Murina jinchui

数据缺乏 DD

数据缺乏 DD	无危 LC	近危 NT	易危 VU	濒危 EN	极危 CR	区域灭绝 RE	野外灭绝 EW	灭绝 EX

分类地位 Taxonomic Status

动物界 Animalia	脊索动物门 Chordata	哺乳纲 Mammalia	翼手目 Chiroptera	蝙蝠科 Vespertilionidae

学名 **Scientific Name**	*Murina jinchui*
命名人 **Species Authority**	Yu, Csorba and Wu, 2020
英文名 **English Name(s)**	Jinchu's Tube-nosed Bat
同物异名 **Synonym(s)**	无 / None
种下单元评估 **Infra-specific Taxa Assessed**	无 / None

评估信息 Assessment Information

评估年份 **Year Assessed**	2021
评定人 **Assessor(s)**	吴毅、蒋志刚 / Yi Wu, Zhigang Jiang
审定人 **Reviewer(s)**	余文华、石艳红、江廷磊 / Wenhua Yu, Yanhong Shi, Tinglei Jiang
其他贡献人 **Other Contributor(s)**	丁晨晨 / Chenchen Ding

理由 Justification: 目前该物种仅发现于模式标本产地——四川卧龙国家级自然保护区，尽管采集到少量标本，然而其种群与分布区仍尚不明确。因此，列为数据缺乏等级 / The Jingchu's Tube-nosed Bat is only found in the Wolong National Nature Reserve of Sichuan, China, where the type specimen was collected. Although a certain number of specimens have been collected, the population and distribution area of this species are not yet known. Thus, it is listed as Data Deficient

地理分布 Geographical Distribution

国内分布 Domestic Distribution

四川 / Sichuan

世界分布 World Distribution

中国 / China

分布标注 Distribution Note

特有种 / Endemic

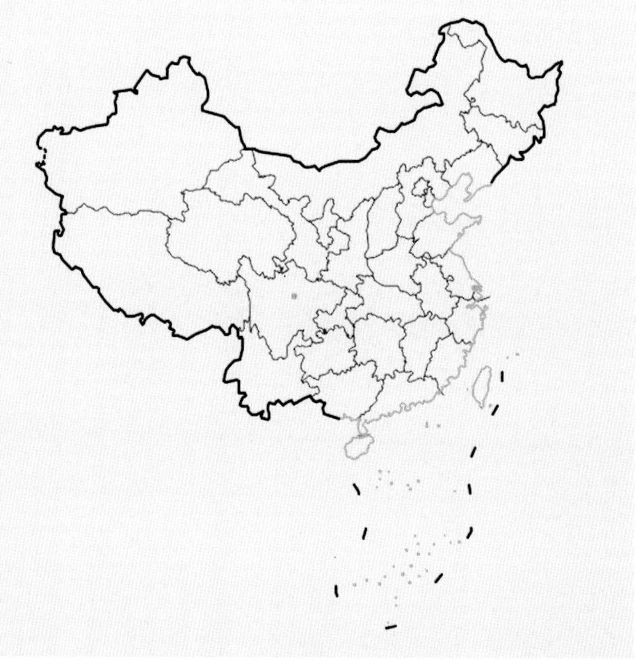

国内分布图
Map of Domestic Distribution

种群 Population

种群数量 Population Size	未知 / Unknown
种群趋势 Population Trend	未知 / Unknown

生境与生态系统 Habitat (s) and Ecosystem (s)

生　　境 Habitat(s)	潮湿常绿落叶阔叶林 / Moist Evergreen and Deciduous Broad-leaved Forest
生态系统 Ecosystem(s)	森林生态系统 / Forest Ecosystem

威胁 Threat (s)

主要威胁 Major Threat(s)	未知 / Unknown

保护级别与保护行动 Protection Category and Conservation Action (s)

国家重点保护野生动物等级 (2021) Category of National Key Protected Wild Animals (2021)	无 / NA
IUCN 红色名录 (2020-2) IUCN Red List (2020-2)	未列入 / NA
CITES 附录 (2019) CITES Appendix (2019)	无 / NA
保护行动 Conservation Action(s)	无 / None

相关文献 Relevant References

Yu *et al*., 2020

锦矗管鼻蝠 *Murina jinchui*　　　　吴毅 摄　By Yi Wu

中国生物多样性红色名录
China's Red List of Biodiversity

荔波管鼻蝠

Murina liboensis

数据缺乏 DD

数据缺乏 DD	无危 LC	近危 NT	易危 VU	濒危 EN	极危 CR	区域灭绝 RE	野外灭绝 EW	灭绝 EX

分类地位 Taxonomic Status

动 物 界 Animalia	脊索动物门 Chordata	哺 乳 纲 Mammalia	翼 手 目 Chiroptera	蝙 蝠 科 Vespertilionidae

学 名 Scientific Name	*Murina liboensis*
命 名 人 Species Authority	Zeng *et al*., 2018
英 文 名 English Name(s)	Libo Tube-nosed Bat
同物异名 Synonym(s)	无 / None
种下单元评估 Infra-specific Taxa Assessed	无 / None

评估信息 Assessment Information

评 估 年 份 Year Assessed	2021
评 定 人 Assessor(s)	周江、蒋志刚 / Jiang Zhou, Zhigang Jiang
审 定 人 Reviewer(s)	余文华、石艳红、江廷磊 / Wenhua Yu, Yanhong Shi, Tinglei Jiang
其他贡献人 Other Contributor(s)	丁晨晨 / Chenchen Ding

理由 Justification: 荔波管鼻蝠最近才由周江团队在贵州荔波茂兰国家级自然保护区发现，目前对其的种群与分布区状况尚不了解。因此，列为数据缺乏等级 / The Libo Tube-nosed Bat was recently discovered by Jiang Zhou's team in their expedition in Maolan National Nature Reserve, Libo, Guizhou. No information on its population and distribution is available so far. Thus, it is listed as Data Deficient

地理分布 Geographical Distribution

国内分布 Domestic Distribution

贵州 / Guizhou

世界分布 World Distribution

中国 / China

分布标注 Distribution Note

特有种 / Endemic

国内分布图
Map of Domestic Distribution

种群 Population

种群数量 **Population Size**	未知 / Unknown
种群趋势 **Population Trend**	未知 / Unknown

生境与生态系统 Habitat (s) and Ecosystem (s)

生　　境 **Habitat(s)**	洞穴、内陆岩石区域、森林 / Cave, Inland Rocky Area, Forest
生态系统 **Ecosystem(s)**	森林生态系统、喀斯特生态系统 / Forest Ecosystem, Karst Ecosystem

威胁 Threat (s)

主要威胁 **Major Threat(s)**	未知 / Unknown

保护级别与保护行动 Protection Category and Conservation Action (s)

国家重点保护野生动物等级 **(2021) Category of National Key Protected Wild Animals (2021)**	无 / NA
IUCN 红色名录 **(2020-2) IUCN Red List (2020-2)**	未列入 / NA
CITES 附录 **(2019) CITES Appendix (2019)**	无 / NA
保护行动 **Conservation Action(s)**	无 / None

相关文献 Relevant References

Zeng *et al*., 2018

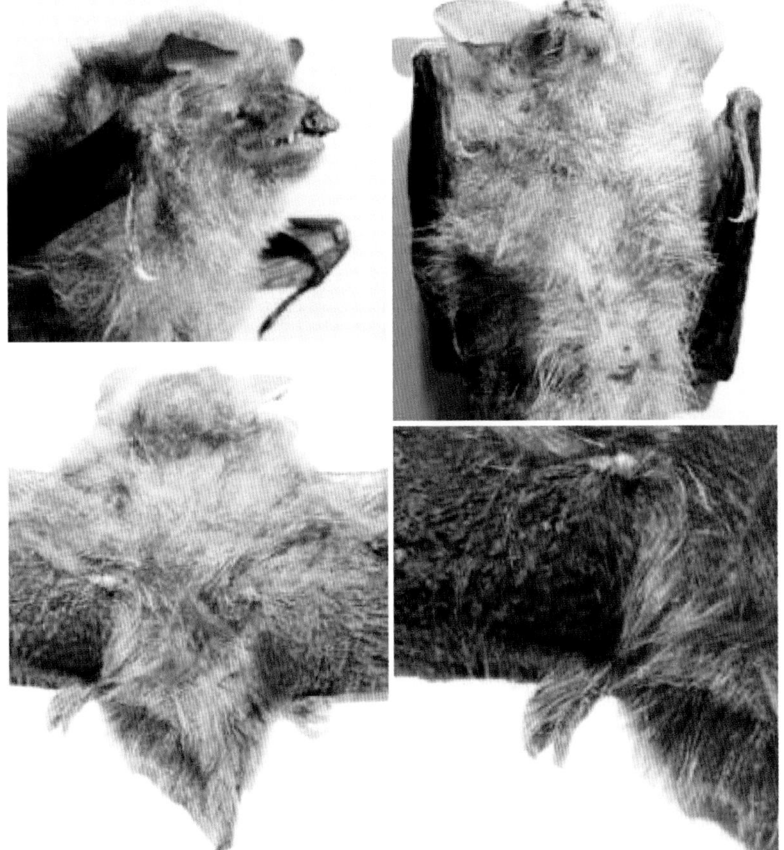

荔波管鼻蝠 *Murina liboensis*　　　周江 摄　By Jiang Zhou

罗蕾莱管鼻蝠

Murina lorelieae

数据缺乏 DD

数据缺乏 DD	无危 LC	近危 NT	易危 VU	濒危 EN	极危 CR	区域灭绝 RE	野外灭绝 EW	灭绝 EX

分类地位 Taxonomic Status

动物界 Animalia	脊索动物门 Chordata	哺乳纲 Mammalia	翼手目 Chiroptera	蝙蝠科 Vespertilionidae

学　名 Scientific Name	*Murina lorelieae*
命名人 Species Authority	Eger and Lim, 2011
英文名 English Name(s)	Lorelie's Tube-nosed Bat
同物异名 Synonym(s)	无 / None
种下单元评估 Infra-specific Taxa Assessed	无 / None

评估信息 Assessment Information

评估年份 Year Assessed	2020
评定人 Assessor(s)	吴毅、蒋志刚 / Yi Wu, Zhigang Jiang
审定人 Reviewer(s)	余文华、石红艳、江廷磊 / Wenhua Yu, Hongyan Shi, Tinglei Jiang
其他贡献人 Other Contributor(s)	李立立、丁晨晨 / Lili Li, Chenchen Ding

理由 Justification: 罗蕾莱管鼻蝠为中国特有种， 仅分布在广西。其种群和生境状况信息很少，因为对其的研究缺乏。因此，列为数据缺乏等级 / Lorelie's Tube-nosed Bat is endemic to China, distributed only in Guangxi. Little information on the population and habitat status of the species are recorded due to deficient studies on it. Thus, it is listed as Data Deficient

地理分布 Geographical Distribution

国内分布 Domestic Distribution

广西 / Guangxi

世界分布 World Distribution

中国 / China

分布标注 Distribution Note

特有种 / Endemic

国内分布图
Map of Domestic Distribution

种群 Population

种群数量 **Population Size**	未知 / Unknown
种群趋势 **Population Trend**	未知 / Unknown

生境与生态系统 Habitat (s) and Ecosystem (s)

生　　境 **Habitat(s)**	森林 / Forest
生态系统 **Ecosystem(s)**	森林生态系统 / Forest Ecosystem

威胁 Threat (s)

主要威胁 **Major Threat(s)**	未知 / Unknown

保护级别与保护行动 Protection Category and Conservation Action (s)

国家重点保护野生动物等级 **(2021) Category of National Key Protected Wild Animals (2021)**	无 / NA
IUCN 红色名录 **(2020-2) IUCN Red List (2020-2)**	数据缺乏 / DD
CITES 附录 **(2019) CITES Appendix (2019)**	无 / NA
保护行动 **Conservation Action(s)**	无 / None

相关文献 Relevant References

Liu and Wu (刘少英和吴毅), 2019; Wilson and Mittermeier, 2019; Burgin *et al*., 2018; Jiang *et al.* (蒋志刚等), 2017; Eger and Lim, 2011

罗蕾莱管鼻蝠 *Murina lorelieae*　　　李锋 摄　By Feng Li

隐姬管鼻蝠
Murina recondita

数据缺乏 DD

数据缺乏 DD	无危 LC	近危 NT	易危 VU	濒危 EN	极危 CR	区域灭绝 RE	野外灭绝 EW	灭绝 EX

分类地位 Taxonomic Status

动物界 Animalia	脊索动物门 Chordata	哺乳纲 Mammalia	翼手目 Chiroptera	蝙蝠科 Vespertilionidae

学　名 Scientific Name	*Murina recondita*
命名人 Species Authority	Kuo, 2009
英文名 English Name(s)	Faint-colored Tube-nosed Bat
同物异名 Synonym(s)	Faint-golden Little Tube-nosed Bat
种下单元评估 Infra-specific Taxa Assessed	无 / None

评估信息 Assessment Information

评估年份 Year Assessed	2020
评定人 Assessor(s)	吴毅、蒋志刚 / Yi Wu, Zhigang Jiang
审定人 Reviewer(s)	余文华、石红艳、江廷磊 / Wenhua Yu, Hongyan Shi, Tinglei Jiang
其他贡献人 Other Contributor(s)	李立立、丁晨晨 / Lili Li, Chenchen Ding

理由 **Justification:** 隐姬管鼻蝠为中国特有种，仅分布在台湾。对其的研究缺乏。因此，列为数据缺乏等级 / Faint-colored Tube-nosed Bat is endemic to China, distributed only in Taiwan. Little information about this species is recorded and studies on it are scarce. Thus, it is listed as Data Deficient

地理分布 Geographical Distribution

国内分布 Domestic Distribution

台湾 / Taiwan

世界分布 World Distribution

中国 / China

分布标注 Distribution Note

特有种 / Endemic

国内分布图
Map of Domestic Distribution

种群 Population

种群数量 **Population Size**	未知 / Unknown
种群趋势 **Population Trend**	未知 / Unknown

生境与生态系统 Habitat (s) and Ecosystem (s)

生　　境 **Habitat(s)**	未知 / Unknown
生态系统 **Ecosystem(s)**	森林生态系统 / Forest Ecosystem

威胁 Threat (s)

主要威胁 **Major Threat(s)**	砍伐 / Logging

保护级别与保护行动 Protection Category and Conservation Action (s)

国家重点保护野生动物等级 **(2021) Category of National Key Protected Wild Animals (2021)**	无 / NA
IUCN 红色名录 (2020-2) IUCN Red List (2020-2)	数据缺乏 / DD
CITES 附录 (2019) CITES Appendix (2019)	无 / NA
保护行动 **Conservation Action(s)**	无 / None

相关文献 Relevant References

Wilson and Mittermeier, 2019; Burgin *et al.*, 2018; Jiang *et al.* (蒋志刚等), 2017; Kuo *et al.*, 2015, 2009

隐姬管鼻蝠 *Murina recondita*　　蒋志刚 绘　Drawn by Zhigang Jiang

榕江管鼻蝠

Murina rongjiangensis

数据缺乏 DD

数据缺乏 DD	无危 LC	近危 NT	易危 VU	濒危 EN	极危 CR	区域灭绝 RE	野外灭绝 EW	灭绝 EX

分类地位 Taxonomic Status

动物界 Animalia	脊索动物门 Chordata	哺乳纲 Mammalia	翼手目 Chiroptera	蝙蝠科 Vespertilionidae

学　名 Scientific Name	*Murina rongjiangensis*
命名人 Species Authority	Cheng *et al*., 2017
英文名 English Name(s)	Rongjiang Tube-nosed Bat
同物异名 Synonym(s)	无 / None
种下单元评估 Infra-specific Taxa Assessed	无 / None

评估信息 Assessment Information

评估年份 Year Assessed	2021
评定人 Assessor(s)	周江、蒋志刚 / Jiang Zhou, Zhigang Jiang
审定人 Reviewer(s)	余文华、石艳红、江廷磊 / Wenhua Yu, Yanhong Shi, Tinglei Jiang
其他贡献人 Other Contributor(s)	丁晨晨 / Chenchen Ding

理由 Justification: 榕江管鼻蝠最近才由周江团队在贵州榕江发现，目前对其种群与分布区尚不了解。因此，列为数据缺乏等级 / The Rongjiang Tube-nosed Bat has been discovered by Jiang Zhou's team in their expedition recently in Rongjiang, Guizhou. No more information is available for its population and distribution so far. Thus, it is listed as Data Deficient

地理分布 Geographical Distribution

国内分布 Domestic Distribution

贵州 / Guizhou

世界分布 World Distribution

中国 / China

分布标注 Distribution Note

特有种 / Endemic

国内分布图
Map of Domestic Distribution

种群 Population

种群数量 **Population Size**	未知 / Unknown
种群趋势 **Population Trend**	未知 / Unknown

生境与生态系统 Habitat (s) and Ecosystem (s)

生　　境 **Habitat(s)**	洞穴 / Cave
生态系统 **Ecosystem(s)**	喀斯特生态系统 / Karst Ecosystem

威胁 Threat (s)

主要威胁 **Major Threat(s)**	未知 / Unknown

保护级别与保护行动 Protection Category and Conservation Action (s)

国家重点保护野生动物等级 **(2021) Category of National Key Protected Wild Animals (2021)**	无 / NA
IUCN 红色名录 **(2020-2) IUCN Red List (2020-2)**	未列入 / NA
CITES 附录 **(2019) CITES Appendix (2019)**	无 / NA
保护行动 **Conservation Action(s)**	无 / None

相关文献 Relevant References

Cheng *et al*., 2017

榕江管鼻蝠 *Murina rongjiangensis*　　周江 摄　By Jiang Zhou

中国生物多样性红色名录
China's Red List of Biodiversity

水甫管鼻蝠
Murina shuipuensis

数据缺乏 DD

数据缺乏 DD	无危 LC	近危 NT	易危 VU	濒危 EN	极危 CR	区域灭绝 RE	野外灭绝 EW	灭绝 EX

分类地位 Taxonomic Status

动物界 Animalia	脊索动物门 Chordata	哺乳纲 Mammalia	翼手目 Chiroptera	蝙蝠科 Vespertilionidae

学　名 **Scientific Name**	*Murina shuipuensis*
命名人 **Species Authority**	Eger and Lim, 2011
英文名 **English Name(s)**	Shuipu's Tube-nosed Bat
同物异名 **Synonym(s)**	无 / None
种下单元评估 **Infra-specific Taxa Assessed**	无 / None

评估信息 Assessment Information

评估年份 **Year Assessed**	2020
评定人 **Assessor(s)**	吴毅、蒋志刚 / Yi Wu, Zhigang Jiang
审定人 **Reviewer(s)**	余文华、石红艳、江廷磊 / Wenhua Yu, Hongyan Shi, Tinglei Jiang
其他贡献人 **Other Contributor(s)**	李立立、丁晨晨 / Lili Li, Chenchen Ding

理由 **Justification:** 水甫管鼻蝠为中国特有种，仅分布在贵州。对其的研究缺乏。因此，列为数据缺乏 / Shuipu's Tube-nosed Bat is endemic to China, distributed only in Guizhou. Studies on this species are deficient. Thus, it is listed as Data Deficient

地理分布 Geographical Distribution

国内分布 **Domestic Distribution**	贵州 / Guizhou
世界分布 **World Distribution**	中国 / China
分布标注 **Distribution Note**	特有种 / Endemic

国内分布图
Map of Domestic Distribution

种群 Population

种群数量 Population Size	未知 / Unknown
种群趋势 Population Trend	未知 / Unknown

生境与生态系统 Habitat (s) and Ecosystem (s)

生　　境 Habitat(s)	森林、喀斯特地貌 / Forest, Karst Landscape
生态系统 Ecosystem(s)	森林生态系统、喀斯特生态系统 / Forest Ecosystem, Karst Ecosystem

威胁 Threat (s)

主要威胁 Major Threat(s)	砍伐、喀斯特洞穴开发 / Logging, Karst Cave Exploitation

保护级别与保护行动 Protection Category and Conservation Action (s)

国家重点保护野生动物等级 (2021) Category of National Key Protected Wild Animals (2021)	无 / NA
IUCN 红色名录 (2020-2) IUCN Red List (2020-2)	数据缺乏 / DD
CITES 附录 (2019) CITES Appendix (2019)	无 / NA
保护行动 Conservation Action(s)	无 / None

相关文献 Relevant References

Wilson and Mittermeier, 2019; Burgin *et al*., 2018; Jiang *et al.* (蒋志刚等), 2017; Eger and Lim, 2011

水甫管鼻蝠 *Murina shuipuensis*　　蒋志刚 绘　Drawn by Zhigang Jiang

乌苏里管鼻蝠

Murina ussuriensis

数据缺乏 DD

数据缺乏 DD	无危 LC	近危 NT	易危 VU	濒危 EN	极危 CR	区域灭绝 RE	野外灭绝 EW	灭绝 EX

分类地位 Taxonomic Status

动物界 Animalia	脊索动物门 Chordata	哺乳纲 Mammalia	翼手目 Chiroptera	蝙蝠科 Vespertilionidae

学名 Scientific Name	*Murina ussuriensis*
命名人 Species Authority	Ognev, 1913
英文名 English Name(s)	Ussuri Tube-nosed Bat
同物异名 Synonym(s)	*Murina silvatica* (Yoshiyuki, 1983)
种下单元评估 Infra-specific Taxa Assessed	无 / None

评估信息 Assessment Information

评估年份 Year Assessed	2020
评定人 Assessor(s)	吴毅、蒋志刚 / Yi Wu, Zhigang Jiang
审定人 Reviewer(s)	余文华、石红艳、江廷磊 / Wenhua Yu, Hongyan Shi, Tinglei Jiang
其他贡献人 Other Contributor(s)	李立立、丁晨晨 / Lili Li, Chenchen Ding

理由 Justification: 乌苏里管鼻蝠的记录较少，对其的研究缺乏。因此，列为数据缺乏等级 / Little information on Ussuri Tube-nosed Bat has been recorded, and studies on it are deficient. Thus, it is listed as Data Deficient

地理分布 Geographical Distribution

国内分布 Domestic Distribution

吉林、黑龙江 / Jilin, Heilongjiang

世界分布 World Distribution

中国；东北亚 / China; Northeast Asia

分布标注 Distribution Note

非特有种 / Non-endemic

国内分布图
Map of Domestic Distribution

种群 Population

种群数量 **Population Size**	未知 / Unknown
种群趋势 **Population Trend**	未知 / Unknown

生境与生态系统 Habitat (s) and Ecosystem (s)

生　　境 **Habitat(s)**	未知 / Unknown
生态系统 **Ecosystem(s)**	森林生态系统 / Forest Ecosystem

威胁 Threat (s)

主要威胁 **Major Threat(s)**	砍伐 / Logging

保护级别与保护行动 Protection Category and Conservation Action (s)

国家重点保护野生动物等级 **(2021) Category of National Key Protected Wild Animals (2021)**	无 / NA
IUCN 红色名录 **(2020-2) IUCN Red List (2020-2)**	数据缺乏 / DD
CITES 附录 **(2019) CITES Appendix (2019)**	无 / NA
保护行动 **Conservation Action(s)**	无 / None

相关文献 Relevant References

Liu and Wu (刘少英和吴毅), 2019; Wilson and Mittermeier, 2019; Burgin *et al*., 2018; Jiang *et al.* (蒋志刚等), 2017; Smith *et al.* (史密斯等), 2009; Simmons, 2005; Maeda, 1980

乌苏里管鼻蝠 *Murina ussuriensis*　　蒋志刚 绘　Drawn by Zhigang Jiang

哈氏彩蝠
Kerivoula hardwickii

数据缺乏 DD

数据缺乏 DD	无危 LC	近危 NT	易危 VU	濒危 EN	极危 CR	区域灭绝 RE	野外灭绝 EW	灭绝 EX

分类地位 Taxonomic Status

动物界 Animalia	脊索动物门 Chordata	哺乳纲 Mammalia	翼手目 Chiroptera	蝙蝠科 Vespertilionidae
学名 **Scientific Name**		*Kerivoula hardwickii*		
命名人 **Species Authority**		Horsfield, 1824		
英文名 **English Name(s)**		Hardwicke's Woolly Bat		
同物异名 **Synonym(s)**		*Kerivoula crypta* (Wroughton and Ryley, 1913); *K. depressa* (Miller, 1906); *K. fusca* Dobson, 1871; *K. hardwickii* (Wroughton and Ryley, 1913) subsp. *crypta*; (转下页)		
种下单元评估 **Infra-specific Taxa Assessed**		无 / None		

评估信息 Assessment Information

评估年份 **Year Assessed**	2020
评定人 **Assessor(s)**	吴毅、蒋志刚 / Yi Wu, Zhigang Jiang
审定人 **Reviewer(s)**	余文华、石红艳、江廷磊 / Wenhua Yu, Hongyan Shi, Tinglei Jiang
其他贡献人 **Other Contributor(s)**	李立立、丁晨晨 / Lili Li, Chenchen Ding

理由 **Justification:** 哈氏彩蝠的记录较少，对其的研究缺乏。因此，列为数据缺乏 / Little information is known about the population and habitat status of Hardwicke's Woolly Bat due to scarce studies on the species. Thus, it is listed as Data Deficient

地理分布 Geographical Distribution

国内分布 Domestic Distribution

四川、广西、云南、福建、重庆、海南 / Sichuan, Guangxi, Yunnan, Fujian, Chongqing, Hainan

世界分布 World Distribution

中国；东南亚 / China; Southeast Asia

分布标注 Distribution Note

非特有种 / Non-endemic

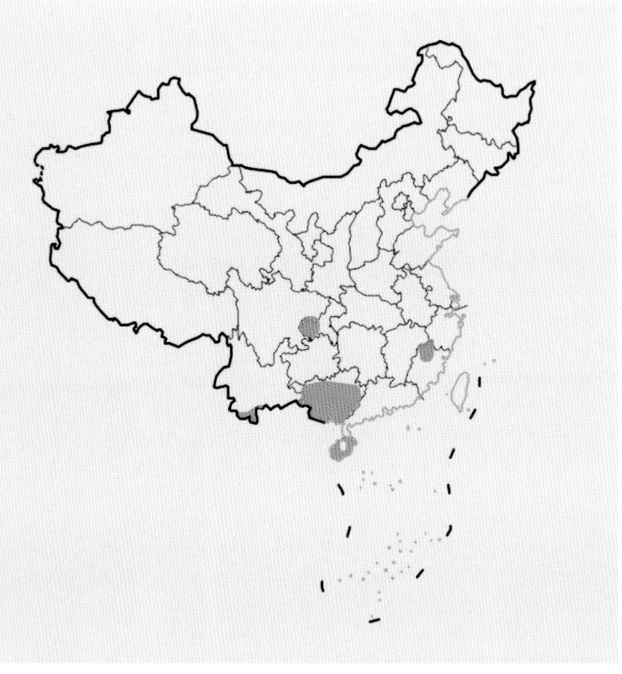

国内分布图
Map of Domestic Distribution

种群 Population

种群数量 Population Size	未知 / Unknown
种群趋势 Population Trend	未知 / Unknown

生境与生态系统 Habitat (s) and Ecosystem (s)

生　　境 Habitat(s)	热带和亚热带湿润低地、山地森林、耕地、人造建筑 / Tropical and Subtropical Moist Lowland, Montane Forest, Arable Land, Building
生态系统 Ecosystem(s)	森林生态系统、农田生态系统、人类聚落生态系统 / Forest Ecosystem, Cropland Ecosystem, Human Settlement Ecosystem

威胁 Threat (s)

主要威胁 Major Threat(s)	砍伐 / Logging

保护级别与保护行动 Protection Category and Conservation Action (s)

国家重点保护野生动物等级 (2021) Category of National Key Protected Wild Animals (2021)	无 / NA
IUCN 红色名录 (2020-2) IUCN Red List (2020-2)	数据缺乏 / DD
CITES 附录 (2019) CITES Appendix (2019)	无 / NA
保护行动 Conservation Action(s)	无 / None

相关文献 Relevant References

Liu and Wu (刘少英和吴毅), 2019; Wilson and Mittermeier, 2019; Burgin *et al*., 2018; Jiang *et al*. (蒋志刚等), 2017; Smith *et al*., 2009; Pan *et al*. (潘清华等), 2007; Wang (王应祥), 2003

(接上页)

K. hardwickii (Miller, 1906) subsp. *depressa*; *K. hardwickii* (Phillips, 1932) subsp. *malpasi*; *K. malpasi* (Phillips, 1932); *Vespertilio hardwickii* (Horsfield, 1824)

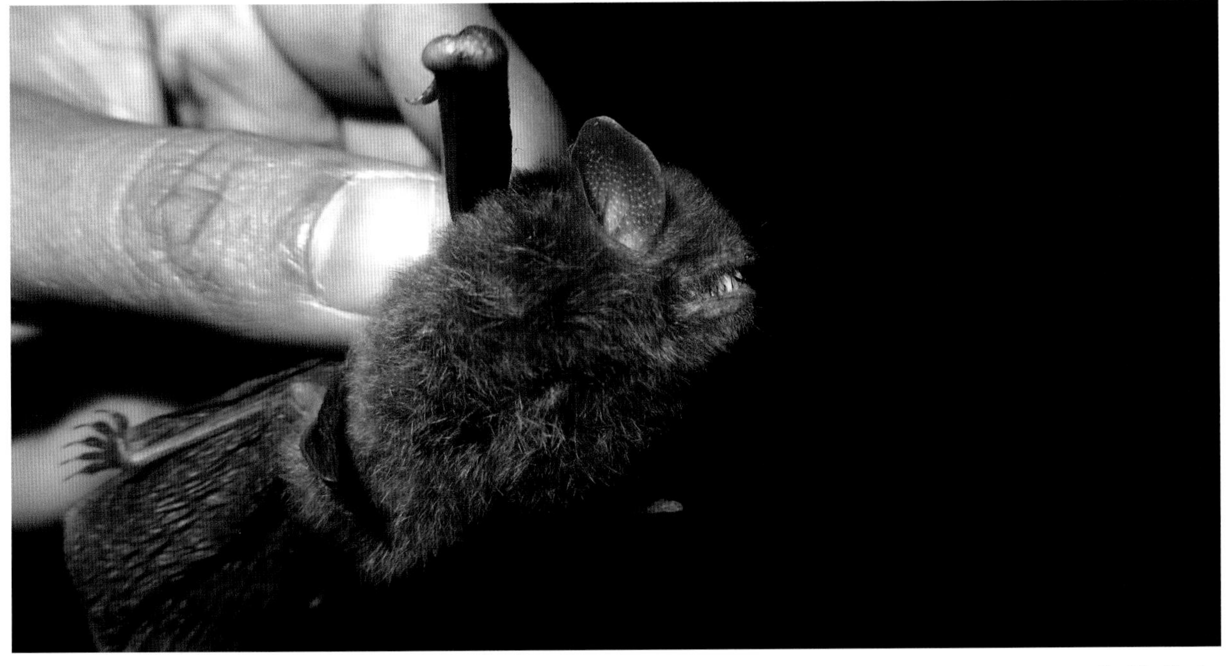

哈氏彩蝠 *Kerivoula hardwickii* By Alice Hughes

数据缺乏 DD	无危 LC	近危 NT	易危 VU	濒危 EN	极危 CR	区域灭绝 RE	野外灭绝 EW	灭绝 EX

分类地位 Taxonomic Status

动物界 Animalia	脊索动物门 Chordata	哺乳纲 Mammalia	食肉目 Carnivora	犬科 Canidae
学名 Scientific Name		*Canis aureus*		
命名人 Species Authority		Linnaeus, 1758		
英文名 English Name(s)		Golden Jackal		
同物异名 Synonym(s)		Asiatic Jackal, Common Jackal		
种下单元评估 Infra-specific Taxa Assessed		无 / None		

评估信息 Assessment Information

评估年份 Year Assessed	2020
评定人 Assessor(s)	蒋志刚 / Zhigang Jiang
审定人 Reviewer(s)	胡慧建、徐爱春 / Huijian Hu, Aichun Xu
其他贡献人 Other Contributor(s)	李立立、丁晨晨 / Lili Li, Chenchen Ding

理由 Justification: 亚洲胡狼分布于中国藏南地区，2018 年在西藏吉隆沟被拍摄到。目前尚缺乏其种群数据。因此，列为数据缺乏 / Golden Jackal is distributed in Zangnan area, China, but there is no accurate population data. Photographs of Golden Jackal were taken in Jilong Valley, Tibet (Xizang), China, in 2018. Thus, it is listed as Data Deficient

地理分布 Geographical Distribution

国内分布 Domestic Distribution

西藏 / Tibet (Xizang)

世界分布 World Distribution

亚洲、欧洲、非洲 / Asia, Europe, Africa

分布标注 Distribution Note

非特有种 / Non-endemic

国内分布图
Map of Domestic Distribution

种群 Population

种群数量 **Population Size**	未知 / Unknown
种群趋势 **Population Trend**	未知 / Unknown

生境与生态系统 Habitat (s) and Ecosystem (s)

生　　境 **Habitat(s)**	从撒哈拉沙漠、常绿森林、半沙漠、中草原和热带草原、森林、红树林、农业区、农村到半城市区域 / From the Sahara Desert to the Evergreen Forest, Semi-desert, Medium to Tropical Grassland and Savanna, Forest, Mangrove, Agricultural Region, Rural and Semi-urban Area
生态系统 **Ecosystem(s)**	森林生态系统、草原生态系统、荒漠生态系统、红树林生态系统、人类聚落生态系统 / Forest Ecosystem, Grassland Ecosystem, Desert Ecosystem, Mangrove Ecosystem, Human Settlement Ecosystem

威胁 Threat (s)

主要威胁 **Major Threat(s)**	未知 / Unknown

保护级别与保护行动 Protection Category and Conservation Action (s)

国家重点保护野生动物等级 **(2021) Category of National Key Protected Wild Animals (2021)**	二级 / Category II
IUCN 红色名录 **(2020-2) IUCN Red List (2020-2)**	无危 / LC
CITES 附录 **(2019) CITES Appendix (2019)**	无 / NA
保护行动 **Conservation Action(s)**	无 / None

相关文献 Relevant References

Burgin *et al.*, 2018; Jiang *et al.* (蒋志刚等), 2017; Mittermeier and Wilson, 2014; Choudhury, 2003

孟加拉狐

Vulpes bengalensis

数据缺乏 DD

数据缺乏 DD	无危 LC	近危 NT	易危 VU	濒危 EN	极危 CR	区域灭绝 RE	野外灭绝 EW	灭绝 EX

分类地位 Taxonomic Status

动 物 界 Animalia	脊索动物门 Chordata	哺 乳 纲 Mammalia	食 肉 目 Carnivora	犬 科 Canidae

学 名 Scientific Name	*Vulpes bengalensis*
命 名 人 Species Authority	Shaw, 1800
英 文 名 English Name(s)	Bengal Fox
同物异名 Synonym(s)	Indian Fox
种下单元评估 Infra-specific Taxa Assessed	无 / None

评估信息 Assessment Information

评 估 年 份 Year Assessed	2020
评 定 人 Assessor(s)	蒋志刚 / Zhigang Jiang
审 定 人 Reviewer(s)	胡慧建、徐爱春 / Huijian Hu, Aichun Xu
其他贡献人 Other Contributor(s)	丁晨晨 / Chenchen Ding

理由 Justification: 孟加拉狐分布于中国藏南地区，目前尚缺乏其种群数据。因此，列为数据缺乏等级 / Bengal Fox is distributed in Zangnan area, China, and its population data is deficient. Thus, it is listed as Data Deficient

地理分布 Geographical Distribution

国内分布 Domestic Distribution

西藏 / Tibet (Xizang)

世界分布 World Distribution

中国；南亚 / China; South Asia

分布标注 Distribution Note

非特有种 / Non-endemic

国内分布图
Map of Domestic Distribution

种群 Population

种群数量 Population Size	未知 / Unknown
种群趋势 Population Trend	未知 / Unknown

生境与生态系统 Habitat (s) and Ecosystem (s)

生　　境 Habitat(s)	易捕猎和挖洞的半干旱的、平坦的和起伏地形的灌木和草原栖息地 / Semi-arid Land, Flat to Undulating Terrain, including Scrub and Grassland habitats where it is easy to hunt and dig hole
生态系统 Ecosystem(s)	灌丛、草地生态系统 / Shrub and Grassland Ecosystem

威胁 Threat (s)

主要威胁 Major Threat(s)	未知 / Unknown

保护级别与保护行动 Protection Category and Conservation Action (s)

国家重点保护野生动物等级 (2021) Category of National Key Protected Wild Animals (2021)	无 / NA
IUCN 红色名录 (2020-2) IUCN Red List (2020-2)	无危 / LC
CITES 附录 (2019) CITES Appendix (2019)	无 / NA
保护行动 Conservation Action(s)	无 / None

相关文献 Relevant References

Burgin *et al*., 2018; Jiang *et al*. (蒋志刚等), 2017; Choudhury, 2003

孟加拉狐 *Vulpes bengalensis*　　By Chinmayisk

越南鼬獾

Melogale cucphuongensis

数据缺乏 DD

数据缺乏 DD	无危 LC	近危 NT	易危 VU	濒危 EN	极危 CR	区域灭绝 RE	野外灭绝 EW	灭绝 EX

分类地位 Taxonomic Status

动 物 界 Animalia	脊索动物门 Chordata	哺 乳 纲 Mammalia	食 肉 目 Carnivora	鼬 科 Mustelidae

学 名 **Scientific Name**	*Melogale cucphuongensis*
命 名 人 **Species Authority**	Nadler *et al*., 2011
英 文 名 **English Name(s)**	Vietnam Ferret-badger
同物异名 **Synonym(s)**	无 / None
种下单元评估 **Infra-specific Taxa Assessed**	无 / None

评估信息 Assessment Information

评 估 年 份 **Year Assessed**	2021
评 定 人 **Assessor(s)**	蒋志刚 / Zhigang Jiang
审 定 人 **Reviewer(s)**	李松 / Song Li
其他贡献人 **Other Contributor(s)**	丁晨晨 / Chenchen Ding

理由 **Justification:** 越南鼬獾是 Nadler *et al.* (2011) 根据在越南发现的两号标本命名的一个鼬科新种，2019 年被李松在中国福建武夷山基于公路上死亡的一个标本发现，是一个中国新记录种。目前缺乏对其种群和栖息地的研究。因此，列为数据缺乏等级 / Vietnam Ferret-badger is described by Nadler *et al.* (2011) and is known from only two specimens. It is a new record species in China identified by Song Li, based on a road killed specimen found in Wuyi Shan, Fujian, China, in 2019. No more information on its population and habitat status in the country has been recorded. Thus, it is listed as Data Deficient

地理分布 Geographical Distribution

国内分布 Domestic Distribution

福建 / Fujian

世界分布 World Distribution

中国、越南 / China, Viet Nam

分布标注 Distribution Note

非特有种 / Non-endemic

国内分布图
Map of Domestic Distribution

种群 Population

种群数量 Population Size	未知 / Unknown
种群趋势 Population Trend	未知 / Unknown

生境与生态系统 Habitat (s) and Ecosystem (s)

生　　境 Habitat(s)	亚热带森林 / Subtropical Forest
生态系统 Ecosystem(s)	森林生态系统 / Forest Ecosystem

威胁 Threat (s)

主要威胁 Major Threat(s)	未知 / Unknown

保护级别与保护行动 Protection Category and Conservation Action (s)

国家重点保护野生动物等级 (2021) Category of National Key Protected Wild Animals (2021)	无 / NA
IUCN 红色名录 (2020-2) IUCN Red List (2020-2)	数据缺乏 / DD
CITES 附录 (2019) CITES Appendix (2019)	无 / NA
保护行动 Conservation Action(s)	无 / None

相关文献 Relevant References

Li *et al*., 2019; Nadler *et al*., 2011

越南鼬獾 *Melogale cucphuongensis*　　蒋志刚 绘　Drawn by Zhigang Jiang

髯海豹

Erignathus barbatus

数据缺乏 DD

数据缺乏 DD	无危 LC	近危 NT	易危 VU	濒危 EN	极危 CR	区域灭绝 RE	野外灭绝 EW	灭绝 EX

分类地位 Taxonomic Status

动 物 界 Animalia	脊索动物门 Chordata	哺 乳 纲 Mammalia	食 肉 目 Carnivora	海 豹 科 Phocidae
学 名 Scientific Name		*Erignathus barbatus*		
命 名 人 Species Authority		Erxleben, 1777		
英 文 名 English Name(s)		Bearded Seal		
同物异名 Synonym(s)		Square Flipper Seal		
种下单元评估 Infra-specific Taxa Assessed		无 / None		

评估信息 Assessment Information

评 估 年 份 Year Assessed	2020
评 定 人 Assessor(s)	周开亚 / Kaiya Zhou
审 定 人 Reviewer(s)	张先锋、王克雄、王丁、祝茜、杨光、蒋志刚 / Xianfeng Zhang, Kexiong Wang, Ding Wang, Qian Zhu, Guang Yang, Zhigang Jiang
其他贡献人 Other Contributor(s)	李立立、丁晨晨 / Lili Li, Chenchen Ding

理由 Justification: 髯海豹为北极及亚北极海域海洋哺乳动物，到达我国海域的个体数不详。因此，列为数据缺乏 / Bearded Seal is sea mammal distributed in the Arctic and sub-Arctic oceans, the number of its individuals reaching China's waters is unknown. Thus, it is listed as Data Deficient

地理分布 Geographical Distribution

国内分布 Domestic Distribution

东海 / East China Sea

世界分布 World Distribution

中国；北极及亚北极海域 / China; Arctic and Sub-Arctic Oceans

分布标注 Distribution Note

非特有种 / Non-endemic

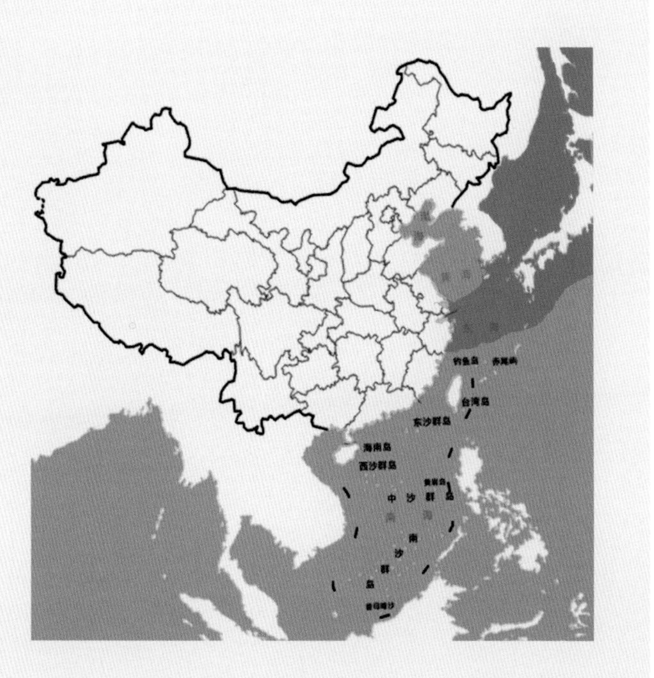

国内分布图
Map of Domestic Distribution

种群 Population

种群数量 **Population Size**	未知 / Unknown
种群趋势 **Population Trend**	稳定 / Stable

生境与生态系统 Habitat (s) and Ecosystem (s)

生　　境 **Habitat(s)**	近海岸海域 / Water Near Coast
生态系统 **Ecosystem(s)**	海洋生态系统 / Ocean Ecosystem

威胁 Threat (s)

主要威胁 **Major Threat(s)**	狩猎及采集、工业污染、气候变化、噪声污染 / Hunting and Collection, Industrial Pollution, Climate Change, Noise Pollution

保护级别与保护行动 Protection Category and Conservation Action (s)

国家重点保护野生动物等级 **(2021) Category of National Key Protected Wild Animals (2021)**	二级 / Category II
IUCN 红色名录 **(2020-2) IUCN Red List (2020-2)**	无危 / LC
CITES 附录 **(2019) CITES Appendix (2019)**	无 / NA
保护行动 **Conservation Action(s)**	无 / None

相关文献 Relevant References

Burgin *et al*., 2018; Jiang *et al*. (蒋志刚等), 2017; Mittermeier and Wilson, 2014; Lau *et al*., 2010; Smith *et al*. (史密斯等), 2009; Wang (王应祥), 2003

髯海豹 *Erignathus barbatus*

渔猫

Prionailurus viverrinus

数据缺乏 DD

数据缺乏 DD	无危 LC	近危 NT	易危 VU	濒危 EN	极危 CR	区域灭绝 RE	野外灭绝 EW	灭绝 EX

分类地位 Taxonomic Status

动物界 Animalia	脊索动物门 Chordata	哺乳纲 Mammalia	食肉目 Carnivora	猫科 Felidae

学名 Scientific Name	*Prionailurus viverrinus*
命名人 Species Authority	Bennett, 1833
英文名 English Name(s)	Fishing Cat
同物异名 Synonym(s)	*Felis viverrinus*
种下单元评估 Infra-specific Taxa Assessed	无 / None

评估信息 Assessment Information

评估年份 Year Assessed	2020
评定人 Assessor(s)	蒋志刚 / Zhigang Jiang
审定人 Reviewer(s)	陈辈乐、李飞 / Bosco P. L. Chan, Fei Li
其他贡献人 Other Contributor(s)	丁晨晨 / Chenchen Ding

理由 Justification: 有关渔猫的记录仅来自云南河口与台湾市场的贸易皮张，这些皮张有可能是经国际贸易流入中国的。分布在中国藏南地区的渔猫，现状未知。因此，列为数据缺乏等级 / The records of skins of Fishing Cat came from the markets of Hekou, Yunnan and Taiwan of China, which were likely to flow into China through international trade, but the status of Fishing Cat in Chinese Zangnan area is unknown. Thus, it is listed as Data Deficient

地理分布 Geographical Distribution

国内分布 Domestic Distribution

西藏 / Tibet (Xizang)

世界分布 World Distribution

中国；南亚 / China; South Asia

分布标注 Distribution Note

非特有种 / Non-endemic

国内分布图
Map of Domestic Distribution

种群 Population

种群数量 Population Size	未知 / Unknown
种群趋势 Population Trend	未知 / Unknown

生境与生态系统 Habitat (s) and Ecosystem (s)

生　　境 Habitat(s)	渔猫广泛分布在包括热带干燥森林在内的多种栖息地，但通常栖息在有芦苇床、潮汐溪流和红树林的沼泽地区；渔猫在较小的、快速移动的水道上的数量不多 / Fishing Cats are widely distributed in a variety of habitats, including Tropical Dry Forest, but they are usually found in Swampy Area with Reed Bed, Tidal Stream, and in Mangrove. Fishing Cats are less common in smaller, fast-moving Waterway
生态系统 Ecosystem(s)	湿地生态系统 / Wetland Ecosystem

威胁 Threat (s)

主要威胁 Major Threat(s)	猎杀 / Hunting

保护级别与保护行动 Protection Category and Conservation Action (s)

国家重点保护野生动物等级 (2021) Category of National Key Protected Wild Animals (2021)	二级 / Category II
IUCN 红色名录 (2020-2) IUCN Red List (2020-2)	易危 / VU
CITES 附录 (2019) CITES Appendix (2019)	无 / NA
保护行动 Conservation Action(s)	未知 / Unknown

相关文献 Relevant References

Burgin *et al.*, 2018; Jiang *et al.* (蒋志刚等), 2017; Jutzeler *et al.*, 2010b; Choudhury, 2003

渔猫 *Prionailurus viverrinus*

数据缺乏 DD	无危 LC	近危 NT	易危 VU	濒危 EN	极危 CR	区域灭绝 RE	野外灭绝 EW	灭绝 EX

分类地位 Taxonomic Status

动物界 Animalia	脊索动物门 Chordata	哺乳纲 Mammalia	偶蹄目 Artiodactyla	鹿科 Cervidae

学名 Scientific Name	*Muntiacus feae*
命名人 Species Authority	Thomas and Doria, 1889
英文名 English Name(s)	Fea's Muntjac
同物异名 Synonym(s)	Tenasserimm Muntjac; *Cervulus feae* (Thomas and Doria, 1889); *Muntiacus feai* (Tortonese in Grubb, 1977)
种下单元评估 Infra-specific Taxa Assessed	无 / None

评估信息 Assessment Information

评估年份 Year Assessed	2020
评定人 Assessor(s)	蒋志刚 / Zhigang Jiang
审定人 Reviewer(s)	陈辈乐、李飞、鲍伟东、鲍毅新、李言阔、丁平、蒋学龙 / Bosco P. L. Chan, Fei Li, Weidong Bao, Yixin Bao, Yankuo Li, Ping Ding, Xuelong Jiang
其他贡献人 Other Contributor(s)	李立立、丁晨晨 / Lili Li, Chenchen Ding

理由 Justification: 林麂曾记录于西藏东南部和云南西北部，但近年来没有任何关于林麂的消息。由于麂属各物种间形态甚为相似，因此，林麂的分类地位尚存争议，有待进一步研究。因此，列为数据缺乏等级 / Fea's Muntjac had been recorded in southeastern Tibet (Xizang) and northwestern Yunnan, but in recent years, there has been no information on the Fea's Muntjac in China. Since all muntjacs are similar, the taxonomy of Fea's Muntjac is still controversial and needs further research. Thus, it is listed as Data Deficient

地理分布 Geographical Distribution

国内分布 Domestic Distribution	
西藏、云南 / Tibet (Xizang), Yunnan	
世界分布 World Distribution	
中国；东南亚 / China; Southeast Asia	
分布标注 Distribution Note	
非特有种 / Non-endemic	

国内分布示意图
Schematic Map of Domestic Distribution

种群 Population

种群数量 **Population Size**	未知 / Unknown
种群趋势 **Population Trend**	未知 / Unknown

生境与生态系统 Habitat (s) and Ecosystem (s)

生　　境 **Habitat(s)**	森林 / Forest
生态系统 **Ecosystem(s)**	森林生态系统 / Forest Ecosystem

威胁 Threat (s)

主要威胁 **Major Threat(s)**	狩猎、开垦 / Hunting, Land Reclaimed for Plantation

保护级别与保护行动 Protection Category and Conservation Action (s)

国家重点保护野生动物等级 **(2021) Category of National Key Protected Wild Animals (2021)**	无 / NA
IUCN 红色名录 **(2020-2) IUCN Red List (2020-2)**	数据缺乏 / DD
CITES 附录 **(2019) CITES Appendix (2019)**	无 / NA
保护行动 **Conservation Action(s)**	未知 / Unknown

相关文献 Relevant References

Burgin *et al*., 2018; Jiang *et al*. (蒋志刚等), 2017; Huang *et al*. (黄薇等), 2008; Wang (王应祥), 2003; Zhang (张词祖等), 1984

数据缺乏 DD	无危 LC	近危 NT	易危 VU	濒危 EN	极危 CR	区域灭绝 RE	野外灭绝 EW	灭绝 EX

分类地位 Taxonomic Status

动物界 Animalia	脊索动物门 Chordata	哺乳纲 Mammalia	偶蹄目 Artiodactyla	鹿科 Cervidae
学名 **Scientific Name**		*Muntiacus putaoensis*		
命名人 **Species Authority**		Amato, Egan and Rabinowitz, 1999		
英文名 **English Name(s)**		Leaf Muntjac		
同物异名 **Synonym(s)**		Putao Muntjac, Leaf Deer		
种下单元评估 **Infra-specific Taxa Assessed**		无 / None		

评估信息 Assessment Information

评估年份 **Year Assessed**	2020
评定人 **Assessor(s)**	蒋志刚 / Zhigang Jiang
审定人 **Reviewer(s)**	蒋学龙、陈辈乐、李飞 / Xuelong Jiang, Bosco P. L. Chan, Fei Li
其他贡献人 **Other Contributor(s)**	李立立、丁晨晨 / Lili Li, Chenchen Ding

理由 Justification: 王应祥 (2003) 曾报道叶麂分布于云南腾冲、梁河、盈江和陇川。后来，再无其他关于叶麂的报告。该种的分类地位及分布状况有待进一步确认。因此，列为数据缺乏等级 / Wang (2003) reported that the Leaf Muntjac was distributed in Tengchong, Lianghe, Yingjiang and Longchuan, Yunnan. However, there have been no other reports about Leaf Muntjac in China since then. The taxonomy status and distribution of this species in China need to be further clarified and researched. Thus, it is listed as Data Deficient

地理分布 Geographical Distribution

国内分布 Domestic Distribution

云南 / Yunnan

世界分布 World Distribution

中国、印度、缅甸 / China, India, Myanmar

分布标注 Distribution Note

非特有种 / Non-endemic

国内分布图
Map of Domestic Distribution

种群 Population

种群数量 **Population Size**	未知 / Unknown
种群趋势 **Population Trend**	下降 / Decreasing

生境与生态系统 Habitat (s) and Ecosystem (s)

生　　境 **Habitat(s)**	森林 / Forest
生态系统 **Ecosystem(s)**	森林生态系统 / Forest Ecosystem

威胁 Threat (s)

主要威胁 **Major Threat(s)**	狩猎、开垦 / Hunting, Land Reclaimed for Plantation

保护级别与保护行动 Protection Category and Conservation Action (s)

国家重点保护野生动物等级 **(2021) Category of National Key Protected Wild Animals (2021)**	无 / NA
IUCN 红色名录 **(2020-2) IUCN Red List (2020-2)**	数据缺乏 / DD
CITES 附录 **(2019) CITES Appendix (2019)**	无 / NA
保护行动 **Conservation Action(s)**	未知 / Unknown

相关文献 Relevant References

Burgin *et al*., 2018; Jiang *et al.* (蒋志刚等), 2017; Pan *et al.* (潘清华等), 2007; Wang (王应祥), 2003; Amato *et al*., 1999

数据缺乏 DD	无危 LC	近危 NT	易危 VU	濒危 EN	极危 CR	区域灭绝 RE	野外灭绝 EW	灭绝 EX

分类地位 Taxonomic Status

动物界 Animalia	脊索动物门 Chordata	哺乳纲 Mammalia	偶蹄目 Artiodactyla	牛科 Bovidae

学　名 **Scientific Name**	*Ovis jubata*
命名人 **Species Authority**	Peters, 1876
英文名 **English Name(s)**	North China Argali
同物异名 **Synonym(s)**	雅布赖盘羊 , Yabulai Argali
种下单元评估 **Infra-specific Taxa Assessed**	无 / None

评估信息 Assessment Information

评估年份 **Year Assessed**	2020
评定人 **Assessor(s)**	蒋志刚 / Zhigang Jiang
审定人 **Reviewer(s)**	初红军、杨维康 / Hongjun Chu, Weikang Yang
其他贡献人 **Other Contributor(s)**	李立立、丁晨晨 / Lili Li, Chenchen Ding

理由 **Justification:** 华北盘羊的生境与种群数据缺乏，这一盘羊模式标本藏于国外博物馆。在山西、河北直至内蒙古大青山一带已经没有盘羊分布，华北盘羊可能已经灭绝。分布在甘肃和内蒙古交界处的雅布赖山的盘羊可能是这一盘羊亚种的残余种群(这一盘羊曾被称为雅布赖盘羊)。因此，列为数据缺乏等级 / Habitat and population data status of the North China Argali are unaware, and its type specimen is in a museum abroad. No argali was sighted in the region from Shanxi, Hebei to Daqing Shan, Inner Mongolia (Nei Mongol) during recent surveys. North China Argali may extinct. However, the argali living in the Yabulai Shan region between Gansu and Inner Mongolia (Nei Mongol) may be the remnant population of the species (that argali species was once called as Yabulai Argali). Thus, it is listed as Data Deficient

地理分布 Geographical Distribution

国内分布 Domestic Distribution

河北 / Hebei

世界分布 World Distribution

中国、蒙古 / China, Mongolia

分布标注 Distribution Note

非特有种 / Non-endemic

国内分布图
Map of Domestic Distribution

种群 Population

种群数量 **Population Size**	未知 / Unknown
种群趋势 **Population Trend**	未知 / Unknown

生境与生态系统 Habitat (s) and Ecosystem (s)

生　　境 **Habitat(s)**	荒漠草原 / Desert Grassland
生态系统 **Ecosystem(s)**	荒漠草原生态系统 / Desert Grassland Ecosystem

威胁 Threat (s)

主要威胁 **Major Threat(s)**	未知 / Unknown

保护级别与保护行动 Protection Category and Conservation Action (s)

国家重点保护野生动物等级 **(2021) Category of National Key Protected Wild Animals (2021)**	无 / NA
IUCN 红色名录 **(2020-2) IUCN Red List (2020-2)**	未列入 / NA
CITES 附录 **(2019) CITES Appendix (2019)**	II
保护行动 **Conservation Action(s)**	列入《迁徙物种保护公约》的"迁徙物种行动计划" / Included in the *Convention on Migratory Species* (CMS)—Species Action Plan

相关文献 Relevant References

Jiang *et al*. (蒋志刚等), 2017; Castelló, 2016; He *et al*. (何志超等), 2015; Wilson and Mittermeier, 2012; Groves and Grubb, 2011

数据缺乏 DD	无危 LC	近危 NT	易危 VU	濒危 EN	极危 CR	区域灭绝 RE	野外灭绝 EW	灭绝 EX

分类地位 Taxonomic Status

动物界 Animalia	脊索动物门 Chordata	哺乳纲 Mammalia	偶蹄目 Artiodactyla	牛科 Bovidae

学名 Scientific Name	*Ovis collium*
命名人 Species Authority	Linnaeus, 1758
英文名 English Name(s)	Kazakhstan Argali
同物异名 Synonym(s)	Karaganda Argali
种下单元评估 Infra-specific Taxa Assessed	无 / None

评估信息 Assessment Information

评估年份 Year Assessed	2020
评定人 Assessor(s)	蒋志刚 / Zhigang Jiang
审定人 Reviewer(s)	初红军、杨维康 / Hongjun Chu,Weikang Yang
其他贡献人 Other Contributor(s)	李立立、丁晨晨 / Lili Li, Chenchen Ding

理由 Justification: 哈萨克盘羊分布在新疆阿勒泰地区，尚缺乏野外调查数据。因此，列为数据缺乏等级 / Kazakhstan Argali is Distributed in the Altay region, Xinjiang. Its field data is deficient. Thus, it is listed as Data Deficient

地理分布 Geographical Distribution

国内分布 Domestic Distribution

新疆 / Xinjiang

世界分布 World Distribution

中国、哈萨克斯坦 / China, Kazakhstan

分布标注 Distribution Note

非特有种 / Non-endemic

国内分布图
Map of Domestic Distribution

种群 Population

种群数量 **Population Size**	未知 / Unknown
种群趋势 **Population Trend**	未知 / Unknown

生境与生态系统 Habitat (s) and Ecosystem (s)

生　　境 **Habitat(s)**	内陆岩石区域、灌丛 / Inland Rocky Area, Shrubland
生态系统 **Ecosystem(s)**	荒漠生态系统 / Desert Ecosystem

威胁 Threat (s)

主要威胁 **Major Threat(s)**	人类干扰、狩猎 / Human Disturbance, Hunting

保护级别与保护行动 Protection Category and Conservation Action (s)

国家重点保护野生动物等级 **(2021) Category of National Key Protected Wild Animals (2021)**	二级 / Category II
IUCN 红色名录 **(2020-2) IUCN Red List (2020-2)**	未列入 / NA
CITES 附录 **(2019) CITES Appendix (2019)**	无 / NA
保护行动 **Conservation Action(s)**	自然保护区内种群得到保护 / Populations in nature reserves are protected

相关文献 Relevant References

Jiang *et al*. (蒋志刚等), 2017; Castelló, 2016; Groves and Grubb, 2011; Wilson and Mittermeier, 2012

哈萨克盘羊 *Ovis collium*　　邸杰 摄　By Jie Di

数据缺乏 DD	无危 LC	近危 NT	易危 VU	濒危 EN	极危 CR	区域灭绝 RE	野外灭绝 EW	灭绝 EX

分类地位 Taxonomic Status

动 物 界 Animalia	脊索动物门 Chordata	哺 乳 纲 Mammalia	偶 蹄 目 Artiodactyla	牛 科 Bovidae

学 名 **Scientific Name**	*Naemorhedus evansi*
命 名 人 **Species Authority**	Lydekker, 1906
英 文 名 **English Name(s)**	Burmese Goral
同物异名 **Synonym(s)**	*Nemorhaedus evansi*
种下单元评估 **Infra-specific Taxa Assessed**	无 / None

评估信息 Assessment Information

评 估 年 份 **Year Assessed**	2020
评 定 人 **Assessor(s)**	蒋志刚 / Zhigang Jiang
审 定 人 **Reviewer(s)**	陈辈乐、李飞 / Bosco P. L. Chan, Fei Li
其他贡献人 **Other Contributor(s)**	李立立、丁晨晨 / Lili Li, Chenchen Ding

理由 **Justification:** Groves 和 Grubb (2011) 报道缅甸斑羚分布于云南西部，但近年来云南西部红外相机所拍摄的斑羚照片无法与中华斑羚区分。云南境内的缅甸斑羚的分类地位及分布状况有待进一步调查。因此，列为数据缺乏等级 / Groves and Grubb (2011) reported that Burmese Goral was distributed in western Yunnan, but the photos of goral taken in recent years by infrared cameras in western Yunnan could not be distinguished from Chinese Goral. The taxonomic status and distribution of Burmese Goral in Yunnan need further investigation. Therefore, it is listed as Data Deficient

地理分布 Geographical Distribution

国内分布 Domestic Distribution

云南 / Yunnan

世界分布 World Distribution

中国；东南亚 / China; Southeast Asia

分布标注 Distribution Note

非特有种 / Non-endemic

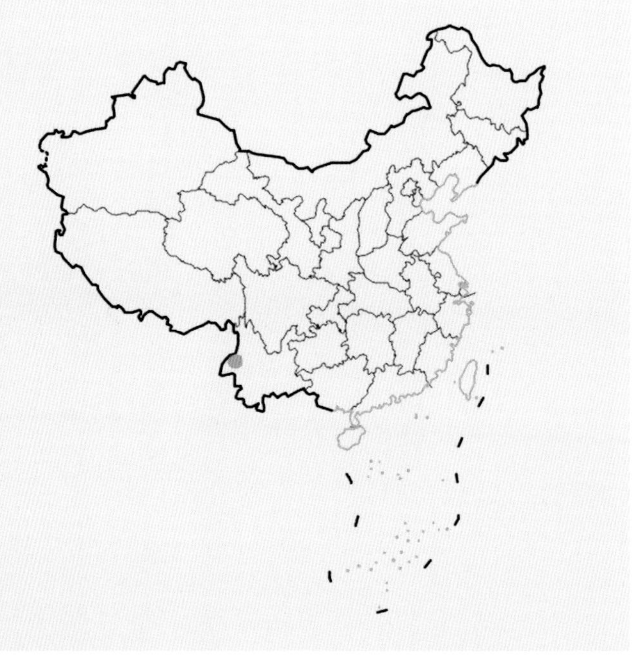

国内分布示意图
Schematic Map of Domestic Distribution

种群 Population

种群数量 **Population Size**	未知 / Unknown
种群趋势 **Population Trend**	未知 / Unknown

生境与生态系统 Habitat (s) and Ecosystem (s)

生　　境 **Habitat(s)**	有森林覆盖的陡峭山崖 / Steep Rocky Mountain Covered by Forest
生态系统 **Ecosystem(s)**	森林生态系统 / Forest Ecosystem

威胁 Threat (s)

主要威胁 **Major Threat(s)**	人类干扰、狩猎 / Human Disturbance, Hunting

保护级别与保护行动 Protection Category and Conservation Action (s)

国家重点保护野生动物等级 **(2021) Category of National Key Protected Wild Animals (2021)**	二级 / Category II
IUCN 红色名录 **(2020-2) IUCN Red List (2020-2)**	未列入 / NA
CITES 附录 **(2019) CITES Appendix (2019)**	无 / NA
保护行动 **Conservation Action(s)**	无 / None

相关文献 Relevant References

Burgin *et al*., 2018; Jiang *et al*. (蒋志刚等), 2017; Castelló, 2016; Wilson and Mittermeier, 2011; Groves and Grubb, 2011

数据缺乏 DD	无危 LC	近危 NT	易危 VU	濒危 EN	极危 CR	区域灭绝 RE	野外灭绝 EW	灭绝 EX

分类地位 Taxonomic Status

动物界 Animalia	脊索动物门 Chordata	哺乳纲 Mammalia	偶蹄目 Artiodactyla	牛科 Bovidae

学名 **Scientific Name**	*Capricornis rubidus*
命名人 **Species Authority**	Blyth, 1863
英文名 **English Name(s)**	Red Serow
同物异名 **Synonym(s)**	*Capricornis sumatraensis* (Blyth, 1863) subsp. *rubidus*
种下单元评估 **Infra-specific Taxa Assessed**	无 / None

评估信息 Assessment Information

评估年份 **Year Assessed**	2020
评定人 **Assessor(s)**	蒋志刚 / Zhigang Jiang
审定人 **Reviewer(s)**	陈辈乐 / Bosco P. L. Chan
其他贡献人 **Other Contributor(s)**	丁晨晨 / Chenchen Ding

理由 Justification: 红鬣羚在中国的分布狭窄，种群数量稀少。2014 年以来，红外相机调查相继在云南腾冲和泸水记录到红鬣羚。但红鬣羚种群与分布区不明。因此，列为数据缺乏等级 / Red Serow has a narrow distribution and small population in China. Since 2014, Red Serow have been recorded by infrared camera surveys in Tengchong and Lushui, Yunnan. But the population and distribution range of Red Serow are still unknown. Thus, it is listed as Data Deficient

地理分布 Geographical Distribution

国内分布 Domestic Distribution

云南 / Yunnan

世界分布 World Distribution

中国、缅甸 / China, Myanmar

分布标注 Distribution Note

非特有种 / Non-endemic

国内分布图
Map of Domestic Distribution

种群 Population

种群数量 **Population Size**	未知 / Unknown
种群趋势 **Population Trend**	未知 / Unknown

生境与生态系统 Habitat (s) and Ecosystem (s)

生　境 **Habitat(s)**	森林、内陆岩石区域、盐碱地、峭壁 / Forest, Inland Rocky Area; Saline, Brackish or Alkaline Land; Cliff
生态系统 **Ecosystem(s)**	森林生态系统 / Forest Ecosystem

威胁 Threat (s)

主要威胁 **Major Threat(s)**	未知 / Unknown

保护级别与保护行动 Protection Category and Conservation Action (s)

国家重点保护野生动物等级 **(2021) Category of National Key Protected Wild Animals (2021)**	二级 / Category II
IUCN 红色名录 **(2020-2) IUCN Red List (2020-2)**	近危 / NT
CITES 附录 **(2019) CITES Appendix (2019)**	无 / NA
保护行动 **Conservation Action(s)**	无 / None

相关文献 Relevant References

Burgin *et al.*, 2018; Castelló, 2016; Groves and Grubb, 2011; Wilson and Mittermeier, 2011

红鬣羚 *Capricornis rubidus*

© Lynx Edicions

数据缺乏 DD	无危 LC	近危 NT	易危 VU	濒危 EN	极危 CR	区域灭绝 RE	野外灭绝 EW	灭绝 EX

分类地位 Taxonomic Status

动 物 界 **Animalia**	脊索动物门 **Chordata**	哺 乳 纲 **Mammalia**	鲸 目 **Cetacea**	须 鲸 科 **Balaenopteridae**

学 名 **Scientific Name**	*Balaenoptera omurai*
命 名 人 **Species Authority**	Wada, Oishi and Yamada, 2003
英 文 名 **English Name(s)**	Omura's Whale
同物异名 **Synonym(s)**	Dwarf Fin Whale
种下单元评估 **Infra-specific Taxa Assessed**	无 / None

评估信息 Assessment Information

评 估 年 份 **Year Assessed**	2020
评 定 人 **Assessor(s)**	周开亚 / Kaiya Zhou
审 定 人 **Reviewer(s)**	张先锋、王克雄、王丁、祝茜、蒋志刚 / Xianfeng Zhang, Kexiong Wang, Ding Wang, Qian Zhu, Zhigang Jiang
其他贡献人 **Other Contributor(s)**	李立立、丁晨晨 / Lili Li, Chenchen Ding

理由 Justification: 大村鲸是 2003 年描述的新物种，目前野外研究很少。因此，列为数据缺乏等级 / Omura's Whale is a new species that was described scientifically in 2003. There have been very few reported field studies. Thus, it is listed as Data Deficient

地理分布 Geographical Distribution

国内分布 Domestic Distribution

东海、台湾海峡、南海 / East China Sea, Taiwan Strait, South China Sea

世界分布 World Distribution

北至日本海，南至澳大利亚，从印度洋西至大西洋毛里塔尼亚的海域 / In the ocean from the Sea of Japan to South Australia, from the Indian Ocean to the ocean region of Mauritania in Atlantic Ocean

分布标注 Distribution Note

非特有种 / Non-endemic

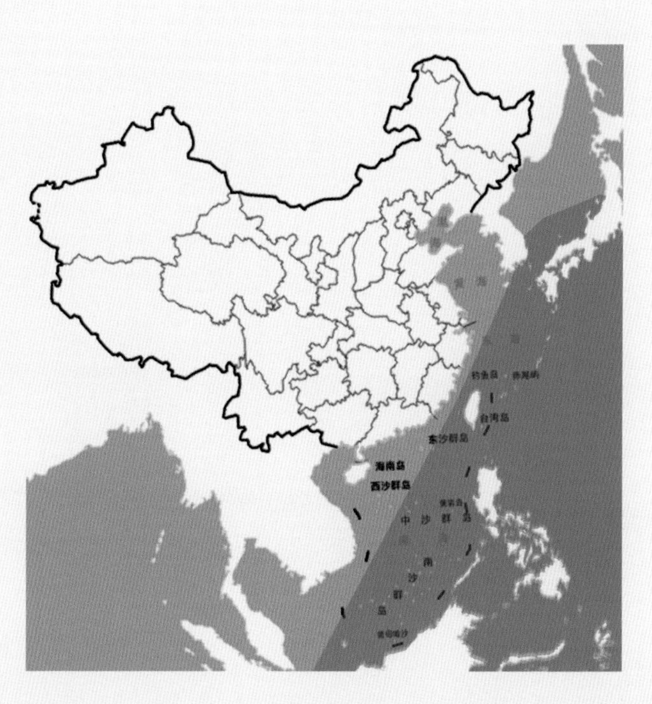

国内分布图
Map of Domestic Distribution

种群 Population

种群数量 **Population Size**	未知 / Unknown
种群趋势 **Population Trend**	未知 / Unknown

生境与生态系统 Habitat (s) and Ecosystem (s)

生　　境 **Habitat(s)**	海洋 / Ocean
生态系统 **Ecosystem(s)**	海洋生态系统 / Ocean Ecosystem

威胁 Threat (s)

主要威胁 **Major Threat(s)**	未知 / Unknown

保护级别与保护行动 Protection Category and Conservation Action (s)

国家重点保护野生动物等级 **(2021) Category of National Key Protected Wild Animals (2021)**	一级 / Category I
IUCN 红色名录 (2020-2) IUCN Red List (2020-2)	数据缺乏 / DD
CITES 附录 (2019) CITES Appendix (2019)	I
保护行动 **Conservation Action(s)**	无 / None

相关文献 Relevant References

Cerchio and Yamada, 2019; Burgin *et al*., 2018; Jiang *et al.* (蒋志刚等), 2015a; Mittermeier and Wilson, 2014; Wang (王丕烈), 2011; Zhou (周开亚), 2004, 2008

大村鲸 *Balaenoptera omurai*

小抹香鲸

Kogia breviceps

数据缺乏 DD

数据缺乏 DD	无危 LC	近危 NT	易危 VU	濒危 EN	极危 CR	区域灭绝 RE	野外灭绝 EW	灭绝 EX

分类地位 Taxonomic Status

动物界 Animalia	脊索动物门 Chordata	哺乳纲 Mammalia	鲸目 Cetacea	抹香鲸科 Physeteridae

学名 **Scientific Name**	*Kogia breviceps*
命名人 **Species Authority**	Blainville, 1838
英文名 **English Name(s)**	Pygmy Sperm Whale
同物异名 **Synonym(s)**	无 / None
种下单元评估 **Infra-specific Taxa Assessed**	无 / None

评估信息 Assessment Information

评估年份 **Year Assessed**	2020
评定人 **Assessor(s)**	周开亚 / Kaiya Zhou
审定人 **Reviewer(s)**	张先锋、王克雄、王丁、祝茜、蒋志刚 / Xianfeng Zhang, Kexiong Wang, Ding Wang, Qian Zhu, Zhigang Jiang
其他贡献人 **Other Contributor(s)**	李立立、丁晨晨 / Lili Li, Chenchen Ding

理由 **Justification:** 中国海域尚无小抹香鲸的数量和趋势的资料。因此，列为数据缺乏 / There is no information on the abundance and population trends of Pygmy Sperm Whale in China's waters. Thus, it is listed as Data Deficient

地理分布 Geographical Distribution

国内分布 Domestic Distribution

黄海、东海、台湾海峡、南海 / Yellow Sea, East China Sea, Taiwan Strait, South China Sea

世界分布 World Distribution

大西洋、太平洋和印度洋热带和温带水域，偶尔出现在如俄罗斯附近的较冷水域 / Tropical and temperate waters of the Atlantic, Pacific Ocean, and Indian Ocean. Occasionally appear in the colder waters, *e.g.* near Russia waters

分布标注 Distribution Note

非特有种 / Non-endemic

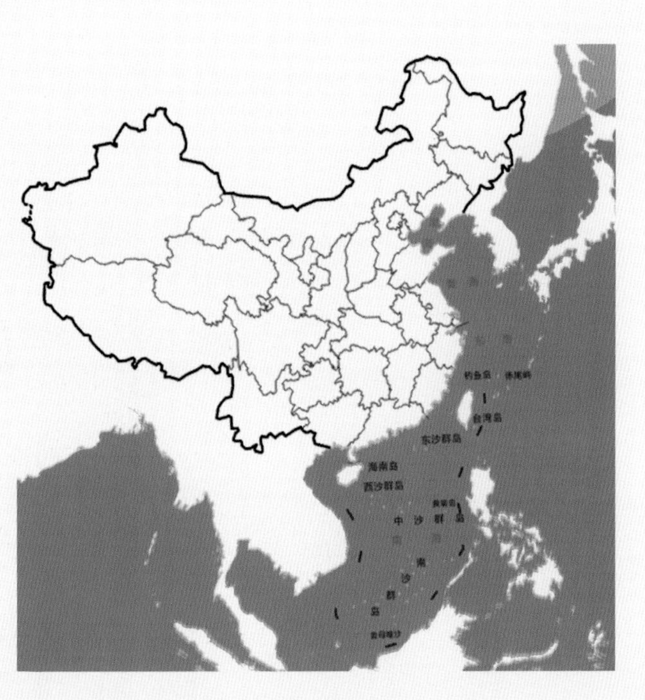

国内分布图
Map of Domestic Distribution

种群 Population

种群数量 **Population Size**	未知 / Unknown
种群趋势 **Population Trend**	未知 / Unknown

生境与生态系统 Habitat (s) and Ecosystem (s)

生　　境 **Habitat(s)**	海洋 / Ocean
生态系统 **Ecosystem(s)**	海洋生态系统 / Ocean Ecosystem

威胁 Threat (s)

主要威胁 **Major Threat(s)**	水下噪声、渔业兼捕 / Under Water Noise, By-catch in Drift-Net Fishery

保护级别与保护行动 Protection Category and Conservation Action (s)

国家重点保护野生动物等级 **(2021) Category of National Key Protected Wild Animals (2021)**	二级 / Category II
IUCN 红色名录 **(2020-2) IUCN Red List (2020-2)**	数据缺乏 / DD
CITES 附录 **(2019) CITES Appendix (2019)**	II
保护行动 **Conservation Action(s)**	法律保护物种 / Legally protected species

相关文献 Relevant References

Burgin *et al*., 2018; Jiang *et al.* (蒋志刚等), 2017, 2015a; Mittermeier and Wilson, 2014; Wang *et al.* (王丕烈等), 2007; Zhou (周开亚), 2008, 2004; Zhou *et al.* (周开亚等), 2001

侏抹香鲸
Kogia sima

数据缺乏 DD

数据缺乏 DD	无危 LC	近危 NT	易危 VU	濒危 EN	极危 CR	区域灭绝 RE	野外灭绝 EW	灭绝 EX

分类地位 Taxonomic Status

动物界 Animalia	脊索动物门 Chordata	哺乳纲 Mammalia	鲸目 Cetacea	抹香鲸科 Physeteridae

学名 **Scientific Name**	*Kogia sima*
命名人 **Species Authority**	Owen, 1866
英文名 **English Name(s)**	Dwarf Sperm Whale
同物异名 **Synonym(s)**	*Kogia simus* (Owen, 1866)
种下单元评估 **Infra-specific Taxa Assessed**	无 / None

评估信息 Assessment Information

评估年份 **Year Assessed**	2020
评定人 **Assessor(s)**	周开亚 / Kaiya Zhou
审定人 **Reviewer(s)**	张先锋、王克雄、王丁、祝茜、蒋志刚 / Xianfeng Zhang, Kexiong Wang, Ding Wang, Qian Zhu, Zhigang Jiang
其他贡献人 **Other Contributor(s)**	李立立、丁晨晨 / Lili Li, Chenchen Ding

理由 **Justification:** 侏抹香鲸数量较多，但尚无全球数量趋势的报道。因此，列为数据缺乏等级 / Population of Dwarf Sperm Whale is fairly abundant, but there is no information about its global population trends. Thus, it is listed as Data Deficient

地理分布 Geographical Distribution

国内分布 Domestic Distribution

黄海、东海、台湾海峡、南海 / Yellow Sea, East China Sea, Taiwan Strait, South China Sea

世界分布 World Distribution

大西洋、太平洋和印度洋热带和温带水域 / Tropical and temperate waters of the Atlantic Ocean, Pacific Ocean and Indian Ocean

分布标注 Distribution Note

非特有种 / Non-endemic

国内分布图
Map of Domestic Distribution

种群 Population

种群数量 **Population Size**	未知 / Unknown
种群趋势 **Population Trend**	未知 / Unknown

生境与生态系统 Habitat (s) and Ecosystem (s)

生　　境 **Habitat(s)**	海洋 / Ocean
生态系统 **Ecosystem(s)**	海洋生态系统 / Ocean Ecosystem

威胁 Threat (s)

主要威胁 **Major Threat(s)**	水下噪声、渔业兼捕、炸药捕鱼 / Under Water Noise, By-catch in Drift-net Fishery, Dynamite Fishing

保护级别与保护行动 Protection Category and Conservation Action (s)

国家重点保护野生动物等级 **(2021) Category of National Key Protected Wild Animals (2021)**	二级 / Category II
IUCN 红色名录 **(2020-2) IUCN Red List (2020-2)**	数据缺乏 / DD
CITES 附录 **(2019) CITES Appendix (2019)**	II
保护行动 **Conservation Action(s)**	法律保护物种 / Legally protected species

相关文献 Relevant References

Burgin *et al*., 2018; Jiang *et al.* (蒋志刚等), 2017, 2015a; Mittermeier and Wilson, 2014; Wang (王丕烈), 2011; Zhou (周开亚), 2008, 2004; Wang (王应祥), 2003

柏氏中喙鲸

Mesoplodon densirostris

数据缺乏 DD

数据缺乏 DD	无危 LC	近危 NT	易危 VU	濒危 EN	极危 CR	区域灭绝 RE	野外灭绝 EW	灭绝 EX

分类地位 Taxonomic Status

动物界 Animalia	脊索动物门 Chordata	哺乳纲 Mammalia	鲸目 Cetacea	喙鲸科 Ziphiidae

学名 Scientific Name	*Mesoplodon densirostris*
命名人 Species Authority	Blainville, 1817
英文名 English Name(s)	Blainville's Beaked Whale
同物异名 Synonym(s)	无 / None
种下单元评估 Infra-specific Taxa Assessed	无 / None

评估信息 Assessment Information

评估年份 Year Assessed	2020
评定人 Assessor(s)	周开亚 / Kaiya Zhou
审定人 Reviewer(s)	张先锋、王克雄、王丁、祝茜、蒋志刚 / Xianfeng Zhang, Kexiong Wang, Ding Wang, Qian Zhu, Zhigang Jiang
其他贡献人 Other Contributor(s)	李立立、丁晨晨 / Lili Li, Chenchen Ding

理由 Justification: 柏氏中喙鲸的全球种群数量信息缺乏，对种群趋势一无所知。因此，列为数据缺乏等级 / There is limited information on global abundance and no information on population trends for this species. Thus, it is listed as Data Deficient

地理分布 Geographical Distribution

国内分布 Domestic Distribution

黄海、东海、台湾海峡、南海 / Yellow Sea, East China Sea, Taiwan Strait, South China Sea

世界分布 World Distribution

所有海洋的热带和温暖水域，且已知在非常高纬度海洋活动 / Tropical and warm waters in all oceans and has been recorded in oceans of very high latitudes

分布标注 Distribution Note

非特有种 / Non-endemic

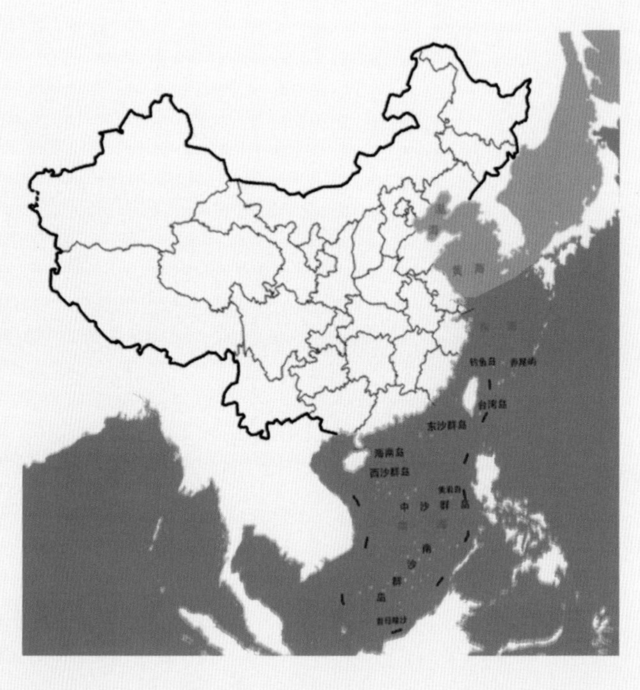

国内分布图
Map of Domestic Distribution

种群 Population

种群数量 **Population Size**	未知 / Unknown
种群趋势 **Population Trend**	未知 / Unknown

生境与生态系统 Habitat (s) and Ecosystem (s)

生　　境 **Habitat(s)**	海洋 / Ocean
生态系统 **Ecosystem(s)**	海洋生态系统 / Ocean Ecosystem

威胁 Threat (s)

主要威胁 **Major Threat(s)**	水下噪声，尤其是海军声呐、渔业兼捕 / Under Water Noise, especially from Naval Sonar, By-catch in Drift-net Fishery

保护级别与保护行动 Protection Category and Conservation Action (s)

国家重点保护野生动物等级 **(2021) Category of National Key Protected Wild Animals (2021)**	二级 / Category II
IUCN 红色名录 **(2020-2) IUCN Red List (2020-2)**	数据缺乏 / DD
CITES 附录 **(2019) CITES Appendix (2019)**	II
保护行动 **Conservation Action(s)**	法律保护物种 / Legally protected species

相关文献 Relevant References

Burgin *et al*., 2018; Jiang *et al.* (蒋志刚等), 2015a; Mittermeier and Wilson, 2014; Wang (王丕烈), 2011; Zhou (周开亚), 2008, 2004; Wang (王应祥), 2003

中国生物多样性红色名录
China's Red List of Biodiversity

银杏齿中喙鲸

Mesoplodon ginkgodens

数据缺乏 DD

数据缺乏 DD	无危 LC	近危 NT	易危 VU	濒危 EN	极危 CR	区域灭绝 RE	野外灭绝 EW	灭绝 EX

分类地位 Taxonomic Status

动物界 Animalia	脊索动物门 Chordata	哺乳纲 Mammalia	鲸目 Cetacea	喙鲸科 Ziphiidae

学名 Scientific Name	*Mesoplodon ginkgodens*
命名人 Species Authority	Nishiwaki and Kamiya, 1958
英文名 English Name(s)	Ginkgo-toothed Beaked Whale
同物异名 Synonym(s)	*Mesoplodon hotaula* (Deraniyagala, 1963)
种下单元评估 Infra-specific Taxa Assessed	无 / None

评估信息 Assessment Information

评估年份 Year Assessed	2020
评定人 Assessor(s)	周开亚 / Kaiya Zhou
审定人 Reviewer(s)	张先锋、王克雄、王丁、祝茜、蒋志刚 / Xianfeng Zhang, Kexiong Wang, Ding Wang, Qian Zhu, Zhigang Jiang
其他贡献人 Other Contributor(s)	李立立、丁晨晨 / Lili Li, Chenchen Ding

理由 Justification: 缺乏对银杏齿中喙鲸的数量估计。因此，列为数据缺乏等级 / There are no estimates of the abundance of Ginkgo-toothed Beaked Whales. Thus, it is listed as Data Deficient

地理分布 Geographical Distribution

国内分布 Domestic Distribution

黄海、东海、台湾海峡、南海 / Yellow Sea, East China Sea, Taiwan Strait, South China Sea

世界分布 World Distribution

印度洋、太平洋热带与温带海域 / Tropical and temperate waters in the Indian Ocean and Pacific Ocean

分布标注 Distribution Note

非特有种 / Non-endemic

国内分布图
Map of Domestic Distribution

种群 Population

种群数量 **Population Size**	未知 / Unknown
种群趋势 **Population Trend**	未知 / Unknown

生境与生态系统 Habitat (s) and Ecosystem (s)

生　　境 **Habitat(s)**	海洋 / Ocean
生态系统 **Ecosystem(s)**	海洋生态系统 / Ocean Ecosystem

威胁 Threat (s)

主要威胁 **Major Threat(s)**	刺网、张网等误捕，地震测量和海军声呐的噪声 / Incidental Mortality in Fishery, including in Gill Net and Set Net; Noise from Seismic Survey and Naval Sonar

保护级别与保护行动 Protection Category and Conservation Action (s)

国家重点保护野生动物等级 (2021) **Category of National Key Protected Wild Animals (2021)**	二级 / Category II
IUCN 红色名录 (2020-2) IUCN Red List (2020-2)	数据缺乏 / DD
CITES 附录 (2019) CITES Appendix (2019)	II
保护行动 **Conservation Action(s)**	法律保护物种 / Legally protected species

相关文献 Relevant References

Burgin *et al*., 2018; Jiang *et al.* (蒋志刚等), 2017, 2015a; Mittermeier and Wilson, 2014; Wang *et al.* (王丕烈等), 2011; Zhou (周开亚), 2008, 2004; Zhou *et al.* (周开亚等), 2001

小中喙鲸

Mesoplodon peruvianus

数据缺乏 DD

数据缺乏 DD	无危 LC	近危 NT	易危 VU	濒危 EN	极危 CR	区域灭绝 RE	野外灭绝 EW	灭绝 EX

分类地位 Taxonomic Status

动物界 Animalia	脊索动物门 Chordata	哺乳纲 Mammalia	鲸目 Cetacea	喙鲸科 Ziphiidae

学名 Scientific Name	*Mesoplodon peruvianus*
命名人 Species Authority	Reyes, Mead and Van Waerebeek, 1991
英文名 English Name(s)	Pygmy Beaked Whale
同物异名 Synonym(s)	无 / None
种下单元评估 Infra-specific Taxa Assessed	无 / None

评估信息 Assessment Information

评估年份 Year Assessed	2020
评定人 Assessor(s)	周开亚 / Kaiya Zhou
审定人 Reviewer(s)	张先锋、王克雄、王丁、祝茜、蒋志刚 / Xianfeng Zhang, Kexiong Wang, Ding Wang, Qian Zhu, Zhigang Jiang
其他贡献人 Other Contributor(s)	李立立、丁晨晨 / Lili Li, Chenchen Ding

理由 **Justification:** 缺乏小中喙鲸种群数量和趋势的信息。因此，列为数据缺乏等级 / Lacking of information on population sizes and trends of Pygmy Beaked Whale. Thus, it is listed as Data Deficient

地理分布 Geographical Distribution

国内分布 Domestic Distribution

东海 / East China Sea

世界分布 World Distribution

太平洋 / Pacific Ocean

分布标注 Distribution Note

非特有种 / Non-endemic

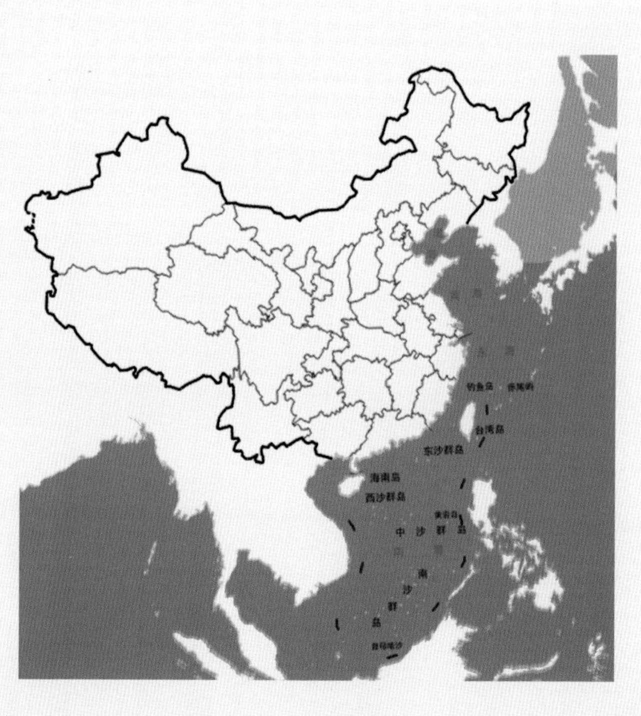

国内分布图
Map of Domestic Distribution

种群 Population

种群数量 **Population Size**	未知 / Unknown
种群趋势 **Population Trend**	未知 / Unknown

生境与生态系统 Habitat (s) and Ecosystem (s)

生　　境 **Habitat(s)**	海洋 / Ocean
生态系统 **Ecosystem(s)**	海洋生态系统 / Ocean Ecosystem

威胁 Threat (s)

主要威胁 **Major Threat(s)**	海军声呐噪声、渔业兼捕 / Naval Sonar Noise, By-catch in Drift-net Fishery

保护级别与保护行动 Protection Category and Conservation Action (s)

国家重点保护野生动物等级 **(2021) Category of National Key Protected Wild Animals (2021)**	二级 / Category II
IUCN 红色名录 **(2020-2) IUCN Red List (2020-2)**	数据缺乏 / DD
CITES 附录 **(2019) CITES Appendix (2019)**	II
保护行动 **Conservation Action(s)**	法律保护物种 / Legally protected species

相关文献 Relevant References

Burgin *et al*., 2018; Jiang *et al.* (蒋志刚等), 2017, 2015a; Mittermeier and Wilson, 2014; Culik, 2011; Wang *et al.* (王丕烈等), 2011

小中喙鲸 *Mesoplodon peruvianus*

贝氏喙鲸

Berardius bairdii

数据缺乏 DD

数据缺乏 DD	无危 LC	近危 NT	易危 VU	濒危 EN	极危 CR	区域灭绝 RE	野外灭绝 EW	灭绝 EX

分类地位 Taxonomic Status

动物界 Animalia	脊索动物门 Chordata	哺乳纲 Mammalia	鲸目 Cetacea	喙鲸科 Ziphiidae

学　名 **Scientific Name**	*Berardius bairdii*
命名人 **Species Authority**	Stejneger, 1883
英文名 **English Name(s)**	Baird's Beaked Whale
同物异名 **Synonym(s)**	无 / None
种下单元评估 **Infra-specific Taxa Assessed**	无 / None

评估信息 Assessment Information

评估年份 **Year Assessed**	2020
评定人 **Assessor(s)**	周开亚 / Kaiya Zhou
审定人 **Reviewer(s)**	张先锋、王克雄、王丁、祝茜、蒋志刚 / Xianfeng Zhang, Kexiong Wang, Ding Wang, Qian Zhu, Zhigang Jiang
其他贡献人 **Other Contributor(s)**	李立立、丁晨晨 / Lili Li, Chenchen Ding

理由 **Justification:** 对贝氏喙鲸全球种群数量和趋势的了解有限。因此，列为数据缺乏等级 / There is only a limited understanding of the global population and trends of Baird's Beaked Whale. Thus, it is listed as Data Deficient

地理分布 Geographical Distribution

国内分布 Domestic Distribution

东海 / East China Sea

世界分布 World Distribution

太平洋 / Pacific Ocean

分布标注 Distribution Note

非特有种 / Non-endemic

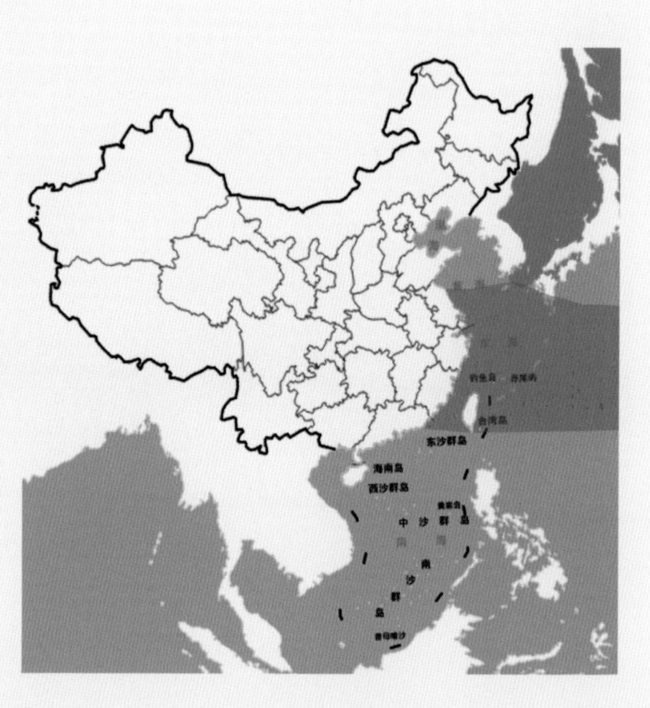

国内分布图
Map of Domestic Distribution

种群 Population

种群数量 **Population Size**	未知 / Unknown
种群趋势 **Population Trend**	未知 / Unknown

生境与生态系统 Habitat (s) and Ecosystem (s)

生　　境 **Habitat(s)**	海洋 / Ocean
生态系统 **Ecosystem(s)**	海洋生态系统 / Ocean Ecosystem

威胁 Threat (s)

主要威胁 **Major Threat(s)**	商业捕鲸、渔业误捕、海军声呐噪声 / Commercial Whaling, Fishery Accidental Catching, Naval Sonar Noise

保护级别与保护行动 Protection Category and Conservation Action (s)

国家重点保护野生动物等级 **(2021) Category of National Key Protected Wild Animals (2021)**	二级 / Category II
IUCN 红色名录 **(2020-2) IUCN Red List (2020-2)**	数据缺乏 / DD
CITES 附录 **(2019) CITES Appendix (2019)**	II
保护行动 **Conservation Action(s)**	法律保护物种 / Legally protected species

相关文献 Relevant References

Burgin *et al.*, 2018; Jiang *et al.* (蒋志刚等), 2017, 2015a; Mittermeier and Wilson, 2014; Wang *et al.* (王丕烈等), 2011; Zhou (周开亚), 2008, 2004; Wang and Wang (王火根和王宇), 1998

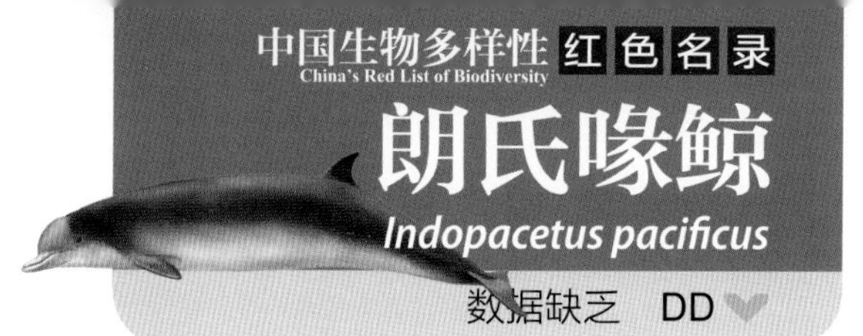

朗氏喙鲸

Indopacetus pacificus

数据缺乏 DD

数据缺乏 DD	无危 LC	近危 NT	易危 VU	濒危 EN	极危 CR	区域灭绝 RE	野外灭绝 EW	灭绝 EX

分类地位 Taxonomic Status

动 物 界 Animalia	脊索动物门 Chordata	哺 乳 纲 Mammalia	鲸 目 Cetacea	喙 鲸 科 Ziphiidae

学 名 **Scientific Name**	*Indopacetus pacificus*
命 名 人 **Species Authority**	Longman, 1926
英 文 名 **English Name(s)**	Indo-Pacific Beaked Whale
同物异名 **Synonym(s)**	Tropical Bottlenose Whale; Longman's Beaked Whale; *Mesoplodon pacificus* (Longman, 1926); *M. pacificus* (Longman, 1926)
种下单元评估 **Infra-specific Taxa Assessed**	无 / None

评估信息 Assessment Information

评 估 年 份 **Year Assessed**	2020
评 定 人 **Assessor(s)**	周开亚 / Kaiya Zhou
审 定 人 **Reviewer(s)**	张先锋、王克雄、王丁、祝茜、蒋志刚 / Xianfeng Zhang, Kexiong Wang, Ding Wang, Qian Zhu, Zhigang Jiang
其他贡献人 **Other Contributor(s)**	李立立、丁晨晨 / Lili Li, Chenchen Ding

理由 Justification: 对全球朗氏喙鲸种群数量和趋势几乎没有了解。因此，列为数据缺乏等级 / There is nearly no information regarding to the global population and trends of Indo-Pacific Beaked Whale. Thus, it is listed as Data Deficient

地理分布 Geographical Distribution

国内分布 Domestic Distribution

东海 / East China Sea

世界分布 World Distribution

大西洋和太平洋的暖温带部分 / Warm to temperate portions of the Atlantic Ocean and Pacific Ocean

分布标注 Distribution Note

非特有种 / Non-endemic

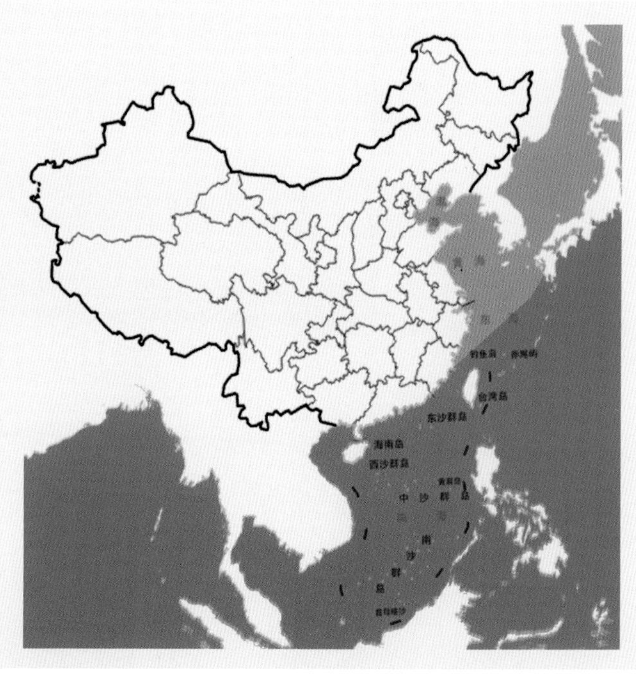

国内分布图
Map of Domestic Distribution

种群 Population

种群数量 **Population Size**	未知 / Unknown
种群趋势 **Population Trend**	未知 / Unknown

生境与生态系统 Habitat (s) and Ecosystem (s)

生　　境 **Habitat(s)**	海洋 / Ocean
生态系统 **Ecosystem(s)**	海洋生态系统 / Ocean Ecosystem

威胁 Threat (s)

主要威胁 **Major Threat(s)**	渔业误捕、海军声呐噪声、塑料碎片 / Fishery Accidental Catching, Naval Sonar Noise, Plastic Debris

保护级别与保护行动 Protection Category and Conservation Action (s)

国家重点保护野生动物等级 **(2021) Category of National Key Protected Wild Animals (2021)**	二级 / Category II
IUCN 红色名录 **(2020-2) IUCN Red List (2020-2)**	数据缺乏 / DD
CITES 附录 **(2019) CITES Appendix (2019)**	II
保护行动 **Conservation Action(s)**	法律保护物种 / Legally protected species

相关文献 Relevant References

Burgin *et al*., 2018; Jiang *et al.* (蒋志刚等), 2017, 2015a; Mittermeier and Wilson, 2014; Culik, 2011; Wang *et al.* (王丕烈等), 2011; Peng *et al.* (彭亚君等), 2009

中国生物多样性红色名录
China's Red List of Biodiversity

飞旋原海豚
Stenella longirostris

数据缺乏 DD

数据缺乏 DD	无危 LC	近危 NT	易危 VU	濒危 EN	极危 CR	区域灭绝 RE	野外灭绝 EW	灭绝 EX

分类地位 Taxonomic Status

动物界 Animalia	脊索动物门 Chordata	哺乳纲 Mammalia	鲸目 Cetacea	海豚科 Delphinidae

学名 Scientific Name	*Stenella longirostris*
命名人 Species Authority	Gray, 1828
英文名 English Name(s)	Spinner Dolphin
同物异名 Synonym(s)	无 / None
种下单元评估 Infra-specific Taxa Assessed	无 / None

评估信息 Assessment Information

评估年份 Year Assessed	2020
评定人 Assessor(s)	周开亚 / Kaiya Zhou
审定人 Reviewer(s)	张先锋、王克雄、王丁、祝茜、蒋志刚 / Xianfeng Zhang, Kexiong Wang, Ding Wang, Qian Zhu, Zhigang Jiang
其他贡献人 Other Contributor(s)	李立立、丁晨晨 / Lili Li, Chenchen Ding

理由 Justification: 对东太平洋的飞旋原海豚种群的研究有多篇文献报道，但对西太平洋、印度洋、大西洋的种群尚无报道。因此，列为数据缺乏等级 / There are several literatures regarding to the populations of the Spinner Dolphin in the Eastern Pacific Ocean. However, there is no report about the populations in the Western Pacific Ocean, Indian Ocean and Atlantic Ocean. Thus, it is listed as Data Deficient

地理分布 Geographical Distribution

国内分布 Domestic Distribution

南海、台湾海峡；福建、台湾、香港、广西 / South China Sea, Taiwan Strait; Fujian, Taiwan, Hong Kong, Guangxi

世界分布 World Distribution

分布在北纬 40° 到南纬 40° 间海域的所有热带和亚热带水域 / All tropical and subtropical waters between 40°N and 40°S

分布标注 Distribution Note

非特有种 / Non-endemic

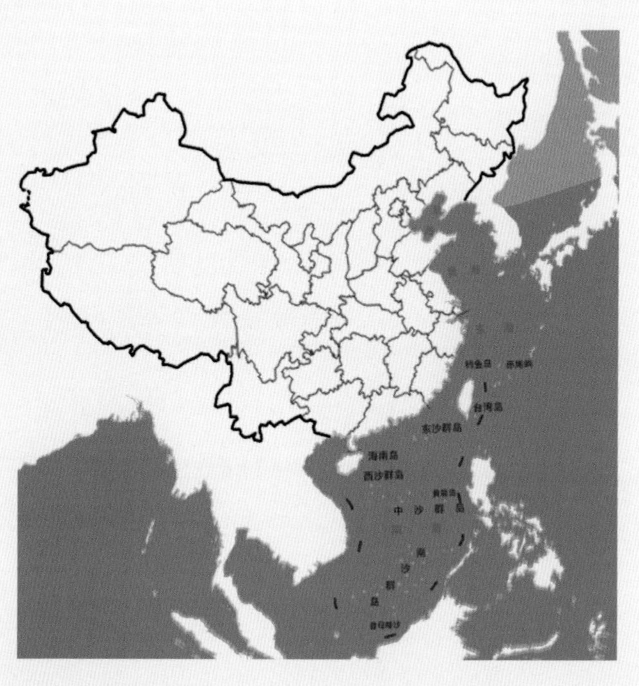

国内分布图
Map of Domestic Distribution

种群 Population

种群数量 Population Size	东太平洋的飞旋原海豚数量合计超过百万头。在其他海洋的飞旋原海豚数量未知 / The numbers of Spinner Dolphin amount to more than a million in the Eastern Pacific Ocean, however its populations in other oceans are unknown
种群趋势 Population Trend	未知 / Unknown

生境与生态系统 Habitat (s) and Ecosystem (s)

生　　境 Habitat(s)	海洋 / Ocean
生态系统 Ecosystem(s)	海洋生态系统 / Ocean Ecosystem

威胁 Threat (s)

主要威胁 Major Threat(s)	定向渔业、兼捕、污染 / Direct Catch in Fishery, By-catch, Pollution

保护级别与保护行动 Protection Category and Conservation Action (s)

国家重点保护野生动物等级 (2021) Category of National Key Protected Wild Animals (2021)	二级 / Category II
IUCN 红色名录 (2020-2) IUCN Red List (2020-2)	数据缺乏 / DD
CITES 附录 (2019) CITES Appendix (2019)	II
保护行动 Conservation Action(s)	法律保护物种 / Legally protected species

相关文献 Relevant References

Burgin *et al*., 2018; Jiang *et al.* (蒋志刚等), 2017, 2015a; Mittermeier and Wilson, 2014; Wang (王丕烈), 2011; Zhou (周开亚), 2008, 2004; Zhu *et al.* (祝茜等), 2007

飞旋原海豚 *Stenella longirostris*

印太瓶鼻海豚

Tursiops aduncus

数据缺乏 DD

数据缺乏 DD	无危 LC	近危 NT	易危 VU	濒危 EN	极危 CR	区域灭绝 RE	野外灭绝 EW	灭绝 EX

分类地位 Taxonomic Status

动物界 Animalia	脊索动物门 Chordata	哺乳纲 Mammalia	鲸目 Cetacea	海豚科 Delphinidae

学名 **Scientific Name**	*Tursiops aduncus*
命名人 **Species Authority**	Ehrenberg, 1833
英文名 **English Name(s)**	Indo-Pacific Bottlenose Dolphin
同物异名 **Synonym(s)**	无 / None
种下单元评估 **Infra-specific Taxa Assessed**	无 / None

评估信息 Assessment Information

评估年份 **Year Assessed**	2020
评定人 **Assessor(s)**	周开亚 / Kaiya Zhou
审定人 **Reviewer(s)**	张先锋、王克雄、王丁、祝茜、蒋志刚 / Xianfeng Zhang, Kexiong Wang, Ding Wang, Qian Zhu, Zhigang Jiang
其他贡献人 **Other Contributor(s)**	李立立、丁晨晨 / Lili Li, Chenchen Ding

理由 **Justification:** 缺乏印太瓶鼻海豚种群数量和趋势的研究报道。因此，列为数据缺乏等级 / No recorded literatures or researches regarding to the population size and trend of Indo-Pacific Bottlenose Dolphin. Thus, it is listed as Data Deficient

地理分布 Geographical Distribution

国内分布 Domestic Distribution

东海、台湾海峡、南海 / East China Sea, Taiwan Strait, South China Sea

世界分布 World Distribution

印度、澳大利亚北部、中国南部、红海和非洲东海岸的水域 / The waters around India, north Australia, south China, the Red Sea, and the eastern coast of Africa

分布标注 Distribution Note

非特有种 / Non-endemic

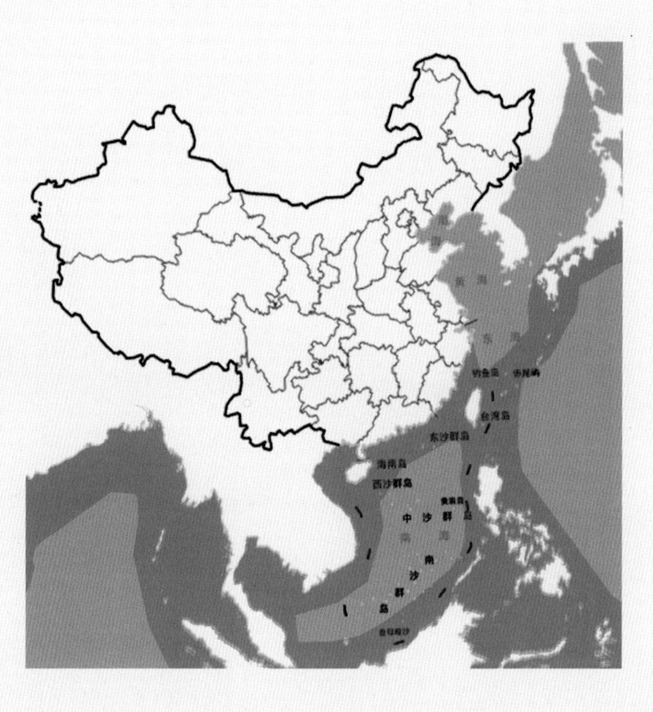

国内分布图
Map of Domestic Distribution

种群 Population

种群数量 Population Size	多个地方种群有上万个个体，但印太瓶鼻海豚的总体数量未知 / There are ten thousands of Indo-Pacific Bottlenose Dolphin in some local populations. However, the total number of its population size is unknown
种群趋势 Population Trend	未知 / Unknown

生境与生态系统 Habitat (s) and Ecosystem (s)

生　　境 Habitat(s)	海洋 / Ocean
生态系统 Ecosystem(s)	海洋生态系统 / Ocean Ecosystem

威胁 Threat (s)

主要威胁 Major Threat(s)	定向捕获、兼捕、栖息地退化、人类干扰 / Direct Catch, By-catch, Habitat Degradation, Human Disturbance

保护级别与保护行动 Protection Category and Conservation Action (s)

国家重点保护野生动物等级 (2021) Category of National Key Protected Wild Animals (2021)	二级 / Category II
IUCN 红色名录 (2020-2) IUCN Red List (2020-2)	数据缺乏 / DD
CITES 附录 (2019) CITES Appendix (2019)	II
保护行动 Conservation Action(s)	法律保护物种 / Legally protected species

相关文献 Relevant References

Burgin *et al*., 2018; Jiang *et al.* (蒋志刚等), 2017, 2015a; Mittermeier and Wilson, 2014; Culik, 2011; Wang (王丕烈), 2011; Zhou (周开亚), 2008, 2004

数据缺乏 DD	无危 LC	近危 NT	易危 VU	濒危 EN	极危 CR	区域灭绝 RE	野外灭绝 EW	灭绝 EX

分类地位 Taxonomic Status

动 物 界 **Animalia**	脊索动物门 **Chordata**	哺 乳 纲 **Mammalia**	鲸 目 **Cetacea**	海 豚 科 **Delphinidae**

学 名 **Scientific Name**	*Orcinus orca*
命 名 人 **Species Authority**	Linnaeus, 1758
英 文 名 **English Name(s)**	Killer Whale
同物异名 **Synonym(s)**	Orca; *Orcinus glacialis* (Berzin and Vladimirov, 1983); *nanus* (Mikhalev *et al.*, 1981)
种下单元评估 **Infra-specific Taxa Assessed**	无 / None

评估信息 Assessment Information

评 估 年 份 **Year Assessed**	2020
评 定 人 **Assessor(s)**	周开亚 / Kaiya Zhou
审 定 人 **Reviewer(s)**	张先锋、王克雄、王丁、祝茜、蒋志刚 / Xianfeng Zhang, Kexiong Wang, Ding Wang, Qian Zhu, Zhigang Jiang
其他贡献人 **Other Contributor(s)**	李立立、丁晨晨 / Lili Li, Chenchen Ding

理由 **Justification:** 虎鲸在世界广泛分布，至少有数以万计的成熟个体。但关于该物种在中国水域分布状况的信息非常少。因此，列为数据缺乏等级 / Killer Whale is widely distributed throughout the world, with at least several tens thousands of mature individuals. However, there is only a little information about this species in China's waters. Thus, it is listed as Data Deficient

地理分布 Geographical Distribution

国内分布 Domestic Distribution

渤海、黄海、东海、南海、台湾海域 / Bohai Sea, Yellow Sea, East China Sea, South China Sea, Taiwan waters

世界分布 World Distribution

全球海域 / All global ocean waters

分布标注 Distribution Note

非特有种 / Non-endemic

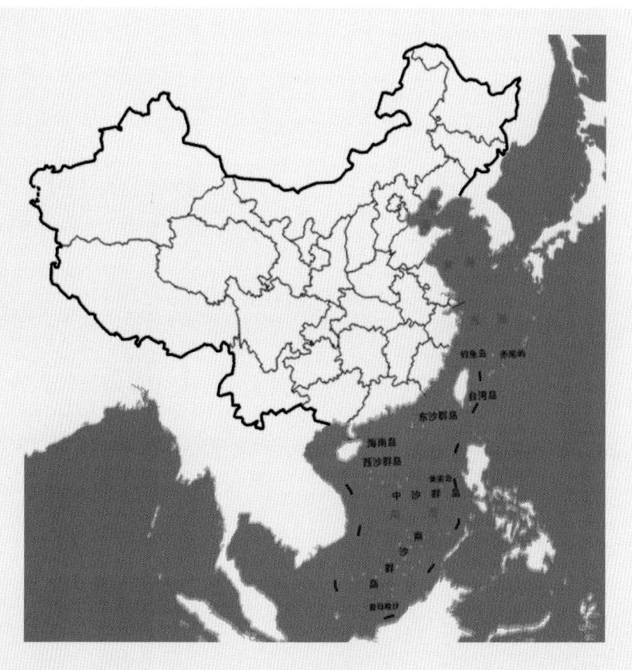

国内分布图
Map of Domestic Distribution

种群 Population

种群数量 Population Size	未知 / Unknown
种群趋势 Population Trend	未知 / Unknown

生境与生态系统 Habitat (s) and Ecosystem (s)

生　　境 Habitat(s)	海洋 / Ocean
生态系统 Ecosystem(s)	海洋生态系统 / Ocean Ecosystem

威胁 Threat (s)

主要威胁 Major Threat(s)	偶然捕获、污染、噪声、栖息地退化 / Incidental Catch, Pollution, Noise, Habitat Degradation

保护级别与保护行动 Protection Category and Conservation Action (s)

国家重点保护野生动物等级 (2021) Category of National Key Protected Wild Animals (2021)	二级 / Category II
IUCN 红色名录 (2020-2) IUCN Red List (2020-2)	数据缺乏 / DD
CITES 附录 (2019) CITES Appendix (2019)	II
保护行动 Conservation Action(s)	法律保护物种 / Legally protected species

相关文献 Relevant References

Burgin *et al*., 2018; Jiang *et al.* (蒋志刚等), 2017, 2015a; Mittermeier and Wilson, 2014; Culik, 2011; Zhou (周开亚), 2008, 2004; Wang and Fan (王火根和范忠勇), 2004

虎鲸 *Orcinus orca*

数据缺乏 DD	无危 LC	近危 NT	易危 VU	濒危 EN	极危 CR	区域灭绝 RE	野外灭绝 EW	灭绝 EX

分类地位 Taxonomic Status

动物界 Animalia	脊索动物门 Chordata	哺乳纲 Mammalia	鲸目 Cetacea	海豚科 Delphinidae
学名 **Scientific Name**		*Pseudorca crassidens*		
命名人 **Species Authority**		Owen, 1846		
英文名 **English Name(s)**		False Killer Whale		
同物异名 **Synonym(s)**		无 / None		
种下单元评估 **Infra-specific Taxa Assessed**		无 / None		

评估信息 Assessment Information

评估年份 **Year Assessed**	2020
评定人 **Assessor(s)**	周开亚 / Kaiya Zhou
审定人 **Reviewer(s)**	张先锋、王克雄、王丁、祝茜、蒋志刚 / Xianfeng Zhang, Kexiong Wang, Ding Wang, Qian Zhu, Zhigang Jiang
其他贡献人 **Other Contributor(s)**	李立立、丁晨晨 / Lili Li, Chenchen Ding

理由 **Justification:** Odell 和 McClune (1999) 曾报道中国和日本沿岸的伪虎鲸种群数约为 16,000 头，但缺乏进一步的研究。因此，列为数据缺乏等级 / Odell and McClune(1999) reported the estimation of 16,000 False Killer Whales in the coastal waters of China and Japan, but lack of further research. Thus, it is listed as Data Deficient

地理分布 Geographical Distribution

国内分布 Domestic Distribution

渤海、黄海、东海、南海、台湾海域 / Bohai Sea, Yellow Sea, East China Sea, South China Sea, Taiwan waters

世界分布 World Distribution

全球海域 / All global ocean waters

分布标注 Distribution Note

非特有种 / Non-endemic

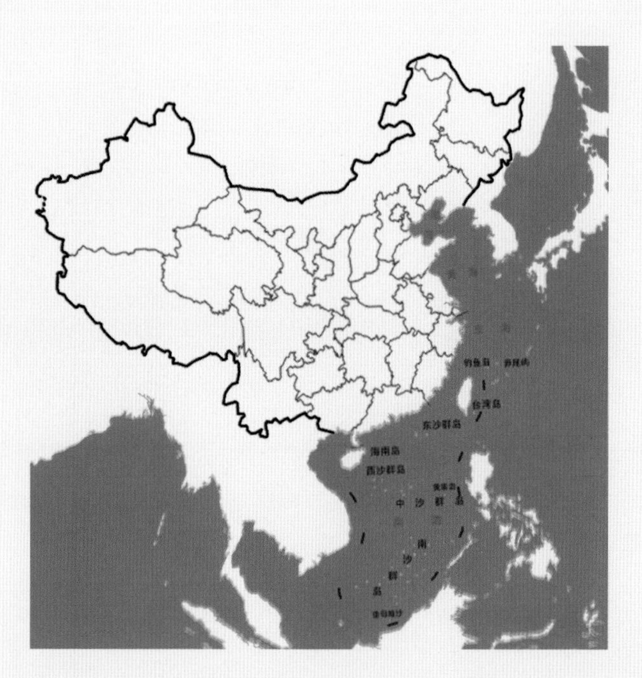

国内分布图
Map of Domestic Distribution

种群 Population

种群数量 **Population Size**	未知 / Unknown
种群趋势 **Population Trend**	未知 / Unknown

生境与生态系统 Habitat (s) and Ecosystem (s)

生　　境 **Habitat(s)**	海洋 / Ocean
生态系统 **Ecosystem(s)**	海洋生态系统 / Ocean Ecosystem

威胁 Threat (s)

主要威胁 **Major Threat(s)**	直接捕杀、渔业兼捕、污染 / Direct Catch, By-catch in Fishery, Pollution

保护级别与保护行动 Protection Category and Conservation Action (s)

国家重点保护野生动物等级 **(2021) Category of National Key Protected Wild Animals (2021)**	二级 / Category II
IUCN 红色名录 **(2020-2) IUCN Red List (2020-2)**	数据缺乏 / DD
CITES 附录 **(2019) CITES Appendix (2019)**	II
保护行动 **Conservation Action(s)**	法律保护物种 / Legally protected species

相关文献 Relevant References

Burgin *et al*., 2018; Jiang *et al*. (蒋志刚等), 2017, 2015a; Mittermeier and Wilson, 2014; Wang (王丕烈), 2011; Zhou (周开亚), 2008, 2004; Odell and McClune, 1999

伪虎鲸 *Pseudorca crassidens*　　© Lynx Edicions

数据缺乏 DD	无危 LC	近危 NT	易危 VU	濒危 EN	极危 CR	区域灭绝 RE	野外灭绝 EW	灭绝 EX

分类地位 Taxonomic Status

动物界 Animalia	脊索动物门 Chordata	哺乳纲 Mammalia	鲸目 Cetacea	海豚科 Delphinidae
学名 Scientific Name		*Feresa attenuata*		
命名人 Species Authority		Gray, 1874		
英文名 English Name(s)		Pygmy Killer Whale		
同物异名 Synonym(s)		无 / None		
种下单元评估 Infra-specific Taxa Assessed		无 / None		

评估信息 Assessment Information

评估年份 Year Assessed	2020
评定人 Assessor(s)	周开亚 / Kaiya Zhou
审定人 Reviewer(s)	张先锋、王克雄、王丁、祝茜、蒋志刚 / Xianfeng Zhang, Kexiong Wang, Ding Wang, Qian Zhu, Zhigang Jiang
其他贡献人 Other Contributor(s)	李立立、丁晨晨 / Lili Li, Chenchen Ding

理由 Justification: 小虎鲸在世界各地的热带和亚热带近海水域呈小群游弋，但缺乏种群数量估计。因此，列为数据缺乏等级 / Pygmy Killer Whale prefers to swim in groups in tropical and subtropical offshore waters around the world. However, there is very little information available on its population size. Thus, it is listed as Data Deficient

地理分布 Geographical Distribution

国内分布 Domestic Distribution

东海、南海、台湾海域 / East China Sea, South China Sea, Taiwan waters

世界分布 World Distribution

全球海域 / All global ocean waters

分布标注 Distribution Note

非特有种 / Non-endemic

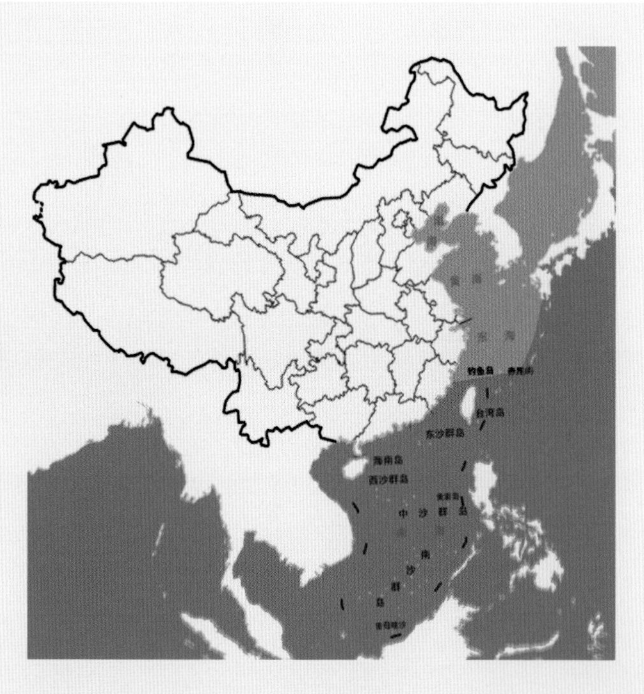

国内分布图
Map of Domestic Distribution

种群 Population

种群数量 **Population Size**	未知 / Unknown
种群趋势 **Population Trend**	未知 / Unknown

生境与生态系统 Habitat (s) and Ecosystem (s)

生　境 **Habitat(s)**	海洋 / Ocean
生态系统 **Ecosystem(s)**	海洋生态系统 / Ocean Ecosystem

威胁 Threat (s)

主要威胁 **Major Threat(s)**	直接捕杀、渔业兼捕、海军声呐噪声、污染 / Direct Catch, By-catch in Fishery, Naval Sonar Noise, Pollution

保护级别与保护行动 Protection Category and Conservation Action (s)

国家重点保护野生动物等级 **(2021) Category of National Key Protected Wild Animals (2021)**	二级 / Category II
IUCN 红色名录 **(2020-2) IUCN Red List (2020-2)**	数据缺乏 / DD
CITES 附录 **(2019) CITES Appendix (2019)**	II
保护行动 **Conservation Action(s)**	法律保护物种 / Legally protected species

相关文献 Relevant References

Burgin *et al.*, 2018; Jiang *et al.* (蒋志刚等), 2017, 2015a; Mittermeier and Wilson, 2014; Wang (王丕烈), 2011; Zhou (周开亚), 2008, 2004

中国生物多样性红色名录
China's Red List of Biodiversity

短肢领航鲸

Globicephala macrorhynchus

数据缺乏 DD

数据缺乏 DD	无危 LC	近危 NT	易危 VU	濒危 EN	极危 CR	区域灭绝 RE	野外灭绝 EW	灭绝 EX

分类地位 Taxonomic Status

动物界 Animalia	脊索动物门 Chordata	哺乳纲 Mammalia	鲸目 Cetacea	海豚科 Delphinidae

学名 Scientific Name	*Globicephala macrorhynchus*
命名人 Species Authority	Gray, 1846
英文名 English Name(s)	Short-finned Pilot Whale
同物异名 Synonym(s)	*Globicephala scammoni* (Bailey, 1936)
种下单元评估 Infra-specific Taxa Assessed	无 / None

评估信息 Assessment Information

评估年份 Year Assessed	2020
评定人 Assessor(s)	周开亚 / Kaiya Zhou
审定人 Reviewer(s)	张先锋、王克雄、王丁、祝茜、蒋志刚 / Xianfeng Zhang, Kexiong Wang, Ding Wang, Qian Zhu, Zhigang Jiang
其他贡献人 Other Contributor(s)	李立立、丁晨晨 / Lili Li, Chenchen Ding

理由 Justification: 尚无中国近海的短肢领航鲸的种群数量报道。因此，列为数据缺乏等级 / The population size of Short-finned Pilot Whale in the coastal waters of China has not been reported. Thus, it is listed as Data Deficient

地理分布 Geographical Distribution

国内分布 Domestic Distribution

黄海、东海、南海、台湾海域 / Yellow Sea, East China Sea, South China Sea, Taiwan waters

世界分布 World Distribution

印度洋、大西洋和太平洋温带和热带海域 / Temperate and tropical regions of the Indian Ocean, Atlantic Ocean, and Pacific Ocean

分布标注 Distribution Note

非特有种 / Non-endemic

国内分布图
Map of Domestic Distribution

种群 Population

种群数量 Population Size	未知 / Unknown
种群趋势 Population Trend	未知 / Unknown

生境与生态系统 Habitat (s) and Ecosystem (s)

生　　境 Habitat(s)	海洋 / Ocean
生态系统 Ecosystem(s)	海洋生态系统 / Ocean Ecosystem

威胁 Threat (s)

主要威胁 Major Threat(s)	捕杀、渔业兼捕、海军噪声、污染 / Catching, By-catch in Fishery, Naval Sonar Noise, Pollution

保护级别与保护行动 Protection Category and Conservation Action (s)

国家重点保护野生动物等级 (2021) Category of National Key Protected Wild Animals (2021)	二级 / Category II
IUCN 红色名录 (2020-2) IUCN Red List (2020-2)	数据缺乏 / DD
CITES 附录 (2019) CITES Appendix (2019)	II
保护行动 Conservation Action(s)	法律保护物种 / Legally protected species

相关文献 Relevant References

Burgin *et al*., 2018; Jiang *et al.* (蒋志刚等), 2017, 2015a; Mittermeier and Wilson, 2014; Wang (王丕烈), 2011; Zhou (周开亚), 2008, 2004; Zhou *et al.* (周开亚等), 2001

数据缺乏 DD	无危 LC	近危 NT	易危 VU	濒危 EN	极危 CR	区域灭绝 RE	野外灭绝 EW	灭绝 EX

分类地位 Taxonomic Status

动 物 界 Animalia	脊索动物门 Chordata	哺 乳 纲 Mammalia	啮 齿 目 Rodentia	松 鼠 科 Sciuridae
学　名 Scientific Name		*Petaurista grandis*		
命 名 人 Species Authority		Swinhoe, 1862		
英 文 名 English Name(s)		Taiwan Flying Squirrel		
同物异名 Synonym(s)		Formosan Giant Flying Squirrel		
种下单元评估 Infra-specific Taxa Assessed		无 / None		

评估信息 Assessment Information

评 估 年 份 Year Assessed	2020
评　定　人 Assessor(s)	蒋志刚、刘少英 / Zhigang Jiang, Shaoying Liu
审　定　人 Reviewer(s)	马勇、鲍毅新 / Yong Ma, Yixin Bao
其他贡献人 Other Contributor(s)	李立立、丁晨晨 / Lili Li, Chenchen Ding

理由 Justification: 缺乏对台湾大鼯鼠的研究。因此，列为数据缺乏等级 / Studies on Taiwan Flying Squirrel is deficient. Thus, it is listed as Data Deficient

地理分布 Geographical Distribution

国内分布 Domestic Distribution

台湾 / Taiwan

世界分布 World Distribution

中国 / China

分布标注 Distribution Note

特有种 / Endemic

国内分布图
Map of Domestic Distribution

种群 Population

种群数量 **Population Size**	未知 / Unknown
种群趋势 **Population Trend**	未知 / Unknown

生境与生态系统 Habitat (s) and Ecosystem (s)

生　　境 **Habitat(s)**	森林 / Forest
生态系统 **Ecosystem(s)**	森林生态系统 / Forest Ecosystem

威胁 Threat (s)

主要威胁 **Major Threat(s)**	未知 / Unknown

保护级别与保护行动 Protection Category and Conservation Action (s)

国家重点保护野生动物等级 **(2021) Category of National Key Protected Wild Animals (2021)**	无 / NA
IUCN 红色名录 **(2020-2) IUCN Red List (2020-2)**	未列入 / NA
CITES 附录 **(2019) CITES Appendix (2019)**	无 / NA
保护行动 **Conservation Action(s)**	无 / None

相关文献 Relevant References

Burgin *et al*., 2018; Jiang *et al.* (蒋志刚等), 2017; Zhang (张荣祖), 1997

台湾大鼯鼠 *Petaurista grandis*　　蒋志刚 绘　Drawn by Zhigang Jiang

海南大鼯鼠

Petaurista hainana

数据缺乏 DD

数据缺乏 DD	无危 LC	近危 NT	易危 VU	濒危 EN	极危 CR	区域灭绝 RE	野外灭绝 EW	灭绝 EX

分类地位 Taxonomic Status

动物界 Animalia	脊索动物门 Chordata	哺乳纲 Mammalia	啮齿目 Rodentia	松鼠科 Sciuridae

学　名 **Scientific Name**	*Petaurista hainana*
命名人 **Species Authority**	Allen, 1925
英文名 **English Name(s)**	Hainan Flying Squirrel
同物异名 **Synonym(s)**	Hainan Giant Flying Squirrel
种下单元评估 **Infra-specific Taxa Assessed**	无 / None

评估信息 Assessment Information

评估年份 **Year Assessed**	2020
评定人 **Assessor(s)**	蒋志刚、刘少英 / Zhigang Jiang, Shaoying Liu
审定人 **Reviewer(s)**	马勇、鲍毅新 / Yong Ma, Yixin Bao
其他贡献人 **Other Contributor(s)**	李立立、丁晨晨 / Lili Li, Chenchen Ding

理由 **Justification:** 对海南大鼯鼠的研究缺乏。因此，列为数据缺乏等级 / Studies on Hainan Flying Squirrel is deficient. Thus, it is listed as Data Deficient

地理分布 Geographical Distribution

国内分布 Domestic Distribution

海南 / Hainan

世界分布 World Distribution

中国 / China

分布标注 Distribution Note

特有种 / Endemic

国内分布图
Map of Domestic Distribution

种群 Population

种群数量 Population Size	未知 / Unknown
种群趋势 Population Trend	未知 / Unknown

生境与生态系统 Habitat (s) and Ecosystem (s)

生　　境 Habitat(s)	森林 / Forest
生态系统 Ecosystem(s)	森林生态系统 / Forest Ecosystem

威胁 Threat (s)

主要威胁 Major Threat(s)	未知 / Unknown

保护级别与保护行动 Protection Category and Conservation Action (s)

国家重点保护野生动物等级 (2021) Category of National Key Protected Wild Animals (2021)	无 / NA
IUCN 红色名录 (2020-2) IUCN Red List (2020-2)	未列入 / NA
CITES 附录 (2019) CITES Appendix (2019)	无 / NA
保护行动 Conservation Action(s)	无 / None

相关文献 Relevant References

Burgin *et al.*, 2018; Jiang *et al.* (蒋志刚等), 2017; Zhang (张荣祖), 1997

海南大鼯鼠 *Petaurista hainana*　　蒋志刚 绘　Drawn by Zhigang Jiang

白颊鼯鼠

Petaurista leucogenys

数据缺乏 DD

数据缺乏 DD	无危 LC	近危 NT	易危 VU	濒危 EN	极危 CR	区域灭绝 RE	野外灭绝 EW	灭绝 EX

分类地位 Taxonomic Status

动物界 Animalia	脊索动物门 Chordata	哺乳纲 Mammalia	啮齿目 Rodentia	松鼠科 Sciuridae

学名 **Scientific Name**	*Petaurista leucogenys*
命名人 **Species Authority**	Temminck, 1827
英文名 **English Name(s)**	Watase's Flying Squirrel
同物异名 **Synonym(s)**	Japanese Giant Flying Squirrel
种下单元评估 **Infra-specific Taxa Assessed**	无 / None

评估信息 Assessment Information

评估年份 **Year Assessed**	2020
评定人 **Assessor(s)**	蒋志刚、刘少英 / Zhigang Jiang, Shaoying Liu
审定人 **Reviewer(s)**	李松、马勇 / Song Li, Yong Ma
其他贡献人 **Other Contributor(s)**	李立立、丁晨晨 / Lili Li, Chenchen Ding

理由 **Justification:** 对白颊鼯鼠的研究缺乏。因此，列为数据缺乏等级 / Studies on Watase's Flying Squirrel is deficient. Thus, it is listed as Data Deficient

地理分布 Geographical Distribution

国内分布 Domestic Distribution

黑龙江 / Heilongjiang

世界分布 World Distribution

中国；东北亚 / China; Northeast China

分布标注 Distribution Note

非特有种 / Non-endemic

国内分布图
Map of Domestic Distribution

种群 Population

种群数量 **Population Size**	未知 / Unknown
种群趋势 **Population Trend**	未知 / Unknown

生境与生态系统 Habitat (s) and Ecosystem (s)

生　　境 **Habitat(s)**	森林 / Forest
生态系统 **Ecosystem(s)**	森林生态系统 / Forest Ecosystem

威胁 Threat (s)

主要威胁 **Major Threat(s)**	未知 / Unknown

保护级别与保护行动 Protection Category and Conservation Action (s)

国家重点保护野生动物等级 **(2021) Category of National Key Protected Wild Animals (2021)**	无 / NA
IUCN 红色名录 **(2020-2) IUCN Red List (2020-2)**	未列入 / NA
CITES 附录 **(2019) CITES Appendix (2019)**	无 / NA
保护行动 **Conservation Action(s)**	未知 / Unknown

相关文献 Relevant References

Burgin *et al*., 2018; Jiang *et al.* (蒋志刚等), 2017; Zhang (张荣祖), 1997

白颊鼯鼠 *Petaurista leucogenys*　　蒋志刚 绘　Drawn by Zhigang Jiang

中国生物多样性红色名录
China's Red List of Biodiversity

云南大鼯鼠

Petaurista yunanensis

数据缺乏 DD

数据缺乏 DD	无危 LC	近危 NT	易危 VU	濒危 EN	极危 CR	区域灭绝 RE	野外灭绝 EW	灭绝 EX

分类地位 Taxonomic Status

动物界 Animalia	脊索动物门 Chordata	哺乳纲 Mammalia	啮齿目 Rodentia	松鼠科 Sciuridae

学名 Scientific Name	*Petaurista yunanensis*
命名人 Species Authority	Anderson, 1875
英文名 English Name(s)	Yunnan Giant Flying Squirrel
同物异名 Synonym(s)	无 / None
种下单元评估 Infra-specific Taxa Assessed	无 / None

评估信息 Assessment Information

评估年份 Year Assessed	2020
评定人 Assessor(s)	蒋志刚、刘少英 / Zhigang Jiang, Shaoying Liu
审定人 Reviewer(s)	马勇、鲍毅新 / Yong Ma, Yixin Bao
其他贡献人 Other Contributor(s)	李立立、丁晨晨 / Lili Li, Chenchen Ding

理由 Justification: 对云南大鼯鼠的研究缺乏。因此，列为数据缺乏等级 / Studies on Yunnan Giant Flying Squirrel are deficient. Thus, it is listed as Data Deficient

地理分布 Geographical Distribution

国内分布 Domestic Distribution

云南、广西、西藏 / Yunnan, Guangxi, Tibet (Xizang)

世界分布 World Distribution

中国 / China

分布标注 Distribution Note

特有种 / Endemic

国内分布图
Map of Domestic Distribution

种群 Population

种群数量 **Population Size**	未知 / Unknown
种群趋势 **Population Trend**	未知 / Unknown

生境与生态系统 Habitat (s) and Ecosystem (s)

生　　境 **Habitat(s)**	森林 / Forest
生态系统 **Ecosystem(s)**	森林生态系统 / Forest Ecosystem

威胁 Threat (s)

主要威胁 **Major Threat(s)**	无 / None

保护级别与保护行动 Protection Category and Conservation Action (s)

国家重点保护野生动物等级 **(2021) Category of National Key Protected Wild Animals (2021)**	无 / NA
IUCN 红色名录 **(2020-2) IUCN Red List (2020-2)**	无危 / LC
CITES 附录 **(2019) CITES Appendix (2019)**	无 / NA
保护行动 **Conservation Action(s)**	无 / None

相关文献 Relevant References

Burgin *et al*., 2018; Jiang *et al*. (蒋志刚等), 2017; Zhang (张荣祖), 1997

云南大鼯鼠 *Petaurista yunanensis*　　　　班鼎盈 摄　By Dingying Ban

绒毛鼯鼠

Eupetaurus cinereus

数据缺乏 DD

数据缺乏 DD	无危 LC	近危 NT	易危 VU	濒危 EN	极危 CR	区域灭绝 RE	野外灭绝 EW	灭绝 EX

分类地位 Taxonomic Status

动物界 Animalia	脊索动物门 Chordata	哺乳纲 Mammalia	啮齿目 Rodentia	松鼠科 Sciuridae

学　　名 Scientific Name	*Eupetaurus cinereus*
命 名 人 Species Authority	Thomas, 1888
英 文 名 English Name(s)	Woolly Flying Squirrel
同物异名 Synonym(s)	无 / None
种下单元评估 Infra-specific Taxa Assessed	无 / None

评估信息 Assessment Information

评估年份 Year Assessed	2020
评　定　人 Assessor(s)	蒋志刚、刘少英 / Zhigang Jiang, Shaoying Liu
审　定　人 Reviewer(s)	马勇、鲍毅新 / Yong Ma, Yixin Bao
其他贡献人 Other Contributor(s)	李立立、丁晨晨 / Lili Li, Chenchen Ding

理由 **Justification:** 对绒毛鼯鼠的研究缺乏。因此，列为数据缺乏等级 / Studies on Woolly Flying Squirrel are deficient. Thus, it is listed as Data Deficient

地理分布 Geographical Distribution

国内分布 Domestic Distribution

云南 / Yunnan

世界分布 World Distribution

中国、巴基斯坦 / China, Pakistan

分布标注 Distribution Note

非特有种 / Non-endemic

国内分布图
Map of Domestic Distribution

种群 Population

种群数量 **Population Size**	未知 / Unknown
种群趋势 **Population Trend**	未知 / Unknown

生境与生态系统 Habitat (s) and Ecosystem (s)

生　　境 **Habitat(s)**	亚热带常绿阔叶林 / Subtropical Evergreen Broad-leaved Forest
生态系统 **Ecosystem(s)**	森林生态系统 / Forest Ecosystem

威胁 Threat (s)

主要威胁 **Major Threat(s)**	无 / None

保护级别与保护行动 Protection Category and Conservation Action (s)

国家重点保护野生动物等级 (2021) **Category of National Key Protected Wild Animals (2021)**	无 / NA
IUCN 红色名录 (2020-2) IUCN Red List (2020-2)	无危 / LC
CITES 附录 (2019) CITES Appendix (2019)	无 / NA
保护行动 **Conservation Action(s)**	无 / None

相关文献 Relevant References

Burgin *et al.*, 2018; Jiang *et al.* (蒋志刚等), 2017; Zheng *et al.* (郑智民等), 2012; Wang (王应祥), 2003; Zahler and Khan, 2003

绒毛鼯鼠 *Eupetaurus cinereus*

中国生物多样性红色名录
China's Red List of Biodiversity

坎氏毛足鼠
Phodopus campbelli

数据缺乏 DD

数据缺乏 DD	无危 LC	近危 NT	易危 VU	濒危 EN	极危 CR	区域灭绝 RE	野外灭绝 EW	灭绝 EX

分类地位 Taxonomic Status

动物界 Animalia	脊索动物门 Chordata	哺乳纲 Mammalia	啮齿目 Rodentia	仓鼠科 Cricetidae

学名 Scientific Name	*Phodopus campbelli*
命名人 Species Authority	Thomas, 1905
英文名 English Name(s)	Campbell's Desert Hamster
同物异名 Synonym(s)	Campbell's Dwarf Hamster; *Phodopus crepidatus* (Hollister, 1912); *tuvinicus* (Orlov and Iskharova, 1974)
种下单元评估 Infra-specific Taxa Assessed	无 / None

评估信息 Assessment Information

评估年份 Year Assessed	2020
评定人 Assessor(s)	蒋志刚、刘少英 / Zhigang Jiang, Shaoying Liu
审定人 Reviewer(s)	马勇、路纪琪、鲍毅新 / Yong Ma, Jiqi Lu, Yixin Bao
其他贡献人 Other Contributor(s)	李立立、丁晨晨 / Lili Li, Chenchen Ding

理由 Justification: 坎氏毛足鼠的研究记录较少，对其的研究缺乏。因此，列为数据缺乏等级 / Little information on Campbell's Desert Hamster has been recorded, and studies on it are deficient. Thus, it is listed as Data Deficient

地理分布 Geographical Distribution

国内分布 Domestic Distribution

内蒙古、河北、辽宁、吉林 / Inner Mongolia (Nei Mongol), Hebei, Liaoning, Jilin

世界分布 World Distribution

中国、哈萨克斯坦、蒙古、俄罗斯 / China, Kazakhstan, Mongolia, Russia

分布标注 Distribution Note

非特有种 / Non-endemic

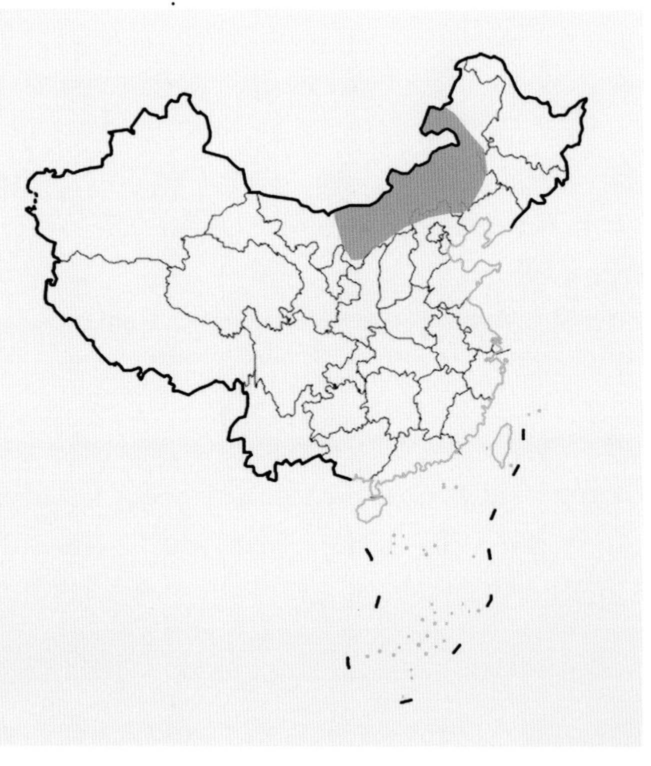

国内分布图
Map of Domestic Distribution

种群 Population

种群数量 **Population Size**	未知 / Unknown
种群趋势 **Population Trend**	稳定 / Stable

生境与生态系统 Habitat (s) and Ecosystem (s)

生　　境 **Habitat(s)**	草地、半荒漠 / Grassland, Semi-desert
生态系统 **Ecosystem(s)**	草地生态系统、荒漠生态系统 / Grassland Ecosystem, Desert Ecosystem

威胁 Threat (s)

主要威胁 **Major Threat(s)**	无 / None

保护级别与保护行动 Protection Category and Conservation Action (s)

国家重点保护野生动物等级 **(2021) Category of National Key Protected Wild Animals (2021)**	无 / NA
IUCN 红色名录 **(2020-2) IUCN Red List (2020-2)**	无危 / LC
CITES 附录 **(2019) CITES Appendix (2019)**	无 / NA
保护行动 **Conservation Action(s)**	无 / None

相关文献 Relevant References

Burgin *et al*., 2018; Jiang *et al.* (蒋志刚等), 2017; Wilson *et al*., 2016; Zheng *et al.* (郑智民等), 2012; Smith *et al.* (史密斯等), 2009; Zheng and Li (郑生武和李保国), 1999; Corbet and Hill, 1991

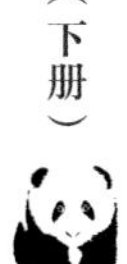

丽江绒鼠
Eothenomys fidelis

数据缺乏 DD

数据缺乏 DD	无危 LC	近危 NT	易危 VU	濒危 EN	极危 CR	区域灭绝 RE	野外灭绝 EW	灭绝 EX

分类地位 Taxonomic Status

动物界 Animalia	脊索动物门 Chordata	哺乳纲 Mammalia	啮齿目 Rodentia	仓鼠科 Cricetidae

学名 Scientific Name	*Eothenomys fidelis*
命名人 Species Authority	Liu *et al.*, 2017
英文名 English Name(s)	Lijiang Black Vole
同物异名 Synonym(s)	无 / None
种下单元评估 Infra-specific Taxa Assessed	无 / None

评估信息 Assessment Information

评估年份 Year Assessed	2020
评定人 Assessor(s)	刘少英、蒋志刚 / Shaoying Liu, Zhigang Jiang
审定人 Reviewer(s)	马勇、路纪琪、鲍毅新 / Yong Ma, Jiqi Lu, Yixin Bao
其他贡献人 Other Contributor(s)	李立立、丁晨晨 / Lili Li, Chenchen Ding

理由 Justification: 丽江绒鼠的种群与栖息地信息缺乏。因此，列为数据缺乏等级 / Information on the population and habitat status of Lijiang Black Vole are deficient. Thus, it is listed as Data Deficient

地理分布 Geographical Distribution

国内分布 Domestic Distribution

云南、四川 / Yunnan, Sichuan

世界分布 World Distribution

中国 / China

分布标注 Distribution Note

特有种 / Endemic

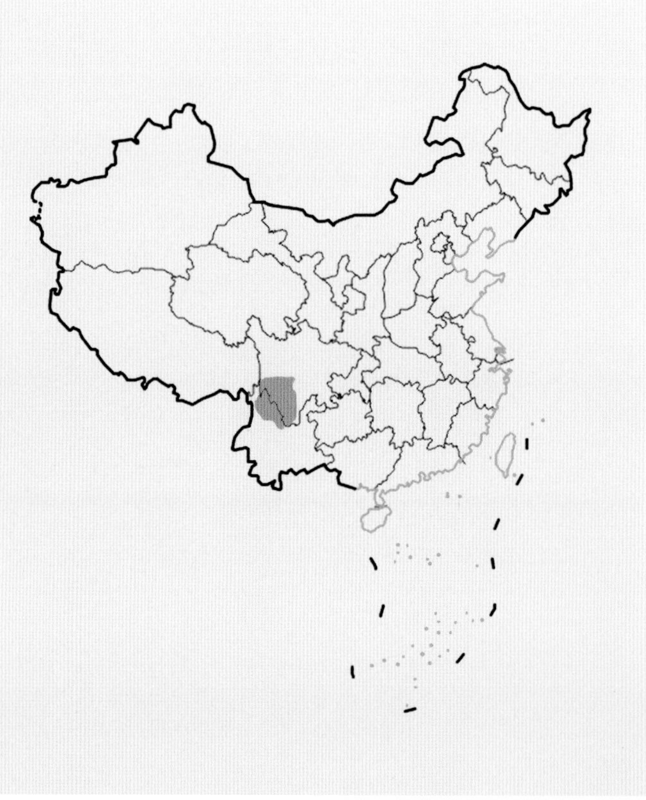

国内分布图
Map of Domestic Distribution

种群 Population

种群数量 **Population Size**	未知 / Unknown
种群趋势 **Population Trend**	未知 / Unknown

生境与生态系统 Habitat (s) and Ecosystem (s)

生　　境 **Habitat(s)**	山地森林 / Montane Forest
生态系统 **Ecosystem(s)**	森林生态系统 / Forest Ecosystem

威胁 Threat (s)

主要威胁 **Major Threat(s)**	无 / None

保护级别与保护行动 Protection Category and Conservation Action (s)

国家重点保护野生动物等级 **(2021) Category of National Key Protected Wild Animals (2021)**	无 / NA
IUCN 红色名录 (2020-2) IUCN Red List (2020-2)	无危 / LC
CITES 附录 (2019) CITES Appendix (2019)	无 / NA
保护行动 **Conservation Action(s)**	无 / None

相关文献 Relevant References

Burgin *et al.*, 2018; Jiang *et al.* (蒋志刚等), 2017; Liu *et al.*, 2017

丽江绒鼠 *Eothenomys fidelis*　　蒋志刚 绘　Drawn by Zhigang Jiang

中国生物多样性 红色名录
China's Red List of Biodiversity

银色高山䶄
Alticola argentatus

数据缺乏 DD

数据缺乏 DD	无危 LC	近危 NT	易危 VU	濒危 EN	极危 CR	区域灭绝 RE	野外灭绝 EW	灭绝 EX

分类地位 Taxonomic Status

动物界 Animalia	脊索动物门 Chordata	哺乳纲 Mammalia	啮齿目 Rodentia	仓鼠科 Cricetidae

学　名 Scientific Name	*Alticola argentatus*
命名人 Species Authority	Severtzov, 1879
英文名 English Name(s)	Silver Mountain Vole
同物异名 Synonym(s)	*Alticola alaica* (Rosanov, 1935); *argurus* (Thomas, 1909); *blanfordi* (Scully, 1880); *gracilis* (Kashkarov, 1923); *lahulius* (Hinton, 1926); *leucurus* (Severtsov, 1873); （转下页）
种下单元评估 Infra-specific Taxa Assessed	无 / None

评估信息 Assessment Information

评估年份 Year Assessed	2020
评定人 Assessor(s)	刘少英、蒋志刚 / Shaoying Liu, Zhigang Jiang
审定人 Reviewer(s)	马勇、路纪琪、鲍毅新 / Yong Ma, Jiqi Lu, Yixin Bao
其他贡献人 Other Contributor(s)	李立立、丁晨晨 / Lili Li, Chenchen Ding

理由 **Justification:** 有关银色高山䶄的记录较少，对其的研究缺乏。因此，列为数据缺乏等级 / Little information on Silver Mountain Vole has been recorded, and studies on it are deficient. Thus, it is listed as Data Deficient

地理分布 Geographical Distribution

国内分布 Domestic Distribution

新疆、甘肃 / Xinjiang, Gansu

世界分布 World Distribution

中国；南亚、中亚 / China; South Asia, Central Asia

分布标注 Distribution Note

非特有种 / Non-endemic

国内分布图
Map of Domestic Distribution

种群 Population

种群数量 Population Size	未知 / Unknown
种群趋势 Population Trend	未知 / Unknown

生境与生态系统 Habitat (s) and Ecosystem (s)

生　　境 Habitat(s)	草地、灌木、内陆岩石区域 / Grassland, Shrubland, Inland Rocky Area
生态系统 Ecosystem(s)	灌丛生态系统、草地生态系统 / Shrubland Ecosystem, Grassland Ecosystem

威胁 Threat (s)

主要威胁 Major Threat(s)	无 / None

保护级别与保护行动 Protection Category and Conservation Action (s)

国家重点保护野生动物等级 (2021) Category of National Key Protected Wild Animals (2021)	无 / NA
IUCN 红色名录 (2020-2) IUCN Red List (2020-2)	无危 / LC
CITES 附录 (2019) CITES Appendix (2019)	无 / NA
保护行动 Conservation Action(s)	无 / None

相关文献 Relevant References

Burgin *et al.*, 2018; Jiang *et al.* (蒋志刚等), 2017; Smith *et al.* (史密斯等), 2009; Pan *et al.* (潘清华等), 2007; Rossolimo *et al.*, 1994

(接上页)
longicauda (Kashkarov, 1923); *longicaudata* (Ognev, 1950); *parvidens* (Schlitter and Setzer, 1973); *phasma* (Miller, 1912); *rosanovi* (Ognev, 1940); *saurica* (Afanasiev and Bazhanov, 1948); *shnitnikovi* (Ognev, 1940); *severtzovi* (Tichomirov and Korchagin, 1889); *subluteus* (Thomas, 1914); *tarasovi* (Rossolimo and Pavlinov, 1992); *villosa* (Kashkarov, 1923); *worthingtoni* (Miller, 1906)

银色高山鼾 *Alticola argentatus*

克氏松田鼠

Neodon clarkei

数据缺乏 DD

数据缺乏 DD	无危 LC	近危 NT	易危 VU	濒危 EN	极危 CR	区域灭绝 RE	野外灭绝 EW	灭绝 EX

分类地位 Taxonomic Status

动物界 Animalia	脊索动物门 Chordata	哺乳纲 Mammalia	啮齿目 Rodentia	仓鼠科 Cricetidae

学名 **Scientific Name**	*Neodon clarkei*
命名人 **Species Authority**	Hinton, 1923
英文名 **English Name(s)**	Clarke's Vole
同物异名 **Synonym(s)**	*Volemys clarkei* (Hinton, 1923)
种下单元评估 **Infra-specific Taxa Assessed**	无 / None

评估信息 Assessment Information

评估年份 **Year Assessed**	2020
评定人 **Assessor(s)**	刘少英、蒋志刚 / Shaoying Liu, Zhigang Jiang
审定人 **Reviewer(s)**	马勇、鲍毅新 / Yong Ma, Yixin Bao
其他贡献人 **Other Contributor(s)**	李立立、丁晨晨 / Lili Li, Chenchen Ding

理由 **Justification:** 克氏松田鼠记录较少，对其的研究缺乏。因此，列为数据缺乏等级 / Information on Clarke's Vole is deficient due to lacking of studies. Thus, it is listed as Data Deficient

地理分布 Geographical Distribution

国内分布 **Domestic Distribution**

云南 / Yunnan

世界分布 **World Distribution**

中国、缅甸 / China, Myanmar

分布标注 **Distribution Note**

非特有种 / Non-endemic

国内分布图
Map of Domestic Distribution

种群 Population

种群数量 Population Size	未知 / Unknown
种群趋势 Population Trend	稳定 / Stable

生境与生态系统 Habitat (s) and Ecosystem (s)

生　　境 Habitat(s)	泰加林、草甸 / Taiga Forest, Meadow
生态系统 Ecosystem(s)	森林生态系统、草地生态系统 / Forest Ecosystem, Grassland Ecosystem

威胁 Threat (s)

主要威胁 Major Threat(s)	无 / None

保护级别与保护行动 Protection Category and Conservation Action (s)

国家重点保护野生动物等级 (2021) Category of National Key Protected Wild Animals (2021)	无 / NA
IUCN 红色名录 (2020-2) IUCN Red List (2020-2)	无危 / LC
CITES 附录 (2019) CITES Appendix (2019)	无 / NA
保护行动 Conservation Action(s)	无 / None

相关文献 Relevant References

Burgin *et al*., 2018; Jiang *et al*. (蒋志刚等), 2017; Wilson *et al*., 2016; Liu *et al*. (刘少英等), 2017

林芝松田鼠
Neodon linzhiensis

数据缺乏 DD

数据缺乏 DD	无危 LC	近危 NT	易危 VU	濒危 EN	极危 CR	区域灭绝 RE	野外灭绝 EW	灭绝 EX

分类地位 Taxonomic Status

动物界 Animalia	脊索动物门 Chordata	哺乳纲 Mammalia	啮齿目 Rodentia	仓鼠科 Cricetidae

学　名 **Scientific Name**	*Neodon linzhiensis*
命名人 **Species Authority**	Liu, 2012
英文名 **English Name(s)**	Linzhi Mountain Vole
同物异名 **Synonym(s)**	无 / None
种下单元评估 **Infra-specific Taxa Assessed**	无 / None

评估信息 Assessment Information

评估年份 **Year Assessed**	2020
评定人 **Assessor(s)**	刘少英、蒋志刚 / Shaoying Liu, Zhigang Jiang
审定人 **Reviewer(s)**	马勇、鲍毅新 / Yong Ma, Yixin Bao
其他贡献人 **Other Contributor(s)**	李立立、丁晨晨 / Lili Li, Chenchen Ding

理由 **Justification:** 林芝松田鼠的种群与栖息地数据缺乏。因此，列为数据缺乏等级 / Information on the population and habitat status of Linzhi Mountain Vole is deficient. Thus, it is listed as Data Deficient

地理分布 Geographical Distribution

国内分布 Domestic Distribution

西藏 / Tibet (Xizang)

世界分布 World Distribution

中国 / China

分布标注 Distribution Note

特有种 / Endemic

国内分布图
Map of Domestic Distribution

种群 Population

种群数量 **Population Size**	未知 / Unknown
种群趋势 **Population Trend**	未知 / Unknown

生境与生态系统 Habitat (s) and Ecosystem (s)

生　　境 **Habitat(s)**	高海拔草地、泰加林 / High Altitude Grassland, Taiga Forest
生态系统 **Ecosystem(s)**	森林生态系统、草地生态系统 / Forest Ecosystem, Grassland Ecosystem

威胁 Threat (s)

主要威胁 **Major Threat(s)**	无 / None

保护级别与保护行动 Protection Category and Conservation Action (s)

国家重点保护野生动物等级 **(2021) Category of National Key Protected Wild Animals (2021)**	无 / NA
IUCN 红色名录 **(2020-2) IUCN Red List (2020-2)**	无危 / LC
CITES 附录 **(2019) CITES Appendix (2019)**	无 / NA
保护行动 **Conservation Action(s)**	无 / None

相关文献 Relevant References

Burgin *et al*., 2018; Jiang *et al.* (蒋志刚等), 2017; Wilson *et al*., 2016; Liu *et al*., 2012a, 2012b

林芝松田鼠 *Neodon linzhiensis*

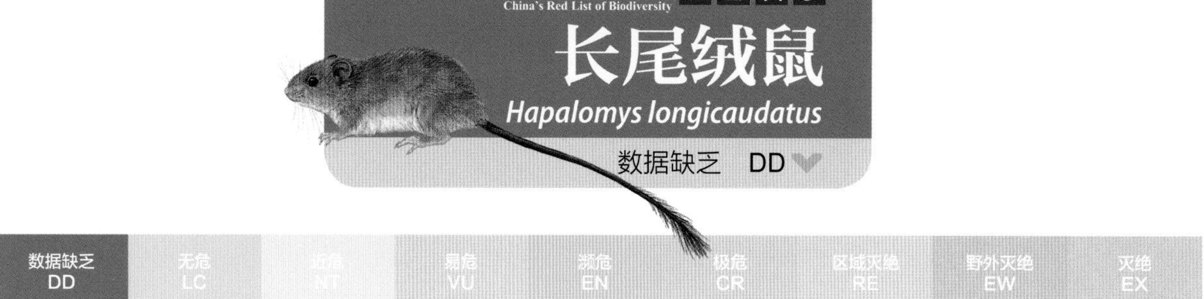

数据缺乏 DD	无危 LC	近危 NT	易危 VU	濒危 EN	极危 CR	区域灭绝 RE	野外灭绝 EW	灭绝 EX

分类地位 Taxonomic Status

动 物 界 Animalia	脊索动物门 Chordata	哺 乳 纲 Mammalia	啮 齿 目 Rodentia	鼠 科 Muridae
学 名 Scientific Name		*Hapalomys longicaudatus*		
命 名 人 Species Authority		Blyth, 1859		
英 文 名 English Name(s)		Long-tailed Marmoset Rat		
同物异名 Synonym(s)		Marmotet Rat; Greater Marmoset Rat		
种下单元评估 Infra-specific Taxa Assessed		无 / None		

评估信息 Assessment Information

评 估 年 份 Year Assessed	2020
评 定 人 Assessor(s)	刘少英、蒋志刚 / Shaoying Liu, Zhigang Jiang
审 定 人 Reviewer(s)	马勇、鲍毅新 / Yong Ma, Yixin Bao
其他贡献人 Other Contributor(s)	李立立、丁晨晨 / Lili Li, Chenchen Ding

理由 Justification: 长尾绒鼠的记录较少，对其的研究缺乏。因此，列为数据缺乏等级 / Little information on the Long-tailed Marmoset Rat has been recorded, and studies on this species are deficient. Thus, it is listed as Data Deficient

地理分布 Geographical Distribution

国内分布 Domestic Distribution

云南 / Yunnan

世界分布 World Distribution

中国；东南亚 / China; Southeast Asia

分布标注 Distribution Note

非特有种 / Non-endemic

国内分布图
Map of Domestic Distribution

种群 Population

种群数量 Population Size	未知 / Unknown
种群趋势 Population Trend	未知 / Unknown

生境与生态系统 Habitat (s) and Ecosystem (s)

生　　境 Habitat(s)	森林 / Forest
生态系统 Ecosystem(s)	森林生态系统 / Forest Ecosystem

威胁 Threat (s)

主要威胁 Major Threat(s)	无 / None

保护级别与保护行动 Protection Category and Conservation Action (s)

国家重点保护野生动物等级 (2021) Category of National Key Protected Wild Animals (2021)	无 / NA
IUCN 红色名录 (2020-2) IUCN Red List (2020-2)	无危 / LC
CITES 附录 (2019) CITES Appendix (2019)	无 / NA
保护行动 Conservation Action(s)	无 / None

相关文献 Relevant References

Burgin *et al.*, 2018; Jiang *et al.* (蒋志刚等), 2017; Wilson *et al.*, 2016; Zheng *et al.* (郑智民等), 2012; Wang (王应祥), 2003

黑姬鼠

Apodemus nigrus

数据缺乏 DD

数据缺乏 DD	无危 LC	近危 NT	易危 VU	濒危 EN	极危 CR	区域灭绝 RE	野外灭绝 EW	灭绝 EX

分类地位 Taxonomic Status

动物界 Animalia	脊索动物门 Chordata	哺乳纲 Mammalia	啮齿目 Rodentia	鼠科 Muridae

学名 Scientific Name	*Apodemus nigrus*
命名人 Species Authority	Ge D, Feijó A and Yang Q, 2019
英文名 English Name(s)	Black Field Mouse
同物异名 Synonym(s)	无 / None
种下单元评估 Infra-specific Taxa Assessed	无 / None

评估信息 Assessment Information

评估年份 Year Assessed	2021
评定人 Assessor(s)	葛德燕 / Deyan Ge
审定人 Reviewer(s)	蒋志刚 / Zhigang Jiang
其他贡献人 Other Contributor(s)	丁晨晨 / Chenchen Ding

理由 **Justification:** 黑姬鼠是 2019 年葛德燕、Anderson Feijó 和杨奇森在贵州和重庆发现的一个新种。目前缺乏对其种群和栖息地的研究。因此，列为数据缺乏等级 / The Black Field Mouse is a new species discovered by Deyan Ge, Feijó Anderson and Qisen Yang in Guizhou and Chongqing in 2019. No information on its population and habitat status has been recorded so far. Thus, it is listed as Data Deficient

地理分布 Geographical Distribution

国内分布 Domestic Distribution

贵州、重庆 / Guizhou, Chongqing

世界分布 World Distribution

中国 / China

分布标注 Distribution Note

特有种 / Endemic

国内分布图
Map of Domestic Distribution

种群 Population

种群数量 Population Size	未知 / Unknown
种群趋势 Population Trend	未知 / Unknown

生境与生态系统 Habitat (s) and Ecosystem (s)

生　　境 Habitat(s)	常绿阔叶林 / Evergreen Broad-leaved Forest
生态系统 Ecosystem(s)	森林生态系统 / Forest Ecosystem

威胁 Threat (s)

主要威胁 Major Threat(s)	未知 / Unknown

保护级别与保护行动 Protection Category and Conservation Action (s)

国家重点保护野生动物等级 (2021) Category of National Key Protected Wild Animals (2021)	无 / NA
IUCN 红色名录 (2020-2) IUCN Red List (2020-2)	未列入 / NA
CITES 附录 (2019) CITES Appendix (2019)	无 / NA
保护行动 Conservation Action(s)	无 / None

相关文献 Relevant References

Ge *et al*., 2019

黑姬鼠 *Apodemus nigrus* 蒋志刚 绘 Drawn by Zhigang Jiang

喜马拉雅姬鼠
Apodemus pallipes

数据缺乏 DD

数据缺乏 DD	无危 LC	近危 NT	易危 VU	濒危 EN	极危 CR	区域灭绝 RE	野外灭绝 EW	灭绝 EX

分类地位 Taxonomic Status

动物界 Animalia	脊索动物门 Chordata	哺乳纲 Mammalia	啮齿目 Rodentia	鼠科 Muridae

学名 Scientific Name	*Apodemus pallipes*
命名人 Species Authority	Barrett-Hamilton, 1900
英文名 English Name(s)	Himalayan Field Mouse
同物异名 Synonym(s)	Ward's Field Mouse; *Apodemus bushengensis* (Zheng, 1979); *pentax* (Wroughton, 1908); *wardi* (Wroughton, 1908)
种下单元评估 Infra-specific Taxa Assessed	无 / None

评估信息 Assessment Information

评估年份 Year Assessed	2020
评定人 Assessor(s)	刘少英、蒋志刚 / Shaoying Liu, Zhigang Jiang
审定人 Reviewer(s)	马勇、鲍毅新 / Yong Ma, Yixin Bao
其他贡献人 Other Contributor(s)	李立立、丁晨晨 / Lili Li, Chenchen Ding

理由 Justification: 关于喜马拉雅姬鼠的记录少，对其种群和栖息地的研究缺乏。因此，列为数据缺乏等级 / Little information on the population and habitat of Himalayan Field Mouse has been recorded, and studies on this species are deficient. Thus, it is listed as Data Deficient

地理分布 Geographical Distribution

国内分布 Domestic Distribution

西藏 / Tibet (Xizang)

世界分布 World Distribution

中国；中亚、南亚 / China; Central Asia, South Asia

分布标注 Distribution Note

非特有种 / Non-endemic

国内分布图
Map of Domestic Distribution

种群 Population

种群数量 **Population Size**	未知 / Unknown
种群趋势 **Population Trend**	未知 / Unknown

生境与生态系统 Habitat (s) and Ecosystem (s)

生　　境 **Habitat(s)**	泰加林 / Taiga Forest
生态系统 **Ecosystem(s)**	森林生态系统 / Forest Ecosystem

威胁 Threat (s)

主要威胁 **Major Threat(s)**	无 / None

保护级别与保护行动 Protection Category and Conservation Action (s)

国家重点保护野生动物等级 **(2021) Category of National Key Protected Wild Animals (2021)**	无 / NA
IUCN 红色名录 **(2020-2) IUCN Red List (2020-2)**	无危 / LC
CITES 附录 **(2019) CITES Appendix (2019)**	无 / NA
保护行动 **Conservation Action(s)**	无 / None

相关文献 Relevant References

Burgin *et al.*, 2018; Jiang *et al.* (蒋志刚等), 2017; Wilson *et al.*, 2016; Zheng *et al.* (郑智民等), 2012; Smith *et al.* (史密斯等), 2009; Pan *et al.* (潘清华等), 2007

喜马拉雅姬鼠 *Apodemus pallipes*

乌拉尔姬鼠
Apodemus uralensis

数据缺乏 DD

数据缺乏 DD	无危 LC	近危 NT	易危 VU	濒危 EN	极危 CR	区域灭绝 RE	野外灭绝 EW	灭绝 EX

分类地位 Taxonomic Status

动物界 Animalia	脊索动物门 Chordata	哺乳纲 Mammalia	啮齿目 Rodentia	鼠科 Muridae
学　名 Scientific Name		*Apodemus uralensis*		
命名人 Species Authority		Pallas, 1811		
英文名 English Name(s)		Herb Field Mouse		
同物异名 Synonym(s)		Ural Field Mouse; Pygmy Field Mouse; *Apodemus baessleri* (Dahl, 1919); *balchanensis* (Kashkarov, 1981); *cimrmani* (Vohralík, 2002); *ciscaucasicus* (Ognev, 1924); (转下页)		
种下单元评估 Infra-specific Taxa Assessed		无 / None		

评估信息 Assessment Information

评估年份 Year Assessed	2020
评　定　人 Assessor(s)	刘少英、蒋志刚 / Shaoying Liu, Zhigang Jiang
审　定　人 Reviewer(s)	马勇、鲍毅新 / Yong Ma, Yixin Bao
其他贡献人 Other Contributor(s)	李立立、丁晨晨 / Lili Li, Chenchen Ding

理由 Justification: 乌拉尔姬鼠的记录较少，对其的研究缺乏。因此，列为数据缺乏等级 / Little information on Herb Field Mouse has been recorded, and studies on this species are deficient. Thus, it is listed as Data Deficient

地理分布 Geographical Distribution

国内分布 Domestic Distribution
新疆 / Xinjiang
世界分布 World Distribution
亚洲、欧洲 / Asia, Europe
分布标注 Distribution Note
非特有种 / Non-endemic

国内分布图
Map of Domestic Distribution

种群 Population

种群数量 **Population Size**	未知 / Unknown
种群趋势 **Population Trend**	稳定 / Stable

生境与生态系统 Habitat (s) and Ecosystem (s)

生　　境 **Habitat(s)**	森林 / Forest
生态系统 **Ecosystem(s)**	森林生态系统 / Forest Ecosystem

威胁 Threat (s)

主要威胁 **Major Threat(s)**	无 / None

保护级别与保护行动 Protection Category and Conservation Action (s)

国家重点保护野生动物等级 **(2021) Category of National Key Protected Wild Animals (2021)**	无 / NA
IUCN 红色名录 (2020-2) IUCN Red List (2020-2)	无危 / LC
CITES 附录 (2019) CITES Appendix (2019)	无 / NA
保护行动 **Conservation Action(s)**	无 / None

相关文献 Relevant References

Burgin *et al*., 2018; Jiang *et al.* (蒋志刚等), 2017; Wilson *et al*., 2016; Zheng *et al.* (郑智民等), 2012; Smith *et al.* (史密斯等), 2009; Wang (王应祥), 2003

(接上页)
kastschenkoi (Kuznetzov, 1932); *major* (Severtsov, 1873); *microps* (Kratochvíl and Rosicky, 1952); *microtis* (Miller, 1912); *mosquensis* (Ognev, 1913); *nankiangensis* (Wang, 1964); *pallidus* (Kashkarov, 1926); *parvulus* (Mosanský, 1994); *tokmak* (Severtzov, 1873)

数据缺乏 DD	无危 LC	近危 NT	易危 VU	濒危 EN	极危 CR	区域灭绝 RE	野外灭绝 EW	灭绝 EX

分类地位 Taxonomic Status

动物界 Animalia	脊索动物门 Chordata	哺乳纲 Mammalia	啮齿目 Rodentia	鼠科 Muridae
学名 Scientific Name		*Rattus rattus*		
命名人 Species Authority		Linnaeus, 1758		
英文名 English Name(s)		Black Rat		
同物异名 Synonym(s)		无 / None		
种下单元评估 Infra-specific Taxa Assessed		无 / None		

评估信息 Assessment Information

评估年份 Year Assessed	2020
评定人 Assessor(s)	蒋志刚 / Zhigang Jiang
审定人 Reviewer(s)	刘少英 / Shaoying Liu
其他贡献人 Other Contributor(s)	丁晨晨 / Chenchen Ding

理由 **Justification:** 黑家鼠全球分布，种群数量大，但在中国为新近发现，种群数量与占有区不详。因此，列为数据缺乏等级 / Black Rat is globally distributed with large populations, but it has only been found in China recently, and its population and habitat status are unknown in China. Thus, it is listed as Data Deficient

地理分布 Geographical Distribution

国内分布 Domestic Distribution

黑龙江、吉林、辽宁、河北、山东、河南、湖北、安徽、江苏、江西、浙江、福建、湖南、广东、广西、重庆、云南、贵州、四川、西藏 / Heilongjiang, Jilin, Liaoning, Hebei, Shandong, Henan, Hubei, Anhui, Jiangsu, Jiangxi, Zhejiang, Fujian, Hunan, Guangdong, Guangxi, Chongqing, Yunnan, Guizhou, Sichuan, Tibet (Xizang)

世界分布 World Distribution

亚洲、欧洲 / Asia, Europe

分布标注 Distribution Note

非特有种 / Non-endemic

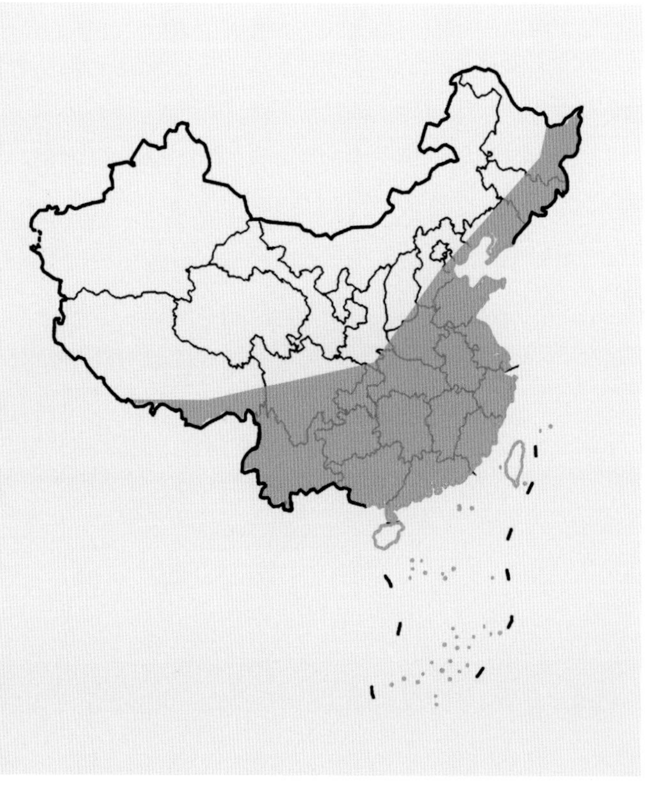

国内分布图
Map of Domestic Distribution

种群 Population

种群数量 **Population Size**	未知 / Unknown
种群趋势 **Population Trend**	下降 / Decreasing

生境与生态系统 Habitat (s) and Ecosystem (s)

生　　境 **Habitat(s)**	主要栖居于人居环境，但也在一些自然或半自然的生境中发现 / Mainly inhabit Human Settlement, but also found in a variety of Natural and Semi-natural Habitats
生态系统 **Ecosystem(s)**	城市生态系统 / Urban Ecosystem

威胁 Threat (s)

主要威胁 **Major Threat(s)**	毒杀、捕捉 / Poisoning, Capturing

保护级别与保护行动 Protection Category and Conservation Action (s)

国家重点保护野生动物等级 **(2021) Category of National Key Protected Wild Animals (2021)**	无 / NA
IUCN 红色名录 **(2020-2) IUCN Red List (2020-2)**	无危 / LC
CITES 附录 **(2019) CITES Appendix (2019)**	无 / NA
保护行动 **Conservation Action(s)**	无 / None

相关文献 Relevant References

Burgin *et al*., 2018; Wilson and Reeder, 2005

黑家鼠 *Rattus rattus*　　　　蒋志刚 绘　Drawn by Zhigang Jiang

剑纹小社鼠

Niviventer gladiusmaculus

数据缺乏 DD

数据缺乏 DD	无危 LC	近危 NT	易危 VU	濒危 EN	极危 CR	区域灭绝 RE	野外灭绝 EW	灭绝 EX

分类地位 Taxonomic Status

动物界 Animalia	脊索动物门 Chordata	哺乳纲 Mammalia	啮齿目 Rodentia	鼠科 Muridae
学名 Scientific Name		*Niviventer gladiusmaculus*		
命名人 Species Authority		Ge *et al.*, 2018		
英文名 English Name(s)		Least White-bellied Rat		
同物异名 Synonym(s)		Least Niviventer		
种下单元评估 Infra-specific Taxa Assessed		无 / None		

评估信息 Assessment Information

评估年份 Year Assessed	2021
评定人 Assessor(s)	葛德燕 / Deyan Ge
审定人 Reviewer(s)	蒋志刚 / Zhigang Jiang
其他贡献人 Other Contributor(s)	丁晨晨 / Chenchen Ding

理由 Justification: 剑纹小社鼠是 2018 年由葛德燕在西藏林芝米林县发现的一个新种，目前尚不了解其种群与栖息地状况。因此，列为数据缺乏等级 / Least White-bellied Rat is a new species discovered by Deyan Ge in Milin County, Linzhi, Tibet (Xizang). No information is available about its population and habitat status. Thus, it is listed as Data Deficient

地理分布 Geographical Distribution

国内分布 Domestic Distribution

西藏 / Tibet (Xizang)

世界分布 World Distribution

中国 / China

分布标注 Distribution Note

特有种 / Endemic

国内分布图
Map of Domestic Distribution

种群 Population

种群数量 **Population Size**	未知 / Unknown
种群趋势 **Population Trend**	未知 / Unknown

生境与生态系统 Habitat (s) and Ecosystem (s)

生　　境 **Habitat(s)**	森林 / Forest
生态系统 **Ecosystem(s)**	森林生态系统 / Forest Ecosystem

威胁 Threat (s)

主要威胁 **Major Threat(s)**	未知 / Unknown

保护级别与保护行动 Protection Category and Conservation Action (s)

国家重点保护野生动物等级 **(2021) Category of National Key Protected Wild Animals (2021)**	无 / NA
IUCN 红色名录 **(2020-2) IUCN Red List (2020-2)**	未列入 / NA
CITES 附录 **(2019) CITES Appendix (2019)**	无 / NA
保护行动 **Conservation Action(s)**	无 / None

相关文献 Relevant References

Ge *et al*., 2018

剑纹小社鼠 *Niviventer gladiusmaculus*　　蒋志刚 绘　Drawn by Zhigang Jiang

数据缺乏 DD	无危 LC	近危 NT	易危 VU	濒危 EN	极危 CR	区域灭绝 RE	野外灭绝 EW	灭绝 EX

分类地位 Taxonomic Status

动物界 Animalia	脊索动物门 Chordata	哺乳纲 Mammalia	啮齿目 Rodentia	鼠科 Muridae

学名 **Scientific Name**	*Niviventer niviventer*
命名人 **Species Authority**	Hodgson, 1836
英文名 **English Name(s)**	Himalayan White-bellied Rat
同物异名 **Synonym(s)**	White-bellied Rat
种下单元评估 **Infra-specific Taxa Assessed**	无 / None

评估信息 Assessment Information

评估年份 **Year Assessed**	2020
评定人 **Assessor(s)**	刘少英、蒋志刚 / Shaoying Liu, Zhigang Jiang
审定人 **Reviewer(s)**	马勇、鲍毅新 / Yong Ma, Yixin Bao
其他贡献人 **Other Contributor(s)**	李立立、丁晨晨 / Lili Li, Chenchen Ding

理由 **Justification:** 白腹鼠的记录较少，对其的研究缺乏。因此，列为数据缺乏等级 / Little information on the Himalayan White-bellied Rat has been recorded, and studies on it are deficient. Thus, it is listed as Data Deficient

地理分布 Geographical Distribution

国内分布 Domestic Distribution

西藏 / Tibet (Xizang)

世界分布 World Distribution

中国；南亚 / China; South Asia

分布标注 Distribution Note

非特有种 / Non-endemic

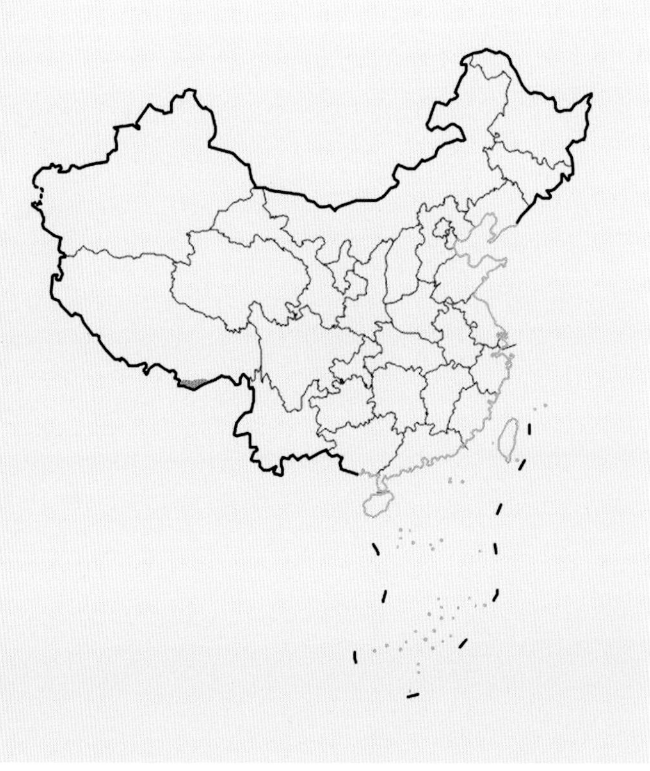

国内分布图
Map of Domestic Distribution

种群 Population

种群数量 **Population Size**	未知 / Unknown
种群趋势 **Population Trend**	未知 / Unknown

生境与生态系统 Habitat (s) and Ecosystem (s)

生　　境 **Habitat(s)**	森林、耕地 / Forest, Arable Land
生态系统 **Ecosystem(s)**	森林生态系统、农田生态系统 / Forest Ecosystem, Cropland Ecosystem

威胁 Threat (s)

主要威胁 **Major Threat(s)**	无 / None

保护级别与保护行动 Protection Category and Conservation Action (s)

国家重点保护野生动物等级 **(2021) Category of National Key Protected Wild Animals (2021)**	无 / NA
IUCN 红色名录 **(2020-2) IUCN Red List (2020-2)**	无危 / LC
CITES 附录 **(2019) CITES Appendix (2019)**	无 / NA
保护行动 **Conservation Action(s)**	无 / None

相关文献 Relevant References

Burgin *et al*., 2018; Jiang *et al*. (蒋志刚等), 2017; Wilson *et al*., 2016; Li *et al*. (李裕冬等), 2007; Choudhury, 2003

缅甸山鼠

Niviventer tenaster

数据缺乏 DD

数据缺乏 DD	无危 LC	近危 NT	易危 VU	濒危 EN	极危 CR	区域灭绝 RE	野外灭绝 EW	灭绝 EX

分类地位 Taxonomic Status

动物界 Animalia	脊索动物门 Chordata	哺乳纲 Mammalia	啮齿目 Rodentia	鼠科 Muridae

学名 **Scientific Name**	*Niviventer tenaster*
命名人 **Species Authority**	Thomas, 1916
英文名 **English Name(s)**	Indochinese Mountain Niviventer
同物异名 **Synonym(s)**	Tenasserim White-bellied Rat; *Niviventer champa* (Robinson and Kloss, 1922); *lotipes* (G. M. Allen, 1926)
种下单元评估 **Infra-specific Taxa Assessed**	无 / None

评估信息 Assessment Information

评估年份 **Year Assessed**	2020
评定人 **Assessor(s)**	刘少英、蒋志刚 / Shaoying Liu, Zhigang Jiang
审定人 **Reviewer(s)**	马勇、鲍毅新 / Yong Ma, Yixin Bao
其他贡献人 **Other Contributor(s)**	李立立、丁晨晨 / Lili Li, Chenchen Ding

理由 **Justification:** 缅甸山鼠的记录较少，对其的研究缺乏。因此，列为数据缺乏等级 / Little information on Indochinese Mountain Niviventer has been recorded, and studies on it are deficient. Thus, it is listed as Data Deficient

地理分布 Geographical Distribution

国内分布 Domestic Distribution

海南 / Hainan

世界分布 World Distribution

中国；东南亚 / China; Southeast Asia

分布标注 Distribution Note

非特有种 / Non-endemic

国内分布图
Map of Domestic Distribution

种群 Population

种群数量 Population Size	未知 / Unknown
种群趋势 Population Trend	未知 / Unknown

生境与生态系统 Habitat (s) and Ecosystem (s)

生　　境 Habitat(s)	热带湿润山地森林 / Tropical Moist Montane Forest
生态系统 Ecosystem(s)	森林生态系统 / Forest Ecosystem

威胁 Threat (s)

主要威胁 Major Threat(s)	无 / None

保护级别与保护行动 Protection Category and Conservation Action (s)

国家重点保护野生动物等级 (2021) Category of National Key Protected Wild Animals (2021)	无 / NA
IUCN 红色名录 (2020-2) IUCN Red List (2020-2)	无危 / LC
CITES 附录 (2019) CITES Appendix (2019)	无 / NA
保护行动 Conservation Action(s)	无 / None

相关文献 Relevant References

Burgin *et al*., 2018; Jiang *et al.* (蒋志刚等), 2017; Wilson *et al*., 2016; Smith *et al.* (史密斯等), 2009; Jing *et al*., 2007; Pan *et al.* (潘清华等), 2007

小泡灰鼠
Berylmys manipulus

数据缺乏 DD

数据缺乏 DD	无危 LC	近危 NT	易危 VU	濒危 EN	极危 CR	区域灭绝 RE	野外灭绝 EW	灭绝 EX

分类地位 Taxonomic Status

动物界 Animalia	脊索动物门 Chordata	哺乳纲 Mammalia	啮齿目 Rodentia	鼠科 Muridae

学名 Scientific Name	*Berylmys manipulus*
命名人 Species Authority	Thomas, 1916
英文名 English Name(s)	Manipur White-toothed Rat
同物异名 Synonym(s)	*Berylmys kekrimus* (Roonwal, 1948)
种下单元评估 Infra-specific Taxa Assessed	无 / None

评估信息 Assessment Information

评估年份 Year Assessed	2020
评定人 Assessor(s)	刘少英、蒋志刚 / Shaoying Liu, Zhigang Jiang
审定人 Reviewer(s)	马勇、鲍毅新 / Yong Ma, Yixin Bao
其他贡献人 Other Contributor(s)	李立立、丁晨晨 / Lili Li, Chenchen Ding

理由 Justification: 小泡灰鼠的记录较少，对其的研究缺乏。因此，列为数据缺乏等级 / Little information on the Manipur White-toothed Rat has been recorded, and studies on it are deficient. Thus, it is listed as Data Deficient

地理分布 Geographical Distribution

国内分布 Domestic Distribution

云南 / Yunnan

世界分布 World Distribution

中国、印度、缅甸 / China, India, Myanmar

分布标注 Distribution Note

非特有种 / Non-endemic

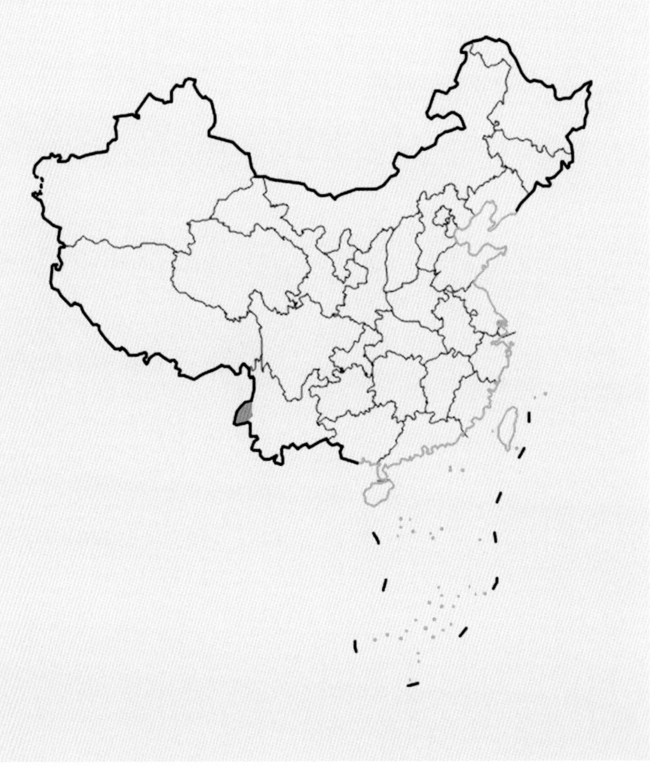

国内分布图
Map of Domestic Distribution

种群 Population

种群数量 Population Size	未知 / Unknown
种群趋势 Population Trend	未知 / Unknown

生境与生态系统 Habitat (s) and Ecosystem (s)

生　　境 Habitat(s)	灌丛、热带湿润山地森林 / Shrubland, Tropical Moist Montane Forest
生态系统 Ecosystem(s)	森林生态系统、灌丛生态系统 / Forest Ecosystem, Shrubland Ecosystem

威胁 Threat (s)

主要威胁 Major Threat(s)	未知 / Unknown

保护级别与保护行动 Protection Category and Conservation Action (s)

国家重点保护野生动物等级 (2021) Category of National Key Protected Wild Animals (2021)	无 / NA
IUCN 红色名录 (2020-2) IUCN Red List (2020-2)	无危 / LC
CITES 附录 (2019) CITES Appendix (2019)	无 / NA
保护行动 Conservation Action(s)	无 / None

相关文献 Relevant References

Burgin *et al.*, 2018; Jiang *et al.* (蒋志刚等), 2017; Wilson *et al.*, 2016; Zheng *et al.* (郑智民等), 2012; Smith *et al.* (史密斯等), 2009; Xing *et al.* (邢雅俊等), 2008

道氏东京鼠

Tonkinomys davovantien

数据缺乏 DD

数据缺乏 DD	无危 LC	近危 NT	易危 VU	濒危 EN	极危 CR	区域灭绝 RE	野外灭绝 EW	灭绝 EX

分类地位 Taxonomic Status

动物界 Animalia	脊索动物门 Chordata	哺乳纲 Mammalia	啮齿目 Rodentia	鼠科 Muridae
学名 Scientific Name		*Tonkinomys davovantien*		
命名人 Species Authority		Musser *et al*., 2006		
英文名 English Name(s)		Daovantien's Limestone Rat		
同物异名 Synonym(s)		无 / None		
种下单元评估 Infra-specific Taxa Assessed		无 / None		

评估信息 Assessment Information

评估年份 Year Assessed	2020
评定人 Assessor(s)	蒋志刚 / Zhigang Jiang
审定人 Reviewer(s)	蒋学龙 / Xuelong Jiang
其他贡献人 Other Contributor(s)	丁晨晨 / Chenchen Ding

理由 Justification: 道氏东京鼠被成市等 (2018) 发现分布在云南麻栗坡狭窄范围内。目前，缺乏其种群与栖息地的数据。因此，列为数据缺乏等级 / Daovantien's Limestone Rat was discovered in a narrow range in Malipo, Yunnan by Cheng *et al*. (2018). However, information on the populations and habitats of this species is deficient. Thus, it is listed as Data Deficient

地理分布 Geographical Distribution

国内分布 Domestic Distribution

云南 / Yunnan

世界分布 World Distribution

中国、越南 / China, Viet Nam

分布标注 Distribution Note

非特有种 / Non-endemic

国内分布图
Map of Domestic Distribution

种群 Population

种群数量 Population Size	未知 / Unknown
种群趋势 Population Trend	未知 / Unknown

生境与生态系统 Habitat (s) and Ecosystem (s)

生　　境 Habitat(s)	常绿阔叶林 / Evergreen Broad-leaved Forest
生态系统 Ecosystem(s)	森林生态系统 / Forest Ecosystem

威胁 Threat (s)

主要威胁 Major Threat(s)	未知 / Unknown

保护级别与保护行动 Protection Category and Conservation Action (s)

国家重点保护野生动物等级 (2021) Category of National Key Protected Wild Animals (2021)	无 / NA
IUCN 红色名录 (2020-2) IUCN Red List (2020-2)	无危 / LC
CITES 附录 (2019) CITES Appendix (2019)	无 / NA
保护行动 Conservation Action(s)	无 / None

相关文献 Relevant References

Cheng *et al.*(成市等), 2018

道氏东京鼠 *Tonkinomys davovantien*　　　　蒋志刚 绘　Drawn by Zhigang Jiang

中国生物多样性红色名录
China's Red List of Biodiversity

武夷山猪尾鼠

Typhlomys cinereus

数据缺乏 DD

数据缺乏 DD	无危 LC	近危 NT	易危 VU	濒危 EN	极危 CR	区域灭绝 RE	野外灭绝 EW	灭绝 EX

分类地位 Taxonomic Status

动物界 Animalia	脊索动物门 Chordata	哺乳纲 Mammalia	啮齿目 Rodentia	刺山鼠科 Platacanthomyidae

学　名 **Scientific Name**	*Typhlomys cinereus*
命名人 **Species Authority**	Chen *et al.*, 2017
英文名 **English Name(s)**	Wuyishan Tree Mouse
同物异名 **Synonym(s)**	无 / None
种下单元评估 **Infra-specific Taxa Assessed**	无 / None

评估信息 Assessment Information

评估年份 **Year Assessed**	2020
评定人 **Assessor(s)**	蒋学龙、蒋志刚 / Xuelong Jiang, Zhigang Jiang
审定人 **Reviewer(s)**	刘少英 / Shaoying Liu
其他贡献人 **Other Contributor(s)**	李立立、丁晨晨 / Lili Li, Chenchen Ding

理由 **Justification:** 武夷山猪尾鼠的记录较少，对其的研究缺乏。因此，列为数据缺乏等级 / Little information on Wuyishan Tree Mouse has been recorded, and the studies on it are deficient. Thus, it is listed as Data Deficient

地理分布 Geographical Distribution

国内分布 Domestic Distribution

安徽、浙江、福建、广西 / Anhui, Zhejiang, Fujian, Guangxi

世界分布 World Distribution

中国 / China

分布标注 Distribution Note

特有种 / Endemic

国内分布图
Map of Domestic Distribution

种群 Population

种群数量 Population Size	未知 / Unknown
种群趋势 Population Trend	未知 / Unknown

生境与生态系统 Habitat (s) and Ecosystem (s)

生　　境 Habitat(s)	次生冷杉林 / Secondary Fir Forest
生态系统 Ecosystem(s)	森林生态系统 / Forest Ecosystem

威胁 Threat (s)

主要威胁 Major Threat(s)	无 / None

保护级别与保护行动 Protection Category and Conservation Action (s)

国家重点保护野生动物等级 (2021) Category of National Key Protected Wild Animals (2021)	无 / NA
IUCN 红色名录 (2020-2) IUCN Red List (2020-2)	无危 / LC
CITES 附录 (2019) CITES Appendix (2019)	无 / NA
保护行动 Conservation Action(s)	无 / None

相关文献 Relevant References

Burgin *et al.*, 2018; Chen *et al.*, 2017b; Jiang *et al.* (蒋志刚等), 2017; Wilson *et al.*, 2016

中国生物多样性红色名录
China's Red List of Biodiversity

大娄山猪尾鼠

Typhlomys daloushanensis

数据缺乏 DD

数据缺乏 DD	无危 LC	近危 NT	易危 VU	濒危 EN	极危 CR	区域灭绝 RE	野外灭绝 EW	灭绝 EX

分类地位 Taxonomic Status

动物界 Animalia	脊索动物门 Chordata	哺乳纲 Mammalia	啮齿目 Rodentia	刺山鼠科 Platacanthomyidae
学　名 **Scientific Name**		*Typhlomys daloushanensis*		
命名人 **Species Authority**		Wang and Li, 1996		
英文名 **English Name(s)**		Daloushan Pygmy Dormouse		
同物异名 **Synonym(s)**		Daloushan Tree Mouse		
种下单元评估 **Infra-specific Taxa Assessed**		无 / None		

评估信息 Assessment Information

评估年份 **Year Assessed**	2020
评定人 **Assessor(s)**	蒋学龙、蒋志刚 / Xuelong Jiang, Zhigang Jiang
审定人 **Reviewer(s)**	刘少英 / Shaoying Liu
其他贡献人 **Other Contributor(s)**	李立立、丁晨晨 / Lili Li, Chenchen Ding

理由 **Justification:** 大娄山猪尾鼠的记录较少，对其的研究缺乏。因此，列为数据缺乏 / Little information on Daloushan Pygmy Dormouse has been recorded, and studies on it are deficient. Thus, it is listed as Data Deficient

地理分布 Geographical Distribution

国内分布 Domestic Distribution

陕西、甘肃、四川、云南、重庆、贵州、广西、湖南、广东 / Shaanxi, Gansu, Sichuan, Yunnan, Chongqing, Guizhou, Guangxi, Hunan, Guangdong

世界分布 World Distribution

中国 / China

分布标注 Distribution Note

特有种 / Endemic

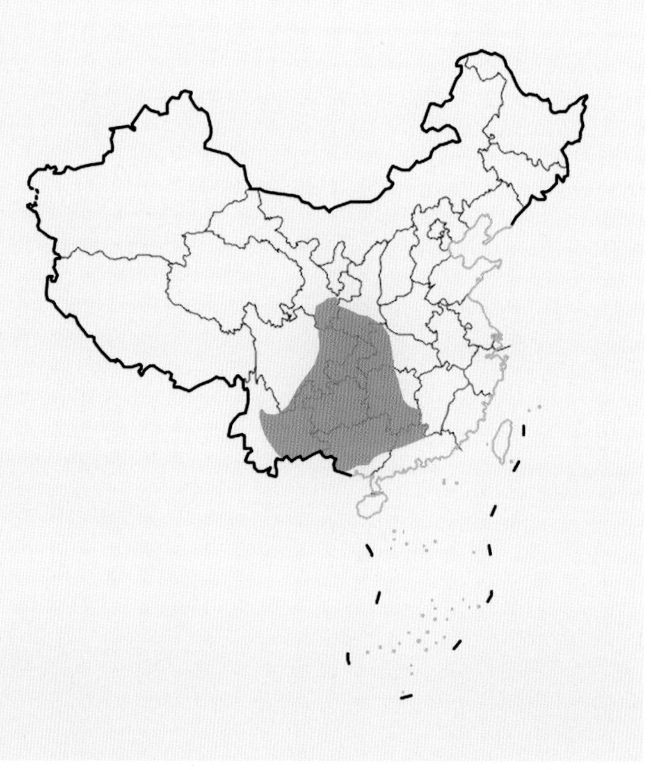

国内分布图
Map of Domestic Distribution

种群 Population

种群数量 Population Size	未知 / Unknown
种群趋势 Population Trend	未知 / Unknown

生境与生态系统 Habitat (s) and Ecosystem (s)

生　　境 Habitat(s)	山地森林 / Montane Forest
生态系统 Ecosystem(s)	森林生态系统 / Forest Ecosystem

威胁 Threat (s)

主要威胁 Major Threat(s)	无 / None

保护级别与保护行动 Protection Category and Conservation Action (s)

国家重点保护野生动物等级 (2021) Category of National Key Protected Wild Animals (2021)	无 / NA
IUCN 红色名录 (2020-2) IUCN Red List (2020-2)	无危 / LC
CITES 附录 (2019) CITES Appendix (2019)	无 / NA
保护行动 Conservation Action(s)	无 / None

相关文献 Relevant References

Burgin *et al*., 2018; Chen *et al*., 2017b; Jiang *et al*. (蒋志刚等), 2017; Wilson *et al*., 2016

小猪尾鼠

Typhlomys nanus

数据缺乏 DD

数据缺乏 DD	无危 LC	近危 NT	易危 VU	濒危 EN	极危 CR	区域灭绝 RE	野外灭绝 EW	灭绝 EX

分类地位 Taxonomic Status

动物界 Animalia	脊索动物门 Chordata	哺乳纲 Mammalia	啮齿目 Rodentia	刺山鼠科 Platacanthomyidae

学 名 Scientific Name	*Typhlomys nanus*
命名人 Species Authority	Chen *et al.*, 2017
英文名 English Name(s)	Nanu Tree Mouse
同物异名 Synonym(s)	无 / None
种下单元评估 Infra-specific Taxa Assessed	无 / None

评估信息 Assessment Information

评估年份 Year Assessed	2020
评 定 人 Assessor(s)	蒋学龙、蒋志刚 / Xuelong Jiang, Zhigang Jiang
审 定 人 Reviewer(s)	刘少英 / Shaoying Liu
其他贡献人 Other Contributor(s)	李立立、丁晨晨 / Lili Li, Chenchen Ding

理由 Justification: 小猪尾鼠的记录较少，对其的研究缺乏。因此，列为数据缺乏等级 / Little information on Nanu Tree Mouse has been recorded, and studies on it are deficient. Thus, it is listed as Data Deficient

地理分布 Geographical Distribution

国内分布 Domestic Distribution

云南 / Yunnan

世界分布 World Distribution

中国 / China

分布标注 Distribution Note

特有种 / Endemic

国内分布图
Map of Domestic Distribution

种群 Population

种群数量 Population Size	未知 / Unknown
种群趋势 Population Trend	未知 / Unknown

生境与生态系统 Habitat (s) and Ecosystem (s)

生　　境 Habitat(s)	山地森林 / Montane Forest
生态系统 Ecosystem(s)	森林生态系统 / Forest Ecosystem

威胁 Threat (s)

主要威胁 Major Threat(s)	无 / None

保护级别与保护行动 Protection Category and Conservation Action (s)

国家重点保护野生动物等级 (2021) Category of National Key Protected Wild Animals (2021)	无 / NA
IUCN 红色名录 (2020-2) IUCN Red List (2020-2)	无危 / LC
CITES 附录 (2019) CITES Appendix (2019)	无 / NA
保护行动 Conservation Action(s)	无 / None

相关文献 Relevant References

Burgin *et al*., 2018; Chen *et al*., 2017b; Jiang *et al*. (蒋志刚等), 2017

小猪尾鼠 *Typhlomys nanus*

小竹鼠

Cannomys badius

数据缺乏 DD

数据缺乏 DD	无危 LC	近危 NT	易危 VU	濒危 EN	极危 CR	区域灭绝 RE	野外灭绝 EW	灭绝 EX

分类地位 Taxonomic Status

动物界 Animalia	脊索动物门 Chordata	哺乳纲 Mammalia	啮齿目 Rodentia	鼹型鼠科 Spalacidae

学　名 Scientific Name	*Cannomys badius*
命名人 Species Authority	Hodgson, 1841
英文名 English Name(s)	Lesser Bamboo Rat
同物异名 Synonym(s)	*Cannomys kastaneus* (Blyth, 1843); *lonnbergi* (Gyldenstolpe, 1917); *minor* (Gray, 1842); *pater* (Thomas, 1915); *plumbescens* (Thomas, 1915)
种下单元评估 Infra-specific Taxa Assessed	无 / None

评估信息 Assessment Information

评估年份 Year Assessed	2020
评定人 Assessor(s)	刘少英、蒋志刚 / Shaoying Liu, Zhigang Jiang
审定人 Reviewer(s)	蒋学龙 / Xuelong Jiang
其他贡献人 Other Contributor(s)	李立立、丁晨晨 / Lili Li, Chenchen Ding

理由 Justification: 小竹鼠记录较少，对其的研究缺乏。因此，列为数据缺乏等级 / Little information on Lesser Bamboo Rat has been recorded, and studies on the species are deficient. Thus, it is listed as Data Deficient

地理分布 Geographical Distribution

国内分布 Domestic Distribution

云南 / Yunnan

世界分布 World Distribution

中国；东南亚 / China; Southeast Asia

分布标注 Distribution Note

非特有种 / Non-endemic

国内分布图
Map of Domestic Distribution

种群 Population

种群数量 Population Size	未知 / Unknown
种群趋势 Population Trend	未知 / Unknown

生境与生态系统 Habitat (s) and Ecosystem (s)

生　　境 Habitat(s)	竹林、灌丛 / Bamboo Grove, Shrubland
生态系统 Ecosystem(s)	森林生态系统、灌丛生态系统 / Forest Ecosystem, Shrubland Ecosystem

威胁 Threat (s)

主要威胁 Major Threat(s)	无 / None

保护级别与保护行动 Protection Category and Conservation Action (s)

国家重点保护野生动物等级 (2021) Category of National Key Protected Wild Animals (2021)	无 / NA
IUCN 红色名录 (2020-2) IUCN Red List (2020-2)	无危 / LC
CITES 附录 (2019) CITES Appendix (2019)	无 / NA
保护行动 Conservation Action(s)	无 / None

相关文献 Relevant References

Burgin *et al*., 2018; Jiang *et al.* (蒋志刚等), 2017; Wilson *et al*., 2016; Zheng *et al.* (郑智民等), 2012; Smith *et al.* (史密斯等), 2009; Wang (王应祥), 2003; He *et al.* (何晓瑞等), 1991

小竹鼠 *Cannomys badius*

暗褐竹鼠

Rhizomys vestitus

数据缺乏 DD

数据缺乏 DD	无危 LC	近危 NT	易危 VU	濒危 EN	极危 CR	区域灭绝 RE	野外灭绝 EW	灭绝 EX

分类地位 Taxonomic Status

动物界 Animalia	脊索动物门 Chordata	哺乳纲 Mammalia	啮齿目 Rodentia	鼹型鼠科 Spalacidae

学名 Scientific Name	*Rhizomys vestitus*
命名人 Species Authority	Thomas, 1921
英文名 English Name(s)	Ward's Bamboo Rat
同物异名 Synonym(s)	无 / None
种下单元评估 Infra-specific Taxa Assessed	无 / None

评估信息 Assessment Information

评估年份 Year Assessed	2020
评定人 Assessor(s)	刘少英、蒋志刚 / Shaoying Liu, Zhigang Jiang
审定人 Reviewer(s)	蒋学龙 / Xuelong Jiang
其他贡献人 Other Contributor(s)	李立立、丁晨晨 / Lili Li, Chenchen Ding

理由 Justification: 暗褐竹鼠的记录较少，对其的研究缺乏。因此，列为数据缺乏等级 / Little information on Ward's Bamboo Rat has been recorded, and studies on it are deficient. Thus, it is listed as Data Deficient

地理分布 Geographical Distribution

国内分布 Domestic Distribution

四川、陕西、甘肃、重庆、湖北、云南、安徽、河南、湖南 / Sichuan, Shaanxi, Gansu, Chongqing, Hubei, Yunnan, Anhui, Henan, Hunan

世界分布 World Distribution

中国 / China

分布标注 Distribution Note

特有种 / Endemic

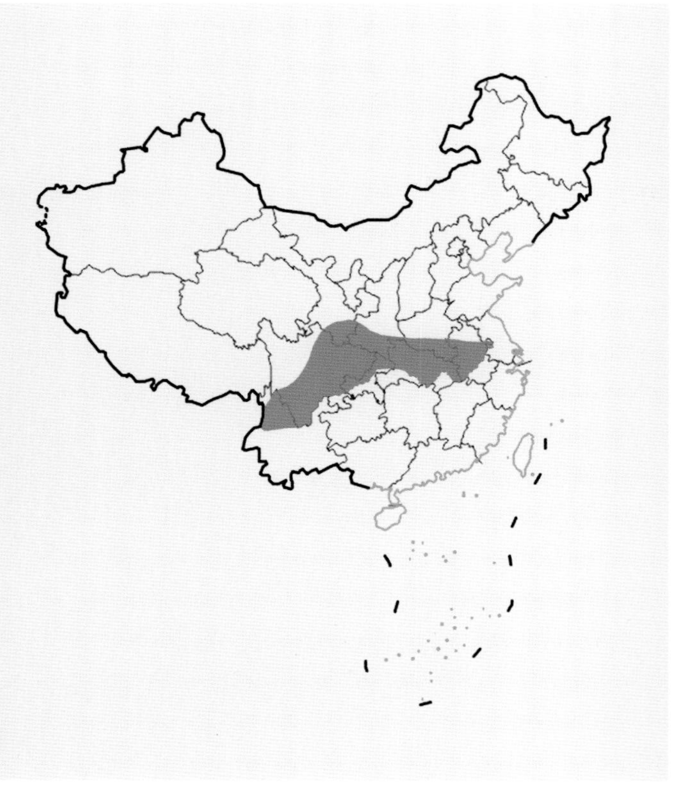

国内分布图
Map of Domestic Distribution

种群 Population

种群数量 **Population Size**	未知 / Unknown
种群趋势 **Population Trend**	未知 / Unknown

生境与生态系统 Habitat (s) and Ecosystem (s)

生　　境 **Habitat(s)**	竹林 / Bamboo Grove
生态系统 **Ecosystem(s)**	森林生态系统 / Forest Ecosystem

威胁 Threat (s)

主要威胁 **Major Threat(s)**	无 / None

保护级别与保护行动 Protection Category and Conservation Action (s)

国家重点保护野生动物等级 **(2021) Category of National Key Protected Wild Animals (2021)**	无 / NA
IUCN 红色名录 **(2020-2) IUCN Red List (2020-2)**	无危 / LC
CITES 附录 **(2019) CITES Appendix (2019)**	无 / NA
保护行动 **Conservation Action(s)**	无 / None

相关文献 Relevant References

Burgin *et al.*, 2018; Jiang *et al.* (蒋志刚等), 2017; Pan *et al.* (潘清华等), 2007; Wang (王应祥), 2003

暗褐竹鼠 *Rhizomys vestitus*　　蒋志刚 绘　Drawn by Zhigang Jiang

甘肃鼢鼠
Myospalax cansus

数据缺乏 DD

数据缺乏 DD	无危 LC	近危 NT	易危 VU	濒危 EN	极危 CR	区域灭绝 RE	野外灭绝 EW	灭绝 EX

分类地位 Taxonomic Status

动物界 Animalia	脊索动物门 Chordata	哺乳纲 Mammalia	啮齿目 Rodentia	鼹型鼠科 Spalacidae

学　名 Scientific Name	*Myospalax cansus*
命名人 Species Authority	Thomas, 1911
英文名 English Name(s)	Gansu Zokor
同物异名 Synonym(s)	*Myospalax baileyi* (Thomas, 1911); *cansus shenseius* (Thomas, 1911); *fontanierii* (Milne-Edwards, 1867); *fontanus* (Thomas, 1912); *kukunoriensis* (Lönnberg, 1926); (转下页)
种下单元评估 Infra-specific Taxa Assessed	无 / None

评估信息 Assessment Information

评估年份 Year Assessed	2020
评定人 Assessor(s)	蒋志刚、刘少英 / Zhigang Jiang, Shaoying Liu
审定人 Reviewer(s)	李保国 / Baoguo Li
其他贡献人 Other Contributor(s)	李立立、丁晨晨 / Lili Li, Chenchen Ding

理由 Justification: 甘肃鼢鼠记录较少，对其的研究缺乏。因此，列为数据缺乏 / Little information on the Gansu Zokor has been recorded, because studies on it are deficient. Thus, it is listed as Data Deficient

地理分布 Geographical Distribution

国内分布 Domestic Distribution

陕西、甘肃、宁夏 / Shaanxi, Gansu, Ningxia

世界分布 World Distribution

中国 / China

分布标注 Distribution Note

特有种 / Endemic

国内分布图
Map of Domestic Distribution

种群 Population

种群数量 **Population Size**	未知 / Unknown
种群趋势 **Population Trend**	未知 / Unknown

生境与生态系统 Habitat (s) and Ecosystem (s)

生　境 **Habitat(s)**	森林、灌丛、草地、耕地 / Forest, Shrubland, Grassland, Arable Land
生态系统 **Ecosystem(s)**	森林生态系统、灌丛生态系统、草地生态系统、农田生态系统 / Forest Ecosystem, Shrubland Ecosystem, Grassland Ecosystem, Cropland Ecosystem

威胁 Threat (s)

主要威胁 **Major Threat(s)**	无 / None

保护级别与保护行动 Protection Category and Conservation Action (s)

国家重点保护野生动物等级 **(2021) Category of National Key Protected Wild Animals (2021)**	无 / NA
IUCN 红色名录 **(2020-2) IUCN Red List (2020-2)**	无危 / LC
CITES 附录 **(2019) CITES Appendix (2019)**	无 / NA
保护行动 **Conservation Action(s)**	无 / None

相关文献 Relevant References

Burgin *et al.*, 2018; Liu *et al.* (刘丽等), 2018; Jiang *et al.* (蒋志刚等), 2017; Zhang *et al.* (张洪峰等), 2006

(接上页)

rufescens (J. A. Allen, 1909)

甘肃鼢鼠 *Myospalax cansus*　　　　蒋志刚 绘　Drawn by Zhigang Jiang

秦岭鼢鼠

Eospalax rufescens

数据缺乏 DD

数据缺乏 DD	无危 LC	近危 NT	易危 VU	濒危 EN	极危 CR	区域灭绝 RE	野外灭绝 EW	灭绝 EX

分类地位 Taxonomic Status

动物界 Animalia	脊索动物门 Chordata	哺乳纲 Mammalia	啮齿目 Rodentia	鼹型鼠科 Spalacidae

学　名 **Scientific Name**	*Eospalax rufescens*
命名人 **Species Authority**	J. Allen, 1909
英文名 **English Name(s)**	Qinling Mountain Zokor
同物异名 **Synonym(s)**	*Myospalas rufecens* (Luo *et al.*, 2000)
种下单元评估 **Infra-specific Taxa Assessed**	无 / None

评估信息 Assessment Information

评估年份 **Year Assessed**	2020
评定人 **Assessor(s)**	蒋志刚、刘少英 / Zhigang Jiang, Shaoying Liu
审定人 **Reviewer(s)**	李保国 / Baoguo Li
其他贡献人 **Other Contributor(s)**	李立立、丁晨晨 / Lili Li, Chenchen Ding

理由 **Justification:** 秦岭鼢鼠记录较少，对其的研究缺乏。因此，列为数据缺乏等级 / Little information on Qinling Mountain Zokor has been recorded, and studies on it are deficient. Thus, it is listed as Data Deficient

地理分布 Geographical Distribution

国内分布 Domestic Distribution

青海、甘肃、宁夏、陕西、四川 / Qinghai, Gansu, Ningxia, Shaanxi, Sichuan

世界分布 World Distribution

中国 / China

分布标注 Distribution Note

特有种 / Endemic

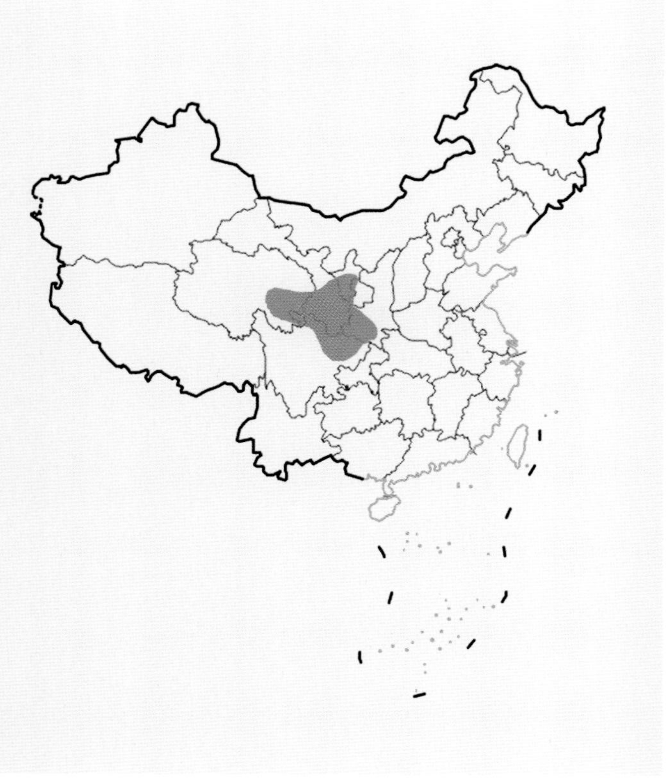

国内分布图
Map of Domestic Distribution

种群 Population

种群数量 **Population Size**	未知 / Unknown
种群趋势 **Population Trend**	未知 / Unknown

生境与生态系统 Habitat (s) and Ecosystem (s)

生　　境 **Habitat(s)**	森林、灌丛、草地、耕地 / Forest, Shrubland, Grassland, Arable Land
生态系统 **Ecosystem(s)**	森林生态系统、灌丛生态系统、草地生态系统、农田生态系统 / Forest Ecosystem, Shrubland Ecosystem, Grassland Ecosystem, Cropland Ecosystem

威胁 Threat (s)

主要威胁 **Major Threat(s)**	无 / None

保护级别与保护行动 Protection Category and Conservation Action (s)

国家重点保护野生动物等级 **(2021) Category of National Key Protected Wild Animals (2021)**	无 / NA
IUCN 红色名录 **(2020-2) IUCN Red List (2020-2)**	无危 / LC
CITES 附录 **(2019) CITES Appendix (2019)**	无 / NA
保护行动 **Conservation Action(s)**	无 / None

相关文献 Relevant References

Burgin *et al*., 2018; Jiang *et al.* (蒋志刚等), 2017; Wilson *et al*., 2016; Zheng *et al.* (郑智民等), 2012; Lu *et al.* (鲁庆彬等), 2010; He *et al.* (何娅等), 2009; Wang (王应祥), 2003

秦岭鼢鼠 *Eospalax rufescens*

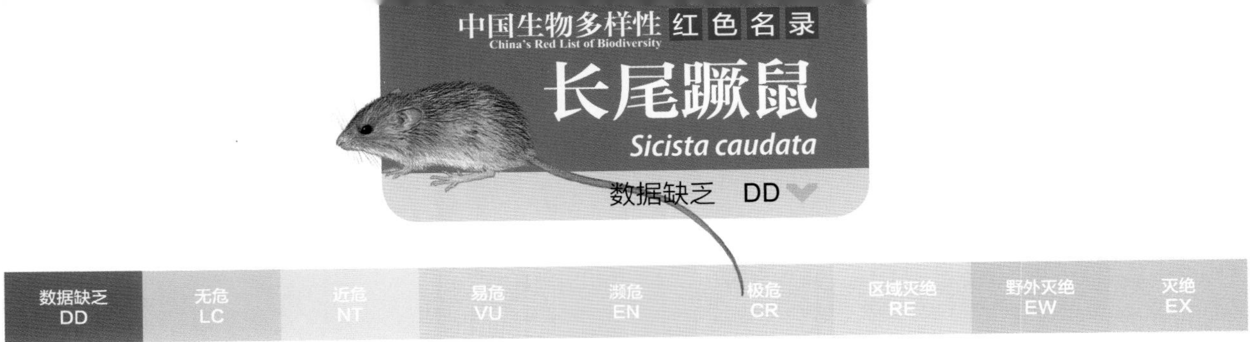

分类地位 Taxonomic Status

动物界 Animalia	脊索动物门 Chordata	哺乳纲 Mammalia	啮齿目 Rodentia	跳鼠科 Dipodidae
学名 Scientific Name		*Sicista caudata*		
命名人 Species Authority		Thomas, 1907		
英文名 English Name(s)		Long-tailed Birch Mouse		
同物异名 Synonym(s)		无 / None		
种下单元评估 Infra-specific Taxa Assessed		无 / None		

评估信息 Assessment Information

评估年份 Year Assessed	2020
评定人 Assessor(s)	蒋志刚、刘少英 / Zhigang Jiang, Shaoying Liu
审定人 Reviewer(s)	马勇、刘伟 / Yong Ma, Wei Liu
其他贡献人 Other Contributor(s)	李立立、丁晨晨 / Lili Li, Chenchen Ding

理由 Justification: 长尾蹶鼠的记录较少，对其的研究缺乏。因此，列为数据缺乏等级 / Little information on the Long-tailed Birch Mouse has been recorded, and studies on it are deficient. Thus, it is listed as Data Deficient

地理分布 Geographical Distribution

国内分布 Domestic Distribution

吉林、黑龙江 / Jilin, Heilongjiang

世界分布 World Distribution

中国、俄罗斯 / China, Russia

分布标注 Distribution Note

非特有种 / Non-endemic

国内分布图
Map of Domestic Distribution

种群 Population

种群数量 **Population Size**	未知 / Unknown
种群趋势 **Population Trend**	未知 / Unknown

生境与生态系统 Habitat (s) and Ecosystem (s)

生　　境 **Habitat(s)**	泰加林、针阔混交林 / Taiga Forest, Coniferous and Broad-leaved Mixed Forest
生态系统 **Ecosystem(s)**	森林生态系统 / Forest Ecosystem

威胁 Threat (s)

主要威胁 **Major Threat(s)**	无 / None

保护级别与保护行动 Protection Category and Conservation Action (s)

国家重点保护野生动物等级 **(2021) Category of National Key Protected Wild Animals (2021)**	无 / NA
IUCN 红色名录 **(2020-2) IUCN Red List (2020-2)**	无危 / LC
CITES 附录 **(2019) CITES Appendix (2019)**	无 / NA
保护行动 **Conservation Action(s)**	无 / None

相关文献 Relevant References

Burgin *et al*., 2018; Jiang *et al.* (蒋志刚等), 2017; Wilson *et al*., 2016; Zheng *et al.* (郑智民等), 2012; Smith *et al.* (史密斯等), 2009; Wang (王应祥), 2003

长尾蹶鼠 *Sicista caudata*

巴里坤跳鼠

Allactaga balikunica

数据缺乏 DD

数据缺乏 DD	无危 LC	近危 NT	易危 VU	濒危 EN	极危 CR	区域灭绝 RE	野外灭绝 EW	灭绝 EX

分类地位 Taxonomic Status

动物界 Animalia	脊索动物门 Chordata	哺乳纲 Mammalia	啮齿目 Rodentia	跳鼠科 Dipodidae

学　名 **Scientific Name**	*Allactaga balikunica*
命名人 **Species Authority**	Hsia and Fang, 1964
英文名 **English Name(s)**	Balikun Jerboa
同物异名 **Synonym(s)**	*Allactaga nataliae* (Sokolov, 1981)
种下单元评估 **Infra-specific Taxa Assessed**	无 / None

评估信息 Assessment Information

评估年份 **Year Assessed**	2020
评定人 **Assessor(s)**	蒋志刚、刘少英 / Zhigang Jiang, Shaoying Liu
审定人 **Reviewer(s)**	马勇、刘伟 / Yong Ma, Wei Liu
其他贡献人 **Other Contributor(s)**	李立立、丁晨晨 / Lili Li, Chenchen Ding

理由 **Justification:** 巴里坤跳鼠记录较少，对其的研究缺乏。因此，列为数据缺乏等级 / Little information on Balikun Jerboa has been recorded, and studies on it are deficient. Thus, it is listed as Data Deficient

地理分布 Geographical Distribution

国内分布 Domestic Distribution

新疆 / Xinjiang

世界分布 World Distribution

中国、蒙古 / China, Mongolia

分布标注 Distribution Note

非特有种 / Non-endemic

国内分布图
Map of Domestic Distribution

种群 Population

种群数量 **Population Size**	未知 / Unknown
种群趋势 **Population Trend**	未知 / Unknown

生境与生态系统 Habitat (s) and Ecosystem (s)

生　　境 **Habitat(s)**	半荒漠草原 / Semi-desert Grassland
生态系统 **Ecosystem(s)**	荒漠生态系统 / Desert Ecosystem

威胁 Threat (s)

主要威胁 **Major Threat(s)**	无 / None

保护级别与保护行动 Protection Category and Conservation Action (s)

国家重点保护野生动物等级 **(2021) Category of National Key Protected Wild Animals (2021)**	无 / NA
IUCN 红色名录 **(2020-2) IUCN Red List (2020-2)**	无危 / LC
CITES 附录 **(2019) CITES Appendix (2019)**	无 / NA
保护行动 **Conservation Action(s)**	无 / None

相关文献 Relevant References

Burgin *et al*., 2018; Jiang *et al.* (蒋志刚等), 2017; Zheng *et al.* (郑智民等), 2012; Smith *et al.* (史密斯等), 2009; Pan *et al.* (潘清华等), 2007; Xia and Fang (夏武平和方喜业), 1964

巴里坤跳鼠 *Allactaga balikunica*　　蒋志刚 绘　Drawn by Zhigang Jiang

数据缺乏 DD	无危 LC	近危 NT	易危 VU	濒危 EN	极危 CR	区域灭绝 RE	野外灭绝 EW	灭绝 EX

分类地位 Taxonomic Status

动物界 Animalia	脊索动物门 Chordata	哺乳纲 Mammalia	兔形目 Lagomorpha	鼠兔科 Ochotonidae
学名 Scientific Name		*Ochotona argentata*		
命名人 Species Authority		Howell, 1928		
英文名 English Name(s)		Helanshan Pika		
同物异名 Synonym(s)		银色鼠兔，Silver Pika		
种下单元评估 Infra-specific Taxa Assessed		无 / None		

评估信息 Assessment Information

评估年份 Year Assessed	2021
评定人 Assessor(s)	蒋志刚 / Zhigang Jiang
审定人 Reviewer(s)	Marc Foggin
其他贡献人 Other Contributor(s)	丁晨晨 / Chenchen Ding

理由 Justification: 贺兰山鼠兔分布在贺兰山局部岩石山区 (Smith *et al*., 1990)。贺兰山鼠兔冬季被毛银浅灰色，夏季则为鲜红色，有强壮的下颌，牙齿和头骨形态特征明显。此外，*O. argentata* 和 *O. pallasi* 的二倍体染色体数目 (2*n*=38) 与 *O. alpine* (2*n*=42) 的不同。它最初被 Howell 描述为 *Ochotona alpina argentata*。马勇在 1985~1987 年采集了 5 具标本。有研究支持该鼠兔是一个独立的物种 (Erbajeva, 1997)。由于缺乏贺兰山鼠兔的种群数据，故将其列为数据缺乏 / The Helanshan Pika is distributed in some rocky mountain areas of Helan Shan (Smith *et al*., 1990). The winter coat of Helanshan Pika is light gray with wool silver, and its summer coat is bright red. The Helanshan Pika has a strong mandibular, prominent teeth and skull shape. Furthermore, the chromosome number of *O. argentata* and *O. pallasi* (2*n*=38) is contrasted with the *O. alpina* (2*n*=42). Originally, it was described by Howell as *Ochotona alpina argentata*. Yong Ma collected five specimens between 1985 and 1987. Studies support that the pika is a unique species (Erbajeva, 1997). Due to lacking of population parameters, it is listed as Data Deficient

地理分布 Geographical Distribution

国内分布 Domestic Distribution

宁夏 / Ningxia

世界分布 World Distribution

中国 / China

分布标注 Distribution Note

特有种 / Endemic

国内分布图
Map of Domestic Distribution

种群 Population

种群数量 **Population Size**	未知 / Unknown
种群趋势 **Population Trend**	未知 / Unknown

生境与生态系统 Habitat (s) and Ecosystem (s)

生　　境 **Habitat(s)**	半荒漠岩石灌丛生境 / Shrubland in Semi-desert Rocky Zone
生态系统 **Ecosystem(s)**	陆生生态系统 / Terrestrial Ecosystem

威胁 Threat (s)

主要威胁 **Major Threat(s)**	无 / None

保护级别与保护行动 Protection Category and Conservation Action (s)

国家重点保护野生动物等级 **(2021) Category of National Key Protected Wild Animals (2021)**	二级 / Category II
IUCN 红色名录 **(2020-2) IUCN Red List (2020-2)**	濒危 / EN
CITES 附录 **(2019) CITES Appendix (2019)**	无 / NA
保护行动 **Conservation Action(s)**	无 / None

相关文献 Relevant References

Burgin *et al*., 2018; Mittermeier and Wilson, 2014; Erbajeva and Ma, 2006; Formozov *et al*., 2004; Yu *et al*., 2000; Erbajeva, 1997; Smith *et al*., 1990

贺兰山鼠兔 *Ochotona argentata*

蒋志刚 绘　Drawn by Zhigang Jiang

黄龙鼠兔
Ochotona huanglongensis

数据缺乏 DD

数据缺乏 DD	无危 LC	近危 NT	易危 VU	濒危 EN	极危 CR	区域灭绝 RE	野外灭绝 EW	灭绝 EX

分类地位 Taxonomic Status

动物界 Animalia	脊索动物门 Chordata	哺乳纲 Mammalia	兔形目 Lagomorpha	鼠兔科 Ochotonidae

学　名 **Scientific Name**	*Ochotona huanglongensis*
命名人 **Species Authority**	Liu *et al.*, 2017
英文名 **English Name(s)**	Huanglong Pika
同物异名 **Synonym(s)**	无 / None
种下单元评估 **Infra-specific Taxa Assessed**	无 / None

评估信息 Assessment Information

评估年份 **Year Assessed**	2020
评定人 **Assessor(s)**	刘少英、蒋志刚 / Shaoying Liu, Zhigang Jiang
审定人 **Reviewer(s)**	苏建平、冯祚建、宗浩、廖继承 / Jianping Su, Zuojian Feng, Hao Zong, Jicheng Liao
其他贡献人 **Other Contributor(s)**	李立立、丁晨晨 / Lili Li, Chenchen Ding

理由 Justification: 黄龙鼠兔是新发现的种，分布区与种群数量数据缺乏。因此，列为数据缺乏等级 / Huanglong Pika is a newly discovered species. Its distribution range and population sizes data are deficient. Thus, it is listed as Data Deficient

地理分布 Geographical Distribution

国内分布 Domestic Distribution	
四川 / Sichuan	
世界分布 World Distribution	
中国 / China	
分布标注 Distribution Note	
特有种 / Endemic	

国内分布图
Map of Domestic Distribution

种群 Population

种群数量 **Population Size**	未知 / Unknown
种群趋势 **Population Trend**	未知 / Unknown

生境与生态系统 Habitat (s) and Ecosystem (s)

生　　境 **Habitat(s)**	针阔混交林、灌丛木、草甸的岩石区域 / Rocky Area of Coniferous and Broad-leaved Mixed Forest, Shrubland, Meadow
生态系统 **Ecosystem(s)**	森林生态系统、灌丛生态系统、草甸生态系统 / Forest Ecosystem, Shrub Ecosystem, Meadow Ecosystem

威胁 Threat (s)

主要威胁 **Major Threat(s)**	无 / None

保护级别与保护行动 Protection Category and Conservation Action (s)

国家重点保护野生动物等级 **(2021) Category of National Key Protected Wild Animals (2021)**	无 / NA
IUCN红色名录**(2020-2) IUCN Red List (2020-2)**	无危 / LC
CITES 附录 **(2019) CITES Appendix (2019)**	无 / NA
保护行动 **Conservation Action(s)**	无 / None

相关文献 Relevant References

Burgin *et al.*, 2018; Jiang *et al.* (蒋志刚等), 2017; Liu *et al.* (刘少英等), 2017

黄龙鼠兔 *Ochotona huanglongensis*　　　　刘洋 摄　By Yang Liu

草原鼠兔

Ochotona pusilla

数据缺乏 DD

数据缺乏 DD	无危 LC	近危 NT	易危 VU	濒危 EN	极危 CR	区域灭绝 RE	野外灭绝 EW	灭绝 EX

分类地位 Taxonomic Status

动 物 界 Animalia	脊索动物门 Chordata	哺 乳 纲 Mammalia	兔 形 目 Lagomorpha	鼠 兔 科 Ochotonidae

学　　名 Scientific Name	*Ochotona pusilla*
命 名 人 Species Authority	Pallas, 1769
英 文 名 English Name(s)	Steppe Pika
同物异名 Synonym(s)	无 / None
种下单元评估 Infra-specific Taxa Assessed	无 / None

评估信息 Assessment Information

评 估 年 份 Year Assessed	2020
评　定　人 Assessor(s)	蒋志刚 / Zhigang Jiang
审　定　人 Reviewer(s)	刘少英 / Shaoying Liu
其他贡献人 Other Contributor(s)	丁晨晨 / Chenchen Ding

理由 Justification: 草原鼠兔目前仅在新疆有报道。2009 年于新疆克拉玛依加依尔山克孜勒乌赞一带的叉子圆柏灌丛中捕获了 2 只草原鼠兔标本。然而，关于该种的种群与生境状况仍知之甚少。因此，列为数据缺乏等级 / Steppe Pika is only reported in Xinjiang, China. Two specimens of Steppe Pika were trapped in *Juniperus sabina* shrub in the Kezilewuzan Area, Jayer Shan Region, Karamay District, Xinjiang in 2009. However, we still know very little information on its population and habitat status. Thus, it is listed as Data Deficient

地理分布 Geographical Distribution

国内分布 Domestic Distribution

新疆 / Xinjiang

世界分布 World Distribution

中国、哈萨克斯坦、吉尔吉斯斯坦、俄罗斯 / China, Kazakhstan, Kyrgyzstan, Russia

分布标注 Distribution Note

非特有种 / Non-endemic

国内分布图
Map of Domestic Distribution

种群 Population

种群数量 **Population Size**	未知 / Unknown
种群趋势 **Population Trend**	未知 / Unknown

生境与生态系统 Habitat (s) and Ecosystem (s)

生　　境 **Habitat(s)**	半荒漠灌丛生境 / Shrubland in Semi-desert Zone
生态系统 **Ecosystem(s)**	陆生生态系统 / Terrestrial Ecosystem

威胁 Threat (s)

主要威胁 **Major Threat(s)**	无 / None

保护级别与保护行动 Protection Category and Conservation Action (s)

国家重点保护野生动物等级 **(2021) Category of National Key Protected Wild Animals (2021)**	无 / NA
IUCN 红色名录 **(2020-2) IUCN Red List (2020-2)**	无危 / LC
CITES 附录 **(2019) CITES Appendix (2019)**	无 / NA
保护行动 **Conservation Action(s)**	无 / None

相关文献 Relevant References

Burgin *et al.*, 2018; Wilson *et al.*, 2016; Shayrave *et al.* (沙依拉吾等), 2009; Sokolov *et al.*, 1977 (Hoffmann and Smith in 2009 translated from Russian)

草原鼠兔 *Ochotona pusilla*　　方红霞 绘　Drawn by Hongxia Fang

参考文献
References

阿不都热合曼·吐尔逊, 余亮, 艾尼瓦尔·吐米尔, 等. 2008. 新疆北部干旱地区鼠类群落空间结构差异的比较研究. 新疆大学学报(自然科学版), 25(2): 211-218. Tursun A, Yu L, Tumur A, *et al*. 2008. Comparative study on the spatial structure differences of the rodent communities from arid areas in northern Xinjiang. *Journal of Xinjiang University* (Natural Science Edition), 25(2): 211-218. (In Chinese with English Abstract)

阿布力米提·阿布都卡迪尔. 2002. 新疆哺乳类(兽纲)名录. 干旱区研究, 19(增刊): 1-75. Abulimiti A. 2002. List of Mammals in Xinjiang. *Arid Zone Research*, 19(Supplementary): 1-75. (In Chinese with English Abstract)

阿布力米提·阿布都卡迪尔, 时琨, 吐尔迅·吐拉克, 等. 2010. 阿克苏拜城天山区盘羊和北山羊的分布与种群资源现状. 野生动物学报, 31(5): 270-275. Abulimiti A, Shi K, Tursun T, *et al*. 2010. Distribution and population resources of Argali Sheep and Ibex in Baicheng of Aksu Tianshan Mountains. *Chinese Journal of Wildlife*, 31(5): 270-275. (In Chinese with English Abstract)

艾尼瓦尔·铁木尔, 张大铭, 苏力旦, 等. 1998. 新疆阿拉山口地区鼠类群落物种多样性的初步研究. 新疆大学学报(自然科学版), 15: 73-76. Anwar T, Zhang D M, Su L D, *et al*. 1998. A preliminary study on species diversity of rodent community in Alashankou, Xinjiang. *Journal of Xinjiang University* (Natural Science Edition), 15: 73-76. (In Chinese with English Abstract)

艾尼瓦尔·吐米尔, 马合木提·哈力克, Airoldi J. 2005. 农田周围生态保留带中普通田鼠的种群生态学: 种群数量动态及结构. 动物学杂志, 40(5): 43-49. Tumur A, Halik M, Airoldi J. 2005. A study of the population ecology of *Microtus arvalis*: population dynamics and structure in the set aside area around the farmland. *Chinese Journal of Zoology*, 40(5): 43-49. (In Chinese with English Abstract)

安冉, 刘斌, 徐艺玫, 等. 2015. 林睡鼠幼鼠的活动规律和行为初步观察. 兽类学报, 35(2): 170-175. An R, Liu B, Xu Y M, *et al*. 2015. Observations of activity patterns and behavior of *Dryomys nitedula* pups. *Acta Theriologica Sinica*, 35(2): 170-175. (In Chinese with English Abstract)

白冰, 周伟, 张庆, 等. 2011. 高黎贡山大塘白眉长臂猿春季栖息地利用及与赧亢的比较. 四川动物, 30(1): 25-30. Bai B, Zhou W, Zhang Q, *et al*. 2011. Habitat utilization by hoolock gibbons (*Hoolock hoolock*) at Datang, Mt. Gaoligong in spring and its comparison with the situation in Nankang. *Sichuan Journal of Zoology*, 30(1): 25-30. (In Chinese with English Abstract)

白德凤, 陈颖, 李俊松, 等. 2018. 西双版纳尚勇自然保护区哺乳动物物种多样性. 生物多样性, 26(1): 75-78. Bai D F, Chen Y, Li J S, *et al*. 2018. Mammal diversity in Shangyong Nature Reserve, Xishuangbanna, Yunnan Province. *Biodiversity Science*, 26(1): 75-78. (In Chinese with English Abstract)

包新康, 杨增武, 赵伟, 等. 2014. 甘肃安西国家级自然保护区脊椎动物20年间的变化. 生物多样性, 22(4): 539-545. Bao X K, Yang Z W, Zhao W, *et al*. 2014. The alteration of the vertebrate resources over the past two decades in Gansu Anxi Extreme Arid National Nature Reserve. *Biodiversity Science*, 22(4): 539-545. (In Chinese with English Abstract)

蔡桂全, 冯祚建. 1982. 高原兔(*Lepus oiostolus*)亚种补充研究——包括两个新亚种. 兽类学报, 2(2): 167-182. Cai G Q, Feng Z J. 1982. A systematic revision of the subspecies of highland hare (*Lepus oiostolus*), including two new subspecies. *Acta Theriologica Sinica*, 2(2): 167-182. (In Chinese with English Abstract)

蔡仁逵, 黄叔怀, 李金良. 1959. 连云港所获长须鲸的鉴定. 南京师范学院学报(自然科学), (3): 1-4. Cai R K, Huang S H, Li J L. 1959. Identification of baleen whales in Lianyungang. *Journal of Nanjing Normal University* (Natural Science), (3): 1-4. (In Chinese with English Abstract)

曹明, 周伟, 白冰, 等. 2010. 滇南勐腊地区威氏小鼷鹿种群生境利用. 动物学研究, 31(3): 303-309. Cao M, Zhou W, Bai B, *et al*. 2010. Habitat use of Williamson's mouse deer (*Tragulus williamsoni*) in Mengla Area, Southern Yunnan. *Zoological Research*, 31(3): 303-309. (In Chinese with English Abstract)

曹伊凡, 林恭华, 卢学峰, 等. 2009. 柯氏鼠兔的食性分析. 动物学杂志, 44(1): 58-62. Cao Y F, Lin G H, Lu X F, *et al*. 2009. Food habits of *Ochotona koslowi*. *Chinese Journal of Zoology*, 44(1): 58-62. (In Chinese with English Abstract)

岑业文, 彭红元. 2010. 广西玉林市翼手类多样性初步调查. 四川动物, 29(5): 609-612. Cen Y W, Peng H Y. 2010. Research on bat diversity in Yulin city, Guangxi. *Sichuan Journal of Zoology*, 29(5): 609-612. (In Chinese with English Abstract)

常勇斌, 贾陈喜, 宋刚, 等. 2018. 西藏错那县发现藏南猕猴. 动物学杂志, 53(2): 243-248. Chang Y B, Jia C X, Song G, *et al*. 2018. Discovery of *Macaca munzala* in Cona, Tibet. *Chinese Journal of Zoology*, 53(2): 243-248. (In Chinese with English Abstract)

陈柏承, 余文华, 吴毅, 等. 2015. 毛翼管鼻蝠在广西和江西分布新纪录及其性二型现象. 四川动物, 34(2): 211-215. Chen B C, Yu W H, Wu Y, *et al*. 2015. New record and sexual dimorphism of *Harpiocephalus harpia* in Guangxi and Jiangxi, China. *Sichuan Journal of Zoology*, 34(2): 211-215. (In Chinese with English Abstract)

陈昌笃, 康利华, 张憬. 2000. 都江堰生物多样性研究与保护. 成都: 四川科学技术出版社: 55-57. Chen C D, Kang L H, Zhang J. 2000. Biodiversity Research and Protection in Dujiangyan. Chengdu: Sichuan Science and Technology Press: 55-57. (In Chinese)

陈道富, 全仁哲, 范喜顺, 等. 2003. 欧亚河狸的生物学特性及其保护与开发. 石河子大学学报(自然科学版), 7(1): 84-86. Chen D F, Quan R Z, Fan X S, *et al*. 2003. Biological characteristics and protection and development of Eurasian beaver. *Journal of Shihezi University* (Natural Science Edition), 7(1): 84-86. (In Chinese with English Abstract)

陈耕, 张林源. 1998. 西藏珠穆朗玛自然保护区野生动物状况与保护. 见: 中国科学院生物多样性委员会, 林业部野生动物和森林植物保护司, 国家环保局自然保护司, 等. 生物多样性与人类未来——第二届全国生物多样性保护与持续利用研讨会论文集. 北京: 中国林业出版社: 155-160. Chen G, Zhang L Y. 1998. The status and protection of wild animals in Qomolangma Nature Reserve in Tibet. *In*: Biodiversity Committee of Chines Academy of Sciences; Department of Wildlife and Forest Plant Protection, Ministry of Forestry; Department of Natural Protection, Ministry of Environmental Protection, *et al*. Biodiversity and the Future of Mankind: Proceedings of the Second National Symposium on Biodiversity Conservation and Sustainable Use. Beijing: China Forestry Press: 155-160. (In Chinese)

陈兼善. 1969. 台湾脊椎动物志(下册). 台北: 台湾商务印书馆. Chen J S. 1969. Vertebrates of Taiwan (Vol. 2). Taipei: Taiwan Commercial Press. (In Chinese)

陈立军, 张文杰, 张小倩, 等. 2014. 典型草原区达乌尔鼠兔繁殖生态学的初步研究. 动物学杂志, 49(5): 649-656. Chen L J, Zhang W J, Zhang X Q, *et al*. 2014. The preliminary study on breeding ecology of *Ochotona dauurica* in typical steppe. *Chinese Journal of Zoology*, 49(5): 649-656. (In Chinese with English Abstract)

陈良, 鲍毅新, 张龙龙, 等. 2010. 九龙山保护区黑麂栖息地选择的季节变化. 生态学报, 30(5): 1227-1237. Chen L, Bao Y X, Zhang L L, *et al*. 2010. Seasonal changes in habitat selection by black muntjac (*Muntiacus crinifrons*) in Jiulong Mountain Nature Reserve. *Acta Ecologica Sinica*, 30(5): 1227-1237. (In Chinese with English Abstract)

陈敏. 2003. 七种蝙蝠回声定位行为生态研究. 长春: 东北师范大学硕士研究生学位论文. Chen M. 2003. Study on Echolocation Behavior of Seven Bats. Changchun: Master's Thesis of Northeast Normal University. (In Chinese with English Abstract)

陈敏, 冯江, 李振新, 等. 2002. 普氏蹄蝠(*Hipposideros pratti*)回声定位声波、形态及捕食策略. 应用生态学报, 13(12): 1629-1632. Chen M, Feng J, Li Z X, *et al*. 2002. Echolocation sound waves, morphological features and foraging strategies in *Hipposideros pratti*. *Journal of Applied Ecology*, 13(12): 1629-1632. (In Chinese with English Abstract)

陈敏杰, 毕超贤, 余梁哥, 等. 2014. 倭蜂猴生物学特性、生存现状及保护对策. 生物学通报, 49(10): 7-10. Chen M J, Bi C X, Yu L G, *et al*. 2014. Biological characteristics, status and conservation policy of pygmy loris. *Biology Bulletin*, 49(10): 7-10. (In Chinese with English Abstract)

陈鹏. 2012. 姬鼠属和白腹鼠属系统发育及谱系地理研究. 北京: 中国科学院研究生院博士研究生学位论文. Chen P. 2012. Systematics and Phylogeography of *Apodemus* and *Niviventer*. Beijing: Doctor's Thesis of Graduate University of Chinese Academy of Sciences. (In Chinese with English Abstract)

陈鹏, 师杜鹃. 2013. 长青自然保护区金猫垂直分布和季节性活动规律研究. 陕西林业科技, (1): 22-24. Chen P, Shi D J. 2013. The vertical distribution and seasonal activity patterns of Asian Golden Cat in Changqing National Nature Reserve. *Shaanxi Forest Science and Technology*, (1): 22-24. (In Chinese with English Abstract)

陈鹏, 王应祥, 林苏, 等. 2014. 中国兽类新纪录——耐氏大鼠*Leopoldamys neilli*. 四川动物, 33(6): 858-864. Chen P, Wang Y X, Lin S, *et al*. 2014. A new record of mammals in China-*Leopoldamys neilli*. *Sichuan Journal of Zoology*, 33(6): 858-864. (In Chinese with English Abstract)

陈水华, 诸葛阳. 1993. 臭鼩染色体的研究. 兽类学报, (2): 8. Chen S H, Zhuge Y. 1993. Chromosome studies of house shrew, *Suncus murinus* (Soricidae). *Acta Theriologica Sinica*, (2): 8. (In Chinese with English Abstract)

陈万里, 谌利民, 马文虎, 等. 2013. 四川羚牛繁殖期集群类型及海拔分布. 四川动物, 32(6): 841-845. Chen W L, Shen L M, Ma W H, *et al*. 2013. Group type and distribution of *Budorcas taxicolor tibetana* in rutting season. *Sichuan Journal of Zoology*, 32(6): 841-845. (In Chinese with English Abstract)

陈小荣, 许大明, 鲍毅新, 等. 2013. G-F指数测度百山祖兽类物种多样性. 生态学杂志, 32(6): 1421-1427. Chen X R, Xu D M, Bao Y X, *et al*. 2013. Mammalian species diversity in Baishanzu National Nature Reserve, Zhejiang Province of East China based on G-F Index. *Chinese Journal of Ecology*, 32(6): 1421-1427. (In Chinese with English Abstract)

陈晓澄, 李文靖. 2009. 西藏东南部灰颈鼠兔(*Ochotona forresti*)一新亚种. 兽类学报, 29(1): 101-105. Chen X C, Li W J. 2009. A new subspecies of *Ochotona forresti* in southeastern Tibet, China. *Acta Theriologica Sinica*, 29(1): 101-105. (In Chinese with English Abstract)

陈延熹, 黄文几, 唐仕敏. 1987. 赣北翼手类区系调查. 兽类学报, 7(1): 13-19. Chen Y X, Huang W J, Tang S M. 1987. The Chiropteran fauna of the north Jiangxi. *Acta Theriologica Sinica*, 7(1): 13-19. (In Chinese with English Abstract)

陈延熹, 黄文几, 唐子英. 1989. 赣南翼手类初步调查. 兽类学报, 9(3): 226-227. Chen Y X, Huang W J, Tang Z Y. 1989. The investigation of Chiroptera in Jiangxi. *Acta Theriologica Sinica*, 9(3): 226-227. (In Chinese with English Abstract)

陈毅, 刘奇, 谭梁静, 等. 2013. 广东省发现南蝠. 动物学杂志, 48(2): 287-291. Chen Y, Liu Q, Tan L J, *et al*. 2013. *Ia io* was dicovered in Guangdong Province. *Chinese Journal of Zoology*, 48(2): 287-291. (In Chinese with English Abstract)

陈永春, 肖林, 李学友. 2013. 白马雪山中华鬣羚种群数量和分布初步调查. 野生动物, 34: 253-255. Chen Y C, Xiao L, Li X Y. 2013. Preliminary study on the population abundance and distribution of Chinese Serow (*Capricornis milneedwardsii*) at Baima Xueshan Reserve. *Chinese Wildlife*, 34: 253-255. (In Chinese with English Abstract)

陈志平, 王应祥, 冯庆, 等. 1996. 云南西双版纳片断热带雨林鼠形啮齿类的物种多样性研究. 动物学研究, 17(4): 451-458. Chen Z P, Wang Y X, Feng Q, *et al*. 1996. The studies on species diversity of myomorpha rodents in the fragmental tropical rainforest in Xishuangbanna, Yunnan. *Zoological Research*, 17(4): 451-458. (In Chinese with English Abstract)

陈志平, 王应祥, 刘瑞清. 1993. 云南兔(*Lepus comus*)的染色体研究. 兽类学报, 13(3): 188-192. Chen Z P, Wang Y X, Liu R Q. 1993. Studies on the chromosomes of Yunnan hare (*Lepus comus*). *Acta Theriologica Sinica*, 13(3): 188-192. (In Chinese with English Abstract)

陈智, 黄乘明, 周歧海, 等. 2008. 白头叶猴 (*Trachypithecus leucocephalus*) 栖息地景观格局的时空变化. 生态学报, 28(2): 587-594. Chen Z, Huang C M, Zhou Q H, *et al*. 2008. Spatial temporal changes of habitat of *Trachypithecus leucocephalus*. *Acta Ecologica Sinica*, 28(2): 587-594. (In Chinese with English Abstract)

陈忠, 蒙以航, 周锋, 等. 2007. 海南岛棕果蝠的活动节律与食性. 兽类学报, 27(2): 112-119. Chen Z, Meng Y H, Zhou F, *et al*. 2007. Activity rhythms and food habits of Leschenault's rousette *Rousettus leschenaulti* on Hainan Island. *Acta Theriologica Sinica*, 27(2): 112-119. (In Chinese with English Abstract)

陈子禧, 吴毅, 余文华, 等. 2018. 哈氏管鼻蝠在中国大陆地区的又一新发现——江西省分布新纪录. 西部林业科学, 47(2): 75-80. Chen Z X, Wu Y, Yu W H, *et al*. 2018. Another Discovery of *Murina harrisoni* in Mainland of China——A New Record from Jiangxi Province. *Journal of West China Forestry Science*, 47(2): 75-80. (In Chinese with English Abstract)

成市, 陈中正, 程峰, 等. 2018. 中国啮齿类一属、种新纪录——道氏东京鼠. 兽类学报, 38(3): 309-314. Cheng S, Chen Z Z, Cheng F, *et al*. 2018. New record of a rodent genus (*Murinae*) in China—*Tonkinomys daovantieni*. *Acta Theriologica Sinica*, 38(3): 309-314. (In Chinese with English Abstract)

程宏毅, 鲍毅新, 陈良, 等. 2008. 黑麂(*Muntiacus crinifrons*)栖息地片断化对种群基因流的影响. 生态学报, 28(3): 1109-1119. Cheng H Y, Bao Y X, Chen L, *et al*. 2008. Effects of habitat fragmentation on gene flow of the black muntjac (*Muntiacus crinifrons*). *Acta Ecologica Sinica*, 28(3): 1109-1119. (In Chinese with English Abstract)

程泽信, 刘武. 2000. 我国野生猪獾资源及其开发利用初探. 经济动物学报, (4): 33-35. Cheng Z X, Liu W. 2000. A preliminary study on wild hog badger resources and their exploitation and utilization in China. *Journal of Economic Animal*, (4): 33-35. (In Chinese with English Abstract)

程志营, 卢贞燕, 梁显堂. 2011. 广西翼手目动物布氏球果蝠新记录. 广西科学, 18(3): 312-313. Cheng Z Y, Lu Z Y, Liang X T. 2011. A new record of *Sphaerias blanfordi* of Chiroptera in Guangxi. *Guangxi Sciences*, 18(3): 312-313. (In Chinese with

English Abstract)

褚新洛. 1989. 云南省志 • 卷六 • 动物志. 昆明: 云南人民出版社: 1-401. Chu X L. 1989. Annals of Yunnan Province. Vol. 6. Fauna. Kunming: Yunnan People's Publishing House: 1-401. (In Chinese)

崔茂欢, 杨国斌, 杨士剑. 2014. 兰坪云岭省级自然保护区兽类资源调查. 大理学院学报, 13(6): 48-53. Cui M H, Yang G B, Yang S J. 2014. Mammals survey of Lanping Yunling Provincial Nature Reserve. *Journal of Dali University*, 13(6): 48-53. (In Chinese with English Abstract)

崔鹏, 徐海根, 吴军, 等. 2014. 中国脊椎动物红色名录指数评估. 生物多样性, 22(5): 589-595. Cui P, Xu H G, Wu J, *et al.* 2014. Assessing the Red List Index for vertebrate species in China. *Biodiversity Science*, 22(5): 589-595. (In Chinese with English Abstract)

崔庆虎, 蒋志刚, 连新明, 等. 2005. 根田鼠栖息地选择的影响因素. 兽类学报, 25(1): 45-51. Cui Q H, Jiang Z G, Lian X M, *et al.* 2005. Factors influencing habitat selection of root voles *Microtus oeconomus*. *Acta Theriologica Sinica*, 25(1): 45-51. (In Chinese with English Abstract)

崔绍朋, 罗晓, 李春旺, 等. 2018. 基于MaxEnt模型预测白唇鹿的潜在分布区. 生物多样性, 26(2): 171-176. Cui S P, Luo X, Li C W, *et al.* 2018. Predicting the potential distribution of white-lipped deer using the Max-Ent model. *Biodiversity Science*, 26(2): 171-176. (In Chinese with English Abstract)

戴强, 袁佐平, 张晋东, 等. 2006. 道路及道路施工对若尔盖高寒湿地小型兽类及鸟类生境利用的影响. 生物多样性, 14(2): 121-127. Dai Q, Yuan Z P, Zhang J D, *et al.* 2006. Road and road construction effects on habitat use of small mammals and birds in Zoige alpine wetland. *Biodiversity Science*, 14(2): 121-127. (In Chinese with English Abstract)

党飞红, 余文华, 王晓云, 等. 2017. 中国渡濑氏鼠耳蝠种名订正. 四川动物, 36(1): 7-13. Dang F H, Yu W H, Wang X Y, *et al.* 2017. Taxonomic Clarification of *Myotis rufoniger* from China. *Sichuan Journal of Zoology*, 36(1): 7-13. (In Chinese with English Abstract)

邓可, 张利周, 李权, 等. 2013. 云南天池自然保护区兽类资源调查. 四川动物, 32(3): 458-463. Deng K, Zhang L Z, Li Q, *et al.* 2013. Mammal survey of Tienchi Nature Reserve, Yunnan Province. *Sichuan Journal of Zoology*, 32(3): 458-463. (In Chinese with English Abstract)

邓庆伟, 刘胜祥, 奚蓉, 等. 2008. 湖北省兽类一新纪录——贵州菊头蝠. 四川动物, 27(3): 411. Deng Q W, Liu S X, Xi R, *et al.* 2008. A new record of mammal in Hubei Province—*Rhinolophus rex*. *Sichuan Journal of Zoology*, 27(3): 411. (In Chinese with English Abstract)

丁晨晨, 胡一鸣, 李春旺, 等. 2018. 印度野牛在中国的分布及其栖息地适宜性分析. 生物多样性, 26(9): 37-47. Ding C C, Hu Y M, Li C W, *et al.* 2018. Distribution and habitat suitability assessment of the gaur *Bos gaurus* in China. *Biodiversity Sciences*, 26(9): 37-47. (In Chinese with English Abstract)

丁铁明, 王作义. 1989. 井冈山自然保护区发现斑蝠. 江西林业科技, (1): 32. Ding T M, Wang Z Y. 1989. *Scotomanes ornatus* found in Jinggangshan Nature Reserve. *Jiangxi Forestry Science and Technology*, (1): 32. (In Chinese with English Abstract)

丁贤明, 钱宝珍, Matsuda J, 等. 2008. 长爪沙鼠的遗传多样性分析. 遗传, 30(7): 877-884. Ding X M, Qian B Z, Matsuda J, *et al.* 2008. Genetic diversity of Mongolian gerbils (*Meriones unguiculatus*). *Hereditas*, 30(7): 877-884. (In Chinese with English Abstract)

董金海, 王广洁, 丁正凰, 等. 1977. 在我国胶州湾内首获成体抹香鲸. 海洋科学, (1): 14-15. Dong J H, Wang G J, Ding Z H, *et al.* 1977. First adult sperm whale in Jiaozhou Bay, China. *Ocean Science*, (1): 14-15. (In Chinese with English Abstract)

董维惠, 侯希贤, 杨玉平. 2006. 内蒙古中西部地区五趾跳鼠种群数量动态研究. 中国媒介生物学及控制杂志, 17(6): 444-446. Dong W H, Hou X X, Yang Y P. 2006. A Study on the population dynamics of *Allactage sibirica* jerboa in the central and western region of Inner Mongolia. *Chinese Journal of Vector Biology and Control*, 17(6): 444-446. (In Chinese with English Abstract)

董维惠, 侯希贤, 杨玉平. 2008. 三趾跳鼠种群数量动态及预测研究. 中华卫生杀虫药械, 14(3): 181-184. Dong W H, Hou X X, Yang Y P. 2008. Population dynamics and prediction for northern three-toed jerbod. *Chinese Journal of Hygienic Insecticides & Equipments*, 14(3): 181-184. (In Chinese with English Abstract)

董聿茂, 诸葛阳, 黄美华. 1989. 浙江动物志. 兽类. 杭州: 浙江科学技术出版社. Dong Y M, Zhuge Y, Huang M H. 1989. Fauna of Zhejiang: Mammal. Hangzhou: Zhejiang Science and Technology Press. (In Chinese)

段海生, 杨振琼, 刘亦仁. 2011. 中国食虫动物名录修定及分布. 华中师范大学学报(自然科学版), 45(3): 466-471. Duan H S, Yang Z Q, Liu Y R. 2011. List revision of insectivorous in China and its distribution. *Journal of Central China Normal University* (Natural Science), 45(3): 466-471. (In Chinese with English Abstract)

段艳芳, 谢朝晖, 胡建业, 等. 2012. 蜂猴、倭蜂猴的现状与保护策略. 生物学通报, 47(7): 4-7. Duan Y F, Xie Z H, Hu J Y, *et al*. 2012. Status and conservation strategies of loris and pygmy loris. *Bulletin of Biology*, 47(7): 4-7. (In Chinese with English Abstract)

鄂晋, 张福顺, 余奕东, 等. 2009. 荒漠区开垦干扰下子午沙鼠种群数量动态与繁殖特征. 内蒙古农业大学学报(自然科学版), 30(2): 140-144. E Jin, Zhang F S, Yu Y D, *et al*. 2009. Population dynamics and reproduction characteristic of mid-day gerbil under farmland disturbance in desert region. *Journal of Inner Mongolia Agricultural University* (Natural Science Edition), 30(2): 140-144. (In Chinese with English Abstract)

樊龙锁, 刘焕金. 1998. 历山自然保护区30种鸟类繁殖特性及成效的研究. 山西林业科技, (3): 27-31. Fan L S, Liu H J. 1998. Reproductive characteristics and effects of 30 species of birds in Lishan Nature Reserve. *Shanxi Forestry Science and Technology*, (3): 27-31. (In Chinese with English Abstract)

范朋飞. 2012. 中国长臂猿科动物的分类和保护现状. 兽类学报, 232(3): 248-258. Fan P F. 2012. Taxonomy and conservation status of gibbons in China. *Acta Theriologica Sinica*, 232(3): 248-258. (In Chinese with English Abstract)

范振鑫, 刘少英, 郭聪, 等. 2009. 林跳鼠亚科的系统学研究述评. 四川动物, 28(1): 157-159. Fan Z X, Liu S Y, Guo C, *et al*. 2009. A review on the phylogenetic study of the subfamily Zapodinae. *Sichuan Journal of Zoology*, 28(1): 157-159. (In Chinese with English Abstract)

范振鑫, 王璐萱, 张修月, 等. 2010. 基于线粒体细胞色素*b*基因对大耳姬鼠和龙姬鼠分类关系的探讨. 四川动物, 29(6): 878-881. Fan Z X, Wang L X, Zhang X Y, *et al*. 2010. The taxonomic relationship between *Apodemus latronum* and *Apodemus draco* based on the Cytochrome *b* gene. *Sichuan Journal of Zoology*, 29(6): 878-881. (In Chinese with English Abstract)

冯江, 李振新, 陈敏, 等. 2002. 同一山洞中五种蝙蝠的回声定位比较及生态位的分化. 生态学报, 22(2): 150-155. Feng J, Li Z X, Chen M, *et al*. 2002. The echolocation comparison and the differentiation of ecology niche of five species bats live in one cave. *Acta Ecologica Sinica*, 22(2): 150-155. (In Chinese with English Abstract)

冯江, 李振新, 陈敏, 等. 2003. 拖网式食鱼蝠——大足鼠耳蝠的形态、回声定位声波及捕食策略. 生态学报, 23(9): 1712-1718. Feng J, Li Z X, Chen M, *et al*. 2003. Morphological features, echolocation calls and foraging strategy in the trawling piscivorous bat: Rickett's big-footed bat *Myotis ricketti*. *Acta Ecologica Sinica*, 23(9): 1712-1718. (In Chinese with English Abstract)

冯利民, 王利繁, 王斌, 等. 2013. 西双版纳尚勇自然保护区野生印支虎及其三种主要有蹄类猎物种群现状调查. 兽类学报, 33(4): 308-318. Feng L M, Wang L F, Wang B, *et al*. 2013. Population status of the Indochinese tiger (*Panthera tigris cobetti*) and density of the three primary ungulate prey species in Shangyong Nature Reserve, Xishuangbanna, China. *Acta Theriologica Sinica*, 33(4): 308-318. (In Chinese with English Abstract)

冯利民, 王志胜, 林柳, 等. 2010. 云南南滚河国家级自然保护区亚洲象种群旱季生境选择及保护策略. 兽类学报, 30(1): 1-10. Feng L M, Wang Z S, Lin L, *et al*. 2010. Habitat selection in dry season of Asian elephant (*Elephas maximus*) and conservation strategies in Nangunhe National Nature Reserve, Yunnan, China. *Acta Theriologica Sinica*, 30(1): 1-10. (In Chinese with English Abstract)

冯庆, 蒋学龙, 李松, 等. 2006. 中国翼手类一属、种新纪录. 动物分类学报, 31(1): 224-230. Feng Q, Jiang X L, Li S, *et al*. 2006. A new record genus *Megaerops* and its two species of bat in China (Chiroptera, Pteropodidae). *Acta Zootaxonomica Sinica*, 31(1): 224-230. (In Chinese with English Abstract)

冯庆, 蒋学龙, 王应祥. 2008. 亚洲南部球果蝠*Sphaerias blanfordi* (Thomas, 1891)的亚种分化. 兽类学报, 28(4): 367-374. Feng Q, Jiang X L, Wang Y X. 2008. Subspecies differentiation for Blanford's fruit bat, *Sphaerias blanfordi* (Pteropodidae, Chiroptera) in southern Asia. *Acta Theriologica Sinica*, 28(4): 367-374. (In Chinese with English Abstract)

冯庆, 王应祥, 林苏. 2007. 中国安氏长舌果蝠的分类记述. 动物学研究, 28(6): 647-653. Feng Q, Wang Y X, Lin S. 2007. Notes of Greater long-tongued fruit bat *Macroglossus sobrinus* in China. *Zoological Research*, 28(6): 647-653. (In Chinese with English Abstract)

冯志勇, 黄秀清, 陈美梨, 等. 1990. 黄毛鼠种群时空动态和近年来鼠害上升的原因的研究. 生态科学, (1): 78-83. Feng Z Y, Huang X Q, Chen M L, *et al*. 1990. Spatio-temporal dynamics of the population of the yellow rats *Rattus losea* and the causes of the increase of rodent damage in recent years. *Ecological Science*, (1): 78-83. (In Chinese with English Abstract)

冯祚建, 蔡桂全, 郑昌琳. 1984. 西藏哺乳类名录. 兽类学报, 4: 341-358. Feng Z J, Cai G Q, Zheng C L. 1984. A checklist of the mammals of Xizang (Tibet). *Acta Theriologica Sinica*, 4: 341-358. (In Chinese with English Abstract)

冯祚建, 蔡桂全, 郑昌林. 1986. 西藏哺乳类. 北京: 科学出版社. Feng Z J, Cai G Q, Zheng C L. 1986. Mammals of Tibet. Beijing: Science Press. (In Chinese)

冯祚建, 郑昌琳. 1985. 中国鼠兔属(*Ochotona*)的研究——分类与分布. 兽类学报, 5: 269-289. Feng Z J, Zheng C L. 1985. Studies on the Pikas (genus *Ochotona*)—Taxonomic notes and distribution. *Acta Theoriologica Sinica*, 5: 269-289. (In Chinese with English Abstract)

符丹凤, 张佑祥, 蒋洵, 等. 2010. 西南鼠耳蝠湖南分布新纪录. 吉首大学学报(自然科学版), 31(3): 106-108. Fu D F, Zhang Y X, Jiang X, *et al*. 2010. Distribution record of *Myotis altarium* Thomas in Hunan Province of China. *Journal of Jishou University* (*Natural Sciences Edition*), 31(3): 106-108. (In Chinese with English Abstract)

符建荣, 雷开明, 孙治宇, 等. 2012. 西藏小型兽类二新纪录. 四川动物, 31(1): 123-124. Fu J R, Lei K M, Sun Z Y, *et al*. 2012. New records of two small mammals in Tibet, China. *Sichuan Journal of Zoology*, 31(1): 123-124. (In Chinese with English Abstract)

付必谦, 陈卫, 王磊, 等. 2008. 北京地区中华姬鼠与大林姬鼠的数量分类研究. 生物数学学报, 23(4): 668-676. Fu B Q, Chen W, Wang L, *et al*. 2008. Quantitative classification of Mice *A. draco* and *A. peninsulae* from Beijing. *Journal of Biomathematics*, 23(4): 668-676. (In Chinese with English Abstract)

付和平, 郭志成, 董清, 等. 2003. 内蒙古阿拉善右旗、额济纳旗啮齿动物区系. 草原与草业, (3): 5-7. Fu H P, Guo Z C, Dong Q, *et al*. 2003. Rodent fauna of Alashan Right Banner and Ejin Banner in Inner Mongolia. *Grassland and Grass Industry*, (3): 5-7. (In Chinese with English Abstract)

付和平, 武晓东, 杨泽龙. 2005. 阿拉善地区不同生境小型兽类群落多样性研究. 兽类学报, 25(1): 32-38. Fu H P, Wu X D, Yang Z L. 2005. Diversity of small mammals communities at different habitats in Alashan region, Inner Mongolia. *Acta Theriologica Sinica*, 25(1): 32-38. (In Chinese with English Abstract)

甘宏协, 胡华斌. 2008. 基于野牛生境选择的生物多样性保护廊道设计: 来自西双版纳的案例. 生态学杂志, 27(12): 2153-2158. Gan H X, Hu H B. 2008. Biodiversity conservation corridor design based on habitat selection of Gaur (*Bos gaurus*): a case study from Xishuangbanna, China. *Chinese Journal of Ecology*, 27(12): 2153-2158. (In Chinese with English Abstract)

高安利. 1991. 江豚(*Neophocaena phocaenoides*)不同种群的形态差异和遗传变异的研究. 南京: 南京师范大学博士研究生学位论文. Gao A L. 1991. Morphological and Geographical Variation of Different Populations of Finless Porpoise. Nanjing: Doctor's Thesis of Nanjing Normal University. (In Chinese with English Abstract)

高行宜, 姚军. 2006. 新疆哈密盆地初冬鹅喉羚的地理分布与种群数量. 干旱区地理, 29(2): 213-218. Gao X Y, Yao J. 2006. Study on the geography distribution and population of *Gazella subgutturosa* in the Hami Basin, Xinjiang in early winter. *Arid Land Geography*, 29(2): 213-218. (In Chinese with English Abstract)

高晶. 2006. 吉林和云南马铁菊头蝠线粒体DNA部分序列变异及其系统发育关系. 长春: 东北师范大学硕士研究生学位论文. Gao J. 2006. Sequence Variation and Phylogenetic Analysis of Greater Horseshoe Bat *Rhinolophus ferrumequinum* in Yunnan and Jilin Provinces Based on Partial Mitochondrial DNA. Changchun: Master's Thesis of Northeast Normal University. (In Chinese with English Abstract)

高倩, 王存富, 薛慧良, 等. 2008. 三个不同地理种群大仓鼠的遗传多样性. 曲阜师范大学学报(自然科学版), 34(2): 93-97. Gao Q, Wang C F, Xue H L, *et al*. 2008. Genetic diversity of cricetulus triton de Winton for three different populations. *Journal of Qufu Normal University* (Natural Science Edition), 34(2): 93-97. (In Chinese with English Abstract)

高武, 陈卫, 傅必谦. 1996. 北京地区翼手类的区系及其分布. 河北大学学报(自然科学版), 16(5): 49-52. Gao W, Chen W, Fu B Q. 1996. The fauna and distribution of Chiropteran in Beijing Area. *Journal of Hebei University* (Natural Science Edition), 16(5): 49-52. (In Chinese with English Abstract)

高耀亭, 冯祚建. 1964. 中国灰尾兔亚种的研究. 动物分类学报, 1(1): 21-32. Gao Y T, Feng Z J. 1964. Subspecies of *Lepus oiostolus* in China. *Acta Zootaxonomica Sinica*, 1(1): 21-32. (In Chinese with English Abstract)

高耀亭, 陆长坤, 张洁, 等. 1962. 云南西双版纳兽类调查报告. 动物学报, 14(7): 180-196. Gao Y T, Lu C K, Zhang J, *et al*. 1962. Mammal survey report of Xishuangbanna, Yunnan. *Acta Zoologica Sinica*, 14(7): 180-196. (In Chinese with English Abstract)

高耀亭, 汪松, 王申裕, 等. 1987. 中国动物志. 兽纲. 第8卷. 食肉目. 北京: 科学出版社. Gao Y T, Wang S, Wang S Y, *et al*. 1987. Fauna Sinica. Mammalia. Vol. 8 Carnivora. Beijing: Science Press. (In Chinese)

高志英. 2008. 吉林省蝙蝠科三种蝙蝠的核型研究. 长春: 吉林农业大学硕士研究生学位论文. Gao Z Y. 2008. Study on Karyotypes of three Species of Vespertilionidae Bats from Jilin. Changchun: Master's Thesis of Jilin Agricultural University. (In Chinese with English Abstract)

高中信. 2006. 中国狼研究进展. 动物学杂志, 41(1): 134-136. Gao Z X. 2006. Review of the research on wolf in China. *Chinese Journal of Zoology*, 41(1): 134-136. (In Chinese with English Abstract)

葛小芳, 孟凡露, 王朋, 等. 2015. 大兴安岭驯鹿(*Rangifer tarandus*)的春季生境选择. 生态学报, 35(15): 5000-5008. Ge X F, Meng F L, Wang P, *et al*. 2015. The spring habitat selection of reindeer (*Rangifer tarandus*) in Great Xing'anling of China. *Acta Ecologica Sinica*, 35(15): 5000-5008. (In Chinese with English Abstract)

龚立新, 顾浩, 孙淙南, 等. 2018. 贵州发现毛翼管鼻蝠和华南菊头蝠及其回声定位声波特征. 动物学杂志, 53(3): 329-338. Gong L X, Gu H, Sun C N, *et al*. 2018. Two New records of the Chiroptera in Guizhou Province (China) and their echolocation calls. *Chinese Journal of Zoology*, 53(3): 329-338. (In Chinese with English Abstract)

龚晓俊, 陈贵春, 刘昭兵, 等. 2013. 贵州省啮齿动物分布及名录. 医学动物防制, 29(1): 1-9. Gong X J, Chen G C, Liu Z B, *et al*. 2013. Rodent distribution and list in Guizhou Province. *Control of Medical Animals*, 29(1): 1-9. (In Chinese with English Abstract)

龚正达, 王应祥, 李章鸿, 等. 2000. 中国鼠兔一新种——片马黑鼠兔. 动物学研究, 21(3): 204-209. Gong Z D, Wang Y X, Li Z H, *et al*. 2000. A new species of pika: Pianma Blacked Pika, *Ochotona nigritia* (Lagomorpha: Ochotonidae) fom Yunnan, China. *Zoological Research*, 21(3): 204-209. (In Chinese with English Abstract)

龚正达, 吴厚永, 段兴德, 等. 2001a. 云南横断山区小型兽类物种多样性与地理分布趋势. 生物多样性, 9(1): 73-79. Gong Z D, Wu H Y, Duan X D, *et al*. 2001a. The species diversity and distribution trends of small mammals in Hengduan Mountains, Yunnan. *Biodiversity Science*, 9(1): 73-79. (In Chinese with English Abstract)

龚正达, 吴厚永, 段兴德, 等. 2001b. 云南玉龙雪山自然保护区小型兽类群落系统聚类分析与区系研究. 疾病预防控制通报, (1): 67-73. Gong Z D, Wu H Y, Duan X D, *et al*. 2001b. Cluster analysis and floristic study of small animal community in Yulong Snow Mountain Nature Reserve, Yunnan. *Chinese Journal of Disease Control and Prevention*, (1): 67-73. (In Chinese with English Abstract)

谷登芝, 周立志, 马勇, 等. 2011. 中国柽柳沙鼠线粒体DNA的地理变异及其亚种分化. 兽类学报, 31(4): 347-357. Gu D Z, Zhou L Z, Ma Y, *et al*. 2011. Geographic variation in mitochondrial DNA sequences and subspecies divergence of the Tamarisk Gerbil (*Meriones tamariscinus*) in China. *Acta Theriologica Sinica*, 31(4): 347-357. (In Chinese with English Abstract)

谷晓明, 路静, 韩建领, 等. 2003a. 蝙蝠科七种蝙蝠的核型. 兽类学报, 23(2): 127-132. Gu X M, Lu J, Han J L, *et al*. 2003a. Karyotypes of Seven Species of Vespertilionidae Bats. *Acta Theriologica Sinica*, 23(2): 127-132. (In Chinese with English Abstract)

谷晓明, 涂云彦, 蒋大池, 等. 2003b. 贵州五种菊头蝠的核型分析. 动物学杂志, 38(1): 18-22. Gu X M, Tu Y Y, Jiang D C, *et al*. 2003b. Karyotype analysis of five *Rhinolophus* species from Guizhou. *Chinese Journal of Zoology*, 38(1): 18-22. (In Chinese with English Abstract)

广东省昆虫研究所动物室, 中山大学生物系. 1983. 海南岛的鸟兽. 北京: 科学出版社: 1-426. Animal Laboratory of Guangdong Institute of Entomology; Department of Biology, Sun Yat-Sen University. 1983. Mammals and Birds on the Hainan Island. Beijing: Science Press: 1-426. (In Chinese)

郭郛, 钱燕文, 马建章. 2004. 中国动物学发展史. 哈尔滨: 东北林业大学出版社. Guo F, Qian Y W, Ma J Z. 2004. The Developing History of China's Zoology. Harbin: Northeast China Forestry University Press. (In Chinese).

郭建荣. 2003. 山西芦芽山自然保护区岩松鼠生态的初步观察. 四川动物, 22(3): 171-172. Guo J R. 2003. Ecology of *Sciurotamias davidianus* in Luya Mountain Nature Reserve, Shanxi Province. *Sichuan Journal of Zoology*, 22(3): 171-172. (In Chinese with English Abstract)

郭世芳, 梁栓柱. 1997. 内蒙古哺乳动物的种类和分布. 内蒙古林业调查设计, (2): 64-69. Guo S F, Liang S Z. 1997. Species and distribution of mammals in Inner Mongolia. *Inner Mongolia Forestry Survey and Design*, (2): 64-69. (In Chinese with English Abstract)

郭宪, 阎萍, 程胜利, 等. 2007. 中国野牦牛遗传资源的保存与利用. 家畜生态学报, 28(5): 96-98. Guo X, Yan P, Cheng S L, *et al*. 2007. Conservation and utilization of genetic Resources of wild yak in China. *Acta Ecologiae Animalis Domastici*, 28(5): 96-98. (In Chinese with English Abstract)

郭延蜀. 2000. 四川梅花鹿的分布、数量及栖息环境的调查. 兽类学报, 20(2): 81-87. Guo Y S. 2000. Distribution, numbers and habitat of Sichuan sika deer (*Cervus nippon sichuanicus*). *Acta Theriologica Sinica*, 20(2): 81-87. (In Chinese with English Abstract)

国家林业局. 2003. 国家重点保护野生动物名录. National Forestry Administration. 2003. National Key Protected Wild Animal Species Checklist. http://www.forestry.gov.cn/ [2018-5-1]. (In Chinese)

国家林业局. 2009. 中国重点陆生野生动物资源调查. 北京: 中国林业出版社. National Forestry Administration. 2009.

Investigation of Key Terrestrial Wild Animal Resource of China. Beijing: China Forestry Press. (In Chinese)

国家林业局. 2015. 中国大熊猫现状. National Forestry Administration. 2015. Status of China's Giant Pandas. http://www.forestry.gov.cn [2018-5-1]. (In Chinese)

韩宝银, 贺红早. 2012. 贵州飞龙洞南蝠的种群研究. 安徽农业科学, 40(25): 558-648. Han B Y, He H Z. 2012. Study of the great evening bats (*Ia io*) populations in Feilong Cave of Guizhou Province. *Journal of Anhui Agricultural Sciences*, 40(25): 558-648. (In Chinese with English Abstract)

郝海邦, 刘少英, 张修月, 等. 2011. 凉山田鼠分子系统进化. 见: 四川省动物学会. 四川省动物学会第九次会员代表大会暨第十届学术研讨会论文集: 66. Hao H B, Liu S Y, Zhang X Y, *et al*. 2011. Molecular phylogeny of *Proedromys liangshanensis*. *In*: Sichuan Animal Society. Proceedings of the 9th Member Congress and 10th Symposium of Sichuan Zoological Society: 66. (In Chinese)

郝玉江, 王克雄, 韩家波, 等. 2011. 中国海兽研究概述. 兽类学报, 31(1): 20-36. Hao Y J, Wang K X, Han J B, *et al*. 2011. Marine mammal researches in China. *Acta Theriologica Sinica*, 31(1): 20-36. (In Chinese with English Abstract)

何锴, 邓可, 蒋学龙. 2012. 中国兽类鼩鼱科一新纪录——高氏缺齿鼩. 动物学研究, 33(5): 542-544. He K, Deng K, Jiang X L. 2012. First record of Van sung's shrew (*Chodsigoa caovansunga*) in China. *Zoological Research*, 33(5): 542-544. (In Chinese with English Abstract)

何向阳, 彭兴文, 梁捷, 等. 2019. 广东丹霞山国家级自然保护区蝙蝠物种多样性调查. 动物学杂志, 54(6): 810-814. He X Y, Peng X W, Liang J, *et al*. 2019. Species diversity surveys of bat species in Danxia Shan National Nature Reserve, Guangdong. *Chinese Journal of Zoology*, 54(6): 810-814. (In Chinese with English Abstract)

何晓瑞, 杨德华. 1982. 我国菲氏叶猴生物学的初步研究. 动物学研究, 3(增刊): 349-354. He X R, Yang D H. 1982. Peliminary studies on the biology of *Presbytis phayrei*. *Zoological Research*, 3(Supplementary): 349-354. (In Chinese with English Abstract)

何晓瑞. 1999. 无尾蹄蝠在云南再次发现. 四川动物, 18(4): 183. He X R. 1999. *Coelops frithi* rediscovered in Yunnan. *Sichuan Journal of Zoology*, 18(4): 183. (In Chinese with English Abstract)

何晓瑞, 杨向东, 李涛. 1991. 中国小竹鼠生态的初步研究. 动物学研究, 12(1): 41-48. He X R, Yang X D, Li T. 1991. A preliminary studies on the ecology of the lesser bamboo rat (*Cannomys badius*) in China. *Zoological Research*, 12(1): 41-48. (In Chinese with English Abstract)

何新焕. 2011. 河南周边省份马铁菊头蝠的种下分类研究. 新乡: 河南师范大学硕士研究生学位论文. He X H. 2011. Subspecies Classification of Horseshoe Bats *Rhinolophus ferrumequinum* in Surrounding Provinces of Henan. Xinxiang: Master's Thesis of Henan Normal University. (In Chinese with English Abstract)

何娅, 周材权, 刘国库, 等. 2012. 斯氏鼢鼠物种地位有效性的探讨. 兽类学报, 37(1): 36-43. He Y, Zhou C Q, Liu G K, *et al*. 2012. Research on the validity of *Eospalax smithii* inferred from molecular and morphological evdience. *Acta Zootaxonomica Sinica*, 37(1): 36-43. (In Chinese with English Abstract)

何娅, 周材权, 潘立. 2009. 四川兽类新纪录——秦岭鼢鼠线粒体cyt *b*基因的克隆及其比较研究. 西华师范大学学报(自然科学版), 30(1): 8-12. He Y, Zhou C Q, Pan L. 2009. New mammalian record in Sichuan Province——*Eospalax Rufesens* mitochondrion Cyt *b* gene's clone and comparative analysis. *Journal of China West Normal University* (Natural Science Edition), 30(1): 8-12. (In Chinese with English Abstract)

何业恒. 1993. 中国珍稀兽类的历史变迁. 长沙: 湖南科学技术出版社. He Y H. 1993. Historical Changes of Rare Animals in China. Changsha: Hunan Science and Technology Press. (In Chinese)

何志超, 毕俊怀, 陈绍勇, 等. 2015. 基于红外相机对额仁淖尔苏木盘羊(*Ovis ammon*)生存现状的研究. 野生动物学报, 36(1): 5-10. He Z C, Bi J H, Chen S Y, *et al*. 2015. Research on argali *Ovis ammon* from the E'ren Nao'er Region of Inner Mongolia using infrared cameras. *Chinese Journal of Wildlife*, 36(1): 5-10. (In Chinese with English Abstract)

贺新平, 卜艳珍, 周会先, 等. 2014. 云南省保山市发现莱氏蹄蝠*Hipposideros lylei*. 四川动物, 33(6): 865-873. He X P, Bu Y Z, Zhou H X, *et al*. 2014. *Hipposideros lylei* found in Baoshan City, Yunnan Province. *Sichuan Journal of Zoology*, 33(6): 865-873. (In Chinese with English Abstract)

洪体玉, 周善义, 叶建平, 等. 2009. 广西发现局部白化中蹄蝠幼仔一例. 动物学杂志, 44(2): 138-140. Hong T Y, Zhou S Y, Ye J P, *et al*. 2009. A partial albino pup of *Hipposideros larvatus* found in Guangxi Province. *Chinese Journal of Zoology*, 44(2): 138-140. (In Chinese with English Abstract)

侯兰新. 1995. 中国的跳鼠家族浅谈. 生物学通报, 30: 14. Hou L X. 1995. A brief discussion on the Chinese jerboa. *Biology Bulletin*, 30: 14. (In Chinese with English Abstract)

侯兰新. 2000. 新疆伊犁地区兽类调查报告. 兽类学报, 20(3): 233-238. Hou L X. 2000. Investigation report on mammals in Yili, Xinjiang. *Acta Zoologica Sinica*, 20(3): 233-238. (In Chinese with English Abstract)

侯兰新, 欧阳霞辉. 2010. 心颅跳鼠亚科(Cardiocraniinae)在中国的分布和分类. 西北民族大学学报(自然科学版), 31(3): 64-67. Hou L X, Ouyang X H. 2010. Distribution and taxonomy of subfamily cardiocraniinae in China. *Journal of Northwest University for Nationalities* (Natural Science Edition), 31: 64-67. (In Chinese with English Abstract)

侯立冰, 丁晶晶, 丁玉华, 等. 2012. 江苏大丰麋鹿种群及管理模式探讨. 野生动物学报, 33(5): 254-257. Hou L B, Ding J J, Ding Y H, *et al*. 2012. Manegement of Père David's deer at Dafeng Milu National Nature Reserve. *Chinese Journal of Wildlife*, 33(5): 254-257. (In Chinese with English Abstract)

侯希贤, 董维惠, 杨玉平. 2003. 鄂尔多斯沙地草场小毛足鼠种群数量动态分析. 中国媒介生物学及控制杂志, 14(3): 177-180. Hou X X, Dong W H, Yang Y P. 2003. Population dynamics analysis of *Phodopus roborovskii* in sandy land of Ordos. *Chinese Journal of Vector Biology and Control*, 14(3): 177-180. (In Chinese with English Abstract)

侯希贤, 董维惠, 周延林, 等. 2000. 鄂尔多斯沙地草场小毛足鼠种群数量动态及预测. 中国媒介生物学及控制杂志, 11(1): 7-10. Hou X X, Dong W H, Zhou Y L. 2000. Analysis on the population dynamics of desert hamster in Ordos Sandland. *Chinese Journal of Vector Biology and Control*, 11(1): 7-10. (In Chinese with English Abstract)

胡刚. 1998. 山蝠、绒山蝠和爪哇伏翼乳酸脱氢酶同工酶的比较研究. 四川师范学院学报(自然科学版), 19: 3-9. Hu G. 1998. A comparative study of tissue lactate dehydrogenase isoenzyme among noctule, villus noctule and Javan pipistrelle. *Journal of Sichuan Teachers College* (Natural Science Edition), 19: 3-9. (In Chinese with English Abstract)

胡刚, 董鑫, 罗洪章, 等. 2011. 过去二十年贵州黑叶猴分布与种群动态及致危因子分析. 兽类学报, 31(3): 306-311. Hu G, Dong X, Luo H Z, *et al*. 2011. The distribution and population dynamics of Francois' langur over the past two decades in Guizhou, China and threats to its survival. *Acta Theriologica Sinica*, 31(3): 306-311. (In Chinese with English Abstract)

胡刚, 杜勇. 2002. 云南省小熊猫(*Ailurus fulgens*)资源分布及保护现状. 西北林学院学报, 17(3): 67-71. Hu G, Du Y. 2002. The current distribution, population and conservation status of *Ailurus fulgens* in Yunnan. *Journal of Northwest Forestry College*, 17(3): 67-71. (In Chinese with English Abstract)

胡锦矗. 2001. 大熊猫研究. 上海: 上海科技教育出版社. Hu J C. 2001. Giant Panda Research. Shanghai: Shanghai Science and Technology Education Press. (In Chinese)

胡锦矗, 胡杰. 2007. 四川兽类名录新订. 西华师范大学学报(自然科学版), 28(3): 165-171. Hu J C, Hu J. 2007. Newly edited catalogue of Sichuan mammals. *Journal of China West Normal University* (Natural Science Edition), 28(3): 165-171. (In Chinese with English Abstract)

胡锦矗, 吴毅. 1993. 四川伏翼属3种蝙蝠新记录. 四川师范学院学报(自然科学版), 14(3): 236-238. Hu J C, Wu Y. 1993. New records of bat's three species of pipistrelle in Sichuan Province. *Journa of China West Normal University* (Natural Sciences), 14(3): 236-238. (In Chinese with English Abstract)

胡开良, 杨剑, 谭梁静, 等. 2012. 同地共栖三种鼠耳蝠食性差异及其生态位分化. 动物学研究, 33: 177-181. Hu K L, Yang J, Tan L J, *et al*. 2012. Dietary differences and niche partitioning in three sympatric *Myotis* species. *Zoological Reseach*, 33: 177-181. (In Chinese with English Abstract)

胡秋波, 吴太平, 蒋洪. 2014. 黄胸鼠生态及防治研究进展. 中华卫生杀虫药械, 20(2): 180-184. Hu Q B, Wu T P, Jiang H. 2014. A review of ecology and control of *Rattus tanezumi*. *Chinese Journal of Hygienic Insecticides & Equipments*, 20(2): 180-184. (In Chinese with English Abstract)

胡诗佳, 彭建军, 于冬梅, 等. 2010. 中华穿山甲的研究及保护现状. 四川动物, 29(4): 673-675. Hu S J, Peng J J, Yu D M, *et al*. 2010. Research and conservation status in Chinese pangolin (*Manis pentadactyla*). *Sichuan Journal of Zoology*, 29(4): 673-675. (In Chinese with English Abstract)

胡一鸣, 李玮琪, 蒋志刚, 等. 2018. 羌塘、可可西里无人区野牦牛种群数量和分布现状. 生物多样性, 26(2): 185-190. Hu Y M, Li W Q, Jiang Z G, *et al*. 2018. A wild yak survey in Chang Tang of Tibet Autonomous Region and Hoh Xili of Qinghai Province. *Biodiversity Science*, 26(2): 185-190. (In Chinese with English Abstract)

胡一鸣, 姚志军, 黄志文, 等. 2014. 西藏珠穆朗玛峰国家级自然保护区哺乳动物区系及其垂直变化. 兽类学报, 34(1): 28-37. Hu Y M, Yao Z J, Huang Z W, *et al*. 2014. Mammalian fauna and its vertical changes in Mt. Qomolangma National Nature Reserve, Tibet, China. *Acta Theriologica Sinica*, 34(1): 28-37. (In Chinese with English Abstract)

胡宜峰, 黎舫, 吴毅, 等. 2018. 海南省蝙蝠新记录——毛翼管鼻蝠. 浙江林业科技, 38(3): 85-88. Hu Y F, Li F, Wu Y, *et al*. 2018. New Record of *Harpiocephalus harpia* in Hainan Province. *Journal Zhejiang Forestry Science Technology*, 38(3): 85-88. (In Chinese with English Abstract)

华朝朗, 杨东, 毕艳玲, 等. 2013. 云南省西黑冠长臂猿现状及保护对策. 林业调查规划, 38(4): 55-60. Hua C L, Yang D, Bi Y L, *et al.* 2013. Status and conservation of western black crested gibbon in Yunnan. *Forest Inventory and Planning*, 38(4): 55-60. (In Chinese with English Abstract)

黄乘明, 周岐海, 李友邦. 2018. 黑头叶猴的行为生态学与保护生物学. 上海: 上海科学技术出版社. Huang C M, Zhou Q H, Li Y B. 2018. Behavioral Ecology and Conservation Biology of François' Langur. Shanghai: Shanghai Science and Technology Press. (In Chinese)

黄辉, 郭宪国, 朱琼蕊. 2013. 我国针毛鼠的研究进展. 医学动物防制, 29(10): 1086-1090. Huang H, Guo X G, Zhu Q R. 2013. Research progress of *Rattus fulvescens* in China. *Control of Medical Animals*, 29(10): 1086-1090. (In Chinese with English Abstract)

黄继荣, 王炎, 李联涛. 2006. 长爪沙鼠生物学特性调查研究. 宁夏农林科技, (6): 36-37. Huang J R, Wang Y, Li L T. 2006. Investigation and study on biological characteristics of *Meriones unguicutatus*. *Ningxia Agroforestry Science and Technology*, (6): 36-37. (In Chinese with English Abstract)

黄继展, 谭梁静, 杨剑, 等. 2013. 澳门翼手类物种多样性调查. 兽类学报, 33(2): 123-132. Huang J Z, Tan L J, Yang J, *et al.* 2013. A recent survey of bat diversity (Mammalia: Chiroptera) in Macau. *Acta Theriologica Sinica*, 33(2): 123-132. (In Chinese with English Abstract)

黄乃伟, 王卓聪, 罗玉梅, 等. 2012. 人类经济活动对长白山自然保护区动物多样性的影响. 北华大学学报(自然科学版), 13(4): 444-450. Huang N W, Wang Z C, Luo Y M, *et al.* 2012. Impact of human economic activities on animal diversity in Changbai Mountain Nature Reserve. *Journal of Beihua University* (Natural Science), 13(4): 444-450. (In Chinese with English Abstract)

黄薇, 夏霖, 冯祚建, 等. 2007. 新疆兽类分布格局及动物地理区划探讨. 兽类学报, 27(4): 325-337. Huang W, Xia L, Feng Z J, *et al.* 2007. Distribution pattern and zoogeographical discussion on mammals in Xinjiang. *Acta Theriologica Sinica*, 27(4): 325-337. (In Chinese with English Abstract)

黄薇, 夏霖, 杨奇森, 等. 2008. 青藏高原兽类分布格局及动物地理区划. 兽类学报, 28(4): 375-394. Huang W, Xia L, Yang Q S, *et al.* 2008. Distribution pattern and zoogeographical division on mammals on the Qinghai-Tibet Plateau. *Acta Theriologica Sinica*, 28(4): 375-394. (In Chinese with English Abstract)

黄翔, 周立志. 2012. 蒙新区子午沙鼠种群的遗传多样性和遗传结构. 兽类学报, 32(3): 179-187. Huang X, Zhou L Z. 2012. Genetic diversity and genetic structure of the mid-day gerbil population in Inner Mongolia-Xinjiang Plateau. *Acta Theriologica Sinica*, 32(3): 179-187. (In Chinese with English Abstract)

黄英, 武晓东. 2004. 内蒙古五趾跳鼠种下数量分类初步研究. 内蒙古农业大学学报(自然科学版), 25(1): 46-52. Huang Y, Wu X D. 2004. Primary study on numerical classification of infer-species of *Allactaga sibirica* in Inner Mongolia. *Neimenggu Nongye Daxue Xuebao* (Natural Science Edition), 25(1): 46-52. (In Chinese with English Abstract)

黄正澜懿, 胡宜峰, 吴华, 等. 2018. 中管鼻蝠在湖北和浙江的分布新纪录. 西部林业科学, (6): 73-77. Huang Z L Y, Hu Y F, Wu H, *et al.* 2018. New Distribution record of *Murina huttoni* in Hubei and Zhejiang Provinces. *Journal of West China Forestry Science*, (6): 73-77. (In Chinese with English Abstract)

黄祥麟. 2012. 沿岸种类鲸豚的系统族群动态研究. 台北: 台湾科学委员会. Huang X L. 2012. Systematic population dynamics of coastal species of whales and dolphins. Taipei: Science Council of Taiwan. (In Chinese)

霍晟, 杨君兴, 向左甫, 等. 2003. 中国灵猫科的支序系统学分析. 动物学研究, 24(6): 413-420. Huo S, Yang J X, Xiang Z F, *et al.* 2003. Cladistic analysis of the family Viverridae (Carnivora) from China. *Zoological Research*, 24(6): 413-420. (In Chinese with English Abstract)

吉晟男, 武晓东, 余奕东, 等. 2009. 荒漠区不同干扰下三趾跳鼠种群数量动态. 内蒙古农业大学学报 (自然科学版), 30(2): 145-150. Ji S N, Wu X D, Yu Y D, *et al.* 2009. Population dynamics of northern three-toed jerboa under different disturbance in desert region. *Journal of Inner Mongolia Agricultural University* (Natural Science Edition), 30(2): 145-150. (In Chinese with English Abstract)

江广华, 肖春芳, 祝友春, 等. 2013. 湖北省啮齿类新纪录——青毛硕鼠. 四川动物, 32(2): 267-268. Jiang G H, Xiao C F, Zhu Y C, *et al.* 2013. A new rodent record in Hubei Province, China: *Berylmys bowersi*. *Sichuan Journal of Zoology*, 32(2): 267-268. (In Chinese with English Abstract)

江廷磊, 冯江. 2011. 中国特有种大卫鼠耳蝠回声定位声波的地理变化: 一个社群适应的案例. 金华: 第七届全国野生动物生态与资源保护学术研讨会. Jiang T L, Feng J. 2011. Geographic changes of echolocation of *Myotis davidii*: a case study of community adaptation. Jinhua: 7th National Symposium on Wildlife Ecology and Resource Conservation. (In Chinese)

江廷磊, 冯江, 朱旭, 等. 2008a. 贵州省发现大足鼠耳蝠分布. 东北师大学报(自然科学版), 40(3): 103-106. Jiang T L, Feng J, Zhu X, *et al*. 2008a. A new record of Rickett's big-footed bat *Myotis ricketti* in Guizhou province. *Journal of Northeast Normal University* (Natural Science Edition), 40(3): 103-106. (In Chinese with English Abstract)

江廷磊, 刘颖, 冯江. 2008b. 中国翼手类一新纪录种. 动物分类学报, 33(1): 212-216. Jiang T L, Liu Y, Feng J. 2008b. A new Chinese record species. *Acta Zootaxonomica Sinica*, 33(1): 212-216. (In Chinese with English Abstract)

姜雪松, 李艳红, 胡杰. 2013. 四川勿角自然保护区的兽类组成与区系. 西华师范大学学报(自然科学版), 34(1): 5-10. Jiang X S, Li Y H, Hu J. 2013. The mammalian fauna and composition in Sichuan Wujiao Nature Reserve. *Journal of China West Normal University* (Natural Science Edition), 34(1): 5-10. (In Chinese with English Abstract)

蒋光藻, 曾录书, 倪健英, 等. 1999. 大足鼠的生物学特性及分布. 西南农业学报, 12(4): 82-85. Jiang G Z, Zeng L S, Ni J Y, *et al*. 1999. Study on the biology and distribution of *Rattus nitidus*. *Southwest China Journal of Agricultural Sciences*, 12(4): 82-85. (In Chinese with English Abstract)

蒋学龙, 李权, 陈中正, 等. 2017. 云南哺乳动物名录. 见: 李德铢. 云南省生物多样性名录. 昆明: 云南人民出版社: 581-588. Jiang X L, Li Q, Chen Z Z, *et al*. 2017. Checklist of mammals in Yunnan Province. *In*: Li D Z. Checklist of Biodiversity in Yunnan Province. Kunming: Yunnan People's Publishing House: 581-588. (In Chinese)

蒋学龙, 王应祥, 马世来. 1993. 中国熊猴的分类整理. 动物学研究, 14(2): 110-117. Jiang X L, Wang Y X, Ma S L. 1993. Taxonomic revision of *Macaca assamensis*. *Zoological Research*, 14(2): 110-117. (In Chinese with English Abstract)

蒋学龙, 王应祥, 王岐山. 1996. 藏酋猴的分类与分布. 动物学研究, 17(4): 361-369. Jiang X L, Wang Y X, Wang Q S. 1996. Taxonomy and distribution of Tibetan macaque (*Macaca thibetana*). *Zoological Research*, 17(4): 361-369. (In Chinese with English Abstract)

蒋志刚. 2004. 普氏野马(*Equus przewalskii*). 动物学杂志, 39(2): 100-101. Jiang Z G. 2004. Wild Horse (*Equus przewalskii*). *Chinese Journal of Zoology*, 39(2): 100-101. (In Chinese with English Abstract)

蒋志刚. 2009. 江西桃红岭梅花鹿国家级自然保护区生物多样性研究. 北京: 清华大学出版社. Jiang Z G. 2009. Biodiversity Study of Sika Deer National Nature Reserve in Taohong Mountain, Jiangxi. Beijing: Tsinghua University Press. (In Chinese)

蒋志刚. 2014. 天际线扫描: 环境与生物多样性保护研究的新方法. 生物多样性, 22(2): 115-116. Jiang Z G. 2014. Horizon Scanning: a new method for environmental and biodiversity conservation. *Biodiversity Science*, 22(2): 115-116. (In Chinese with English Abstract)

蒋志刚. 2016a. 中国脊椎动物生存现状研究. 生物多样性, 24(5): 495-499. Jiang Z G. 2016a. Assessing the surviving status of vertebrates in China. *Biodiversity Science*, 24(5): 495-499. (In Chinese with English Abstract)

蒋志刚. 2016b. 地球上有多少物种? 科学通报, 61(21): 2337-2343. Jiang Z G. 2016b. How many species are there on the Earth? *Chinese Science Bulletin*, 61(21): 2337-2343. (In Chinese with English Abstract)

蒋志刚, 江建平, 王跃招, 等. 2020. 国家濒危物种红色名录的生物多样性保护意义. 生物多样性, 28(5): 558-565. Jiang Z G, Jiang J P, Wang Y Z. 2020. Significance of the country red list of endangered species in conserving biodiversity. *Biodiversity Science*, 28(5): 558-565. (In Chinese with English Abstract)

蒋志刚, 樊恩源. 2003. 关于物种濒危等级标准之探讨——对IUCN物种濒危等级的思考. 生物多样性, 11(5): 383-392. Jiang Z G, Fan E Y. 2003. Exploring the endangered species criteria: rethinking the IUCN Red List Criteria. *Biodiversity Science*, 11(5): 383-392. (In Chinese with English Abstract)

蒋志刚, 雷润华, 刘丙万, 等. 2003. 普氏原羚研究概述. 动物学杂志, 38(6): 129-132. Jiang Z G, Lei R H, Liu B W, *et al*. 2003. A review on the researches of Przewalski's gazelle. *Chinese Journal of Zoology*, 38(6): 129-132. (In Chinese with English Abstract)

蒋志刚, 李立立, 胡一鸣, 等. 2018. 青藏高原有蹄类动物多样性和特有性: 演化与保护. 生物多样性, 26(2): 158-170. Jiang Z G, Li L L, Hu Y M, *et al*. 2018. Diversity and endemism of ungulates on the Qinghai-Tibetan Plateau: Evolution and conservation. *Biodiversity Science*, 26(2): 158-170. (In Chinese with English Abstract)

蒋志刚, 罗振华. 2012. 物种受威胁状况评估: 研究进展与中国的案例. 生物多样性, 20(5): 612-622. Jiang Z G, Luo Z H. 2012. Assessing species endangerment status: progress in research and an example from China. *Biodiversity Science*, 20(5): 612-622. (In Chinese with English Abstract)

蒋志刚, 马勇, 吴毅, 等. 2015a. 中国哺乳动物多样性. 生物多样性, 23(3): 351-364. Jiang Z G, Ma Y, Wu Y, *et al*. 2015a. China's mammalian diversity. *Biodiversity Science*, 23(3): 351-364. (In Chinese with English Abstract)

蒋志刚, 马勇, 吴毅, 等. 2015b. 中国哺乳动物多样性及地理分布. 北京: 科学出版社. Jiang Z G, Ma Y, Wu Y, *et al*. 2015b. China's Mammal Diversity and Geographic Distribution. Beijing: Science Press. (In both Chinese and English)

蒋志刚, 张林源, 杨戎生, 等. 2001. 中国麋鹿种群密度制约现象与发展策略. 动物学报, 47(1): 53-58. Jiang Z G, Zhang L Y, Yang R S, *et al*. 2001. Density dependent growth and population management strategy for Père David's deer in China. *Acta Zoologica Sinica*, 47(1): 53-58. (In Chinese with English Abstract)

蒋志刚, 刘少英, 吴毅, 等. 2017. 中国哺乳动物多样性(第2版). 生物多样性, 25(8): 886-895. Jiang Z G, Liu S Y, Wu Y, *et al*. 2017. China's mammal diversity (2nd edition). *Biodiversity Science*, 25(8): 886-895. (In Chinese with English Abstract)

蒋志刚, 马克平. 2014. 保护生物学原理. 北京: 科学出版社. Jiang Z G, Ma K P. 2014. Principles of Conservation Biology. Beijing: Science Press. (In Chinese)

金崑, 刘世荣, 顾志宏, 等. 2005. 我国川金丝猴的重要栖息地及自然保护区. 乌鲁木齐: 中国林学会2005 年学术年会. Jin K, Liu S R, Gu Z H, *et al*. 2005. Important habitat and nature reserve of golden snub-nosed monkey in China. Urumqi: Annual Meeting of China Association for Science and Technology, China Forestry Society. (In Chinese)

金崑, 马建章. 2004. 中国黄羊资源的分布、数量、致危因素及保护. 东北林业大学学报, 32(2): 104-106. Jin K, Ma J Z. 2004. Distribution and quantity threatening factors and protection of Mongolian gazelle. *Journal of Northeast Forestry University*, 32(2): 104-106. (In Chinese with English Abstract)

金一, 魏世宝, 苗婷婷, 等. 2007. 中华鼯鼠的分类与分布. 经济动物学报, 11(3): 175-178. Jin Y, Wei S B, Miao T T, *et al*. 2007. Classification and distribution of Chinese Flying Squirrel. *Journal of Economic Animal*, 11(3): 175-178. (In Chinese with English Abstract)

金子之史. 1992. シリーズ 日本の哺乳類 各論編, 日本の哺乳類 17 スミスネズミ. 哺乳類科学, 32(1): 39-54. Kaneko. 1992. Series on mammals of Japan, 17 Smith mice of mammals of Japan. *Mammalian Sciences*, 32(1): 39-54. (In Japanese with English Abstract)

鞠丹, 杨娇, 施路一. 2013. 大兴安岭猞猁冬季生境选择. 林业科技, 38(4): 56-58. Ju D, Yang J, Shi L Y. 2013. Habitat selection of *Felis lynx* during winter in the Daxing'anling. *Forestry Science & Technology*, 38(4): 56-58. (In Chinese with English Abstract)

孔令雪, 张虹, 任娟, 等. 2011. 繁殖期不同时段赤腹松鼠巢域的变化. 兽类学报, 31(3): 251-256. Kong L X, Zhang H, Ren J, *et al*. 2011. Variations in home range of *Callosciurus erythraeus* during different breeding periods. *Acta Theriologica Sinica*, 31(3): 251-256. (In Chinese with English Abstract)

乐佩琦, 陈宜瑜. 1998. 中国濒危动物红皮书 • 鱼类. 北京: 科学出版社. Yue P Q, Chen Y Y. 1998. Red Book of Endangered Animals in China: Fishes. Beijing: Science Press. (In Chinese)

雷俊宏. 1991. 棕熊亚种一新纪录. 八一农学院学报, (2): 10-12. Lei J H. 1991. A new record of subspecies of brown bear in China. *Journal of Bayi College of Agriculture*, (2): 10-12. (In Chinese with English Abstract)

雷伟, 李玉春. 2008. 水獭的研究与保护现状. 生物学杂志, 25(1): 47-50. Lei W, Li Y C. 2008. Study and conservation status of otters. *Journal of Biology*, 25(1): 47-50. (In Chinese with English Abstract)

黎舫, 王晓云, 余文华, 等. 2017. 罗蕾莱管鼻蝠在模式产地外的发现——云南分布新纪录. 动物学杂志, 52(5): 727-736. Li F, Wang X Y, Yu W H, *et al*. 2017. Discovery of *Murina lorelieae* Beyond Its Type Locality—a New *Murina* Record from Yunnan, China. *Chinese Journal of Zoology*, 52(5): 727-736. (In Chinese with English Abstract).

黎运喜, 张泽钧, 孙宜然, 等. 2011. 四川唐家河自然保护区黑腹绒鼠对夏季生境的选择. 四川动物, 30(2): 161-165. Li Y X, Zhang Z J, Sun Y R, *et al*. 2011. Summer habitat selection by *Eothenomys melanogaster* in Tangjiahe Nature Reserve, Sichuan Province. *Sichuan Journal of Zoology*, 30(2): 161-165. (In Chinese with English Abstract)

黎运喜, 张泽钧, 孙宜然, 等. 2012. 唐家河自然保护区高山姬鼠和中华姬鼠夏季生境选择的比较. 生态学报, 32(4): 1241-1248. Li Y X, Zhang Z J, Sun Y R, *et al*. 2012. A comparison of summer habitats selected by sympatric *Apodemus chevrieri* and *A. draco* in Tangjiahe Nature Reserve, China. *Acta Ecologic Sinica*, 32(4): 1241-1248. (In Chinese with English Abstract)

李保国, 陈服官. 1989. 鼢鼠属凸颅亚属(*Eospalax*) 的分类研究及一新亚种. 动物学报, 35(1): 89-95. Li B G, Chen F G. 1989. A taxonomic study and new subspecies of the subgenus *Eospalax*, genus Myospalax. *Acta Zoologica Sinica*, 35(1): 89-95. (In Chinese with English Abstract)

李成涛. 2011. 达赉湖保护区赤狐(*Vulpes vulpes*)的生境选择和景观特征分析. 曲阜: 曲阜师范大学硕士研究生学位论文. Li C T. 2011. Habitat selection and landscape feature analysis of *Vulpes vulpes* in Dalai Lake Reserve. Qufu: Master's Dissertation of Qufu Normal University. (In Chinese with English Abstract)

李春旺, 蒋志刚, 周嘉樀, 等. 2002. 内蒙古巴彦淖尔盟蒙古野驴的数量, 分布和保护对策. 兽类学报, 22(1): 1-6. Li C W, Jiang Z G, Zhou J D, *et al*. 2002. Distribution, numbers and conservation of Mongolian wild ass *Equus hemionus hemionus* in

west Inner Mongolia. *Acta Theriologica Sinica*, 22(1): 1-6. (In Chinese with English Abstract)

李德浩, 王祖祥, 吴翠珍. 1989. 青海经济动物志. 西宁: 青海人民出版社. Li D H, Wang Z X, Wu C Z. 1989. Economic Animals of Qinghai. Xining: Qinghai People's Publishing House. (In Chinese)

李德伟, 尹锋, 曾玉, 等. 2010. 海南岛翼手类地理分布格局的聚类分析. 生物学杂志, 27(2): 16-20. Li D W, Yin F, Zeng Y, *et al*. 2010. Cluster analysis on the distribution patterns of Chiroptera on Hainan Island. *Journal of Biology*, 27(2): 16-20. (In Chinese with English Abstract)

李飞, 郑玺, 张华荣, 等. 2017. 广东省珠海市近海诸岛水獭现状与保护建议. 生物多样性, 25(8): 840-846. Li F, Zheng X, Zhang H R, *et al*. 2017. The current status and conservation of otters on the coastal islands of Zhuhai, Guangdong Province, China. *Biodiversity Science*, 25(8): 840-846. (In Chinese with English Abstract)

李飞虹, 杨奇森, 温知新, 等. 2020. 安氏白腹鼠的形态分化与分布范围修订. 兽类学报, 40(3): 209-230. Li F H, Yang Q S, Wen Z X, *et al*. 2020. A study on morphological variation and geographical range of Anderson's white bellied rat. *Acta Theriological Sinica*, 40(3): 209-230. (In Chinese with English Abstract)

李国红. 2010. 贵州马铁菊头蝠群体遗传结构的微卫星分析. 贵州师范大学学报(自然科学版), 28(1): 19-21. Li G H. 2010. Genetic structure of a population of *Rhinolophus ferrumequinum* in Guizhou using microsatellite markers. *Journal of Guizhou Normal University* (Natural Science Edition), 28(1): 19-21. (In Chinese with English Abstract)

李国军, 石杲, 李保荣. 2013. 内蒙古啮齿目松鼠科种类鉴别与分类探讨. 医学动物防制, (2): 198-199. Li G J, Shi G, Li B R. 2013. Identification and classification of Sciuridae, Rodentia in Inner Mongolia. *Control of Medical Animals*, (2): 198-199. (In Chinese with English Abstract)

李国松, 杨显明, 张宏雨, 等. 2011. 云南新平哀牢山西黑冠长臂猿分布与群体数量. 动物学研究, 32(6): 675-683. Li G S, Yang X M, Zhang H Y, *et al*. 2011. Population and distribution of western black crested gibbon (*Nomascus concolor*) at Ailao Mountain, Xinping, Yunnan. *Zoological Research*, 32(6): 675-683. (In Chinese with English Abstract)

李健雄, 王应祥. 1992. 中国橙腹长吻松鼠种下分类的探讨. 动物学研究, 13(3): 235-244. Li J X, Wang Y X. 1992. Taxonomic study on subspecies of *Dremomys lokriah* (Sciuridae, Rodent) from Southwest China—Note with a new subspecies. *Zoological Research*, 13(3): 235-244. (In Chinese with English Abstract)

李俊, 阿布力米提 • 阿不都卡迪尔, 等, 2007. 红尾沙鼠 (*Meriones libycus*) 的年龄鉴定及种群年龄组成. 干旱区研究, 24(1): 43-48. Li J, Abdukadir A. 2007. Study on the age identification and the population age composition of *Meriones libycus*. *Arid Zone Research,* 24(1): 43-48. (In Chinese with English Abstract)

李俊生, 吴建平, 姜兆文. 2001. 呼伦贝尔草原黄羊体况的初步评价. 兽类学报, 21(2): 81-87. Li J S, Wu J P, Jiang Z W. 2001. A preliminary value on body condition in Mongolian gazelle (*Procapra gutturosa*) in Hulunbeier grassland. *Acta Theriologica Sinica*, 21(2): 81-87. (In Chinese with English Abstract)

李秋阳, 赵秀兰, 杨滨. 2013. 云南省沧源县黄胸鼠种群年龄组的划分及分析. 中国媒介生物学及控制杂志, 24(1): 39-42. Li Q Y, Zhao X L, Yang B. 2013. *Rattus tanezumi* age divisions and population analysis in Cangyuan County, Yunnan Province, China. *Chinese Journal of Vector Biology and Control*, 24(1): 39-42. (In Chinese with English Abstract)

李世斌, 陈安国, 李波, 等. 1993. 洞庭平原褐家鼠年龄分组及种群年龄动态分析. 兽类学报, 13(2): 123-130. Li S B, Chen A G, Li B, *et al*. 1993. Age determination and age composition of *Rattus norvegicus* population on Dongting plain. *Acta Theriologica Sinica*, 13(2): 123-130. (In Chinese with English Abstract)

李思华. 1989. 1949-1988年我国兽类新种、新亚种暨新纪录. 兽类学报, 9(1): 71-77. Li S H. 1989. The new species new subspecies and new records of mammals in China during 1949-1988. *Acta Theriologica Sinica*, 9(1): 71-77. (In Chinese with English Abstract)

李松, 杨君兴, 蒋学龙, 等. 2008. 中国巨松鼠*Ratufa bicolor* (Sciuridae: Ratufinae) 头骨形态的地理学变异. 兽类学报, 28(2): 201-206. Li S, Yang J X, Jiang X L, *et al*. 2008. Geographic variation in giant squirrels *Ratufa bicolor* (Sciuridae: Ratufinae) from China based on cranial measurable variables. *Acta Theriologica Sinica*, 28(2): 201-206. (In Chinese with English Abstract)

李文靖, 曲家鹏, 陈晓澄. 2009. 青海省翼手目类一新纪录——东方蝙蝠. 四川动物, 28(5): 738. Li W J, Qu J P, Chen X C. 2009. A new record of Chiroptera in Qinghai Province: *Vespertilio sinensis*. *Sichuan Journal of Zoology*, 28(5): 738. (In Chinese with English Abstract)

李晓晨, 王廷正. 1995. 攀鼠的分类商榷. 动物学研究, 16: 325-328. Li X C, Wang T Z. 1995. Discussion of taxonomy of vernaya's climbing mouse. *Zoological Research*, 16: 325-328. (In Chinese with English Abstract)

李言阔, 单继红, 李佳, 等. 2013. 獐(*Hydropotes inermis*)生态学研究进展. 野生动物学报, 34(5): 270-273. Li Y K, Shan J H, Li

J, *et al*. 2013. Research advances on the ecology of Chinese water deer (*Hydropotes inermis*). *Chinese Wildlife*, 34(5): 270-273. (In Chinese with English Abstract)

李艳红, 关进科, 黎大勇, 等. 2013. 白马雪山自然保护区灰头小鼯鼠的巢址特征. 生态学报, 33(19): 6035-6040. Li Y H, Guan J K, Li D Y, *et al*. 2013. Nest site characteristics of *Petaurista caniceps* in Baima Snow Mountain Nature Reserve. *Acta Ecologica Sinica*, 33(19): 6035-6040. (In Chinese with English Abstract)

李艳红, 吴攀文, 胡杰. 2007. 四川栗子坪自然保护区的兽类区系与资源. 四川动物, 26(4): 841-845. Li Y H, Wu P W, Hu J. 2007. Mammalian fauna and resources in Liziping Nature Reserve, Sichuan. *Sichuan Journal of Zoology*, 26(4): 841-845. (In Chinese with English Abstract)

李艳丽, 张佑祥, 刘志霄, 等. 2012. 湖南省翼手目新纪录——大耳菊头蝠. 四川动物, 31(5): 825-827. Li Y L, Zhang Y X, Liu Z X, *et al*. 2012. A new record of *Rhinolophus macrotis* in Hunan Province. *Sichuan Journal of Zoology*, 31(5): 825-827. (In Chinese with English Abstract)

李义明, 李典谟. 1994. 人为活动对舟山群岛大中型兽的影响——大中型兽受威胁状态分析. 生物多样性, 2(4): 140-145. Li Y M, Li D M. 1994. The effects of human activities on large and middle mammals on Zhoushan Islands-Analysis of threatened status of large and middle mammals. *Chinese Biodiversity*, 2(4): 140-145. (In Chinese with English Abstract)

李义明, 廖明尧, 喻杰, 等. 2005. 社群大小的年变化、气候和人类活动对神农架自然保护区川金丝猴日移动距离的影响. 生物多样性, 13(5): 432-438. Li Y M, Liao M Y, Yu J, *et al*. 2005. Affects of annual change in group size, human disturbances and weather on daily travel distance of a group in Sichuan snub-nosed monkey (*Rhinopithecus roxellana*) in Shennongjia Nature Reserve, China. *Biodiversity Science*, 13(5): 432-438. (In Chinese with English Abstract)

李瑛. 1997. *Eospalax*亚属的地理分布变迁. 陕西师范大学学报(自然科学版)(s1): 42-47. Li Y. 1997. Changes of the geographical distribution of subgenus *Eospalax*. *Journal of Shaanxi Normal University* (Natural Science Edition), (s1): 42-47. (In Chinese with English Abstract)

李永项. 2012. 山羊寨中更新世食虫类及其动物地理与环境变迁研究. 西安: 西北大学博士研究生学位论文: 83. Li Y X. 2012. Study on the Biogeography and Environmental Changes of Middle Pleistocene Insectivores in Yangzhai. Xi'an: Doctor's Thesis of Northwestern University: 83. (In Chinese with English Abstract)

李友邦, 黄乘明, 韦振逸, 等. 2009. 广西猕猴分布数量及其保护. 广西师范大学学报(自然科学版), 27(1): 79-83. Li Y B, Huang C M, Wei Z Y, *et al*. 2009. Distribution and protection of macaques in Guangxi. *Journal of Guangxi Normal University* (Natural Science Edition), 27(1): 79-83. (In Chinese with English Abstract)

李友邦, 韦振逸. 2012. 广西扶绥弄邓黑叶猴种群数量和保护. 安徽农业科学, 40(26): 12952-12953. Li Y B, Wei Z Y. 2012. Survey on distribution and population of *Trachypithecus francoisi* in Nongdeng, Fusui of Guangxi. *Journal of Anhui Agricultural Sciences*, 40(26): 12952-12953. (In Chinese with English Abstract)

李玉春, 吴毅, 陈忠. 2006. 海南岛发现大足鼠耳蝠分布新记录. 兽类学报, 26(2): 211-212. Li Y C, Wu Y, Chen Z. 2006. A new record of Rickett's big-footed bat *Myotis ricketti* on Hainan Island of China. *Acta theriologica Sinica*, 26(2): 211-212. (In Chinese with English Abstract)

李裕冬, 刘少英, 曾宗永. 2007. 白腹鼠属几个相似种的差异探讨. 四川动物, 26(1): 41-45. Li Y D, Liu S Y, Zeng Z Y. 2007. Discussion about different characters of four species in *Niviventer*. *Sichuan Journal of Zoology*, 26(1): 41-45. (In Chinese with English Abstract)

李月辉, 胡志斌, 冷文芳, 等. 2007. 大兴安岭呼中区紫貂生境格局变化及采伐的影响. 生物多样性, 15(3): 232-240. Li Y H, Hu Z B, Leng W F, *et al*. 2007. Habitat pattern change of *Martes zibellina* and the impact of timber harvest in Huzhong Area in Greater Ching'an Mountains, Northeast China. *Biodiversity Science*, 15(3): 232-240. (In Chinese with English Abstract)

李云秀, 潘莉, 曾涛, 等. 2012. 四川木里鸭咀自然保护区兽类资源调查. 四川林业科技, 33(4): 56-60. Li Y X, Pan L, Zeng T, *et al*. 2012. Surveys of mammals in Muli Yazui Nature Reserve, Sichuan Province. *Journal of Sichuan Forestry Science and Technology*, 33(4): 56-60. (In Chinese with English Abstract)

李枝林, 韩建芳. 1988. 羽尾跳鼠自然繁殖情况的初步观察. 四川动物, 7(2): 19. Li Z L, Han J F. 1988. A preliminary observation on the natural reproduction of *Scirtopoda telum*. *Sichuan Journal of Zoology*, 7(2): 19. (In Chinese with English Abstract)

李志刚, 魏辅文, 周江. 2010. 海南长臂猿线粒体D-loop区序列分析及种群复壮. 生物多样性, 18(5): 523-527. Li Z G, Wei F W, Zhou J. 2010. Mitochondrial DNA D-loop sequence analysis and population rejuvenation of Hainan gibbons (*Nomascus hainanus*). *Biodiversity Science*, 18(5): 523-527. (In Chinese with English Abstract)

李致祥. 1981. 中国麝一新种的记述. 动物学研究, 2(2): 157-161. Li Z X. 1981. On a new species of musk-deer from China.

Zoological Research, 2(2): 157-161. (In Chinese with English Abstract)
李致祥, 林正玉. 1983. 云南灵长类的分类和分布. 动物学研究, 4(2): 3-12. Li Z X, Lin Z Y. 1983. Classification and distribution of living primates in Yunnan China. *Zoological Research*, 4(2): 3-12. (In Chinese with English Abstract)
梁仁济, 董永文. 1984. 皖南地区翼手类初步研究. 兽类学报, 4(4): 440-442. Liang R J, Dong Y W. 1984. Bat from south Anhui. *Acta Theriologica Sinica*, 4(4): 440-442. (In Chinese with English Abstract)
梁仁济, 董永文. 1985. 绒山蝠生态的初步调查. 兽类学报, 5(1): 11-15. Liang R J, Dong Y W. 1985. On the ecology of *Nyctalus velutinus*. *Acta Theriologica Sinica*, 5(1): 11-15. (In Chinese with English Abstract)
梁仁济, 李炳华, 陈菲菲, 等. 1983. 安徽省翼手类新记录. 安徽师大学报(自然科学版), (1): 58-63. Liang R J, Li B H, Chen F F, *et al*. 1983. New records of bats in Anhui Province. *Journal of Anhui Normal University* (Natural Science Edition), (1): 58-63. (In Chinese with English Abstract)
梁艺于, 胡杰, 杨志松, 等. 2009. 四川甘洛马鞍山自然保护区兽类初步调查. 西华师范大学学报(自然科学版), 30(3): 246-252. Liang Y Y, Hu J, Yang Z S, *et al*. 2009. A preliminary survey of mammals in Maanshan Nature Reserve in Ganluo, Sichuan, China. *Journal of China West Normal University* (Natural Sciences Edition), 30(3): 246-252. (In Chinese with English Abstract)
廖锐, 郭光普, 刘洋, 等. 2015. 西藏墨脱县小型兽类多样性研究. 四川林业科技, 36(1): 6-10. Liao R, Guo G P, Liu Y, *et al*. 2015. Biodiversity of small mammals in Mêdog, Tibet of China. *Sichuan Forestry Science and Technology*, 36(1): 6-10. (In Chinese with English Abstract)
廖炎发. 1988. 青海荒漠猫的一些生物学资料. 兽类学报, 8(2): 128-131. Liao Y F. 1988. Some biological information of desert cat in Qinghai. *Acta Theriologica Sinica*, 8(2): 128-131. (In Chinese with English Abstract)
林爱青, 王磊, 刘森, 等. 2009. 江苏省蝙蝠新纪录——皮氏菊头蝠. 动物学杂志, 44(3): 113-117. Lin A Q, Wang L, Liu S, *et al*. 2009. A new record of *Rhinolophus pearsoni* in Jiangsu Province. *Chinese Journal of Zoology*, 44(3): 113-117. (In Chinese with English Abstract)
林洪军, 尹皓, 齐彤辉, 等. 2012. 高颅鼠耳蝠回声定位声波特征与分析. 四川动物, 31: 6-9. Lin H J, Yin H, Qi T H, *et al*. 2012. Characteristics and analysis of echolocation calls by *Myotis siligorensis*. *Sichuan Journal of Zoology*, 31: 6-9. (In Chinese with English Abstract)
林纪春, 张渝疆, 张兰英. 1989. 长尾黄鼠年龄鉴定及其种群年龄组成的研究. 兽类学报, 9(3): 216-220. Lin J C, Zhang Y J, Zhang L Y. 1989. Age determination and composition in a population of *Citellus undulatus*. *Acta Theriologica Sinica*, 9(3): 216-220. (In Chinese with English Abstract)
林杰, 徐文轩, 杨维康, 等. 2011. 亚洲野驴生态生物学研究现状. 生态学杂志, 30(10): 2351-2358. Lin J, Xu W X, Yang W K, *et al*. 2011. Present situation of eco-biological study on *Equus hemionus*. *Chinese Journal of Ecology*, 30(10): 2351-2358. (In Chinese with English Abstract)
林良恭. 2000. 台湾陆域哺乳动物多样性与保育. 生物多样性(季刊), (1): 106-115. Lin L G. 2000. Terrestrial mammal diversity and conservation in Taiwan. *Biodiversity* (Quarterly), (1): 106-115. (In Chinese)
林良恭. 2002. 台湾外来种脊椎动物现状. 全球变迁通讯杂志, 33: 8-13. Lin L G. 2002. Current status of ecdemic vertebrates from Taiwan. *Journal of Global Change*, 33: 8-13. (In Chinese with English Abstract)
林良恭, 李玲玲, 郑锡奇. 1997. 台湾的蝙蝠. 台中: 自然科学博物馆. Lin L G, Li L L, Zheng X Q. 1997. Bats in Taiwan. Taizhong: Museum of Natural Science. (In Chinese)
林柳, 张龙田, 罗爱东, 等. 2011. 尚勇保护区亚洲象种群数量动态、种群结构及季节分布格局. 兽类学报, 31(3): 226-234. Lin L, Zhang L T, Luo A D, *et al*. 2011. Population dynamics, structure and seasonal distribution pattern of Asian elephant (*Elephas maximus*) in Shangyong Protected Area, Yunnan, China. *Acta Theriologica Sinica*, 31(3): 226-234. (In Chinese with English Abstract)
刘丰, 宋雨, 颜识涵, 等. 2009. 大趾鼠耳蝠线粒体DNA控制区结构及变异. 动物学杂志, 44(4): 19-27. Liu F, Song Y, Yan S H, *et al*. 2009. Structure and Sequence Variation of the Mitochondrial DNA Control Region in *Myotis macrodactylus*. *Chinese Journal of Zoology*, 44(4): 19-27. (In Chinese with English Abstract)
刘昊, 石红艳, 王刚. 2010. 中华鼠耳蝠的分布及研究现状. 绵阳师范学院学报, 29(11): 66-73. Liu H, Shi H Y, Wang G. 2010. Distribution and Research Progress of *Myotis chinensis*. *Journal of Mianyang Normal University*, 29(11): 66-73. (In Chinese with English Abstract)
刘鹤, 李乐, 马强, 等. 2011. 野猪研究进展. 四川动物, 30(2): 310-314. Liu H, Li L, Ma Q, *et al*. 2011. Review on wild boar research. *Sichuan Journal of Zoology*, 30(2): 310-314. (In Chinese with English Abstract)

刘丽, 周延山, 楚彬, 等. 2018. 基于线粒体基因、形态学和栖息地指标的两种鼢鼠分类研究. 兽类学报, 38(4): 402-410. Liu L, Zhou Y S, Chu B, *et al.* 2018. Classification of two zokor species based on mitochondrial gene, morphological and habitat indices. *Acta Theriologica Sinica*, 38(4): 402-410. (In Chinese with English Abstract)

刘奇, 陈珉, 陈毅, 等. 2014. 湖北省和江苏省发现尼泊尔鼠耳蝠. 动物学杂志, 49(4): 483-489. Liu Q, Chen M, Chen Y, *et al.* 2014. *Myotis nipalensis* discovered in Hubei and Jiangsu Provinces, China. *Journal of Zoology*, 49(4): 483-489. (In Chinese with English Abstract)

刘仁华, 陈曦, 高从政, 等. 1989. 东北鼢鼠种群结构及繁殖初步研究. 齐齐哈尔师范学院学报(自然科学版), (2): 13-20. Liu R H, Chen X, Gao C Z, *et al.* 1989. A preliminary study on population structure and propagation of *Myospalax psilurus*. *Journal of Qiqihar Normal University* (Natural Science Edition), (2): 13-20. (In Chinese with English Abstract)

刘瑞玉. 2008. 中国海洋生物名录. 北京: 科学出版社. Liu R Y. 2008. Checklist of marine biota of China seas. Beijing: Science Press. (In Chinese)

刘森, 江廷磊, 施利民, 等. 2008. 无尾蹄蝠的回声定位声波特征及分析. 动物学研究, 29(1): 95-98. Liu S, Jiang T L, Shi L M, *et al.* 2008. Characteristics and analysis of echolocation calls by *Coelops frithi*. *Zoological Research*, 29(1): 95-98. (In Chinese with English Abstract)

刘少英, 冉江洪, 林强, 等. 2001. 三峡工程重庆库区翼手类研究. 兽类学报, 21(2): 123-131. Liu S Y, Ran J H, Lin Q, *et al.* 2001. Bats in Three Gorges Reservoir area, Chongqing. *Acta Theriologica Sinica*, 21(2): 123-131. (In Chinese with English Abstract)

刘少英, 孙治宇, 冉江洪, 等. 2005. 四川九寨沟自然保护区兽类调查. 兽类学报, 25(3): 273-281. Liu S Y, Sun Z Y, Ran J H, *et al.* 2005. Mammalian survey of Jiuzhaigou National Nature Reserve, Sichuan Province. *Acta Theriologica Sinica*, 25(3): 273-281. (In Chinese with English Abstract)

刘少英, 张明, 孙治宇. 2011. 西藏工布自然保护区生物多样性. 重庆: 西南师范大学出版社. Liu S Y, Zhang M, Sun Z Y. 2011. Biodiversity in the Gongbujiangda Nature Reserve. Chongqing: Southwest China Normal University Press. (In Chinese)

刘少英, 吴毅. 2019. 中国兽类图鉴. 厦门: 海峡书局. Liu S Y, Wu Y. 2019. Handbooks of Mammals of China. Xiamen: The Strait Publishing and Distributing Group. (In Chinese)

刘少英, 靳伟, 廖锐, 等. 2017. 基于Cyt *b*基因和形态学的鼠兔属系统发育研究及鼠兔属1新亚属5新种描述. 兽类学报, 37(1): 1-43. Liu S Y, Jin W, Liao R, *et al.* 2017. Phylogenetic study of *Ochotona* based on mitochondrial Cyt *b* and morphology with a description of one new subgenus and five new species. *Acta Theriologica Sinica*, 37(1): 1-43. (In Chinese with English Abstract)

刘姝, 初红军, 王渊, 等. 2013. 普氏野马(*Equus przewalskii*)重引入区域的社区保护意识调查分析. 干旱区研究, 30(1): 135-143. Liu S, Chu H J, Wang Y, *et al.* 2013. Survey and analysis of the awareness of nomads in the peripheral communities in protecting the wild-back(*Equus przewalski*). *Arid Zone Research*, 30(1): 135-143. (In Chinese with English Abstract)

刘伟. 2012. 太行山南段洞栖蝙蝠研究. 新乡: 河南师范大学硕士研究生学位论文. Liu W. 2012. Cave Dwelling Bats in South Section of Taihang Mountain. Xinxiang: Master's Thesis of Henan Normal University. (In Chinese with English Abstract)

刘伟石, 胡德夫, 郜二虎. 2007. 甘肃省豹的生存现状调查. 四川动物, 26(1): 86-88. Liu W S, Hu D F, Gao E H. 2007. Surviving status of leopard (*Panthera pardus*) in Gansu Province. *Sichuan Journal of Zoology*, 26(1): 86-88. (In Chinese with English Abstract)

刘文超. 2009. 普氏蹄蝠(*Hipposideros pratti*)微卫星位点的筛选及交叉种扩增. 上海: 华东师范大学硕士研究生学位论文. Liu W C. 2009. Selection and Cross-species Amplification of *Hipposideros pratti* Microsatellite Loci. Shanghai: Master's Thesis of East China Normal University. (In Chinese with English Abstract)

刘文华, 佟建明. 2005. 中国的麝资源及其保护与利用现状分析. 中国农业科技导报, 7(4): 28-32. Liu W H, Tong J M. 2005. Analysis on protection and utilization of musk deer resources in China. *Review of China Agricultural Science and Technology*, 7(4): 28-32. (In Chinese with English Abstract)

刘务林, 乔治・B・夏勒. 2003. 野牦牛的分布和现状. 西藏科技, (11): 17-23. Liu W L, Schaller G B. 2003. Distribution and status of wild yak. *Tibet Science and Technology*, (11): 17-23. (In Chinese with English Abstract)

刘曦庆, 彭建军, 高赛飞, 等. 2011. 穿山甲的走私贸易概况、物种鉴定与形态比较. 林业科技通讯, (5): 11-14. Liu X Q, Peng J J, Gao S F, *et al.* 2011. Overview of pangolin smuggling trade, species identification and morphological comparison. *Practical Forestry Technology*, (5): 11-14. (In Chinese with English Abstract)

刘晓明, 魏辅文, 李明, 等. 2002. 中国姬鼠属的系统学研究述评. 兽类学报, 22(1): 46-52. Liu X M, Wei F W, Li M, *et al.* 2002. A review of the phylogenetic study on the Genus *Apodemus* of China. *Acta Theriologica Sinica*, 22(1): 46-52. (In Chinese

with English Abstract)

刘鑫, 王政昆, 肖治术. 2011. 小泡巨鼠和社鼠对珍稀濒危植物红豆树种子的捕食和扩散作用. 生物多样性, 19(1): 93-96. Liu X, Wang Z K, Xiao Z S. 2011. Patterns of seed predation and dispersal of an endangered rare plant *Ormosia hosiei* by Edward's long-tailed rats and Chinese white-bellied rats. *Biodiversity Science*, 19(1): 93-96. (In Chinese with English Abstract)

刘延德, 周昭敏, 周材权, 等. 2006. 四川产中菊头蝠喜马拉雅亚种和马铁菊头蝠日本亚种外部形态及头骨的比较. 动物学杂志, 41(1): 103-107. Liu Y D, Zhou Z M, Zhou C Q, *et al*. 2006. Comparison of morphological and skull of *Rhinolophus affinis himalayanus* and *R. ferrumequinum nippon*. *Chinese Journal of Zoology*, 41(1): 103-107. (In Chinese with English Abstract)

刘艳华, 张明海. 2009. 黑龙江省不同山系狍种群遗传多样性分析. 动物学研究, 30(2): 113-120. Liu Y H, Zhang M H. 2009. Population genetic diversity of roe deer (*Capreolus pygargus*) in mountains of Heilongjiang Province. *Zoological Research*, 30(2): 113-120. (In Chinese with English Abstract)

刘洋, 刘少英, 孙治宇, 等. 2011. 山西省兽类一新纪录——川西缺齿鼩鼱. 四川动物, 30(6): 967-968. Liu Y, Liu S Y, Sun Z Y, *et al*. 2011. A new record of *Chodsigoa hypsibia* in Shanxi Province. *Sichuan Journal of Zoology*, 30(6): 967-968. (In Chinese with English Abstract)

刘洋, 刘少英, 孙治宇, 等. 2013a. 鼩鼹亚科 (Talpidae: Uropsilinae) 一新种. 兽类学报, 33(2): 113-122. Liu Y, Liu S Y, Sun Z Y, *et al*. 2013a. A new species of Uropsilus (Talpidae: Uropsilinae) from Sichuan China. *Acta Theriologica Sinica*, 33(2): 113-122. (In Chinese with English Abstract)

刘洋, 王昊, 刘少英. 2010. 苔原鼩鼱(*Sorex tundrensis*)在中国分布的首次证实. 兽类学报, 30(4): 439-443. Liu Y, Wang H, Liu S Y. 2010. First confirmation of the distribution of tundra shrew (*Sorex tundrensis*) in China. *Acta Theriologica Sinica*, 30(4): 439-443. (In Chinese with English Abstract)

刘洋, 张惠, 刘应雄, 等. 2013b. 四川卡莎湖自然保护区兽类资源调查. 四川林业科技, 34(6): 39-43. Liu Y, Zhang H, Liu Y X, *et al*. 2013b. A preliminary survey of mammal fauna of Kasha Lake Nature Reserve, Sichuan Province. *Journal of Sichuan Forestry Science and Technology*, 34(6): 39-43. (In Chinese with English Abstract)

刘颖, 冯江, 陈敏, 等. 2003. 毛腿鼠耳蝠回声定位声波的分析. 东北师大学报(自然科学版), 35: 113-116. Liu Y, Feng J, Chen M, *et al*. 2003. The analysis on echolocation calls of *Myotis fimbriatus* (Chiroptera: Vespertilionidae). *Journal of Northeast Normal University* (Natural Science Edition), 35: 113-116. (In Chinese with English Abstract)

刘应雄, 张惠, 刘洋. 2014. 四川千佛山自然保护区兽类资源调查. 四川林业科技, 35(3): 65-69. Liu Y X, Zhang H, Liu Y. 2014. A survey of animal resources in Qianfo Shan Nature Reserve in Sichuan Province. *Journal of Sichuan Forestry Science and Technology*, 35(3): 65-69. (In Chinese with English Abstract)

刘长乐, 邹琦, 郜二虎, 等. 2009. 福建省豹的分布调查. 林业科技, 34(2): 35-37. Liu C L, Zou Q, Gao E H, *et al*. 2009. The distribution of leopard (*Panthera pardus*) in Fujian Province. *Forestry Science & Technology*, 34(2): 35-37. (In Chinese with English Abstract)

刘正祥, 洪梅, 杨桂荣, 等. 2013. 香格里拉县小型兽类垂直空间生态位初步研究. 动物学杂志, 48(4): 619-625. Liu Z X, Hong M, Yang G R, *et al*. 2013. Preliminary study on vertical spatial niche of small mammals in Shangrila County of Yunnan Province. *Chinese Journal of Zoology*, 48(4): 619-625. (In Chinese with English Abstract)

刘志霄, 盛和林. 1998. 栖息地片段化与隔离对兽类种群的影响. 生物学通报, 33: 18-20. Liu Z X, Sheng H L. 1998. Effects of habitat fragmentation and isolation on animal populations. *Chinese Journal of Biology*, 33: 18-20. (In Chinese with English Abstract)

刘志霄, 盛和林. 2000. 我国麝的生态研究与保护问题概述. 动物学杂志, 35(3): 54-57. Liu Z X, Sheng H L. 2000. Ecological research and protection of musk deer in China. *Chinese Journal of Zoology*, 35(3): 54-57. (In Chinese with English Abstract)

刘志霄, 张佑祥, 张礼标. 2013. 中国翼手目动物区系分类与分布研究进展、趋势与前景. 动物学研究, 34(6): 687-693. Liu Z X, Zhang Y X, Zhang L B. 2013. Research perspectives and achievements in taxonomy and distribution of bats in China. *Zoological Research*, 34(6): 687-693. (In Chinese with English Abstract)

刘铸, 张隽晟, 白薇, 等. 2019. 中国东北地区鼩鼱科动物分类与分布. 兽类学报, 39(1): 8-26. Liu Z, Zhang J S, Bai W, *et al*. 2019. Classification and distribution of Soricidae in Northeastern China. *Acta Theriologica Sinica*, 39(1): 8-26. (In Chinese with English Abstract)

刘铸, 解瑞雪, 刘欢, 等. 2016. 黑龙江省横道河子地区发现细鼩鼱(食虫目: 鼩鼱科). 动物学杂志, 51(3): 487-491. Liu Z, Xie R X, Liu H. 2016. The Slender Shrew (*Sorex gracillimus* Thomas, 1907; Insetivora: Soricidae) was found in Hengdaohezi of

Heilongjiang Province, China. *Chinese Journal of Zoology*, 51(3): 487-491. (In Chinese with English Abstract)

卢学理, 袁喜才, 彭建军, 等. 2008. 海南坡鹿种群发展动态与保护建议. 四川动物, 27(1): 138-141. Lu X L, Yuan X C, Peng J J, *et al.* 2008. Dynamics and Conservation Suggestions of Hainan Eld's Deer. *Sichuan Journal of Zoology*, 27(1): 138-141. (In Chinese with English Abstract)

鲁庆彬, 张阳, 周材权. 2010. 秦岭鼢鼠的洞穴选择与危害防控. 生态学报, 31(7): 1993-2001. Lu Q B, Zhang Y, Zhou C Q. 2010. Cave-site selection of Qinling zokors with their prevention and control. *Acta Ecologica Sinica*, 31(7): 1993-2001. (In Chinese with English Abstract)

陆雪, 袁兴勤, 余依建, 等. 2007. 鬣羚分类与分布的初步研究. 四川动物, 26(4): 929-930. Lu X, Yuan X Q, Yu Y J, *et al.* 2007. Preliminary analysis of *Capriconis sumatraensis* classification and distribution. *Sichuan Journal of Zoology*, 26(4): 929-930. (In Chinese with English Abstract)

陆长坤, 王宗祎, 全国强, 等. 1965. 云南西部临沧地区兽类的研究. 动物分类学报, 2(4): 279-295. Lu C K, Wang Z W, Quan G Q, *et al.* 1965. Mammals in Lincang, western Yunnan. *Acta Zootaxonomica Sinica*, 2(4): 279-295. (In Chinese with English Abstract)

路纪琪, 刘彬. 2008. 河南省哺乳动物分布新纪录——小泡巨鼠. 四川动物, 27(3): 435. Lu J Q, Liu B. 2008. A new record of mammals in Henan Province—*Leopoldamys edwardsi*. *Sichuan Journal of Zoology*, 27(3): 435. (In Chinese with English Abstract)

罗键, 高红英. 2002. 重庆市翼手类调查及保护建议. 四川动物, 21(1): 45-46. Luo J, Gao H Y. 2002. Investigation on bats in Chongqing. *Sichuan Journal of Zoology*, 21(1): 45-46. (In Chinese with English Abstract)

罗键, 高红英. 2006. 在重庆和辽宁发现绯鼠耳蝠*Myotis formosus*. 四川动物, 25(1): 131-132. Luo J, Gao H Y. 2006. *Myotis formosus*, a record new of Chiroptera in Chongqing and Liaoning. *Sichuan Journal of Zoology*, 25(1): 131-132. (In Chinese with English Abstract)

罗金红, 颜识涵, 宋雨, 等. 2009. 大趾鼠耳蝠回声定位声波特征与分析. 动物学杂志, 44(1): 133-138. Luo J H, Yan S H, Song Y, *et al.* 2009. Characters of echolocation calls in *Myotis macrodactylus*. *Chinese Journal of Zoology*, 44(1): 133-138. (In Chinese with English Abstract)

罗丽. 2011. 基于微卫星标记的中国马铁菊头蝠种群遗传多样性与遗传结构研究. 长春: 东北师范大学硕士研究生学位论文. Luo L. 2011. Genetic Diversity and Population Structure of *Rhinolophus ferrumequinum* in China Based on Microsatellites. Changchun: Master's Thesis of Northeast Normal University. (In Chinese with English Abstract)

罗丽, 卢冠军, 罗金红, 等. 2011. 湖南省蝙蝠新纪录——大足鼠耳蝠. 动物学杂志, 46(2): 148-152. Luo L, Lu G J, Luo J H, *et al.* 2011. *Myotis ricketti*—a new bat record of Hunan Province. *Chinese Journal of Zoology*, 46(2): 148-152. (In Chinese with English Abstract)

罗蓉, 等. 1993. 贵州兽类志. 贵阳: 贵州科技出版社: 1-422. Luo R, *et al.* 1993. Mammals in Guizhou. Guiyang: Guizhou Science and Technology Press: 1-422. (In Chinese)

罗晓, 李峰, 陈静, 等. 2016. 青海湖地区狗獾分类地位和狗獾属进化历史探讨. 生物多样性, 24(6): 694-700. Luo X, Li F, Chen J, *et al.* 2016. The taxonomic status of badgers in the Qinghai Lake area and evolutionary history of Meles. *Biodiversity Science*, 24(6): 694-700. (In Chinese with English Abstract)

罗一宁. 1987. 我国兽类新记录——缺齿鼠耳蝠. 兽类学报, 7(2): 159. Luo Y N. 1987. A new record of mammal in China—*Myotis annectans* in Yunnan. *Acta Theriologica Sinica*, 7(2): 159. (In Chinese with English Abstract)

罗泽珣. 2000. 中国动物志 • 兽纲 • 第六卷 • 啮齿目 下册 仓鼠科. 北京: 科学出版社. Luo Z X. 2000. Fauna Sinica, Mammalia Vol. 6 Rodentia II Cricetidae. Beijing: Science Press. (In Chinese)

罗泽珣, 李振营. 1982. 我国雪兔的分类研究. 东北林业大学学报, 10(2): 159-167. Luo Z X, Li Z Y. 1982. A systematic review of the Chinese varying hare, *Lepus timidus* Linnaeus. *Journal of Northeast Forestry University*, 10(2): 159-167. (In Chinese with English Abstract)

罗忠华. 2011. 云南无量山国家级自然保护区西部黑冠长臂猿景东亚种的群体数量与分布调查. 四川动物, 30(2): 283-287. Luo Z H. 2011. Survey on populations and distribution of western black crested Gibbons (*Nomascus concolor jingdongensis*) from Wuliang Shan National Nature Reservge. *Sichuan Journal of Zoology*, 30(2): 283-287. (In Chinese with English Abstract)

麻应太, 王西峰. 2008. 秦岭羚牛资源现状与保护. 陕西林业科技, (2): 80-83. Ma Y T, Wang X F. 2008. Current status and protection measures of Golden Takin (*Budorcas taxicolor bedfordi*) in Qinling Mountain Ranges. *Shaanxi Forest Science and Technology*, (2): 80-83. (In Chinese with English Abstract)

马建章, 李津友. 1979. 西伯利亚旱獭生态调查研究. 东北林业大学学报, 7(1): 63-71. Ma J Z, Li J Y. 1979. An ecological investigation of Siberian marmot. *Journal of Northeast Forestry University*, 7(1): 63-71. (In Chinese with English Abstract)

马建章, 戎可, 宗诚. 2008. 松鼠生态学研究现状与展望. 动物学杂志, 43(1): 159-164. Ma J Z, Rong K, Zong C. 2008. The ecology of Eurasian red squirrels: recent advances and future prospects. *Chinese Journal of Zoology*, 43(1): 159-164. (In Chinese with English Abstract)

马杰, 梁冰, 张劲硕, 等. 2005. 北京地区大足鼠耳蝠主要食物及其食性组成的季节变化. 动物学报, 51(1): 7-11. Ma J, Liang B, Zhang J S, *et al*. 2005. Major item and seasonal variation in the diet of Rickett's big-footed bat *Myotis ricketti* in Beijing. *Acta Zoologica Sinica*, 51(1): 7-11. (In Chinese with English Abstract)

马强, 苏化龙. 2004. 黑叶猴(*Trachypithecus francoisi*). 动物学杂志, 39(3): 32. Ma Q, Su H L. 2004. Francois's Leaf Monkey(*Trachypithecus francoisi*). *Chinese Journal of Zoology*, 39(3): 32. (In Chinese with English Abstract)

马瑞俊, 蒋志刚. 2006. 青海湖流域环境退化对野生陆生脊椎动物的影响. 生态学报, 26(9): 3066-3073. Ma R J, Jiang Z G. 2006. Impacts of environmental degradation on wild vertebrates in the Qinghai Lake drainage. *Acta Ecologica Sininca*, 26(9): 3066-3073. (In Chinese with English Abstract)

马世来, 王应祥, 施立明. 1990. 麂属(*Muntiacus*)一新种. 动物学研究, 11(1): 47-53. Ma S L, Wang Y X, Shi L M. 1990. A new species of the genus *Muntiacus* from Yunnan, China. *Zoological Research*, 11(1): 47-53. (In Chinese with English Abstract)

马晓婷, 黄玲, 刘玉静, 等. 2014. 社鼠(*Niviventer confucianus*)线粒体基因组全序列分析. 中国细胞生物学学报, 36(8): 1084-1091. Ma X T, Huang L, Liu Y J, *et al*. 2014. Sequence Analysis of the Complete Mitochondrial Genome of *Niviventer confucianus*. *Chinese Journal of Cell Biology*, 36(8): 1084-1091. (In Chinese with English Abstract)

马逸清, 胡锦矗. 1998. 中国熊类资源数量估计及保护对策. 生命科学研究, (2): 205-211. Ma Y Q, Hu J C. 1998. On the resources and conservation of bears in China. *Life Science Research*, (2): 205-211. (In Chinese with English Abstract)

马勇, 王逢桂, 金善科, 等. 1982. 新疆黄兔尾鼠的分布及其生态习性的初步观察. 兽类学报 2(1): 81-88. Ma Y, Wang F G, Jin S K, *et al*. 1982. On Distribution and ecology of Yellow Steppe Lemming (*Lagurus luteus*) of Xinjiang. *Acta Theriologica Sinica*, 2(1): 81-88. (In Chinese with English Abstract)

马勇, 杨奇森, 周立志. 2008. 中国啮齿动物分类学与地理分布. 见: 郑智民, 姜志宽, 陈安国. 啮齿动物学. 上海: 上海交通大学出版社: 35-42. Ma Y, Yang Q S, Zhou L Z. 2008. Taxonomy and geographical distribution of Chinese rodents. *In*: Zheng Z M, Jiang Z K, Chen A G. Rodent Zoology. Shanghai: Shanghai Jiaotong University Press: 35-42. (In Chinese)

马勇. 1964. 山西短棘猬属的一个新种. 动物分类学报, 1(1): 31-36. Ma Y. 1964. A new species of hedgehog from Shansi Province, *Hemiechinus sylvaticus* sp. nov. *Acta Zootaxonomica Sinica*, 1(1): 31-36. (In Chinese with English Abstract)

马勇, 李思华. 1979. 长耳跳鼠一新亚种. 动物分类学报, 4(3): 109-111. Ma Y, Li S H. 1979. A new subspecies of the long-eared jerboa from Xinjiang. *Acta Zootaxonomica Sinica*, 4(3): 109-111. (In Chinese with English Abstract)

马勇, 王逢桂, 金善科, 等. 1981. 新疆北部地区啮齿动物(Glires)的分类研究. 兽类学报, 11(1): 177-188. Ma Y, Wang F G, Jin S K, *et al*. 1981. The taxonomic research of glires in Northern Xinjiang. *Acta Theriologica Sinica*, 11(1): 177-188. (In Chinese with English Abstract)

马志强, 韩家波, 姜大为, 等. 2007. 渤海虎平岛周围海域的斑海豹种群动态初步调查. 水产科学, 26(8): 455-457. Ma Z Q, Han J B, Jiang D W, *et al*. 2007. Population dynamics of spotted seals in the waters around Huping island in the Bohai Sea. *Fisheries Science*, 26(8): 455-457. (In Chinese with English Abstract)

买尔旦·吐尔干. 2006. 吐鲁番盆地鼠类群落结构与多样性研究. 乌鲁木齐: 中国科学院新疆生态与地理研究所硕士研究生学位论文. Turkan M. 2006. Rodent Community Structure and Diversity in Turpan Basin. Urmqi: Master's Thesis of Xinjiang Institute of Ecology and Geography, Chinese Academy Of Sciences. (In Chinese with English Abstract)

毛秀光. 2010. 皮氏菊头蝠与云南菊头蝠系统地理学研究. 上海: 华东师范大学博士研究生学位论文. Mao X G. 2010. Phylogeography of *Rhinolophus pearsoni* and *Rhinolophus yunnanensis*. Shanghai: Doctor's Thesis of East China Normal University. (In Chinese with English Abstract)

门兴元, 郭宪国, 董文鸽, 等. 2006. 珀氏长吻松鼠和赤腹松鼠在保护区与非保护区各年龄松林内的种群动态. 动物学研究, 27(1): 29-33. Men X Y, Guo X G, Dong W G, *et al*. 2006. Population dynamics of *Dremomys pernyi* and *Callosciurus erythraeus* in protective and non-protective pine forests at different ages. *Frontiers of Biology in China*, 27(1): 29-33. (In Chinese with English Abstract)

孟超, 张洪海, 陈玉才. 2008. 中国狼(*Canis lupus chanco*)线粒体全基因组序列分析. 中国生物化学与分子生物学报, 24(12): 1170-1176. Meng C, Zhang H H, Chen Y C. 2008. Sequencing and analysis of mitochondrial genome of Chinese Grey Wolf (*Canis lupus chanco*). *Chinese Journal of Biochemistry and Molecular Biology*, 24(12): 1170-1176. (In Chinese with

English Abstract)

孟玉萍, 胡德夫, 何东阳, 等. 2009. 中国新疆放归普氏野马的繁殖状况. 生物学通报, 44(5): 1-4. Meng Y P, Hu D F, He D Y, *et al*. 2009. Breeding status of *Equus przewalskii* released from Xinjiang, China. *Bulletin of Biology*, 44(5): 1-4. (In Chinese with English Abstract)

米景川, 于少祥, 潘井坤. 2003. 达乌尔黄鼠的种群年龄动态及其生命表研究. 医学动物防制, 19(5): 264-267. Mi J C, Yu S X, Pan J K. 2003. Population age dynamics and life table of *Spermophilus dauricus*. *Control of Medical Animals*, 19(5): 264-267. (In Chinese with English Abstract)

娜日苏, 苏和, 武晓东. 2009. 五趾跳鼠的植物性食物选择与其栖息地植被的关系. 草地学报, 17(3): 383-388. Na R S, Su H, Wu X D. 2009. Botanic food preference of *Allactaga sibirica* Forster and its relationship with the vegetation conditions of their habitat. *Acta Agrestia Sinica*, 17(3): 383-388. (In Chinese with English Abstract)

牛红星. 2008. 河南省翼手类区系分布与系统学研究. 石家庄: 河北师范大学硕士研究生学位论文. Niu H X. 2008. Distribution and Systematics of Chiroptera in Henan Province. Shijiazhuang: Master's Thesis of Hebei Normal University. (In Chinese with English Abstract)

牛红星, 张学成, 马惠霞. 2008. 河南省菊头蝠科1新纪录——皮氏菊头蝠 *Rhinolophus pearsoni*. 河南师范大学学报(自然科学版), 36(1): 147-148. Niu H X, Zhang X C, Ma H X. 2008. New record of bat in Henan Province—*Rhinolophus pearsoni*. *Journal of Henan Normal University* (Natural Science Edition), 36(1): 147-148. (In Chinese with English Abstract)

潘会, 周显明, 杨再学, 等. 2012. 关岭县锡金小家鼠种群生态特征初步探讨. 山地农业生物学报, 31(5): 381-384. Pan H, Zhou X M, Yang Z X, *et al*. 2012. Preliminary investigation in ecological characteristics of *Mus pahari* population in Guanling County. *Journal of Mountain Agriculture and Biology*, 31(5): 381-384. (In Chinese with English Abstract)

潘清华, 王应祥, 岩崑. 2007. 中国哺乳动物彩色图鉴. 北京: 中国林业出版社. Pan Q H, Wang Y X, Yan K. 2007. A Field Guide to Mammals of China. Beijing: China Forestry Press. (In Chinese)

裴俊峰. 2011. 陕西省翼手类新纪录——大菊头蝠. 动物学杂志, 46(6): 130-133. Pei J F. 2011. A new record of woolly horseshoe bat (*Rhinolophus luctus*) in Shaanxi Province. *Chinese Journal of Zoology*, 46(6): 130-133. (In Chinese with English Abstract)

裴俊峰. 2012. 陕西省翼手类新纪录——西南鼠耳蝠. 四川动物, 31(2): 290-292. Pei J F. 2012. A new record of *Myotis altarium* in Shaanxi Province. *Sichuan Journal of Zoology*, 31(2): 290-292. (In Chinese with English Abstract)

裴俊峰, 冯祁君. 2014. 陕西省发现大足鼠耳蝠. 动物学杂志, 49(3): 443-446. Pei J F, Feng Q J. 2014. *Myotis ricketti* Found in Shaanxi Province. *Chinese Journal of Zoology*, 49(3): 443-446. (In Chinese with English Abstract)

彭红元, 陈伟才, 张修月. 2010. 麝的系统发育学研究进展. 四川动物, 29(5): 666-671. Peng H Y, Chen W C, Zhang X Y. 2010. A review on the phylogenetic study of musk deer. *Sichuan Journal of Zoology*, 29(5): 666-671. (In Chinese with English Abstract)

彭鸿授, 高耀亭, 陆长坤, 等. 1962. 四川西南和云南西北部兽类的分布研究. 动物学报, 14(Suppl.): 105-132. Peng H S, Gao Y T, Lu C K, *et al*. 1962. Report on mammals from south-western szechwen and northwestern Yunnan. *Acta Zoologica Sinica*, 14(Suppl.): 105-132. (In Chinese with English Abstract)

彭基泰, 周华明, 刘伟. 2007. 青藏高原东南横断山脉甘孜地区哺乳动物调查及区系研究报告. 四川动物, 25(4): 747-753. Peng J T, Zhou H M, Liu W. 2007. Investigation on mammal and fauna in Ganzi Prefecture in Hengduan Mountains, southeast of Qinghai—Tibet Plateau. *Sichuan Journal of Zoology*, 25(4): 747-753. (In Chinese with English Abstract)

彭培英, 郭宪国. 2014. 社鼠的研究现状及进展. 四川动物, 33(5): 792-800. Peng P Y, Guo X G. 2014. The research status and progresses of *Niviventer confucianus*. Sichuan *Journal of Zoology*, 33(5): 792-800. (In Chinese with English Abstract)

彭亚君, 王以凡, 钱周兴, 等. 2009. 石浦海域齿鲸类一新纪录种记述. 海洋学研究, 27(4): 117-120. Peng Y J, Wang Y F, Qian Z X, *et al*. 2009. Description of a new record species of whales from Chinese coastal waters. *Journal of Marine Sciences*, 27(4): 117-120. (In Chinese with English Abstract)

彭燕章, 叶智彰, 张耀平, 等. 1988. 金丝猴分类及系统发育关系. 动物学研究, 9(3): 239-248. Peng Y Z, Ye Z Z, Zhang Y P, *et al*. 1988. The classification and phylogeny of snub-nosed monkey (*Rhinopithecus* spp.) based on gross morphological characters. *Zoological Research*, 9(3): 239-248. (In Chinese with English Abstract)

朴龙国, 王绍先, 朴正吉. 2013. 长白山兽类. 长春: 吉林科学技术出版社. Piao L G, Wang S X, Piao Z J. 2013. Mammals of Changbai Mountains. Changchun: Science and Technology Publishing House of Jilin. (In Chinese)

朴仁峰, 俞曙林. 1990. 长白瀑布水流下游首次发现水鼩鼱. 延边农学院学报, (1): 58-60. Piao R F, Yu S L. 1990. Water shrews were first found in the lower reaches of Changbai Waterfall. *Journal of Yanbian Agricultural University*, (1): 58-60. (In

Chinese with English Abstract)

朴正吉, 睢亚臣, 崔志刚, 等. 2011. 长白山自然保护区猫科动物种群数量变化及现状. 动物学杂志, 46(3): 78-84. Piao Z J, Sui Y C, Cui Z G, *et al*. 2011. The history and current status of felid population in Changbai Mountain Nature Reserve. *Chinese Journal of Zoology*, 46(3): 78-84. (In Chinese with English Abstract)

平晓鸽, 李春旺, 李春林, 等. 2018. 普氏原羚分布、种群和保护现状. 生物多样性, 26(2): 177-184. Ping X G, Li C W, Li C L, *et al*. 2018. The distribution, population and conservation status of Przewalski's gazelle, *Procapra przewalskii*. *Biodiversity Science*, 26(2): 177-184. (In Chinese with English Abstract)

乔洪海, 刘伟, 杨维康, 等. 2011. 大沙鼠行为生态学研究现状. 生态学杂志, 30(3): 603-610. Qiao H H, Liu W, Yang W K, *et al*. 2011. Behavior ecology of great gerbil *Rhombomys opimus*: A review. *Chinese Journal of Ecology*, 30(3): 603-610. (In Chinese with English Abstract)

秦岭, 孟祥明, Kryukov A, 等. 2007. 陕西秦岭平河梁自然保护区小型兽类的组成与分布. 动物学研究, 3(3): 231-242. Qin L, Meng X M, Korablev A, *et al*. 2007. Species and distribution patterns of small mammals in the Pingheliang Nature Reserve of Qinling Mountain, Shaanxi. *Zoological Research*, 3(3): 231-242. (In Chinese with English Abstract)

秦瑜, 张明海. 2009. 中国马鹿的研究现状及展望. 野生动物学报, 30(2): 100-104. Qin Y, Zhang M H. 2009. Review of researches of red deer (*Cervus elaphus*) and perspects in China. *Chinese Journal of Wildlife*, 30(2): 100-104. (In Chinese with English Abstract)

秦长育. 1985. 阿拉善黄鼠数量分布及有关生态学调查分析. 动物学杂志, 20(6): 14-18. Qin C Y. 1985. Population distribution and related ecological investigation and analysis of *Spermophilus alaschanicus*. *Chinese Journal of Zoology*, 20(6): 14-18. (In Chinese with English Abstract)

秦长育. 1991. 宁夏啮齿动物区系及动物地理区划. 兽类学报, 11(2): 143-151. Qin C Y. 1991. On the faunistics and regionalization of glires in Ningxia Autonomous Region, China. *Acta Theriologica Sinica*, 11(2): 143-151. (In Chinese with English Abstract)

邱广龙, 周浩郎, 覃秋荣, 等. 2013. 海草生态系统与濒危海洋哺乳动物儒艮的相互关系. 海洋环境科学, 32(6): 970-974. Qiu G L, Zhou H L, Qin Q R, *et al*. 2013. Interactions between seagrass ecosystem and the endangered marine mammal dugong. *Marine Environmental Science*, 32(6): 970-974. (In Chinese with English Abstract)

裘丽, 冯祚建. 2004. 青藏公路沿线白昼交通运输等人类活动对藏羚羊迁徙的影响. 动物学报, 50(4): 669-674. Qiu L, Feng Z J. 2004. Effects of traffic during daytime and other human activities on the migration of Tibetan Antelope along the Qinghai-Tibet highway, Qinghai-Tibet Plateau. *Acta Zoologica Sinica*, 50(4): 669-674. (In Chinese with English Abstract)

任宝平, 李明, 魏辅文, 等. 2004. 滇金丝猴(*Rhinopithecus bieti*). 动物学杂志, 39(5): 111. Ren B P, Li M, Wei F W, *et al*. 2004. The Yunnan snub-nosed monkey (*Rhinopithecus bieti*). *Chinese Journal of Zoology*, 39(5): 111. (In Chinese with English Abstract)

任梦非, 黄海娇. 2009. 完达山东部林区冬季东北兔的生境选择. 野生动物学报, 30(6): 302-304. Ren M F, Huang H J. 2009. Winter habitat selection of Manchurian Hare in forest region of East Wandashan. *Chinese Journal of Wildlife*, 30(6): 302-304. (In Chinese with English Abstract)

沙依拉吾, 穆晨, 倪亦菲, 等. 2009. 新疆加依尔山发现草原鼠兔. 动物学杂志, 44(4): 152-154. Shayrave, Mu C, Ni Y F, *et al*. 2009. New record of *Ochotona pusilla* in Jayer Mountain, Xinjiang. *Chinese Journal of Zoology*, 44(4): 152-154. (In Chinese with English Abstract)

沙依拉吾, 武什肯. 1996. 社田鼠生物学特性的观察. 动物学杂志, 31(4): 25-27. Shayrave, Vashkent. 1996. Observations on the biological characteristics of *Microtus socialis*. *Chinese Journal of Zoology*, 31(4): 25-27. (In Chinese with English Abstract)

莎莉, 郭凤清. 1999. 内蒙古大兴安岭林区啮齿动物名录. 中国地方病防治杂志, (3): 174-175. Sha L, Guo F Q. 1999. Rodent list in the Greater Khingan range of Inner Mongolia. *Chinese Journal of Endemic Disease Control*, (3): 174-175. (In Chinese with English Abstract)

单文娟, 马合木提·哈力克. 2013. 塔里木兔种群遗传多样性初探. 生物技术, (3): 46-49. Shan W J, Haric M. 2013. Genetic diversity of *Lepus yarkandensis* population. *Biotechnology*, (3): 46-49. (In Chinese with English Abstract)

邵伟伟, 华攀玉, 周善义, 等. 2007. 棕果蝠微卫星位点的筛选及其对近缘种的通用性. 兽类学报, 27(4): 385-388. Shao W W, Hua P Y, Zhou S Y, *et al*. 2007. The isolation of new microsatellite loci in *Rousettus leschenaulti* and their applicability in closely related species. *Acta Theriologica Sinica*, 27(4): 385-388. (In Chinese with English Abstract)

生态环境部. 2018. 中国生态环境公报. 北京: 生态环境部. http://www.cenews.com.cn/ [2018-12-20] Minstry of Ecological Environment. 2018. Bulletin of China's Ecological Environment. Beijing: Minstry of Ecological Environment. http://www.

cenews.com.cn/ [2018-12-20] (In Chinese)
盛和林. 1983. 哺乳动物学概论. 上海: 华东师范大学出版社. Sheng H L. 1983. Introduction to Mammology. Shanghai: East China Normal University Press. (In Chinese)
盛和林, 陆厚基. 1982. 黄鼬的产仔环境和鼬巢密度调查. 兽类学报, 2(1): 29-34. Sheng H L, Lu H J. 1982. The Environment preference of nesting and nest density of the Female Weaseis (*Mustela sibirica*). *Acta Theriologica Sinica*, 2(1): 29-34. (In Chinese with English Abstract)
盛和林, 大泰司纪之, 陆厚基. 1998. 中国野生哺乳动物. 北京: 中国林业出版社. Sheng H L, Datai S J Z, Lu H J. 1998. Wild Mammals of China. Beijing: China Forestry Press. (In Chinese)
师蕾, 陈新文, 敬凯, 等. 2013. 云南省德钦县兽类区系调查. 云南师范大学学报(自然科学版), 33(5): 64-70. Shi L, Chen X W, Jing K, *et al*. 2013. Investigation of animal fauna in Deqin County of Yunnan Province. *Journal of Yunnan Normal University* (Natural Science Edition), 33(5): 64-70. (In Chinese with English Abstract)
施白南, 赵尔宓. 1980. 四川资源动物志. 成都: 四川人民出版社. Shi B N, Zhao E M. 1980. Animal Resources of Sichuan. Chengdu: Sichuan People's Publishing House. (In Chinese)
施立明, 陈玉泽. 1989. 鼷鹿云南亚种(*Tragulus javanicus williamsoni*) 的核型分析. 动物学报, 35: 41-44. Shi L M, Chen Y Z. 1989. The karyotype analysis of Yunnan Mouse Deer (*Tragulus javanicus williamsoni*). *Acta Zoologica Sinica*, 35: 41-44. (In Chinese with English Abstract)
施银柱, 边疆晖, 王权业, 等. 1991. 高寒草甸地区小哺乳动物群落多样性的初步研究. 兽类学报, 11(4): 279-284. Shi Y Z, Bian J H, Wang Q Y, *et al*. 1991. Studies on species diversity of small mammal community at alpine meadow. *Acta Theriologica Sinica*, 11(4): 279-284. (In Chinese with English Abstract)
施友仁, 王秀玉. 1978. 我国黄海北部发现的黑露脊鲸. 水产科技, (1): 25-27. Shi Y R, Wang X Y. 1978. Black right whale found in northern Yellow Sea of China. *Aquatic Science and Technology*, (1): 25-27. (In Chinese with English Abstract)
石红艳, 吴毅, 胡锦矗. 2000. 中华山蝠的研究进展及保护对策. 四川动物, 19(1): 39-40. Shi H Y, Wu Y, Hu J C. 2000. Research advances and conservation strategy on *Nyctalus velutinus*. *Sichuan Journal of Zoology*, 19(1): 39-40. (In Chinese with English Abstract)
石玉林, 于贵瑞, 王浩, 等. 2015. 中国生态环境安全态势分析与战略思考. 资源科学, 37(7): 1305-1313. Shi Y L, Yu G R, Wang H, *et al*. 2015. Assessment of China's ecological environment and strategic thinking. *Resources Science*, 37(7): 1305-1313. (In Chinese with English Abstract)
史密斯 A T, 解焱, 吉玛 F. 2009. 中国兽类野外手册. 长沙: 湖南教育出版社. Smith A T, Xie Y, Gemma F. 2009. A Guide to the Mammals of China. Changsha: Hunan Education Publishing House. (In Chinese)
史荣耀, 郎彩琴. 2000. 长尾仓鼠生态的观察. 四川动物, 19(1): 33-34. Shi R R, Lang C Q. 2000. Observation on ecology of *Cricetulus longicandatus*. *Sichuan Journal of Zoology*, 19(1): 33-34. (In Chinese with English Abstract)
史文博, 王慧, 朱立峰, 等. 2010. 晚更新世气候波动及长江阻隔对小麂皖南种群和大别山种群遗传分化与基因流模式的影响. 兽类学报, 30(4): 390-399. Shi W B, Wang H, Zhu L F, *et al*. 2010. The genetic divergence and gene flow pattern of two muntjac deer (*Muntiacus reevesii*) populations, Wannan and Dabie Mountains, from the effect of Yangtze River and the late Pleistocene glacial oscillations. *Acta Theriologica Sinica*, 30(4): 390-399. (In Chinese with English Abstract)
寿振黄. 1962. 中国经济动物志: 兽类. 北京: 科学出版社. Shou Z H. 1962. China Economic Animals: Mammals. Beijing: Science Press. (In Chinese)
寿振黄, 汪松, 陆长坤, 等. 1966. 海南岛的兽类调查. 动物分类学报, 3(3): 260-276. Shou Z H, Wang S, Lu C K, *et al*. 1966. Mammals on Hainan Island. *Acta Zootaxonomica Sinica*, 3(3): 260-276. (In Chinese with English Abstract)
寿振黄, 汪松. 1959. 海南食虫目(Insectivora)之一新属新种, 海南新毛猬(*Neohylomys hainanensis* gen. *et* sp. nov.). 动物学报 11(3): 422-426. Shou Z H, Wang S. 1959. New genus and new species of Hainan Insectivora: *Neohylomys hainanensis* gen. *et* sp. nov. *Acta Zoologica Sinica*, 11(3): 422-426. (In Chinese with English Abstract)
寿振黄, 张洁. 1958. 大竹鼠的初步调查. 生物学通报, (2): 28-30. Shou Z H, Zhang J. 1958. Preliminary investigation on giant bamboo rats. *Bulletin of Biology*, (2): 28-30. (In Chinese with English Abstract)
帅凌鹰, 宋延龄, 李俊生, 等. 2006. 黑河流域中游地区荒漠——绿洲景观区啮齿动物群落结构. 生物多样性, 14(6): 525-533. Shuai L Y, Song Y L, Li J S, *et al*. 2006. Rodent community structure of desert-oasis landscape in the middle reaches of the Heihe River. *Biodiversity Science*, 14(6): 525-533. (In Chinese with English Abstract)
宋华. 2009. 基于线粒体16SrDNA的贵州菊头蝠属(翼手目: 菊头蝠科)的分子系统进化关系. 贵阳: 贵州师范大学硕士研究生学位论文. Song H. 2009. Molecular Phylogenetic Relationship of the Genus Rhinolophus (Chiroptera: Rhinolophidae)

Based on Mitochondrial 16SrDNA. Guiyang: Master's Thesis of Guizhou Normal University. (In Chinese with English Abstract)

宋先华, 陈建, 周江. 2014. 贵州省发现高鞍菊头蝠. 动物学杂志, 49(1): 126-131. Song X H, Chen J, Zhou J. 2014. *Rhinolophus paradoxolophus* discovered in Guizhou Province. *Chinese Journal of Zoology*, 49(1): 126-131. (In Chinese with English Abstract)

宋延龄, 李俊生, 曾治高, 等. 2002. 甘肃河西走廊不同生境中鼠类群落结构初步研究. 生物多样性, 10(4): 386-392. Song Y L, Li J S, Zeng Z G, *et al*. 2002. Diversity of rodents communities in different habitats in Hexi Corridor, Gansu Province. *Biodiversity Science*, 10(4): 386-392. (In Chinese with English Abstract)

苏旭坤, 董世魁, 刘世梁, 等. 2014. 阿尔金山自然保护区土地利用/覆被变化对藏野驴栖息地的影响. 生态学杂志, 33(1): 141-148. Su X K, Dong S K, Liu S L, *et al*. 2014. Effects of land use/land cover change (LUCC) on habitats of Tibetan wild donkey in Aerjin Mountain National Nature Reserve. *Chinese Journal of Ecology*, 33(1): 141-148. (In Chinese with English Abstract)

粟海军, 蔡静芸, 冉景丞, 等. 2013. 贵州佛顶山自然保护区兽类资源及其特征分析. 四川动物, 32(1): 137-142. Su H J, Cai J Y, Ran J C, *et al*. 2013. A field survey and analysis on mammal resources of Fodingshan Nature Reserve in Guizhou Province. *Sichuan Journal of Zoology*, 32(1): 137-142. (In Chinese with English Abstract)

孙崇烁, 高耀亭. 1976. 我国猫科新纪录——云猫 (*Pardofelis marmorata*). 动物学报, 3: 15. Sun C S, Gao Y T. 1976. A new record of Felidae in China: *Pardofelis marmorata*. *Acta Zoologica Sinica*, 3: 15. (In Chinese with English Abstract)

孙国政, 倪庆永, 黄蓓, 等. 2012. 西黑冠长臂猿的种群数量、分布与现状. 林业建设, (1): 38-44. Sun G Z, Ni Q Y, Huang B, *et al*. 2012. Population, distribution and status of the west black crested gibbon. *Forestry Construction*, (1): 38-44. (In Chinese with English Abstract)

孙克萍, 冯江, 金龙如, 等. 2006. 依据回声定位声波参数判别同域栖息的蝙蝠种类. 东北师范大学学报(自然科学版), 38(3): 109-114. Sun K P, Feng J, Jin L R, *et al*. 2006. Identification of sympatric bat species by the echolocation calls. *Journal of Northeast Normal University* (Natural Science Edition), 38(3): 109-114. (In Chinese with English Abstract)

孙孟军, 鲍毅新. 2001. 浙江省獐的分布与资源调查. 浙江林业科技, 21(6): 20-24. Sun M J, Bao Y X. 2001. Investigation on distribution and resources of *Hydropotes inermis* in Zhejiang Province. *Journal of Zhejiang Forestry Science and Technology*, 21(6): 20-24. (In Chinese with English Abstract)

孙铭娟, 高行宜, 邵明勤. 2002. 鹅喉羚(*Gazella subgutturosa*)研究动态. 干旱区研究, 19(3): 75-80. Sun M J, Gao X Y, Shao M Q. 2002. Study Trends about *Gazella subgutturosa*. *Arid Zone Research*, 19(3): 75-80. (In Chinese with English Abstract)

孙平, 魏万红, 赵亚军, 等. 2005. 局部环境增温对根田鼠冬季种群的影响. 兽类学报, 25(3): 261-268. Sun P, Wei W H, Zhao Y J, *et al*. 2005. Effects of locally environmental warming on root vole population in winter. *Acta Theriologica Sinica*, 25(3): 261-268. (In Chinese with English Abstract)

孙涛, 王博石, 刘志瑾, 等. 2010. 近缘种扩增法对黑叶猴微卫星位点的筛选及特征分析. 兽类学报, 30(3): 351-353. Sun T, Wang B S, Liu Z J, *et al*. 2010. Identification and characterization of microsatellite markers via cross-species amplification from François' langur (*Trachypithecus francoisi*). *Acta Theriologica Sinica*, 30(3): 351-353. (In Chinese with English Abstract)

孙治宇, 刘少英, 郭延蜀, 等. 2013. 二郎山小型兽类区系及分布格局. 兽类学报, 33(1): 82-89. Sun Z Y, Liu S Y, Guo Y S, *et al*. The faunal composition and distribution of small mammals in Erlang Mountains. *Acta Theriologica Sinica*, 33(1): 82-89. (In Chinese with English Abstract)

谈建文, 冯顺柏, 张淑君, 等. 2005. 神农架野生猕猴及其生境现状. 湖北林业科技, (5): 27-29. Tan J W, Feng S B, Zhang S J, *et al*. 2005. Current status of wild *Macaca mulatta* and its habitat in Shennongjia. *Hubei Forestry Science and Technology*, (5): 27-29. (In Chinese with English Abstract)

谭邦杰. 1955. 哺乳类动物图鉴. 北京: 科学出版社. Tan B J. 1955. An illustrated handbook of Mammals. Beijing: Science Press. (In Chinese)

谭邦杰. 1992. 哺乳动物分类名录. 北京: 中国医药科技出版社. Tan B J. 1992. Classification Lists of Mammals. Beijing: China Medical Science and Technology Press. (In Chinese)

谭敏, 朱光剑, 洪体玉, 等. 2009. 中国翼手类新记录——小蹄蝠. 动物学研究, 30(2): 204-208. Tan M, Zhu G J, Hong T Y, *et al*. 2009. New record of a bat species from China, *Hipposideros cineraceus* (Blyth, 1853). *Zoological Research*, 30(2): 204-208. (In Chinese with English Abstract)

唐华兴, 陈天波, 刘晟源, 等. 2011. 广西弄岗自然保护区黑叶猴的种群动态. 四川动物, 30(1): 136-140. Tang H X, Chen T B,

Liu S Y, *et al*. 2011. The population dynamics of Francois Langur *Trachypithecus francoisi* in Nonggang Nature Reserve, Guangxi, China. *Sichuan Journal of Zoology*, 30(1): 136-140. (In Chinese with English Abstract)

唐占辉, 盛连喜, 曹敏, 等. 2005. 西双版纳地区犬蝠和棕果蝠食性的初步研究. 兽类学报, 25(4): 367-372. Tang Z H, Sheng L X, Cao M, *et al*. 2005. Diet of *Cynopterus sphinx* and *Rousettus leschenaulti* in Xishuangbanna. *Acta Theriologica Sinica*, 25(4): 367-372. (In Chinese with English Abstract)

唐中海, 彭波, 游章强, 等. 2009. 中华竹鼠的洞穴结构及其生境利用特征. 动物学杂志, 44(6): 36-40. Tang Z H, Peng B, You Z Q, *et al*. 2009. Habitation Selection and Den Structure Characteristics of *Rhizomys sinensis* in Piankou Natural Reserve. *Chinese Journal oF Zoology*, 44(6): 36-40. (In Chinese with English Abstract)

滕丽微, 刘振生, 宋延龄, 等. 2005. 海南大田保护区内赤麂的种群数量和特征. 兽类学报, 25(2): 138-142. Teng L W, Liu Z S, Song Y L, *et al*. 2005. Population size and characteristics of Indian Muntjac (*Muntiacus muntjak*) at Hainan Datian National Nature Reserve. *Acta Theriologica Sinica*, 25(2): 138-142. (In Chinese with English Abstract)

田贵全, 宋沿东, 刘强, 等. 2012. 山东省濒危物种多样性调查与评价. 生态环境学报, 21(1): 27-32. Tian G Q, Song Y D, Liu Q, *et al*. 2012. Investigation and evaluation of endangered species diversity in Shandong Province. *Ecology and Environment Sciences*, 21(1): 27-32. (In Chinese with English Abstract)

田瑜, 邬建国, 寇晓军, 等. 2009. 东北虎种群的时空动态及其原因分析. 生物多样性, 17(3): 211-225. Tian Y, Wu J G, Kou X J, *et al*. 2009. Spatiotemporal pattern and major causes of the Amur tiger population dynamics. *Biodiversity Science*, 17(3): 211-225. (In Chinese with English Abstract)

仝磊, 路纪琪. 2010a. 黄胸鼠对假海桐和截头石栎种子的贮藏和取食. 兽类学报, 30(3): 270-277. Tong L, Lu J Q. 2010a. Hoarding and consumption on seeds of *Pittosporopsis kerrii* and *Lithocarpus truncates* by Buff-breasted rat (*Rattus flavipectus*). *Acta Theriologica Sinica*, 30(3): 270-277. (In Chinese with English Abstract)

仝磊, 路纪琪. 2010b. 西双版纳地区小型哺乳动物群落结构及其季节变动. 生态学杂志 29(9): 1770-1776. Tong L, Lu J Q. 2010b. Community structure and its seasonal variation of small mammals in Xishuangbanna of Yunnan, China. *Chinese Journal of Ecology*, 29(9): 1770-1776. (In Chinese with English Abstract)

涂飞云, 韩卫杰, 刘晓华, 等. 2014. 江西哺乳动物组成及区系研究. 江西农业大学学报, 36(4): 848-854. Tu F Y, Han W J, Liu X H, *et al*. 2014. A study of mammal species composition and fauna in Jiangxi Province, China. *Journal of Jiangxi Agricultural University*, 36(4): 848-854. (In Chinese with English Abstract)

汪松. 1958. 食虫目. 见: 中国科学院动物研究所兽类组. 东北兽类调查报告. 北京: 科学出版社. Wang S. 1958. Insectivora. *In*: Mamamal Group of Institute of Zoology, Chinese Academy of Sciences. Northeast Mammals Survey Report. Beijing: Science Press. (In Chinese)

汪松. 1959. 东北兽类补遗. 动物学报, 11(3): 344-348. Wang S. 1959. Addendum to the mammals of northeast China. *Acta Zoologica Sinica*, 11(3): 344-348. (In Chinese with English Abstract)

汪松. 1998. 中国濒危动物红皮书 • 兽类. 北京: 科学出版社. Wang S. 1998. Red Book of Endangered Animals in China • Mammals. Beijing: Science Press. (In Chinese)

汪松, 解焱. 2004. 中国物种红色名录(第一卷) 红色名录. 北京: 高等教育出版社. Wang S, Xie Y. 2004. China Species Red List Vol. I Red List. Beijing: Higher Education Press. (In Chinese)

汪松, 解焱. 2009. 中国物种红色名录(第二卷) 脊椎动物. 北京: 高等教育出版社. Wang S, Xie Y. 2009. China Species Red List Vol. II Vertebrates. Beijing: Higher Education Press. (In Chinese)

王德忠, 罗宁, 谷景和, 等. 1998. 赛加羚羊(*Saiga tatarica*)在我国原产地的引种驯养. 生物多样性, 6: 309-311. Wang D Z, Luo N, Gu J H, *et al*. 1998. The introduction and domestication of Saiga (*Saiga tatarica*) in its original distribution area of China. *Biodiversity Sciences*, 6: 309-311. (In Chinese with English Abstract)

王定国. 1988. 额济纳旗和肃北马鬃山北部边境地区啮齿动物调查. 动物学杂志, 23(6): 24-27. Wang D G. 1988. Rodent survey in the northern border area of Ejinaqi and Mazong mountain, Subei. *Chinese Journal of Zoology*, 23(6): 24-27. (In Chinese with English Abstract)

王东风. 1993. 黑龙江省兽类新纪录——水鼩鼱. 野生动物, (4): 22-25. Wang D F. 1993. New animal record of Heilongjiang province—water shrew. *Chinese Journal of Wildlife*, (4): 22-25. (In Chinese with English Abstract)

王福麟, 王小非. 1995. 中国的复齿鼯鼠. 生物学通报, (7): 11-13. Wang F L, Wang X F. 1995. *Trogopterus xanthipes* in China. *Bulletin of Biology*, (7): 11-13. (In Chinese with English Abstract)

王好峰, 路纪琪, 汤发友, 等. 2008. 太行山猕猴自然保护区金钱豹资源现状及其保护. 河南林业科技, 28(2): 94-95. Wang H F, Lu J Q, Tang F Y, *et al*. 2008. Resources status and protection of *Pantera pardus* in Taihangshan Mountains National

Reseve, Jiyuan, China. *Journal of Henan Forestry Science & Technology*, 28(2): 94-95. (In Chinese with English Abstract)

王昊, 李松岗, 潘文石. 2002. 秦岭野生大熊猫(*Ailuropoda melanoleuca*)的种群存活力分析. 北京大学学报(自然科学版), 38(6): 756-761. Wang H, Li S G, Pan W S. 2002. Population Viability Analysis of giant panda (*Ailuropoda melanoleuca*) in Qinling Mountains. *Journal of Peking University* (Natural Science), 38(6): 756-761. (In Chinese with English Abstract)

王红娜. 2010. 河流不同生境中大趾鼠耳蝠回声定位声波研究. 长春: 东北师范大学硕士研究生学位论文. Wang H N. 2010. Echolocation of *Myotis macrodactylus* in Different Habitats of Rivers. Changchun: Master's Thesis of Northeast Normal University. (In Chinese with English Abstract)

王红愫. 2008. 大足鼠(*Rattus nitidus*)种群动态和繁殖特性研究. 云南大学学报(自然科学版), S1: 180-183, 201. Wang H S. 2008. Population dynamics and reproductive characteristics of *Rattus nitidus*. *Journal of Yunnan University* (Natural Science Edition), S1: 180-183, 201. (In Chinese with English Abstract)

王会, 李娜, 熬磊, 等. 2009. 毛腿鼠耳蝠的核型、G-带和C-带研究. 贵州师范大学学报(自然科学版), 27(2): 12-14. Wang H, Li N, Ao L, *et al*. 2009. A study on karyotypes, G-bands and C-bands of *Myotis fimbriatus*. *Journal of Guizhou Normal University* (Natural Sciences Edition), 27(2): 12-14. (In Chinese with English Abstract)

王火根, 范忠勇. 2004. 浙江海兽及其分布. 动物学杂志, 39(1): 60-63. Wang H G, Fan Z Y. 2004. Marine mammals and their distributions in coastal waters of Zhejiang. *Chinese Journal of Zoology*, 39(1): 60-63. (In Chinese with English Abstract)

王火根, 王宇. 1998. 东海发现的贝氏喙鲸. 水产科学, (5): 11-13. Wang H G, Wang Y. 1998. A Baird's beaked whale from the East China Sea. *Aquatic Science*, (5): 11-13. (In Chinese with English Abstract)

王健, 刘群秀, 唐登奎, 等. 2009. 湖北后河自然保护区果子狸栖息地选择的初步研究. 兽类学报, 29(2): 216-222. Wang J, Liu Q X, Tang D K, *et al*. 2009. Habitat selection of masked palm civet in Houhe Nature Reserve, Hubei. *Acta Theriologica Sinica*, 29(2): 216-222. (In Chinese with English Abstract)

王静. 2009. 多空间尺度下马铁菊头蝠生境选择与空间分布预测. 长春: 东北师范大学博士研究生学位论文. Wang J. 2009. Habitat selection and spatial distribution prediction of *Rhinolophus ferrumequinum* on multi-spatial scales. Changchun: Doctor's Thesis of Northeast Normal University. (In Chinese with English Abstract)

王静, 王新华, 江廷磊, 等. 2010. 马铁菊头蝠捕食活动与猎物资源的关系. 兽类学报, 30(2): 157-162. Wang J, Wang X H, Jiang T L, *et al*. 2010. Relationships between foraging activity of greater horseshoe bat (*Rhinolophus ferrumequinum*) and prey resources. *Acta Theriologica Sinica*, 30(2): 157-162. (In Chinese with English Abstract)

王静, Tiunov M P, 江廷磊, 等. 2009. 吉林省新纪录东方蝙蝠 *Vespertilio sinensis* (Peters, 1880)的回声定位声波特征与分析. 兽类学报, 29(3): 321-325. Wang J, Tiunov M P, Jiang T L, *et al*. 2009. Spectrum analysis of the echolocation calls of a new record species *Vespertilio sinensis* (Peters, 1880) from Jilin Province, China. *Acta Theriologica Sinica*, 29(3): 321-325. (In Chinese with English Abstract)

王君, 时坤, Riordan P. 2012. 新疆塔什库尔干岩羊和北山羊种群密度调查. 野生动物学报, 33(3): 113-117. Wang J, Shi K, Riordan P. 2012. Study on population density of ungulates in Taxkorgan, Xinjiang, China. *Chinese Journal of Wildlife*, 33(3): 113-117. (In Chinese with English Abstract)

王开锋, 靳铁治, 齐晓光, 等. 2010. 甘肃马鬃山发现小地兔. 动物学杂志, 45(6): 145-148. Wang K F, Jin T Z, Qi X G, *et al*. 2010. Little Earth Hare (*Pygeretmus pumilio*)was found in Mazongshan, Gansu Province. *Chinese Journal of Zoology*, 45(6): 145-148. (In Chinese with English Abstract)

王兰萍, 耿荣庆, 常洪, 等. 2009. 大额牛起源与系统地位的遗传学分析. 云南农业大学学报(自然科学版), 24(2): 231-234. Wang L P, Geng R Q, Chang H, *et al*. 2009. Genetic analysis on origin and phylogenetic status of gayal (*Bos frontalis*). *Journal of Yunnan Agricultural University* (Natural Science Edition), 24(2): 231-234. (In Chinese with English Abstract)

王磊, 江廷磊, 孙克萍, 等. 2010. 东亚水鼠耳蝠形态描述与分类. 动物分类学报, 35(2): 360-365. Wang L, Jiang T L, Sun K P, *et al*. 2010. Morphological description and taxonomical status of *Myotis petax*. *Acta Zootaxonomica Sinica*, 35(2): 360-365. (In Chinese with English Abstract)

王力军, 邢志刚, 汪继超, 等. 2010. 海南文昌椰林湾发现儒艮的尸体及死亡原因分析. 兽类学报, 30(3): 354-356. Wang L J, Xing Z G, Wang J C, *et al*. 2010. The recovered carcass of a dugong (*Dugong dugon*) in Yelin Bay of Wenchang City, Hainan Province and its cause of death. *Acta Theriologica Sinica*, 30(3): 354-356. (In Chinese with English Abstract)

王丕烈. 2011. 中国鲸类. 北京: 化学工业出版社. Wang P L. 2011. Chinese Cetaceans. Beijing: Chemical Industry Press. (In Chinese)

王丕烈, 韩家波, 马志强. 2008. 黄渤海斑海豹种群现状调查. 野生动物学报, 29: 29-31. Wang P L, Han J B, Ma Z Q. 2008. Status survey of spotted seals (*Phoca largha*) in Bohai and Yellow Sea. *China Journal of Wildlife*, 29: 29-31. (In Chinese with English Abstract)

王丕烈, 韩家波. 2007. 中国水域中华白海豚种群分布现状与保护. 海洋环境科学, 26(5): 484-487. Wang P L, Han J B. 2007. Present status of distribution and protection of Chinese white dolphin (*Sousa chinensis*) population in Chinese waters. *Marine Environmental Science*, 26(5): 484-487. (In Chinese with English Abstract)

王丕烈, 鹿志创. 2009. 中国水域灰鲸种群历史记录和现状分析. 水产科学, 28(12): 767-771. Wang P L, Lu Z C. 2009. Historical records and current status of Western Gray Whale in China's waters. *Fisheries Science*, 28(12): 767-771. (In Chinese with English Abstract)

王丕烈, 童慎汉, 袁红梅. 2007. 福建漳浦搁浅的小抹香鲸. 水产科学, 26(12): 671-674. Wang P L, Tong S H, Yuan H M. 2007. Stranding of pygmy sperm whale in Zhangpu, Fujian Province. *Fisheries Science*, 26(12): 671-674. (In Chinese with English Abstract)

王岐山, 李进华, 杨兆芬. 1994a. 中国的短尾猴. 生物学通报, 29(6): 5-7. Wang Q S, Li J H, Yang Z F. 1994a. Stump-tailed macaque in China. *Bulletin of Biology*, 29(6): 5-7. (In Chinese with English Abstract)

王岐山, 李进华, 李明. 1994b. 短尾猴种群生态学研究 I . 短尾猴种群动态及分析. 兽类学报, 14(3): 161-165. Wang Q S, Li J H, Li M. 1994b. Studies on population ecology of Tibetan monkeys (*Macaca thibetana*) I. Population dynamics and analysis of Tibetan monkeys. *Acta Theriologica Sinica*, 14(3): 161-165. (In Chinese with English Abstract)

王思博, 孙玉珍. 1997. 巨泡五趾跳鼠*Allactaga bullata* Allen分布区范围及界限. 疾病预防控制通报, (2): 87-92. Wang S B, Sun Y Z. 1997. On the areal limits and bounds of bullae enlarged jeboa *Allactaga bullata* Allen. *Endemic Diseases Bulletin*, (2): 87-92. (In Chinese with English Abstract)

王思博, 黎唯, 蒋卫, 等. 2000. 新疆啮齿动物新种新亚种新记录种与某些鼠种的新分布. 干旱区研究, 17(4): 23-26. Wang S B, Li W, Jiang W, *et al*. 2000. New species, new subspecies and newly recorded species of rodents and new distribution of some mice species of the Xinjiang in the recent years. *Arid Zone Research*, 17(4): 23-26. (In Chinese with English Abstract)

王廷正, 许文贤. 1993. 陕西啮齿动物志. 西安: 陕西师范大学出版社. Wang T Z, Xu W X. 1993. Rodents in Shaanxi. Xi'an: Shaanxi Normal University Press. (In Chinese)

王婉莹. 2007. 普氏蹄蝠、鲁氏菊头蝠、长翼蝠的生态、形态及耳蜗结构和听觉功能的研究. 西安: 陕西师范大学硕士研究生学位论文. Wang W Y. 2007. Ecology, Morphology, Cochlear Structure and Auditory Function of *Hipposideros pratti*, *Rhinolophus rouxii*, *Miniopterus Schrebersi*. Xi'an: Master's Thesis of Shaanxi Normal University. (In Chinese with English Abstract)

王先艳, 吴福星, 妙星, 等. 2013. 福建平潭一头误捕灰鲸的部分形态学记录. 兽类学报, 33(1): 18-27. Wang X Y, Wu F X, Miao X, *et al*. 2013. Partial morphological records of a gray whale (*Eschrichtius robustus*) incidentally caught at Pingtan, Fujian Province, China. *Acta Theriologica Sinica*, 33(1): 18-27. (In Chinese with English Abstract)

王香亭. 1991. 甘肃脊椎动物志. 兰州: 甘肃科学技术出版社. Wang X T. 1991. Vertebrates in Gansu. Lanzhou: Gansu Science and Technology Press. (In Chinese)

王晓云, 张秋萍, 郭伟健, 等. 2016. 水甫管鼻蝠在模式产地外的发现——广东和江西省新纪录. 兽类学报, 36(1): 118-122. Wang X Y, Zhang Q P, Guo W J, *et al*. 2016. Discovery of *Murina shuipuensis* outside of its type locality—new record from Guangdong and Jiangxi Provinces, China. *Acta Theriologica Sinica*, 36(1): 118-122. (In Chinese with English Abstract)

王新华. 2008. 马铁菊头蝠活动与其食物资源关系研究. 长春: 吉林农业大学硕士研究生学位论文. Wang X H. 2008. Study on the Relationship Between the Activities of *Rhinolophus ferrumequinum* and Their Food Resources. Changchun: Master's Thesis of Jilin Agricultural University. (In Chinese with English Abstract)

王亚明, 薛亚东, 夏友福. 2011. 滇西北滇金丝猴栖息地景观格局分析及其破碎化评价. 林业调查规划, 36(2): 34-37. Wang Y M, Xue Y D, Xia Y F. 2011. Landscape pattern and its fragmentation evaluation of habitat of *Rhinopithecus bieti* in Northwest Yunnan. *Forest Inventory and Planning*, 36(2): 34-37. (In Chinese with English Abstract)

王延校, 王芳, 高伶丽, 等. 2012. 山西省菊头蝠科1新纪录——大耳菊头蝠*Rhinolophus macrotis*. 河南师范大学学报(自然科学版), 40(2): 147-148. Wang Y X, Wang F, Gao L L, *et al*. 2012. New record of Rhinolophidae in Shanxi Province—*Rhinolophus macrotis*. *Journal of Henan Normal University* (Natural Science Edition), 40(2): 147-148. (In Chinese with English Abstract)

王延校. 2012. 云南南部洞栖蝙蝠初步调查. 新乡: 河南师范大学. Wang Y X. 2012. Preliminary Investigation on Cave Dwelling Bats in Southern Yunnan. Xinxiang: Master's Thesis of Henan Normal University. (In Chinese with English Abstract)

王应祥, 蒋学龙, 冯庆. 1999. 中国叶猴类的分类、现状与保护. 动物学研究, 20(4): 306-315. Wang Y X, Jiang X L, Feng Q.

1999. Taxonomy, status and conservation of Leaf Monkeys in China. *Zoological Research*, 20(4): 306-315. (In Chinese with English Abstract)

王应祥, 蒋学龙, 冯庆, 等. 1997. 云南豹猫资源量的可持续利用与保护. 兽类学报, 17: 31-42. Wang Y X, Jiang X L, Feng Q, *et al*. 1997. Abundance, sustainable utilization and conservation of leopard cat in Yunnan. *Acta Theriologica Sinica*, 17: 31-42. (In Chinese with English Abstract)

王应祥, 李崇云, 陈志平. 1996. 猪尾鼠的分类、分布与分化. 兽类学报, 16: 54-66. Wang Y X, Li C Y, Chen Z P. 1996. Taxonomy, distribution and differentiation on *Typhlomys cinereus* (Platacant homyidae, Mammalia). *Acta Theriologica Sinica*, 16: 54-66. (In Chinese with English Abstract)

王应祥. 2003. 中国哺乳动物种和亚种分类名录与分布大全. 北京: 中国林业出版社. Wang Y X. 2003. A Complete Checklist of Mammal Species and Subspecies in China－A Taxonomic and Geographic Reference. Beijing: China Forestry Publishing House. (In Chinese with English Abstract)

王应祥, 罗泽珣, 冯祚建. 1985. 云南兔*Lepus comus* G. Allen的分类订正——包括两个新亚种的描记. 动物学研究, 6(1): 802-804. Wang Y X, Luo Z X, Feng Z J. 1985. Taxonomic revision of Yunnan Hare, *Lepus comus* G. Allen with description of two new subspecies. *Zoological Research*, 6(1): 802-804. (In Chinese with English Abstract)

王酉之. 1985. 睡鼠科一新属新种——四川毛尾睡鼠. 兽类学报, 5: 67-75. Wang Y Z. 1985. A new genus and species of Gliridae—*Chaetocauda sichuanensis* gen. et sp. nov. *Acta Theriologica Sinica*, 5: 67-75. (In Chinese with English Abstract)

王酉之, 胡锦矗. 1999. 四川兽类原色图鉴. 北京: 中国林业出版社. Wang Y Z, Hu J C. 1999. Coloured Field Guide of Mammals in Sichuan. Beijing: China Forestry Press. (In Chinese)

王于玫, 刘泽昕, 张闻捷, 等. 2014. 短尾鼩江西省分布新纪录及其地理分布范围的探讨. 兽类学报, 34(2): 200-204. Wang Y M, Liu Z X, Zhang W J, *et al*. 2014. A new record of *Anourosorex squamipes* in Jiangxi Province with a discussion of its geographical range. *Acta Theriologica Sinica*, 34(2): 200-204. (In Chinese with English Abstract)

王玉玺, 张淑云. 1993. 中国兽类分布名录(一). 野生动物, 14(4): 12-17. Wang Y X, Zhang S Y. 1993. List of mammals' distribution in China (I). *Chinese Journal of Wildlife*, 14(4): 12-17.(In Chinese)

王淯, 胡锦矗, 谌利民, 等. 2006. 唐家河自然保护区小哺乳动物空间生态位初步研究. 兽类学报, 25(4): 379-384. Wang Y, Hu J C, Chen L M, *et al*. 2005. Preliminary study on spatial niches of small mammals in Tangjiahe Nature Reserve. *Acta Theriologica Sinica*, 25(4): 379-384. (In Chinese with English Abstract)

王渊, 初红军, 韩丽丽, 等. 2016. 野放普氏野马(*Equus przewalskii*)家域面积及其影响因素. 生态学报, 36(2): 545-553. Wang Y, Chu H J, Han L L, *et al*. 2016. Factors affecting the home range of reintroduced *Equus przewalskii* in the Mt. Kalamaili Ungulate Nature Reserve. *Acta Ecologica Sinica*, 36(2): 545-553. (In Chinese with English Abstract)

王渊, 刘务林, 刘锋, 等. 2019. 西藏墨脱县孟加拉虎种群数量调查. 兽类学报, 39(5): 504-513. Wang Y, Liu W L, Liu F, *et al*. 2019. Investigation on the population of wild Bengal tiger (*Panthera tigris tigris*) in Mêdog, Tibet. *Acta Theriologica Sinica*, 39(5): 504-513. (In Chinese with English Abstract)

王正寰, 王小明, 鲁庆斌. 2004. 四川省石渠县藏狐昼间行为特征观察. 兽类学报, 24(4): 357-360. Wang Z H, Wang X M, Lu Q B. 2004. Observation on the daytime behaviour of Tibetan for (*Vulpes ferrilata*) in Shiqu county, Sichuan Province, China. *Acta Theriologica Sinica*, 24(4): 357-360. (In Chinese with English Abstract)

王志伟, 谢梦洁, 赵言文. 2010. 江苏省生物多样性及其保育. 江苏农业科学, (4): 341-344. Wang Z W, Xie M J, Zhao Y W. 2010. Biodiversity and its conservation in Jiangsu Province. *Jiangsu Agricultural Science*, (4): 341-344. (In Chinese with English Abstract)

王宗祎, 汪松. 1962. 青海发现的大狐蝠(*Pteropus giganteus* Brünnich). 动物学报, (4): 63. Wang Z W, Wang S. 1962. The discovery of a flying fox (*Pteropus giganteus Brünnich*) from Chin-Hai Province, Northwestern China. *Acta Zoologica Sinica*, (4): 63. (In Chinese with English Abstract)

韦力. 2007. 黑髯墓蝠的食性、回声定位信号特征及其出飞时间的研究. 桂林: 广西师范大学硕士研究生学位论文. Wei L. 2007. Study on the Feeding Habits, Echolocation Signal Characteristics and Flight-out Time of *Taphozous melanopogon*. Guilin: Master's Thesis of Guangxi Normal University. (In Chinese with English Abstract)

韦力, 甘雨满, 李周全, 等. 2011. 六种共栖蝙蝠的回声定位信号及翼型特征的比较. 兽类学报, 31(2): 155-163. Wei L, Gan Y M, Li Z Q, *et al*. 2011. Comparisons of echolocation calls and wing morphology among six sympatric bat species. *Acta Theriologica Sinica*, 31(2): 155-163. (In Chinese with English Abstract)

韦力, 周善义, 张礼标, 等. 2006. 三种共栖蝙蝠的回声定位信号特征及其夏季食性的比较. 动物学研究, 27(3): 235-241. Wei L, Zhou S Y, Zhang L B, *et al*. 2006. Characteristics of echolocation calls and summer diet of three sympatric insectivorous bat species. *Zoological Research*, 27(3): 235-241. (In Chinese with English Abstract)

魏学文. 2013. 基于核基因的中国大耳菊头蝠复合体(*Rhinolophus macrotis* complex) 比较分子系统地理学研究. 长春: 东北师范大学硕士研究生学位论文. Wei X W. 2013. Comparative Molecular Phylogeography of *Rhinolophus macrotis* Complex in China Based on Nuclear Genes. Changchun: Master's Thesis of Northeast Normal University. (In Chinese with English Abstract)

温立嘉, 时坤, 黄建, 等. 2014. 西藏墨脱鸟兽红外相机监测初报. 生物多样性, 22(6): 798-799. Wen L J, Shi K, Huang J, *et al*. 2014. Preliminary analysis of mammal and bird diversity monitored with camera traps in Mêdog, Tibet. *Biodiversity Science*, 22(6): 798-799. (In Chinese with English Abstract)

温知新, 尹三军, 冉江洪, 等. 2010. 四川洪雅县赤腹松鼠巢址选择研究. 四川动物, 29(5): 540-545. Wen Z X, Yin S J, Ran J H, *et al*. 2010. Nest-site selection by the red-bellied squirrel (*Callosciueus erythraeus*) in Hongya County, Sichuan Province. *Sichuan Journal of Zoology*, 29(5): 540-545. (In Chinese with English Abstract)

文榕生. 2016. 再探历史时期新疆分布的虎. 四川动物, 35(2): 311-320. Wen R S. 2016. Historical distribution of tiger in Xinjiang Uygur Autonomous Region. *Sichuan Journal of Zoology*, 35(2): 311-320. (In Chinese with English Abstract)

吴爱国. 2001. 板齿鼠的种群数量变动. 中国媒介生物学及控制杂志, 12(1): 14-15. Wu A G. 2001. Variation of the population density of *Bandicota indica*. *Chinese Journal of Vector Biology and Control*, 12(1): 14-15. (In Chinese with English Abstract)

吴爱国. 2002. 卡氏小鼠的种群数量变动. 中国媒介生物学及控制杂志, 13(4): 255-256. Wu A G. 2002. Variation of the population density of *Mus caroli*. *Chinese Journal of Vector Biology and Control*, 13(4): 255-256. (In Chinese with English Abstract)

吴德林, 奉勇, 窦秦川, 等. 1995. 卡氏小鼠种群数量变动特征及其与环境因子的关系. 兽类学报, 15(1): 60-64. Wu D L, Feng Y, Dou Q C, *et al*. 1995. The population fluctuation characteristics and relations to environmental factors in *Mus caroli*. *Acta Theriologica Sinica*, 15(1): 60-64. (In Chinese with English Abstract)

吴家炎, 裴俊峰. 2007. 白唇鹿的研究现状及保护策略. 野生动物学报, 28(5): 36-39. Wu J Y, Pei J F. 2007. Present status of research of white-lipped deer and its conservation strategy. *China Journal of Wildlife*, 28(5): 36-39. (In Chinese with English Abstract)

吴家炎, 裴俊峰. 2011. 秦岭和大巴山区翼手类及其动物地理分布特征. 兽类学报, 31(4): 358-370. Wu J Y, Pei J F. 2011. Bats (Chiroptera) and their zoogeographic distribution characteristics in the Qinling and Daba Mountain ranges. *Acta Theriologica Sinica*, 31(4): 358-370. (In Chinese with English Abstract)

吴家炎, 王伟. 2006. 中国麝类. 北京: 中国林业出版社. Wu J Y, Wang W. 2006. Chinese Musk Deer. Beijing: China Forestry Press. (In Chinese)

吴建国, 吕佳佳. 2009. 气候变化对滇金丝猴分布的潜在影响. 气象与环境学报, 25(6): 1-10. Wu J G, Lyu J J. 2009. Potential effects of climate change on the distributions of Yunnan snub-nosed monkey (*Pygathrix bieti*) in China. *Journal of Meteorology and Environment*, 25(6): 1-10. (In Chinese with English Abstract)

吴鹏举, 张恩迪. 2006. 西藏慈巴沟自然保护区羚牛栖息地选择. 兽类学报, 26(2): 152-158. Wu P J, Zhang E D. 2006. Habitat selection of takin (*Budorcas taxicolor*) in Cibagou Nature Reserve of Tibet, China. *Acta Theriologica Sinica*, 26(2): 152-158. (In Chinese with English Abstract)

吴诗宝, 王应祥, 冯庆. 2005. 中国兽类一新纪录——爪哇穿山甲. 动物分类学报, 30(2): 440-443. Wu S B, Wang Y X, Feng Q. 2005. A new record of Mammalia in China: *Manis javanica*. *Acta Zootaxonomica Sinica*, 30(2): 440-443. (In Chinese with English Abstract)

吴逸群, 刘科科. 2010. 川金丝猴生态生物学研究进展. 陕西林业科技, (6): 42-44. Wu Y Q, Liu K K. 2010. Ecological biology of Sichuan snub-nosed Monkey. *Shaanxi Forest Science And Technology*, (6): 42-44. (In Chinese with English Abstract)

吴毅, 本川雅治, 李玉春, 等. 2011. 广东省二种兽类新纪录——鼩猬 (*Neotetracus sinensis*) 和短尾鼩 (*Anourosorex squamipes*). 兽类学报, 31(3): 317-319. Wu Y, Motokawa M, Li Y C, *et al*. 2011. New records of shrew gymnure (*Neotetracus sinensis*) and Chinese mole shrew (*Anourosorex squamipes*) from Guangdong Province. *Acta Theriologica Sinica*, 31(3): 317-319. (In Chinese with English Abstract)

吴毅, 陈子禧, 王晓云, 等. 2017. 哈氏管鼻蝠在广东的新发现及南岭树栖蝙蝠物种多样性. 广州大学学报(自然科学版), 16(3): 1-7. Wu Y, Chen Z X, Wang X Y, et al. 2017. New record of Murina harissoni in Guangdong and diversity of arboreal bats in Nanling National Nature Reserve. *Journal of Guangzhou University*, 16(3): 1-7. (In Chinese with English Abstract)

吴毅, 侯万儒, 胡锦矗, 等. 2000. 四川地区大蹄蝠某些年龄特征的比较研究. 兽类学报, 20(4): 284-288. Wu Y, Hou W R, Hu

J C, *et al*. 2000. Comparative study on some age characteristic of *Hipposideros armiger* in Sichuan area. *Acta Theriologica Sinica*, 20(4): 284-288. (In Chinese with English Abstract)

吴毅, 胡锦矗, 张国修, 等. 1988. 四川省兽类新纪录. 四川动物, 7(3): 39. Wu Y, Hu J C, Zhang G X, *et al*. 1988. New mammal records in Sichuan. *Sichuan Journal of Zoology*, 7(3): 39. (In Chinese)

吴毅, 梁颖华, 尤君丽, 等. 2001. 广东省蝙蝠三新记录. 四川动物, 20(2): 91. Wu Y, Liang Y H, You J L, *et al*. 2001. Three new records of bats in Guangdong Province. *Sichuan Journal of Zoology*, 20(2): 91. (In Chinese)

吴毅, 彭洪源. 2005. 广东省蝙蝠(Chiroptera)二新记录. 四川动物, 24(2): 176-177. Wu Y, Peng H Y. 2005. Two new records of the Chiroptera in Guangdong Province. *Sichuan Journal of Zoology*, 24(2): 176-177. (In Chinese with English Abstract)

吴毅, 魏辅文, 袁重桂, 等. 1990. 两种纹背鼩鼱鉴别特征的探讨. 四川动物, 1(1): 26. Wu Y, Wei F W, Yuan Z G, *et al*. 1990. Identification of two species of stripe-backed shrew. *Sichuan Journal of Zoology*, 1(1): 26. (In Chinese)

吴毅, 杨奇森, 夏霖, 等. 2004a. 中国蝙蝠新记录——马氏菊头蝠. 动物学杂志, 39(5): 109-110. Wu Y, Yang Q S, Xia L, *et al*. 2004a. New record of Chinese bats: *Rhinolophus marshalli*. *Chinese Journal of Zoology*, 39(5): 109-110. (In Chinese with English Abstract)

吴毅, 余嘉明, 曾凡, 等. 2014. 广东兽类新纪录——褐扁颅蝠及其中国的地理分布. 广州大学学报(自然科学版), 13(6): 23-27. Wu Y, Yu J M, Zeng F, *et al*. 2014. First record of *Tylonycteris robustula* in Guangdong Province with a discussion of its geographical range in China. *Journal of Guangzhou University* (Natural Science Edition), 13(6): 23-27. (In Chinese with English Abstract)

吴毅, 余文华, 李小琼. 2005. 小黄蝠(*Scotophilus kuhlii*)生态的初步研究. 哈尔滨: 第二届全国野生动物生态与资源保护学术讨论会. Wu Y, Yu W H, Li X Q. 2005. A preliminary study on the ecology of *Scotophilus kuhlii*. Harbin: The 2nd National Symposium on Wildlife Ecology and Resource Conservation. (In Chinese)

吴毅, 张成菊, 梁智文, 等. 2007. 广州市区翼手类物种多样性的研究. 广州大学学报(自然科学版), 6(2): 14-17. Wu Y, Zhang C J, Liang Z W, *et al*. 2007. Study on species diversity of Chiroptera in Guangzhou. *Journal of Guangzhou University* (Natural Science Edition), 6(2): 14-17. (In Chinese with English Abstract)

吴毅, 张成菊, 余文华, 等. 2006. 广州市蝙蝠的多样性及在农业生态环境中的作用. 华南农业大学学报, 27(4): 47-51. Wu Y, Zhang C J, Yu W H, *et al*. 2006. Research on bat diversity and its agricultural eco-environment function in Guangzhou City. *Journal of South China Agricultural University*, 27(4): 47-51. (In Chinese with English Abstract)

吴毅, 郑福军, 李艳, 等. 2004b. 广州地区濒危物种扁颅蝠*Tylonycteris pachypus*的种群数量变化与环境因素的关系. 中山大学学报(自然科学版), 43(5): 91-94. Wu Y, Zheng F J, Li Y, *et al*. 2004b. The relationship of quantitative change of *Tylonycteris pachypus* population with factors of environment in Guangzhou. *Journal of Sun Yat-Sen University* (Natural Science Edition), 43(5): 91-94. (In Chinese with English Abstract)

武明录, 王秀辉, 安春林, 等. 2006. 河北省兽类资源调查. 河北林业科技, (2): 20-23. Wu M L, Wang X H, An C L, *et al*. 2006. Animals resources investigation of Hebei province. *The Journal of Hebei Forestry Science and Technology*, (2): 20-23. (In Chinese with English Abstract)

武文华, 付和平, 武晓东, 等. 2007. 应用马尔可夫链模型预测长爪沙鼠和黑线仓鼠种群数量. 动物学杂志 42(6): 69-78. Wu W H, Fu H P, Wu X D, *et al*. 2007. Forecasting the population dynamics of *Meriones unguiculatus* and *Cricetulus barabansis* by applying Markov model. *Chinese Journal oF Zoology*, 42(6): 69-78. (In Chinese with English Abstract)

武晓东, 傅和平, 苏吉安, 等. 2002. 两种小型兽类在我国的新分布区. 动物学杂志, 37(2): 67-68. Wu X D, Fu H P, Su J A, *et al*. 2002. The new area of distribution of two small mammals in China. *Chinese Journal of Zoology*, 37(2): 67-68. (In Chinese with English Abstract)

武晓东, 傅和平, 庄光辉, 等. 2003. 内蒙古阿拉善地区啮齿动物的地理分布及区划. 动物学杂志, 38(2): 27-31. Wu X D, Fu H P, Zhuang G H, *et al*. 2003. Geographical distribution of rodents in the Alashan Region of Inner Mongolia. *Chinese Journal of Zoology*, 38(2): 27-31. (In Chinese with English Abstract)

夏霖, 杨奇森, 魏辅文, 等. 2004. 马麝诸种群地理分化初步探讨. 兽类学报, 24(1): 1-5. Xia L, Yang Q S, Wei F W, *et al*. 2004. Study on geographical division of alpine musk deer (*Moschus sifanicus*). *Acta Theriologica Sinica*, 24(1): 1-5. (In Chinese with English Abstract)

夏武平. 1964. 中国动物图谱: 兽类. 北京: 科学出版社. Xia W P. 1964. Album of China's Animals: Mammals. Beijing: Science Press. (In Chinese)

夏武平, 方喜业. 1964. 巨泡五趾跳鼠(跳鼠科)之一新亚种. 动物分类学报, 1(1): 18-20. Xia W P, Fang X Y. 1964. A new subspecies of *Allactaga bullata* (Diplodidae). *Acta Zootaxonomica Sinica*, 1(1): 18-20. (In Chinese with English Abstract)

夏武平, 张荣祖. 1995. 灵长类研究与保护. 北京: 中国林业出版社. Xia W P, Zhang R Z. 1995. Primates: Research and Conservation. Beijing: China Forestry Publishing House. (In Chinese)

夏亚军. 2011. 雾灵山动物垂直分布. 河北林业科技, (1): 29-30. Xia Y J. 2011. Study on vertical distribution of the animals in Wuling mountain. *Hebei Forestry Science and Technology*, (1): 29-30. (In Chinese with English Abstract)

肖红, 侯兰新, 刘坪, 等. 2003. 伊犁地区兽类区系调查. 陕西师范大学学报(自然科学版), (s2): 14-20. Xiao H, Hou L X, Liu P, *et al*. 2003. Fauna survey of Yili area. *Journal of Shaanxi Normal University* (Natural Science Edition), (s2): 14-20. (In Chinese with English Abstract)

谢文华, 杨锡福, 李俊年, 等. 2014. 八大公山自然保护区地栖性小兽多样性初步研究. 生物多样性, 22(2): 216-222. Xie W H, Yang X F, Li J N, *et al*. 2014. A preliminary study of the biodiversity of ground-dwelling small mammals in Badagongshan National Nature Reserve, Hunan Province. *Biodiversity Science*, 22(2): 216-222. (In Chinese with English Abstract)

辛景禧, 邱梦辞. 1990. 臭鼩鼱(*Suncus murinus*)的繁殖生物学初步研究. 生态科学, (1): 129-140. Xin J X, Qiu M C. 1990. Preliminary study on reproductive biology of *Suncus murinus*. *Ecological Science*, (1): 129-140. (In Chinese with English Abstract)

邢雅俊, 周立志, 马勇. 2008. 中国湿润半湿润地区啮类动物的分布格局. 动物学杂志, 43(5): 51-61. Xing Y J, Zhou L Z, Ma Y. 2008. Geographical distribution pattern of glires species in the humid and semi-humid Region of China. *Chinese Journal of Zoology*, 43(5): 51-61. (In Chinese with English Abstract)

徐爱春, 斯幸峰, 王彦平, 等. 2014. 千岛湖片段化栖息地地栖哺乳动物的红外相机监测及最小监测时长. 生物多样性, 22(6): 764-772. Xu A C, Si X F, Wang Y P, *et al*. 2014. Camera traps and the minimum trapping effort for ground-dwelling mammals in fragmented habitats in the Thousand Island Lake, Zhejiang Province. *Biodiversity Science*, 22(6): 764-772. (In Chinese with English Abstract)

徐纯柱, 张洪海, 马建章. 2010. 紫貂线粒体基因组全序列结构及其进化. 北京林业大学学报, 32(1): 82-88. Xu C Z, Zhang H H, Ma J Z. 2010. Organization of the complete mitochondrial genome and its evolution in sable. *Journal of Beijing Forestry University*, 32(1): 82-88. (In Chinese with English Abstract)

徐峰, 马鸣, 吴逸群. 2011. 新疆托木尔峰国家级自然保护区雪豹的种群密度. 兽类学报, 31(2): 205-210. Xu F, Ma M, Wu Y Q. 2011. Population density of snow leopards (*Panthera uncia*) in Tomur National Nature Reserve of Xinjiang, China. *Acta Theriologica Sinica*, 31(2): 205-210. (In Chinese with English Abstract)

徐海龙. 2012. 大蹄蝠、大菊头蝠及红白鼯鼠线粒体全基因组序列分析. 雅安: 四川农业大学硕士研究生学位论文. Xu H L. 2012. Mitochondrial Genome Sequence Analysis of *Hipposideros armiger*, *Rhinolophus luctus* and *Petaurista alborufus*. Ya'an: Master's Thesis of Sichuan Agricultural University. (In Chinese with English Abstract)

徐剑, 邹佩贞, 温彩燕, 等. 2002. 广东省大陆翼手目动物区系与地理区划. 中山大学学报(自然科学版), 41(3): 77-80. Xu J, Zou P Z, Wen C Y, *et al*. 2002. A study on the fauna and geographic distribution of Chiroptera in continent of Guangdong Province. *Journal of Sun Yat-Sen University* (Natural Science Edition), 41(3): 77-80. (In Chinese with English Abstract)

徐金会, 王琳琳, 薛慧良, 等. 2009. 喜马拉雅旱獭种群微卫星变异及遗传多样性. 动物学杂志, 44(2): 34-40. Xu J H, Wang L L, Xue H L, *et al*. 2009. Microsatellite variation and genetic diversity in *Marmota himalayana*. *Chinese Journal of Zoology*, 44(2): 34-40. (In Chinese with English Abstract)

徐龙辉. 1984. 花白竹鼠(*Rhizomys pruinosus*)的生物学研究. 兽类学报, 4(2): 99-105. Xu L H. 1984. Biological study on *Rhizomys pruinosus*. *Acta Theriologica Sinica*, 4(2): 99-105. (In Chinese with English Abstract)

徐伟霞, 周昭敏, 章敬旗, 等. 2005. 中华菊头蝠头骨形态特征地理差异的研究. 四川动物, 24(4): 31-34. Xu W X, Zhou Z M, Zhang J Q, *et al*. 2005. Study on geographical variations of skull morphology of *Rhinolophus sinicus*. *Sichuan Journal of Zoology*, 24(4): 31-34. (In Chinese with English Abstract)

徐文轩, 乔建芳, 刘伟, 等. 2008. 鹅喉羚生态生物学研究现状. 生态学杂志, 27(2): 257-262. Xu W X, Qiao J F, Liu W, *et al*. 2008. Ecology and biology of *Gazella subgutturosa*: Current situation of studies. *Chinese Journal of Ecology*, 27(2): 257-262. (In Chinese with English Abstract)

徐学良. 1975. 分布在黑龙江省的白鼬. 动物学杂志, 10(3): 28-29. Xu X L. 1975. White Stoats distributed in Heilongjiang Province. *Chinese Journal of Zoology*, 10(3): 28-29. (In Chinese with English Abstract)

徐亚君, 程炳功, 方德安, 等. 1984. 安徽省徽州地区翼手类及其越冬生态的初步观察. 兽类学报, 5(2): 86-94. Xu Y J, Chen B G, Fang D A, *et al*. 1984. A preliminary observation on Chiroptera in Huizhou region, Anhui province and their overwintering ecology. *Acta Theriologica Sinica*, 5(2): 86-94. (In Chinese with English Abstract)

徐亚君, 程炳功, 方德安, 等. 1982. 宽耳犬吻蝠(*Tadarida teniotis* Rafinesoque)在安徽的发现. 兽类学报, 1(2): 200. Xu Y J,

Cheng B G, Fang D A, *et al.* 1982. On discovering the *Tadarida teniotis* Rafinesque in Anhui. *Acta Theriologica Sinica*, 1(2): 200. (In Chinese)

徐肇华, 黄文几. 1982. 仓鼠属三个种的核型分析. 兽类学报, 2(2): 201-210. Xu Z H, Huang W J. 1982. On the karyotypes of three species of *Cricetulus*. *Acta Theriologica Sinica*, 2(2): 201-210. (In Chinese with English Abstract)

徐忠鲜, 余文华, 吴毅, 等. 2013. 江西省翼手目一新纪录——无尾蹄蝠. 四川动物, 32(2): 263-266, 268. Xu Z X, Yu W H, Wu Y, *et al.* 2013. A new record bat of *Coelops frithi* in Jiangxi Province, China. *Sichuan Journal of Zoology*, 32(2): 263-266, 268. (In Chinese with English Abstract)

徐忠鲜, 余文华, 吴毅, 等. 2014. 艾氏管鼻蝠种群遗传结构初步研究及其分类探讨. 兽类学报, 34(3): 270-277. Xu Z X, Yu W H, Wu Y, *et al.* 2014. Preliminary study on population genetic structure and taxonomy of Elery's tubenosed bat (*Murina eleryi*). *Acta Theriologica Sinica*, 34(3): 270-277. (In Chinese with English Abstract)

许立杰, 冯江, 刘颖, 等. 2008. 小菊头蝠和单角菊头蝠分类地位的探讨. 东北师大学报(自然科学版), 40(1): 95-99. Xu L J, Feng J, Liu Y, *et al.* 2008. Taxonomic status of *Rhinolophus blythi* and *R. monoceros*. *Journal of Northeast Normal University* (Natural Science Edition), 40(1): 95-99. (In Chinese with English Abstract)

许凌, 范宇, 蒋学龙, 等. 2013. 树鼩进化分类地位的分子证据. 动物学研究, 34(2): 70-76. Xu L, Fan Y, Jiang X L, *et al.* 2013. Molecular evidence on the phylogenetic position of tree shrews. *Zoological Research*, 34(2): 70-76. (In Chinese with English Abstract)

许再富. 2000. 历史上向“天朝”上贡对滇南犀牛灭绝和亚洲象濒危过程的影响. 生物多样性, 8(1): 112-119. Xu Z F. 2000. The effects of paying tribute to the imperial court in the history on rhinoceros' extinction and elephant's endangerment in Southern Yunnan. *Biodiversity Sciences*, 8(1): 112-119. (In Chinese with English Abstract)

严旬. 2005. 野生大熊猫现状、面临的挑战及展望. 兽类学报, 25(4): 402-406. Yan X. 2005. Status, challenge and prospect of wild giant pandas. *Acta Theriologica Sinica*, 25(4): 402-406. (In Chinese with English Abstract)

严志堂, 钟明明. 1984. 小家鼠(*Mus musculus*)种群动态预测及机制的探讨. 兽类学报, 4(2): 139-146. Yan Z T, Zhong M M. 1984. The prediction to fluctuations in Home Mouse (*Mus musculus*) population. *Acta Theriologica Sinica*, 4(2): 139-146. (In Chinese with English Abstract)

杨道德, 马建章, 何振, 等. 2007. 湖北石首麋鹿国家级自然保护区麋鹿种群动态. 动物学报, 53(6): 947-952. Yang D D, Ma J Z, He Z, *et al.* 2007. Population dynamics of the Père David's deer *Elaphurus davidianus* in Shishou Milu National Nature Reserve, Hubei Province, China. *Acta Zoologica Sinica*, 53(6): 947-952. (In Chinese with English Abstract)

杨德华, 等. 1993. 西双版纳动物志. 昆明: 云南大学出版社: 1-299. Yang D H, *et al.* 1993. Fauna of Xishuangbanna. Kunming: Yunnan University Press: 1-299. (In Chinese)

杨德华, 张家银, 李纯. 1988. 云南野牛的数量分布. 动物学杂志, 23(1): 39-41, 57. Yang D H, Zhang J Y, Li C. 1988. Population distribution of wild ox in Yunnan. *China Journal of Zoology*, 23(1): 39-41, 57. (In Chinese with English Abstract)

杨光, 周开亚. 1996. 误捕及其对海兽种群的影响. 应用生态学报, 7(3): 326-331. Yang G, Zhou K Y. 1996. Incidental catch and its impact on marine mammal populations. *Chinese Journal of Applied Ecology*, 7(3): 326-331. (In Chinese with English Abstract)

杨光荣, 王应祥. 1987. 休氏壮鼠(*Hadromys humei*)一新亚种. 兽类学报, 7(1): 46-50. Yang G R, Wang Y X. 1987. A new subspecies of *Hadromys humei* (Muridae, Mammalia) from Yunnan, China. *Acta Theriologica Sinica*, 7(1): 46-50. (In Chinese with English Abstract)

杨光荣, 杨学时. 1985. 大绒鼠的生物学资料. 动物学杂志, 20: 38-44. Yang G R, Yang X S. 1985. Biological information of *Eothenomys miletus*. *Chinese Journal of Zoology*, 20: 38-44. (In Chinese with English Abstract)

杨光照. 2007. 云南省的猕猴资源现状及其保护与开发利用. 西部林业科学, 36(2): 147-149. Yang G Z. 2007. Present resource situation of *Macaca mulatta* and its protection and utilization. *Western Forestry Science*, 36(2): 147-149. (In Chinese)

杨海龙, 李迪强, 朵海瑞, 等. 2010. 梵净山国家级自然保护区植被分布与黔金丝猴生境选择. 林业科学研究, 23(3): 393-398. Yang H L, Li D Q, Duo H R, *et al.* 2010. Vegetation distribution in Fanjing Mountain National Nature Reserve and habitat selection of Guizhou Golden Monkey. *Forest Science and Technology Research*, 23(3): 393-398. (In Chinese with English Abstract)

杨鸿嘉. 1976. 台湾产鲸类之研究. 台湾博物馆科学季刊, 19: 131-178. Yang H J. 1976. Research on cetaceans from Taiwan. *Taiwan Museum Science Quarterly*, 19: 131-178. (In Chinese)

杨锐, 孙俊杰, 王福勋. 2010. 不同地区大菊头蝠的比较与发现. 见: 中国声学学会. 2010年中国西部声学学术交流会. 云南腾冲. Yang R, Sun J J, Wang F X. 2010. Comparison and discovery of greater horseshoe bats in different regions. *In*:

Acoustical Society of China. Proceedings of 2010 Western China Acoustics Academic Exchange Conference. Tengchong, Yunnan. (In Chinese)

杨士剑, 诸葛阳. 1989. 臭鼩的繁殖和种群年龄结构. 兽类学报, 9(3): 195-210. Yang S J, Zhuge Y. 1989. Studies on reproduction and population age structure of *Suncus murinus*. *Acta Theriologica Sinica*, 9(3): 195-210. (In Chinese with English Abstract)

杨业勤, 雷孝平, 杨传东, 等. 2002. 黔金丝猴的野外生态. 贵阳: 贵州科技出版社. Yang Y Q, Lei X P, Yang C D, *et al*. 2002. Ecology of the Wild Guizhou Snub-nosed Monkey. Guiyang: Guizhou Science and Technology Press. (In Chinese)

杨跃敏, 曾宗永, 罗明澍, 等. 1994. 大足鼠种群动态的非线性模型及逐步回归分析. 兽类学报, 14(2): 130-137. Yang Y M, Zeng Z Y, Luo M S, *et al*. 1994. Nonlinear model and stepwise regression of population dynamics of *Rattus nitidus*. *Acta Theriologica Sinica*, 14(2): 130-137. (In Chinese with English Abstract)

杨再学, 雷邦海, 金星, 等. 2013. 凯里市黑腹绒鼠种群数量变动规律. 中国农学通报, 29(36): 378-381. Yang Z X, Lei B H, Jin X, *et al*. 2013. Fluctuation law of population quantity of *Eothenomys melanogaster* in Kaili City. *Chinese Agricultural Science Bulletin*, 29(36): 378-381. (In Chinese with English Abstract)

杨再学, 雷邦海, 金星, 等. 2014. 针毛鼠的形态及其种群生态特征. 四川动物, 33(3): 393-398. Yang Z X, Lei B H, Jin X, *et al*. 2014. Morphology of *Niviventer fulvescens* and its population ecological characteristics. *Sichuan Journal of Zoology*, 33(3): 393-398. (In Chinese with English Abstract)

杨再学, 郑元利, 金星. 2007. 黑线姬鼠(*Apodemus agrarius*)的种群繁殖参数及其地理分异特征. 生态学报, 27(6): 2425-2434. Yang Z X, Zheng Y L, Jin X. 2007. Species reproductive parameters and the comparison of geography variation in *Apodemus agrarius*. *Acta Ecologica Sinica*, 27(6): 2425-2434. (In Chinese with English Abstract)

姚积生. 2009. 甘肃安南坝野骆驼国家级自然保护区野骆驼现状及其保护对策. 甘肃林业科技, 34(2): 46-49. Yao J S. 2009. Resources protection countermeasures for *Camelus ferus* national nature reserve in Annan Dam in Gansu. Gansu Forestry Science and Technology, 34(2): 46-49. (In Chinese with English Abstract)

叶生荣, 雷刚. 2010. 新疆呼图壁县发现天山蹶鼠. 疾病预防控制通报, (1): 24. Ye S R, Lei G. 2010. *Sicista tianshanica* found in Hutubi County, Xinjiang. *Chinese Journal of Disease Control and Prevention*, (1): 24. (In Chinese with English Abstract)

叶晓堤, 马勇, 张津生, 等. 2002. 绒鼠类系统学研究(啮齿目: 仓鼠科: 田鼠亚科). 动物分类学报, 27(1): 173-182. Ye X D, Ma Y, Zhang J S, *et al*. 2002. A summary of *Eothenomi* (Rodentia: Cricetidae: Microtinae). *Acta Zootaxonomica Sinica*, 27(1): 173-182. (In Chinese with English Abstract)

殷宝法, 于智勇, 杨生妹, 等. 2007. 青藏公路对藏羚羊、藏原羚和藏野驴活动的影响. 生态学杂志, 26(6): 810-816. Yin B F, Yu Z Y, Yang S M, *et al*. 2007. Effects of Qinghai-Tibetan highway on the activities of *Pantholops hodgsoni*, *Procapra picticaudata* and *Equus kiang*. *Chinese Journal of Ecology*, 26(6): 810-816. (In Chinese with English Abstract)

尹皓, 林洪军, 齐彤辉, 等. 2011. 大卫鼠耳蝠回声定位声波、翼型特征及夏季食性分析. 动物学杂志, 46(6): 34-39. Yin H, Lin H J, Qi T H, *et al*. 2011. Echolocation calls, wing shape and summer diet of *Myotis davidii*. *Chinese Journal of Zoology*, 46(6): 34-39. (In Chinese with English Abstract)

由玉岩. 2013. 栖息洞穴干扰对特有种大卫鼠耳蝠种群数量和基因丰富度的影响. 生物学杂志, 30(2): 28-32. You Y Y. 2013. Influence on endemic bat *Myotis davidii* population size and genetic richness by the habitat cave interference. *Journal of Biology*, 30(2): 28-32. (In Chinese with English Abstract)

由玉岩, 杜江峰. 2011. 中国特有蝙蝠大卫鼠耳蝠种群长距离殖民事件. 应用生态学报, 22(3): 773-778. You Y Y, Du J F. 2011. A long distance colonization event of Chinese endemic bat *Myotis davidii*. *The Journal of Applied Ecology*, 22(3): 773-778. (In Chinese with English Abstract)

由玉岩, 刘森, 王磊, 等. 2009. 山东省翼手目一新纪录——宽耳犬吻蝠. 动物学杂志, 44(3): 122-126. You Y Y, Liu S, Wang L, *et al*. 2009. A new record of the Chiroptera in Shandong Province—*Tadarida teniotis*. *Chinese Journal of Zoology*, 44(3): 122-126. (In Chinese with English Abstract)

游章强, 唐中海, 杨远斌, 等. 2014. 察青松多白唇鹿国家级自然保护区白唇鹿对夏季生境的选择. 兽类学报, 34(1): 46-53. You Z Q, Tang Z H, Yang Y B, *et al*. 2014. Summer habitat selection by white-lipped deer (*Cervus albirostris*) in Chaqingsongduo White-lipped Deer National Nature Reserve. *Acta Theriologica Sinica*, 34(1): 46-53. (In Chinese with English Abstract)

于晓东, 罗天宏, 伍玉明, 等. 2006. 长江流域兽类物种多样性的分布格局. 动物学研究, 27(2): 121-143. Yu X D, Luo T H, Wu Y M, *et al*. 2006. A large-scale pattern in species diversity of mammals in the Yangtze River Basin. *Zoological Research*, 27(2): 121-143. (In Chinese with English Abstract)

余国睿, 陈浒, 周江, 等. 2014. 施秉喀斯特兽类物种多样性价值与保护. 贵州师范大学学报(自然科学版), 32(4): 29-33. Yu G R, Chen H, Zhou J, *et al*. 2014. Value and conservation on mammal species diversity in Shibing Karst. *Journal of Guizhou Normal University* (Natural Science), 32(4): 29-33. (In Chinese with English Abstract)

余梁哥, 陈敏杰, 杨士剑, 等. 2013. 利用红外相机调查屏边县大围山倭蜂猴、蜂猴及同域兽类. 四川动物, 32(6): 814-818. Yu L G, Chen M J, Yang S J, *et al*. 2013. Camera Ttrapping survey of Nyticebus *pygmaeus*, *Nyticebus coucang* and Other sympatric mammals at Dawei Mountain, Yunnan. *Sichuan Journal of Zoology*, 32(6): 814-818. (In Chinese with English Abstract)

余文华, 胡宜锋, 郭伟健, 等. 2017. 毛翼管鼻蝠在湖南的新发现及中国适生分布区预测. 广州大学学报(自然科学版), 16(3): 16-20. Yu W H, Hu Y F, Guo W J, *et al*. 2017. New discovery of *Harpiocephalus harpia* in Hunan Province and its potential distribution area in China. *Journal of Guangzhou University* (Natural Science Edition), 16(3): 16-20. (In Chinese with English Abstract)

余文华, 吴毅, 李玉春, 等. 2008. 海南岛发现褐扁颅蝠(*Tylonycteris robustula*)分布新纪录. 广州大学学报(自然科学版), 7(5): 30-33. Yu W H, Wu Y, Li Y C, *et al*. 2008. A new record of greater bamboo bat *Tylonycteris robustula* on the Hainan Island. *Journal of Guangzhou University* (Natural Science Edition), 7(5): 30-33. (In Chinese with English Abstract)

余子寒, 吴倩倩, 石胜超, 等. 2018. 湖南省衡东县发现长指鼠耳蝠. 动物学杂志, 53(5): 701-708. Yu Z H, Wu Q Q, Shi S C, *et al*. 2018. The Kashmir Cave *Myotis* (*Myotis longipes*) Was Found in Hengdong County Hunan Province, China. *Chinese Journal of Zoology*, 53(5): 701-708. (In Chinese with English Abstract)

袁帅, 武晓东, 付和平, 等. 2011. 不同干扰下荒漠啮齿动物群落多样性的多尺度分析. 生态学报, 31(7): 1982-1992. Yuan S, Wu X D, Fu H P, *et al*. 2011. Multi-scales analysis on diversity of desert rodent communities under different disturbances. *Acta Ecologica Sinica*, 31(7): 1982-1992. (In Chinese with English Abstract)

岳阳, 胡宜峰, 雷博宇, 等. 2019. 毛翼管鼻蝠性二型特征及其在湖北和浙江的分布新纪录. 兽类学报, 39(2): 142-154. Yue Y, Hu Y F, Lei B Y, *et al*. 2019. Sexual Dimorphism in *Harpiocephalus harpia* and its New Records from Hubei and Zhejiang, China. *Acta Theriologica Sinica*, 39(2): 142-154. (In Chinese with English Abstract)

曾峰. 2012. 翼手目六种蝙蝠毛发分类特征的研究. 长沙: 湖南师范大学硕士研究生学位论文. Zeng F. 2012. A Study on the Hair Classification of Six Species of Bats. Changsha: Master's Thesis of Hunan Normal University. (In Chinese with English Abstract)

曾国仕, 郑合勋, 邓天鹏. 2010. 河南伏牛山北坡果子狸夏季巢穴生境特征. 生态学报, 20(2): 498-503. Zeng G S, Zheng H X, Deng T P. 2010. A preliminary analysis on summer caves selection of Masked palm civet (*Paguma larvata*) on the north slope of Funiu Mountain. *Acta Ecologica Sinica*, 20(2): 498-503. (In Chinese with English Abstract)

曾治高, 宋延龄. 2008. 秦岭羚牛的生态与保护对策. 生物学通报, 43(8): 1-4. Zeng Z G, Song Y L. 2008. Ecology and conservation strategies of Golden Takin. *Bulletin of Biology*, 43(8): 1-4. (In Chinese with English Abstract)

查木哈, 袁帅, 张晓东, 等. 2013. 不同干扰下阿拉善荒漠啮齿动物群落格局的变动特征. 生态环境学报, (12): 1879-1886. Zha M H, Yuan S, Zhang X D, *et al*. 2013. Research on the rodent community patterns variation features under different disturbances in Alashan Desert. *Ecology and Environment Sciences*, (12): 1879-1886. (In Chinese with English Abstract)

扎史其, 陈新文, 敬凯, 等. 2014. 维西县哺乳动物区系调查. 林业调查规划, 39(2): 73-77. Zha S Q, Chen X W, Jing K, *et al*. 2014. Investigation of mammal fauna in Weixi County. *Forest Inventory and Planning*, 39(2): 73-77. (In Chinese with English Abstract)

张斌, 王昊, 周华明, 等. 2014. 九龙北部贡嘎山区小型兽类调查. 四川林业科技, 35(3): 79-80. Zhang B, Wang H, Zhou H M, *et al*. 2014. Survey of small mammals in the Gongga Mountain Area in Northern Jiulong County. *Sichuan Forestry Science and Technology*, 35(3): 79-80. (In Chinese with English Abstract)

张婵, 王艳梅, 牛红星. 2013. 河南省栾川县伏牛山发现翼手目物种大菊头蝠. 动物学杂志, 48(4): 650-654. Zhang C, Wang Y M, Niu H X. 2013. *Rhinolophus luctus* found in Funiu Mountain, Luanchuan County, Henan Province. *Chinese Journal of Zoology*, 48(4): 650-654. (In Chinese with English Abstract)

张词祖, 盛和林, 陆厚基. 1984. 我国西藏的菲氏麂(*Muntiacus feae*). 兽类学报, 4(2): 88. Zhang C Z, Sheng H L, Lu H J. 1984. On the Fea's muntjak from Xizang (Tibet), China. *Acta Theriologica Sinica*, 4(2): 88. (In Chinese with English Abstract)

张冬冬, 朱洪强, 姜春艳, 等. 2014. 原麝生境选择影响因子的研究概况. 经济动物学报, 18(1): 44-46. Zhang D D, Zhu H Q, Jiang C Y, *et al*. 2014. Research Situation of Impact Factors on Habitat Selection of Siberian *Moschus moschiferus*. *Journal of Economic Animal*, 18(1): 44-46. (In Chinese with English Abstract)

张峰, 胡德夫, 李凯, 等. 2009. 普氏野马繁殖群在组建和放归初期的争斗行为与社群等级建立. 动物学杂志, 44(4): 58-

63. Zhang F, Hu D F, Li K, *et al*. 2009. The agonistic behavior and hierarchical formation of the *Equus przewalskii* herd in the individual coalition and initial releasing period. *Chinese Journal of Zoology*, 44(4): 58-63. (In Chinese with English Abstract)

张孚允, 杨若莉. 1980. 甘肃南部的胡兀鹫. 动物学报, 11(1): 86-90. Zhang F Y, Yang R L. 1980. Bearded vulture from southern Gansu. *Acta Zoologica Sinica*, 11(1): 86-90. (In Chinese with English Abstract)

张海龙, 吴建平, 刘永志, 等. 2008. 大兴安岭原麝夏季的生境选择. 生态学杂志, 27(8): 1313-1316. Zhang H L, Wu J P, Liu Y Z, *et al*. 2008. Habitat selection by *Moschus moschiferus* in summer in Daxing'an Mountains. *Chinese Journal of Ecology*, 27(8): 1313-1316. (In Chinese with English Abstract)

张洪峰, 王开锋, 靳铁治. 2006. 宁夏宁南山区红庄林场甘肃鼢鼠分布密度与危害研究. 四川动物, 25(4): 870-872. Zhang H F, Wang K F, Jin T Z. 2006. Density of distribution and harm of Gansu zoker in the Hongzhuang tree farm of Ningnan mountain area. *Sichuan Journal of Zoology*, 25(4): 870-872. (In Chinese with English Abstract)

张洪海, 窦华山, 翟红昌, 等. 2006. 三种犬科动物春季洞穴特征. 生态学报, 26(12): 3980-3988. Zhang H H, Dou H S, Zhai H C, *et al*. 2006. Characteristics of dens in spring of three species of canids. *Acta Ecologica Sinica*, 26(12): 3980-3988. (In Chinese with English Abstract)

张洪海, 马建章. 2000. 紫貂春季和夏季生境选择的初步研究. 动物学报, 46(4): 399-406. Zhang H H, Ma J Z. 2000. Preliminary research on the habitat selection of sable in spring and summer. *Acta Zoologica Sinica*, 46(4): 399-406. (In Chinese with English Abstract)

张洪亮, 李芝喜, 王人潮, 等. 2000. 基于GIS的贝叶斯统计推理技术在印度野牛生境概率评价中的应用. 遥感学报, 4(1): 66-70. Zhang H L, Li Z X, Wang R C, *et al*. 2000. Application of Bayesian statistics inference techniques based on GIS to the evaluation of habitat probabilities of *Bos gaurus* Readei. *Journal of Remote Sensing*, 4(1): 66-70. (In Chinese with English Abstract)

张建军. 2000. 黄喉貂生态特性的初步观察. 河北林果研究, (S1): 195-196. Zhang J J. 2000. Preliminary observation on ecological characteristics of yellow-throated marten. *Hebei Journal of Forestry and Orchard Research*, (S1): 195-196. (In Chinese with English Abstract)

张杰, 于洪伟. 2005. 黑龙江兽类物种多样性研究. 国土与自然资源研究, (3): 77-78. Zhang J, Yu H W. 2005. Research species diversity of mammals in northeast of China. *Territory & Natural Resources Study*, (3): 77-78. (In Chinese with English Abstract)

张进. 2014. 中国犬科动物线粒体基因组研究及系统发育分析. 曲阜: 曲阜师范大学硕士研究生学位论文. Zhang J. 2014. Mitochondrial Genome Research and Phylogenetic Analysis of Canidae Species in China. Qufu: Master's Thesis of Qufu Normal University. (In Chinese with English Abstract)

张劲硕, 张礼标, 赵辉华, 等. 2005. 中国翼手类新记录——小褐菊头蝠. 动物学杂志, 40(2): 96-98. Zhang J S, Zhang L B, Zhao H H, *et al*. 2005. First record of Chinese bats: *Rhinolophus stheno*. *Chinese Journal of Zoology*, 40(2): 96-98. (In Chinese with English Abstract)

张劲硕, Lynch E, Krejca K H, 等. 2009. 重庆翼手类一新纪录——三叶蹄蝠. 动物学杂志, 44(1): 46. Zhang J S, Lynch E, Krejca K H, *et al*. 2009. A new record for bats in Chongqing: *Aselliscus stoliczkanus*. *Chinese Journal of Zoology*, 44(1): 46. (In Chinese with English Abstract)

张君, 黄小富, 周材权. 2010. 四川东阳沟自然保护区兽类区系调查. 西华师范大学学报(自然科学版), 31(4): 327-332. Zhang J, Huang X F, Zhou C Q. 2010. Preliminary Report on Mammal Fauna of Dongyanggou Nature Reserve. *Journal of China West Normal University* (Natural Science Edition), 31(4): 327-332. (In Chinese with English Abstract)

张礼标, 巩艳艳, 朱光剑, 等. 2010. 中国翼手目新记录——马来假吸血蝠. 动物学研究, 31: 328-332. Zhang L B, Gong Y Y, Zhu G J, *et al*. 2010. New record of a bat species from China, *Megaderma spasma* (Linnaeus, 1758). *Zoological Research*, 31: 328-332. (In Chinese with English Abstract)

张礼标, 洪体玉, 韦力, 等. 2011. 扁颅蝠的扩散行为研究. 兽类学报, 31(3): 244-250. Zhang L B, Hong T Y, Wei L, *et al*. 2011. Dispersal behaviour of the lesser flat-headed bat, *Tylonycteris pachypus* (Chiroptera: Vespertilionidae). *Acta Theriologica Sinica*, 31(3): 244-250. (In Chinese with English Abstract)

张礼标, 刘奇, 沈琪琦, 等. 2014. 广东省蝙蝠新纪录——大黑伏翼. 兽类学报, 34(3): 292-297. Zhang L B, Liu Q, Shen Q Q, *et al*. 2014. New bat record from Guangdong Province in China—*Arielulus circumdatus* (Temminck, 1840). *Acta Theriologica Sinica*, 34(3): 292-297. (In Chinese with English Abstract)

张礼标, 龙勇诚, 张劲硕, 等. 2005. 中国翼手类新记录——马氏菊头蝠. 兽类学报, 25(1): 77-80. Zhang L B, Long Y C, Zhang

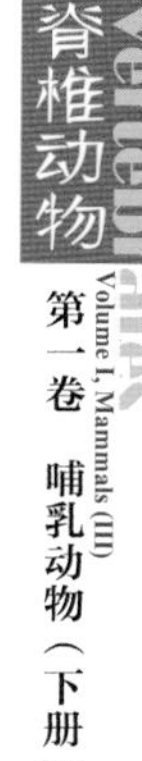

J S, *et al*. 2005. New record of bat species—*Rhinolophus marshalli* from China. *Acta Theriologica Sinica*, 25(1): 77-80. (In Chinese with English Abstract)

张礼标, 张劲硕, 梁冰, 等. 2004. 中国翼手类新纪录——小巨足蝠. 动物学研究, 25(6): 556-559. Zhang L B, Zhang J S, Liang B, *et al*. 2004. New record of a bat species from China, *Myotis hasseltii* (Temminck, 1840). *Zoological Research*, 25(6): 556-559. (In Chinese with English Abstract)

张礼标, 张伟, 张树义. 2007. 印度假吸血蝠捕食鼠耳蝠. 动物学研究, 28(1): 104-105. Zhang L B, Zhang W, Zhang S Y. 2007. Indian False Vampire Bat feeding on *Myotis*. *Zoological Research*, 28(1): 104-105. (In Chinese with English Abstract)

张礼标, 朱光剑, 于冬梅, 等. 2008. 海南、贵州和四川三省翼手类新纪录——褐扁颅蝠. 兽类学报, 28(3): 316-320. Zhang L B, Zhu G J, Yu D M, *et al*. 2008. New record of *Tylonycteris robustula* (Chiroptera: Vespertilionidae) from Hainan, Guizhou, Sichuan Province. *Acta Theriologica Sinica*, 28(3): 316-320. (In Chinese with English Abstract)

张立. 2006. 中国亚洲象现状及研究进展. 生物学通报, 41(11): 1-3. Zhang L. 2006. Current conservation status and research progress on Asian elephants in China. *Bulletin of Biology*, 41(11): 1-3. (In Chinese with English Abstract)

张立, 李麒麟, 孙戈, 等. 2010. 穿山甲种群概况及保护. 生物学通报, 45(9): 1-4. Zhang L, Li Q L, Sun G, *et al*. 2010. Population and conservation status of pangolins. *Bulletin of Biology*, 45(9): 1-4. (In Chinese with English Abstract)

张立志, 葛玉祥, 何蒙德, 等, 2011. 红花尔基樟子松林国家级自然保护区松鼠的生境选择特征. 野生动物, 32(3): 123-125. Zhang L Z, Ge Y X, He M D, *et al*. 2011. Habitat slection by Eurasian Red Squirrels in Honghuaerji Nature Reserve. *Chinese Journal of Wildlife*, 32(3): 123-125. (In Chinese with English Abstract)

张明海. 2002. 黑龙江省熊类资源现状及其保护对策. 动物学杂志, 37(6): 47-52. Zhang M H. 2002. Status and conservation strategies of bear resources in Heilongjiang Province. *Chinese Journal of Zoology*, 37(6): 47-52. (In Chinese with English Abstract)

张琴, 刘奇, 杨昌腾, 等. 2017. 广东省发现长指鼠耳蝠及其回声定位声波特征. 动物学杂志, 52(3): 521-529. Zhang Q, Liu Q, Yang C T, *et al*. 2017. Discovery of Kashmir Cave *Myotis*, *Myotis longipes* in Guangdong Province (China) and its Echolocation Calls. *Chinese Journal of Zoology*, 52(3): 521-529. (In Chinese with English Abstract)

张琼, 曾治高, 孙丽凤, 等. 2009. 海南坡鹿的起源、进化及保护. 兽类学报, 29(4): 365-371. Zhang Q, Zeng Z G, Sun L F, *et al*. 2009. The origin and phylogenetics of Hainan Eld's deer and implications for Eld's deer conservation. *Acta Theriologica Sinica*, 29(4): 365-371. (In Chinese with English Abstract)

张秋萍, 余文华, 吴毅, 等. 2014. 江西省蝙蝠新纪录——褐扁颅蝠及其核型报道. 四川动物, 33(5): 746-749. Zhang Q P, Yu W H, Wu Y, *et al*. 2014. A new record of *Tylonycteris robustula* in Jiangxi Province, China and its karyotype. *Sichuan Journal of Zoology*, 33(5): 746-749. (In Chinese with English Abstract)

张荣祖. 1979. 中国自然地理: 动物地理. 北京: 科学出版社. Zhang R Z. 1979. China's Physical Geography: Zoological Geography. Beijing: Science Press. (In Chinese)

张荣祖. 1999. 中国动物地理. 北京: 科学出版社. Zhang R Z. 1999. China's Zoological Geography. Beijing: Science Press. (In Chinese)

张荣祖. 1997. 中国哺乳动物分布. 北京: 中国林业出版社: 1-280. Zhang R Z. 1997. Distribution of Mammals in China. Beijing: China Forestry Press: 1-280. (In Chinese)

张荣祖. 2002. 中国地质事件与哺乳动物的分布. 动物学报, 48(2): 141-153. Zhang R Z. 2002. Geological events and mammalian distribution in China. *Acta Zoologica Sinica*, 48(2): 141-153. (In Chinese with English Abstract)

张三亮, 刘荣堂, 寇明君, 等. 2008. 甘肃省鼢鼠亚科动物形态学标记多样性研究. 中国森林病虫, 27(5): 1-3. Zhang S L, Liu R T, Kou M J, *et al*. 2008. An experiment with several rodenticides on zokor control. *Chinese Journal of Forest Pests*, 27(5): 1-3. (In Chinese with English Abstract)

张世炎, 张涛, 陈安. 2012. 廉江市2006—2010年板齿鼠种群动态和繁殖特性. 医学动物防制, (2): 119-120. Zhang S Y, Zhang T, Chen A. 2012. Study on the population dynamics and breeding characteristics of Bandicota in 2006-2010 at Lianjiang. *Control of Medical Animals*, (2): 119-120. (In Chinese with English Abstract)

张树义, 赵辉华, 冯江, 等. 2000. 长尾鼠耳蝠飞行状态下的回声定位叫声. 科学通报, 45(5): 526-528. Zhang S Y, Zhao H H, Feng J, *et al*. 2000. Echolocation calls of *Myotis frater* in flight. *Chinese Science Bulletin*, 45(5): 526-528. (In Chinese with English Abstract)

张维道. 1984. 中华鼠耳蝠(*Myotis chinensis* Tomes)和绒鼠耳蝠(*Myotis laniger* Peters)的染色体分析. 安徽师大学报(自然科学版), (2): 42-47. Zhang W D. 1984. Chromosome analysis of *Myotis chinensis* Tomes and *Myotis laniger* Peters. *Journal of Anhui Normal University* (Natural Science Edition), (2): 42-47. (In Chinese with English Abstract)

张维道. 1985. 宽耳犬吻蝠(*Tadarida teniotis* Insignis)和普氏蹄蝠(*Hipposideros pratti*)染色体组型分析. 兽类学报, 5(3): 189-193. Zhang W D. 1985. A study on karyotypes of the bats *Tadarida teniotis* Insignis blyth and *Hipposideros pratti* Thomas. *Acta Theriologica Sinica*, 5(3): 189-193. (In Chinese with English Abstract)

张维道, 宛敏, 周立新. 1983. 毛腿鼠耳蝠和折翼蝠染色体分析. 遗传, 5(6): 40-41. Zhang W D, Wan M, Zhou L X. 1983. Chromosome analysis of *Myotis fimbriatus* and *Miniopterus schreibersi*. *Hereditas*, 5(6): 40-41. (In Chinese with English Abstract)

张伟. 2008. 犬蝠的栖息地、社群结构及其嗅觉在觅食行为中的作用研究. 桂林: 广西师范大学硕士研究生学位论文. Zhang W. 2008. Habitat, Social Structure and Olfactory Function in Foraging Behavior of *Cynopterus sphinx*. Guilin: Master's Thesis of Guangxi Normal University. (In Chinese with English Abstract)

张先锋, 刘仁俊, 赵庆中, 等. 1993. 长江中下游江豚种群现状评价. 兽类学报, 13(4): 260-270. Zhang X F, Liu R J, Zhao Q Z, *et al*. 1993. The population of finless porpoise in the middle and lower reaches of Yangtze River. *Acta Theriologica Sinica*, 13(4): 260-270. (In Chinese with English Abstract)

张显理, 于有志. 1994. 宁夏哺乳动物区系与地理区划研究. 兽类学报, 15(2): 128-136. Zhang X L, Yu Y Z. 1994. Study on the fauna and the zoogeographical division of mammals in Ningxia. *Acta Theriologica Sinica*, 15(2): 128-136. (In Chinese with English Abstract)

张小龙, 张恩迪. 2002. 江苏大丰麋鹿自然保护区内獐在冬季各种生境中分布的初步研究. 四川动物, 21(1): 19-22. Zhang X L, Zhang E D. 2002. Distribution pattern of *Hydropotes inermis* in various habitats in Jiangsu Dafeng Père David's Deer State Nature Reserve. *Sichuan Journal of Zoology*, 21(1): 19-22. (In Chinese with English Abstract)

张晓东, 武晓东, 付和平, 等. 2013. 荒漠破碎化生境中长爪沙鼠集合种群野外验证研究. 动物学杂志, 48(6): 834-843. Zhang X D, Wu X D, Fu H P, *et al*. 2013. The Mongolian gerbils meta-population in habitat fragmentation in Alxa Desert: A field verification study. *Chinese Journal of Zoology*, 48(6): 834-843. (In Chinese with English Abstract)

张晓华, 王广仁. 2002. 科尔沁草原达乌尔黄鼠的生态调查. 中国地方病防治杂志, 17(3): 188. Zhang X H, Wang G R. 2002. Ecological investigation of *Spermophilus dauricus* in Horqin grassland. *Chinese Journal of Endemic Disease Control*, 17(3): 188. (In Chinese with English Abstract)

张旭, 鲍毅新, 刘军, 等. 2013. 千岛湖岛屿社鼠的种群数量动态特征. 生态学报, 33(15): 4665-4673. Zhang X, Bao Y X, Liu J, *et al*. 2013. Population dynamics of *Niviventer confucianus* on Thousand Island Lake. *Acta Ecologica Sinica*, 33(15): 4665-4673. (In Chinese with English Abstract)

张燕均, 邓柏生, 李玉春, 等. 2010. 西南鼠耳蝠广东新纪录及其核型. 兽类学报, 30(4): 460-464. Zhang Y J, Deng B S, Li Y C, *et al*. 2010. A new record of *Myotis altarium* and its karyotype in Guangdong, China. *Acta Theriologica Sinica*, 30(4): 460-464. (In Chinese with English Abstract)

张阳, 周材权, 鲁庆彬, 等. 2011. 山西隰县中华鼢鼠洞址选择. 四川动物, 30(4): 607-611. Zhang Y, Zhou C Q, Lu Q B, *et al*. 2011. Cave-site Selection of *Eospalax fontanieri* in Xi County, Shanxi Province. *Sichuan Journal of Zoology*, 30(4): 607-611. (In Chinese with English Abstract)

张英, 陈鹏. 2013. 陕西长青国家级自然保护区林麝种群分布与保护. 陕西林业科技, (3): 28-30. Zhang Y, Chen P. 2013. Distribution and conservation of *Moschus berezovskii* in Changqing National Nature Reserve. *Shaanxi Forest Science and Technology*, (3): 28-30. (In Chinese with English Abstract)

张佑祥, 刘志霄, 胡开良, 等. 2008. 大菊头蝠在湖南省分布新纪录. 动物学杂志, 43(2): 141-144. Zhang Y X, Liu Z X, Hu K L, *et al*. 2008. A new record of woolly horseshoe bat *Rhinolophus luctus* in Hunan Province. *Chinese Journal of Zoology*, 43(2): 141-144. (In Chinese with English Abstract)

张于光, 何丽, 朵海瑞, 等. 2009. 基于粪便DNA的青海雪豹种群遗传结构初步研究. 兽类学报, 29(3): 310-315. Zhang Y G, He L, Duo H R, *et al*. 2009. A preliminary study on the population genetic structure of snow leopard (*Unica unica*) in Qinghai Province utilizing fecal DNA. *Acta Theriologica Sinica*, 29(3): 310-315. (In Chinese with English Abstract)

张渝疆, 张富春, 孙素荣, 等. 2004. 准噶尔盆地东南缘草原兔尾鼠(*Lagurus lagurus*)种群空间分布研究. 新疆大学学报(自然科学版), 21(3): 300-303. Zhang Y J, Zhang F C, Sun S R, *et al*. 2004. Study on population spatial distribution of *Lagurus lagurus* in southeastern Dzungaria Basin of Xinjiang. *Journal of Xinjiang University* (Natural Science Edition), 21(3): 300-303. (In Chinese with English Abstract)

张云智, 龚正达, 冯锡光, 等. 2002. 云南白草岭鼠形小兽群落结构及垂直分布. 动物学杂志, 37(2): 63-66. Zhang Y Z, Gong Z D, Feng X G, *et al*. 2002. The community structure and vertical distribution of small mammal in Mt. Baicaoling, Yunnan Province, China. *Chinese Journal of Zoology*, 37(2): 63-66. (In Chinese with English Abstract)

张桢珍, 江廷磊, 李振新, 等. 2008. 吉林省发现长尾鼠耳蝠. 动物学杂志, 43(3): 150-153. Zhang Z Z, Jiang T L, Li Z X, *et al*. 2008. A new record of *Myotis frater* in Jilin Province. *Chinese Journal of Zoology*, 43(3): 150-153. (In Chinese with English Abstract)

章敬旗, 周友兵, 徐伟霞, 等. 2004. 几种麝分类地位的探讨. 西华师范大学学报(自然科学版), 25(3): 251-255. Zhang J Q, Zhou Y B, Xu W X, *et al*. 2004. Discussion about musk-deer's classification. *Journal of Sichuan Teachers College* (Natural Science Edition), 25(3): 251-255. (In Chinese with English Abstract)

赵尔宓. 1998. 中国濒危动物红皮书——两栖爬行类. 北京: 科学出版社. Zhao E M. 1998. *Red Book of Endangered Animals in China: Amphibians and Reptiles*. Beijing: Science Press. (In Chinese)

赵辉华, 张树义, 周江, 等. 2002. 中国翼手类新记录——高鞍菊头蝠. 兽类学报, 22(1): 74-76. Zhao H H, Zhang S Y, Zhou J, *et al*. 2002. New record of bats from China: *Rhinolophus paradoxolophus*. *Acta Theriologica Sinica*, 22(1): 74-76. (In Chinese with English Abstract)

赵景辉, 赵伟刚, 陈玉山, 等. 2005. 中国的河狸资源考察. 特产研究, 27(3): 38-41. Zhao J H, Zhao W G, Chen Y S, *et al*. 2005. Observation and study on natural resource in Chinese beaver. *Special Wild Economic Animal and Plant Research*, 27(3): 38-41. (In Chinese with English Abstract)

赵肯堂. 1984. 蒙古黄兔尾鼠的生态观察. 兽类学报, 4(3): 217-222. Zhao K T. 1984. Ecological observation of *Eolagurus przewalskii*. *Acta Zoologica Sinica*, 4(3): 217-222. (In Chinese with English Abstract)

赵肯堂. 1977. 五趾心颅跳鼠的生态调查. 内蒙古大学学报(自然科学版), (1): 64-71. Zhao K T. 1977. Ecological investigation of *Cardiocranius paradoxus*. *Journal of Inner Mongolia University* (Natural Science Edition), (1): 64-71. (In Chinese)

赵肯堂. 1981. 内蒙古啮齿动物. 呼和浩特: 内蒙古人民出版社. Zhao K T. 1981. Inner Mongolia Rodents. Huhehaote: Inner Mongolia People's Publishing House. (In Chinese)

赵启龙, 黄蓓, 郭光, 等. 2016. 云南临沧邦马山西黑冠长臂猿种群历史及现状. 四川动物, 35(1): 1-8. Zhao Q L, Huang B, Guo G, *et al*. 2016. Population Status of *Nomascus concolor* in Bangma Mountain, Lincang, Yunnan. *Sichuan Journal of Zoology*, 35(1): 1-8. (In Chinese with English Abstract)

赵天飙, 杨持, 周立志, 等. 2005. 中国大沙鼠生态学研究进展. 内蒙古大学学报(自然科学版), 36: 591-596. Zhao T B, Yang C, Zhou L Z, *et al*. 2005. Advance of ecological study on great gerbil (*Rhombomys opimus*) in China. *Journal of Inner Mongolia University* (Natural Science Edition), 36: 591-596. (In Chinese with English Abstract)

赵天飙, 张忠兵, 李新民, 等. 2001. 大沙鼠和子午沙鼠的种群生态位. 兽类学报, 21(1): 76-79. Zhao T B, Zhang Z B, Li X M, *et al*. 2001. Studies on the spatial patterns of the populations of *Rhombomys opimus* and *Meriones meridianus*. *Acta Theriologica Sinica*, 21(1): 76-79. (In Chinese with English Abstract)

赵正阶. 1999. 中国东北地区珍稀濒危动物志. 北京: 中国林业出版社. Zhao Z J. 1999. Rare and Endangered Animals in Northeast China. Beijing: China Forestry Press. (In Chinese)

郑昌琳. 1986. 中国兽类之种数. 兽类学报, 6(1): 78-80. Zheng C L. 1986. The number of mammalian species in China. *Acta Theriologica Sinica*, 6(1): 78-80. (In Chinese with English Abstract)

郑昌琳, 汪松. 1980. 白尾松田鼠分类志要. 动物分类学报, 5(1): 108-114. Zheng C L, Wang S. 1980. On the toxonomic status of *Pitymys leucurus* Blyth. *Acta Zootaxonomica Sinica*, 5(1): 108-114. (In Chinese with English Abstract)

郑光美, 王岐山. 1998. 中国濒危动物红皮书——鸟类. 北京: 科学出版社. Zheng G M, Wang Q S. 1998. Red Book of Endangered Animals in China: Birds. Beijing: Science Press. (In Chinese)

郑生武. 1994. 中国西北地区珍稀濒危动物志. 北京: 中国林业出版社. Zheng S W. 1994. Rare and Endangered Animals in Northwest China. Beijing: China Forestry Press. (In Chinese)

郑生武, 李保国. 1999. 中国西北地区脊椎动物系统检索与分布. 西安: 西北大学出版社. Zheng S W, Li B G. 1999. Key and Distribution of Vertebrates in Northwest China. Xi'an: Northwestern University Press. (In Chinese)

郑涛, 张迎梅. 1990. 甘肃省啮齿动物区系及地理区划的研究. 兽类学报, 10(2): 137-144. Zheng T, Zhang Y M. 1990. The fauna and geographical division on glires of Gansu Province. *Acta Theriologica Sinica*, 10(2): 137-144. (In Chinese with English Abstract)

郑伟成, 刘军, 潘成椿, 等. 2012. 中国特有动物黑麂的研究. 野生动物学报, 33(5): 283-288. Zheng W C, Liu J, Pan C C, *et al*. 2012. Review of research on black muntjac (*Muntiacus crinifrons*), an endemic species in China. *Chinese Journal of Wildlife*, 33(5): 283-288. (In Chinese with English Abstract)

郑锡奇. 2010. 台湾蝙蝠图鉴. 台北: 农业委员会特有生物研究保育中心. Zheng X Q. 2010. Bat Atlas of Taiwan. Taipei: Center for Endemic Biology Research and Conservation, Council of Agriculture. (In Chinese)

郑永烈. 1981. 我国兽类新纪录——缅甸鼬獾. 兽类学报, 1: 158. Zheng Y L. 1981. New record of ferret badger Iin China. *Acta Theriologica Sinica*, 1: 158. (In Chinese with English Abstract)

郑智民, 姜志宽, 陈安国. 2012. 啮齿动物学(第2版). 上海: 上海交通大学出版社. Zheng Z M, Jiang Z K, Chen A G. 2012. Rodent Zoology (2nd ed). Shanghai: Shanghai Jiaotong University Press. (In Chinese)

郑作新. 1952. 脊椎动物分类学. 北京: 农业出版社. Zheng Z X. 1952. Taxonomy of Vertebrate. Beijing: Agricultural Press. (In Chinese)

郑作新. 1963. 脊椎动物分类学(第2版). 北京: 农业出版社. Zheng Z X. 1963. Taxonomy of Vertebrates (2nd ed.). Beijing: Agricultural Press. (In Chinese)

郑作新. 1982. 脊椎动物分类学(第3版). 北京: 农业出版社. Zheng Z X. 1982. Taxonomy of Vertebrates (3rd ed.). Beijing: Agricultural Press. (In Chinese)

中国科学院动物研究所兽类研究组. 1958. 东北兽类调查报告. 北京: 科学出版社. Mammal Research Group, Institute of Zoology, Chinese Academy of Sciences. 1958. Animal Survey in Northeast China. Beijing: Science Press. (In Chinese)

中国科学院内蒙古草原生态系统定位站. 1988. 白音锡勒地区的兽类区系特征. 草原生态系统研究. 北京: 科学出版社. Inner Mongolia Grassland Ecosystem Location Station, Chinese Academy of Sciences. 1988. Fauna in Baiyin Xile Area, Grassland Ecosystem Research. Beijing: Science Press. (In Chinese)

中国科学院青海甘肃综合考察队. 1987. 青海甘肃兽类调查报告. 北京: 科学出版社: 1-80. Qinghai and Gansu Investigation Team of Chinese Academy of Sciences. 1987. Qinghai and Gansu Animal Survey Report. Beijing: Science Press: 1-80. (In Chinese)

中国科学院西北高原生物研究所. 1989. 青海经济动物志. 西宁: 青海人民出版社: 1-735. Northwest Institute of Plateau Biology, Chinese Academy of Sciences. 1989. Qinghai Economic Animals. Xining: Qinghai People's Publishing House: 1-735. (In Chinese)

中华人民共和国濒危物种进出口管理办公室, 中华人民共和国濒危物种科学委员会. 2017. 濒危野生动植物国际贸易公约附录I、附录II和附录III. 第17届缔约国大会通过. http://www.cites.org.cn/ [2018-12-11]. General Office of Endangered Species Importing and Exporting of People's Republic of China, Endangered Species Scientific Commission of People's Republic of China. 2017. Appendix I, II and III of Convention on International Trade of Endangered Species of Wild Fauna and Flora. Adapted at COP-17 of CITES. http://www.cites.org.cn/ [2018-12-11]. (In Chinese)

中华人民共和国农业部. 2014. 农业部关于进一步加强长江江豚保护管理工作的通知. http://www.moa.gov.cn/govpublic/CJB/201410/t20141014_4104286.htm [2014-10-14] Ministry of Agriculture, People's Republic of China. 2014. Notification of the Ministry of Agriculture on Further Strengthening the Conservation Management of Fresh Dolphins in the Yangtze River. http://www.moa.gov.cn/govpublic/CJB/201410/t20141014_4104286.htm [2014-10-14] (In Chinese)

钟福生. 2001. 小灵猫的资源、开发利用现状与分布. 湖南生态科学学报, 7(2): 24-26. Zhong F S. 2001. Status and distribution of exploitation and utilization in Lesser Civits' resource. *Journal of Hunan Environment-biological Polytechnic*, 7(2): 24-26. (In Chinese with English Abstract)

钟宇, 刘奇, 刘超飞, 等. 2014. 褐家鼠挖掘活动对上海九段沙湿地植物群落与土壤水盐的影响. 兽类学报, 34(1): 62-70. Zhong Y, Liu Q, Liu C F, *et al*. 2014. The impacts of burrowing activities of introduced *Rattus norvegicus* on plant communities and moisture content and salinity of topsoil in Jiuduan-Sha Wetland, Shanghai. *Acta Theriologica Sinica*, 34(1): 62-70. (In Chinese with English Abstract)

周朝霞, 艾祯仙, 陆小欢, 等. 2009. 三都县褐家鼠种群数量动态与繁殖规律. 贵州农业科学, 37(7): 83-85. Zhou Z X, Ai Z X, Lu X H, *et al*. 2009. Law of population dynamics and reproduction of *Rattus norvegicus* in Sandu County. *Guizhou Agricultural Sciences*, 37(7): 83-85. (In Chinese with English Abstract)

周江. 2001. 贵州省七种蝙蝠空间生态位及种间关系研究. 贵阳: 贵州师范大学硕士研究生学位论文. Zhou J. 2001. Study on Spatial Niche and Interspecific Relationship of Seven Bats in Guizhou Province. Guiyang: Master's Thesis of Guizhou Normal University. (In Chinese with English Abstract)

周江, 谢家骅, 戴强, 等. 2002. 皮氏菊头蝠夏季的捕食行为对策. 动物学研究, 23(2): 120-128. Zhou J, Xie J Y, Dai Q, *et al*. 2002. Feeding behavioral strategy of *Rhinolophus pearsoni* in summer. *Zoological Research*, 23(2): 120-128. (In Chinese with English Abstract)

周江, 杨天友, 侯秀发. 2011. 贵州省发现侏伏翼. 动物学杂志, 46(1): 115-119. Zhou J, Yang T Y, Hou X F. 2011. The least pipistrelle (*Pipistrellus tenuis*) was discovered in Guizhou Province. *Chinese Journal of Zoology*, 46(1): 115-119. (In Chinese with English Abstract)

周江, 杨天友. 2009. 贵州省蝙蝠科二新纪录. 四川动物, 28(6): 925. Zhou J, Yang T Y. 2009. Two New records of vespertilionidae in Guizhou Province. *Sichuan Journal of Zoology*, 28(6): 925. (In Chinese with English Abstract)

周江, 杨天友. 2010. 贵州省松桃县东部地区翼手目物种多样性. 动物学杂志, 45(2): 52-59. Zhou J, Yang T Y. 2010. The Chiroptera species diversity in eastern of Songtao, Guizhou Province. *Chinese Journal of Zoology*, 45(2): 52-59. (In Chinese with English Abstract)

周江, 杨天友. 2012a. 贵州省翼手目一新纪录——大山蝠. 动物学杂志, 47(1): 119-123. Zhou J, Yang T Y. 2012a. A new record of *Nyctalus aviator* in Guizhou Province. *Chinese Journal of Zoology*, 47(1): 119-123. (In Chinese with English Abstract)

周江, 杨天友. 2012b. 贵州省鼠耳蝠属一新纪录——狭耳鼠耳蝠. 四川动物, 31(1): 120-123. Zhou J, Yang T Y. 2012b. A new record of *Myotis blythii* in Guizhou Province. *Sichuan Journal of Zoology*, 31(1): 120-123. (In Chinese with English Abstract)

周开亚. 1982. 关于白鱀豚的保护. 南京师范学院学报(自然科学版), (4): 71-74. Zhou K Y. 1982. Protection of Baiji dolphin. *Journal of Nanjing Normal University* (Natural Science Edition), (4): 71-74. (In Chinese)

周开亚. 2004. 中国动物志 • 兽纲 • 第九卷 鲸目、食肉目 海豹总科、海牛目. 北京: 科学出版社. Zhou K Y. 2004. Fauna Sinica, Mammalia Vol. 9 Cetacea, Carnivora, Phocoidea and Sirenia. Beijing: Science Press. (In Chinese)

周开亚. 2008. 脊椎动物: 鲸类、食肉类和海牛类. 见: 刘瑞玉. 中国海洋生物名录. 北京: 科学出版社: 1082-1085. (In Chinese and English)

周开亚, 解斐生, 黎德伟, 等. 2001. 联合国粮农组织物种鉴定手册: 中国的海兽. 罗马: 联合国粮农组织. Zhou K Y, Xie F S, Li D W, *et al*. 2001. FAO Species Identification Manual: Marine Mammals of China. Rome: FAO. (In Chinese)

周开亚, 杨光, 高安利, 等. 1998. 南京—湖口江段长江江豚的种群数量和分布特点. 南京师范大学学报(自然科学版), 21(2): 91-98. Zhou K Y, Yang G, Gao A L, *et al*. 1998. Population and distribution characteristics of the finless porpoise in the Yangtze river from Nanjing to Hukou. *Journal of Nanjing Normal University* (Natural Science Edition), 21(2): 91-98. (In Chinese)

周立志, 马勇, 李迪强. 2000. 大沙鼠在中国的地理分布. 动物学报, 46(2): 130-137. Zhou L Z, Ma Y, Li D Q. 2000. Distribution of great gerbil (*Rhombomys opimus*) in China. *Acta Zoologica Sinica*, 46(2): 130-137. (In Chinese with English Abstract)

周立志, 马勇, 叶晓堤. 2002. 中国干旱地区啮齿动物物种分布的区域分异. 动物学报, 48(2): 183-194. Zhou L Z, Ma Y, Ye X D. 2002. Distribution of glires in arid regions of China. *Chinese Journal of Zoology*, 48(2): 183-194. (In Chinese with English Abstract)

周全, 吴毅, 肖玲, 等. 2005. 扁颅蝠的栖息地及最北分布. 动物学杂志, 40(6): 114-116. Zhou Q, Wu Y, Xiao L, *et al*. 2005. The habitat and the northernmost distribution of *Tylonycteris pachypus*. *Chinese Journal of Zoology*, 40(6): 114-116. (In Chinese with English Abstract)

周全, 徐忠鲜, 余文华, 等. 2014. 广东省南岭发现毛翼管鼻蝠及其核型与回声定位声波特征. 动物学杂志, 49(1): 41-45. Zhou Z, Xu Z X, Yu W H, *et al*. 2014. The Occurrence of Bat *Harpiocephalus harpia* from Nanlin, Guangdong and its Karyotypes, Echolocation Calls. *Chinese Journal of Zoology*, 49(1): 41-45. (In Chinese with English Abstract)

周全, 张燕均, 本川雅治, 等. 2011. 广东省南岭新纪录种中管鼻蝠的形态测量、核型及超声波数据. 动物学杂志, 46(1): 109-114. Zhou Q, Zhang Y J, Motokawa M, *et al*. 2011. A new record bat *Murina huttoni* from Guangdong, China and its morphology, karyotypes, echolocation calls. *Chinese Journal of Zoology*, 46(1): 109-114. (In Chinese with English Abstract)

周全, 张燕均, 杨平, 等. 2012. 广东省蝙蝠新纪录种——大墓蝠. 四川动物, 31(2): 287-289. Zhou Q, Zhang Y J, Yang P, *et al*. 2012. A new record bat of *Taphozous theobaldi* in Guangdong, China. *Sichuan Journal of Zoology*, 31(2): 287-289. (In Chinese with English Abstract)

周树武, 梁江明, 曾竣, 等. 2007. 合浦县啮齿动物种类及其分布的研究. 中华卫生杀虫药械, 13(4): 275-277. Zhou S W, Liang J M, Zeng J, *et al*. 2007. Species and distribution of rodent in Hepu county. *Chinese Journal of Hygienic Insecticides & Equipments*, 13(4): 275-277. (In Chinese with English Abstract)

周现召. 2012. 河北省珍稀濒危动物分布格局的研究. 石家庄: 河北师范大学硕士研究生学位论文. Zhou X Z. 2012. Research on the Distribution Pattern of Rare and Endangered Animals in Hebei Province. Shijiazhuang: Master's Thesis of Hebei Normal University. (In Chinese with English Abstract)

周旭东, 黄健, 张永军, 等. 2005a. 新疆甘家湖自然保护区啮齿动物群落结构与空间格局的研究. 四川动物, 24(2): 138-142. Zhou X D, Huang J, Zhang Y J, *et al*. 2005a. Research on rodent community structure and space pattern at Ganjiahu National Nature Reserve in Xinjiang. *Sichuan Journal of Zoology*, 24(2): 138-142. (In Chinese with English Abstract)

周昭敏, 徐伟霞, 吴毅, 等. 2005b. 中菊头蝠中国三亚种的形态特征比较. 动物学研究, 26(6): 645-651. Zhou Z M, Xu W X,

Wu Y, *et al*. 2005b. Morphometric characteristics of three subspecies of *Rhinolophus affinis* in China. *Zoological Research*, 26(6): 645-651. (In Chinese with English Abstract)

周昭敏, 赵宏, 张忠旭, 等. 2012. 中国穿山甲与爪哇穿山甲甲片异速生长分析及其在司法鉴定中的应用. 动物学研究, 33(3): 271-275. Zhou Z M, Zhao H, Zhang Z X, *et al*. 2012. Allometry of scales in Chinese pangolins (*Manis pentadactyla*) and Malayan pangolins (*Manis javanica*) and application in judicial expertise. *Zoological Research*, 33(3): 271-275. (In Chinese with English Abstract)

朱斌良. 2008. 海南岛翼手目(Chiroptera)物种多样性初步调查与保护对策研究. 海口: 海南师范大学硕士研究生学位论文. Zhu B L. 2008. Preliminary Investigation and Protection of Species Diversity of Chiroptera, Hainan Island. Haikou: Master's Thesis of Hainan Normal University. (In Chinese with English Abstract)

朱光剑. 2007. 海南岛犬蝠食性、栖息地类型和棕果蝠活动规律的研究. 海口: 海南师范大学硕士研究生学位论文. Zhu G J. 2007. Feeding Habits, Habitat Types and Activity Patterns of Rousettus leschenaultii on Hainan Island. Haikou: Master's Thesis of Hainan Normal University. (In Chinese with English Abstract)

朱光剑, 韩乃坚, 洪体玉, 等. 2008a. 海南属种新纪录——中华山蝠的回声定位信号、栖息地及序列分析. 动物学研究, 29(4): 447-451. Zhu G J, Han N J, Hong T Y, *et al*. 2008a. Echolocation call, roost and sequence analysis of new record of *Nyctalus plancyi* (Chiroptera: Vespertilionidae) on Hainan Island. *Zoological Research*, 29(4): 447-451. (In Chinese with English Abstract)

朱光剑, 李德伟, 叶建平, 等. 2008b. 南蝠海南岛分布新纪录、回声定位信号和ND1分析. 动物学杂志, 43(5): 69-75. Zhu G J, Li D W, Ye J P, *et al*. 2008b. New Record of *Ia io* in Hainan Island, and its echolocation pulses and ND1 analysis. *Chinese Journal of Zoology*, 43(5): 69-75. (In Chinese with English Abstract)

朱红艳, 曾涛, 刘洋, 等. 2010. 四川黄龙自然保护区兽类资源调查. 四川林业科技, 31(5): 83-88. Zhu H Y, Zeng T, Liu Y, *et al*. 2010. Investigations of mammal resources in Huanglong Nature Reserve in Sichnan. *Sichuan Forestry Science and Technology*, 31(5): 83-88. (In Chinese with English Abstract)

朱琼蕊, 郭宪国, 黄辉. 2014. 小家鼠的研究现状. 热带医学杂志, 14(3): 392-396. Zhu Q R, Guo X G, Huang H. 2014. Current research of house mice. *Journal of Tropical Medicine*, 14(3): 392-396.

朱曦, 曹炜斌, 王军. 2010. 舟山普陀山岛兽类区系及分布. 浙江农林大学学报, 27(1): 110-115. Zhu X, Cao W B, Wang J. 2010. Mammalian fauna and distribution of Putuoshan Island in Zhoushan. *Journal of Zhejiang A & F University*, 27(1): 110-115. (In Chinese with English Abstract)

朱旭, 刘颖, 施利民, 等. 2009. 大棕蝠江南亚种回声定位声波特征与分析. 四川动物, 28(1): 59-63. Zhu X, Liu Y, Shi L M, *et al*. 2009. Characteristics and analysis of echolocation calls by *Eptesicus serotinus andersoni*. *Sichuan Journal of Zoology*, 28(1): 59-63. (In Chinese with English Abstract)

朱妍, 李波, 张伟, 等. 2011. 俄罗斯与中国紫貂保护利用现状的比较. 经济动物学报, 15(4): 198-202. Zhu Y, Li B, Zhang W, *et al*. 2011. Current status comparison of sable conservation and utilization in Russia and China. *Journal of Economic Animal*, 15(4): 198-202. (In Chinese with English Abstract)

祝茜, 姜波, 汤庭耀. 2000. 中国海洋哺乳动物的种类、分布及其保护对策. 海洋科学, 24(9): 35-39. Zhu Q, Jiang B, Tang T Y. 2000. Species, distribution, and protection of marine mammals in the Chinese coastal waters. *Marine Science*, 24(9): 35-39. (In Chinese with English Abstract)

祝茜, 李响, 马牧, 等. 2007. 长吻飞旋海豚的一些生物学测量数据. 海洋科学, 31(8): 15-17. Zhu Q, Li X, Ma M, *et al*. 2007. Some measurements of the Spinner dolphin *Stenella longirostris*. *Marine Sciences*, 31(8): 15-17. (In Chinese with English Abstract)

庄炜. 1991. 中国貉(*Nyctereutes procyonoides*)线粒体DNA多态性及其与亚种分化的关系. 北京: 中国科学院研究生院硕士研究生学位论文. Zhuang W. 1991. Mitochondrial DNA Polymorphism of *Nyctereutes procyonoides* and its Relationship with Subspecies Differentiation. Beijing: Master's Thesis of Graduate School, Chinese Academy of Sciences. (In Chinese with English Abstract)

邹波, 王庭林, 宁振东, 等. 2012. 山西省鼠类主要天敌——鼬类的分布与数量估测. 农业技术与装备, (18): 7-9. Zou B, Wang T L, Ning Z D, *et al*. 2012. Estimation of distribution and population of weasels, the main natural enemies of rodents in Shanxi Province. *Agricultural Technology and Equipment*, (18): 7-9. (In Chinese with English Abstract)

Abdukadir A, Khan B, Masuda R, *et al*. 2010. Asiatic wild cat (*Felis silvestris ornata*) is no more a 'Least Concern' species in Xinjiang, China. *Pakistan Journal of Wildlife*, 1(2): 57-63.

Abdukadir A, Khan B. 2013. Status of Asiatic wild cat and its habitat in Xinjiang Tarim Basin, China. *Open Journal of Ecology*,

3(8): 551.

Abe H. 1995. Revision of the Asian moles of the genus *Mogera*. *Journal of the Mammalogical Society of Japan*, 20: 51-68.

Abramov A. 2002. Variation of the baculum structure of the *Palaearctic badger* (Carnivora, Mustelidae, *Meles*). *Russian Journal of Theriology*, 1(1): 57-60.

Abramov A V, Duckworth J W, Wang Y X, *et al.* 2008. The stripe-backed weasel *Mustela strigidorsa*: taxonomy, ecology, distribution and status. *Mammal Review*, 38(4): 247-266.

Abramov A V, Meschersky I G, Rozhnov V V. 2009. On the taxonomic status of the harvest mouse *Micromys minutus* (Rodentia: Muridae) from Vietnam. *Zootaxa*, 2199(1): 58-68.

Adler G H. 1996. Habitat relations of two endemic species of highland forest rodents in Taiwan. *Zoological Studies* (Taipei), 35(2): 105-110.

Ai H, He K, Chen Z, *et al.* 2018. Taxonomic revision of the genus *Mesechinus* (Mammalia: Erinaceidae) with description of a new species. *Zoological Research*, 39(5): 335-347.

Allen G M. 1923. New Chinese insectivores. *American Museum Novitates*, 100: 1-11.

Allen G M. 1938-1940. The Mammals of China and Mongolia. Vol. XI. Part. I, II. New York: American Museum (Natural History).

Amano M. 2018. Finless porpoise. *In*: Würsig B, Thewissen J G M, Kovacs K M. Encyclopedia of Marine Mammals. 3rd ed. San Diego: Academic Press/ Elsevier: 372-375.

Amato G, Egan M G, Rabinowitz A. 1999. A new species of muntjac, *Muntiacus putaoensis* (Artiodactyla: Cervidae) from northern Myanmar. *Animal Conservation Forum*, 2(1): 1-7.

American Society of Mammalogists. 2019. Mammal Diversity Database. http://www.mammaldiversity.org/. [2019-2-20]

Apagow P M. 2007. Species: Demarcation and diversity. *In*: Purvis A, Gittleman J L, Brooks T. Phylogeny and Conservation. Cambridge: Cambridge University Press: 19-56.

Argyropulo A I. 1948. A review of Recent species of the family Lagomyidae Lilljeb., 1886 (Lagomorpha, Mammalia). *Trudy Zoologicheskogo Instituta Akademii Nauk SSR* (*Proceedings of the Zoological Institute of the USSR*), 7: 124-128. (In Russian)

Baklushinskaya I Y, Formozov N A. 1999. New species of the 38-chromosomal pikas. *In*: Abstract of the VI Meeting of the Theriological Society of RAS. Moscow, 17. (In Russian)

Banaszek A, Bogomolov P, Feoktistova N, *et al.* 2020. *Cricetus cricetus*. The IUCN Red List of Threatened Species 2020. https://dx.doi.org/10.2305/IUCN.UK.2020-2.RLTS.T5529A111875852.en. [2020-9-10]

Bao W. 2010. *Eurasian lynx* in China-present status and conservation challenges. *Cat News* (Special Issue), 5: 22-25.

Barnosky A D, Matzke N, Tomiya S, *et al.* 2011. Has the Earth's sixth mass extinction already arrived? *Nature*, 471(7336): 51-57.

Bar-Ona Y M, Phillips R, Milo R. 2018. The biomass distribution on Earth. http://www.pnas.org/cgi/doi/10.1073/pnas.1711842115. [2019-4-30]

Bleisch W V, Buzzard P J, Zhang H, *et al.* 2009. Surveys at a Tibetan antelope *Pantholops hodgsonii* calving ground adjacent to the Arjinshan Nature Reserve, Xinjiang, China: Decline and recovery of a population. *Oryx*, 43(2): 191-196.

Brook S M, Donnithorne-Tait D, Lorenzini R, *et al.* 2017. *Cervus hanglu* (amended version of 2017 assessment). The IUCN Red List of Threatened Species 2017: e.T4261A120733024. https://dx.doi.org/10.2305/IUCN.UK.2017-3.RLTS.T4261A120733024.en. [2020-7-17]

Bryant J V, Gottelli D, Zeng X, *et al.* 2016. Assessing current genetic status of the *Hainan gibbon* using historical and demographic baselines: implications for conservation management of species of extreme rarity. *Molecular Ecology*, 25(15): 3540-3556.

Burgin C J, Colella J P, Kahn P L, *et al.* 2018. How many species of mammals are there? *Journal of Mammalogy*, 99: 1-14.

Butchart S H M, Walpole M, Collen B, *et al.* 2010. Global biodiversity: indicators of recent declines. *Science*, 328(5982): 1164-1168.

Buzzard P J, Zhang H, Xü D, *et al.* 2010. A globally important wild yak *Bos mutus* population in the Arjinshan Nature Reserve, Xinjiang, China. *Oryx*, 44(4): 577-580.

Cao L, Wang X, Fang S. 2003. A molecular phylogeny of Bharal and dwarf blue sheep based on mitochondrial cytochrome b gene sequences. *Acta Zoologica Sinica*, 49(2): 198-204.

Castelló J R. 2016. Bovids of the World. Princeton: Princeton University Press.

Catalogue of Life. 2019. http://www.catalogueoflife.org/annual-checklist/2018/. [2019-1-15]

Cerchio S, Yamada T K. 2018. Omura's whale. *In*: Würsig B, Thewissen J G M, Kovacs K M. Encyclopedia of Marine Mammals. 3rd ed. San Diego: Academic Press/ Elsevier: 656-659.

Chakraborty S, Srinivasulu C, Srinivasulu B, *et al*. 2004. Checklist of insectivores (Mammalia: Insectivora) of South Asia. *Zoos' Print Journal*, 19: 1361-1371.

Challender D W S, Nash H C, Waterman C. 2019. *Pangolins: Science, Society and Conservation*. New York: Academic Press.

Chan B P L, Tan X, Tan W. 2008. Rediscovery of the critically endangered eastern black crested gibbon *Nomascus nasutus* (Hylobatidae) in China, with preliminary notes on population size, ecology and conservation status. *Asian Primates Journal*, 1(1): 17-25.

Chan B P L, Lo Y F P, Mo Y. 2020. New hope for the Hainan gibbon: formation of a new group outside its known range. *Oryx*, 54(3): 296-298.

Chen J, Liu T, Deng H, *et al*. 2017a. A new species of *Murina* bats was discovered in Guizhou Province, China. *Cave Research*, 2: 1

Chen P, Hua Y. 1989. Distribution, population size and protection of *Lipotes vexillifer*. *In*: Perrin W F, Brownell Jr R L, Zhou K, *et al*. Biology and Conservation of the River Dolphins. Switzerland: IUCN Species Survival Commission Occasional Paper(No. 3): 81-85.

Chen S, Liu S, Liu Y, *et al*. 2012. Molecular phylogeny of Asiatic short-tailed shrews, genus *Blarinella* Thomas, 1911 (Mammalia: Soricomorpha: Soricidae) and its taxonomic implications. *Zootaxa*, 3250: 43-53.

Chen Y. 2009. Distribution patterns and faunal characteristic of mammals on Hainan Island of China. *Folia Zoologica*, 58(4): 372-384.

Chen Y, Peng H. 2010. Research on bat diversity in Yulin City, Guangxi. *Sichuan Journal of Zoology*, 29: 609-612.

Chen Z, He K, Huang C, *et al*. 2017b. Integrative systematic analyses of the genus *Chodsigoa* (Mammalia: Eulipotyphla: Soricidae), with descriptions of new species. *Zoological Journal of the Linnaean Society*, 180(3): 694-713.

Cheng F, He K, Chang Z, *et al*. 2017. Phylogeny and systematic revision of the genus *Typhlomys* (Rodentia, Platacanthomyidae): with description of a new species. *Journal of Mammalogy*, 98(3): 731-743.

Cheng H C, Fang Y P, Chou C H. 2015. A Photographic Guide to the Bats of Taiwan. Nantou: Endemic Species Research Institute.

China Species Information Service. 2018. *Myotis pequinius*. Available at: http://www.chinabiodiversity.com; http://www.baohu.org. [2018-5- 22]

Choudhury. 2003. The Mammals of Arunachal Pradesh. New Deli: Daya Books.

Christiansen P, Kitchener A. 2011. A neotype of the clouded leopard (*Neofelis nebulosa* Griffith, 1821). *Mammalian Biology*, 76(3): 325-331.

Chu H, Jiang Z. 2009. Distribution and conservation of Sino-Mongolian beaver, *Castor fiber birulai*. *Oryx*, 43(2): 197-202.

Chung K P, Corlett R T. 2006. Rodent diversity in a highly degraded tropical landscape: Hong Kong, South China. *Biodiversity and Conservation*, 15: 4521-4532.

Clark H O, Murdoch J D, Newman D P, *et al*. 2009. *Vulpes corsac* (Carnivora: Canidae). *Mammalian Species*, 832: 1-8.

Clark H O, Newman D P, Murdoch J D, *et al*. 2008. *Vulpes ferrilata* (Carnivora: Canidae). *Mammalian Species*, 821: 1-6.

Corbet G B. 1978. The Mammals of the Palaearctic Region: a Taxonomic Review. London, New York: British Museum (Natural History) and Cornell University Press.

Corbet G B. 1988. The family Erinaceidae: a synthesis of its taxonomy, phylogeny, ecology and zoogeography. *Mammal Review*, 18: 117-172.

Corbet G B, Hill J E. 1986. A World List of Mammalian Species. 2nd ed. London: British Museum (Natural History).

Corbet G B, Hill J E. 1992. Mammals of the Indo-Malayan Region: A Systematic Review. Oxford: Oxford University Press.

Corbet G B, Hill J E. 1991. A World List of Mammalian Species. 3rd ed. London: British Museum (Natural History).

Corlett R T. 2014. The Ecology of Tropical East Asia. 2nd ed. Oxford: Oxford University Press.

Crandall K A, Agapow P M, Bininda-Emonds O R P, *et al*. 2004. The impact of species concept on biodiversity studies. *Quarterly Review of Biology*, 79(2): 161-179.

Cronquist A. 1978. Once again, what is a species? *In*: Knutson L V. Biosystematics in Agriculture. Chicago: The University of Chicago Press: 3-20.

Csorba G, Bates P J J. 2005. Description of a new species of *Murina* from Cambodia (Chiroptera: Vespertilionidae: Murininae). *Acta Chiropterologica*, 7(1): 1-7.

Csorba G, Lee L L. 1999. A new species of vespertilionid bat from Taiwan and a revision of the taxonomic status of Arielulus and Thainycteris (Chiroptera: Vespertilionidae). *Journal of Zoology* (London), 248: 361-367.

Csorba G P, Ujhelyi P, Thomas N. 2003. Horseshoe Bats of the World. Shropshire: Alana Books.

Cui S, Milner-Gulland E J, Singh N J, *et al.* 2017. Historical range, extirpation and prospects for reintroduction of Saigas in China. *Scientific Reports*, 7: 44200.

Culik B G. 2011. Odontocetes–the toothed whales. Bonn: UNEP/CMS/ASCOBANS Secretariat: 311.

Dahal N, Lissovsky A A, Lin Z, *et al.* 2017a. Corrigendum to "Genetics, morphology and ecology reveal a cryptic pika lineage in the Sikkim Himalaya" [*Mol. Phylogenet. Evol.* 106: 55-60]. *Molecular Phylogenetics and Evolution*, 107: 645.

Dahal N, Lissovsky A A, Lin Z, *et al.* 2017b. Genetics, morphology and ecology reveal a cryptic pika lineage in the Sikkim Himalaya. *Molecular Phylogenetics and Evolution*, 106: 55-60.

Darwin C. 1859. On the Origin of Species by Means of Natural Selection, or the Preservation of Favoured Races in the Struggle for Life. London: John Murray.

Ding L, Chen M, Pan T, *et al.* 2014. Complete mitochondrial DNA sequence of *Lepus sinensis* (Leporidae: *Lepus*). *DNA Sequence*, 27: 1711-1712.

Eger J L, Lim B K. 2011. Three new species of *Murina* from southern China (Chiroptera: Vespertilionidae). *Acta Chiropterologica*, 13(2): 227-243.

Elleman J R, Morrison T C S. 1951. Checklist of Palaeartic and Indian Mammals. London: Birtish Musem (Nature History).

Erbajeva M A. 1988. *Pischukhi kainozoya* (taxonomia, systematica, filogenia) [Cenozoic pikas taxonomy, systematics, phylogeny)]. Moscow: Nauka. (In Russian)

Erbajeva M A. 1997. The history and systematics of the genus *Ochotona*. *Proceeding of XIIth Lagomorph Workshop*, *Game and Wildlife*, 17(3): 505.

Erbajeva M A. 2003. Late Miocene ochotonids (Mammalia, Lagomorpha) from Central Mongolia. *Neues Jahrbuch für Geologie und Paläeontologie Monatshefte*, 4: 212-222.

Erbajeva M A, Ma Y. 2006. A new look at the taxonomic status of *Ochotona argentata* Howell, 1928. *Acta Zoologica Cracoviensia*, 49A(1-2): 135-149.

Escobar L E, Awan M N, Qiao H. 2015. Anthropogenic disturbance and habitat loss for the red-listed Asiatic black bear (*Ursus thibetanus*): Using ecological niche modeling and nighttime light satellite imagery. *Biological Conservation*, 191: 400-407.

Fan P, He K, Chen X, *et al.* 2017. Description of a new species of *Hoolock* gibbon (Primates: Hylobatidae) based on integrative taxonomy. *American Journal Primatology*, 79(5): 10.1002/ajp.22631.

Fan P, Wen X, Huo S, *et al.* 2011. Distribution and conservation status of the Vulnerable eastern hoolock gibbon *Hoolock leuconedys* in China. *Oryx*, 45(6): 129-134.

Fan Z, Liu S, Liu Y, *et al.* 2009. Molecular phylogeny and taxonomic reconsideration of the subfamily Zopodinae (Rodentia: Dipodidae), with an emphasis on Chinese species. *Molecular Phylogenetics and Evolution*, 51(3): 447-453.

Fang Y, Lee L. 2002. Re-evaluation of the Taiwanese white-toothed shrew, *Crocidura tadae* Tokuda and Kano, 1936 (Insectivora: Soricidae) from Taiwan and two offshore islands. *Journal of Zoology* (London), 257: 145-154.

Fellowes J R, Chan B P L, Zhou J, *et al.* 2008. Current status of the Hainan gibbon (*Nomascus hainanus*): progress of population monitoring and other priority actions. *Asian Primates Journal*, 1(1): 2-9.

Feng L, Jutzeler E. 2010. Clouded leopard. Cat News(Special Issue), 5: 34-36.

Feng L, Lin L, Zhang L, *et al.* 2008a. Evidence of wild tigers in southwest China—a preliminary survey of the Xishuangbanna National Nature Reserve. Cat News, 48: 4-6.

Feng Q, Li S, Wang Y. 2008b. A new species of bamboo bat (Chiroptera: Vespertilionidae: *Tylonycteris*) from Southwestern China. *Zoological Science* (Tokyo), 25: 225-234.

Formozov N A. 1997. Pikas (*Ochotona*) of the world: Systematics and Conservation. *Proceeding of XIIth Lagomorph Workshop*, *Game and Wildlife*, 17(3): 506-507.

Formozov N A, Baklushinskaya I, Ma Y. 2004. Taxonomic status of the Helan-Shan pika, *Ochotona argentata*, from the Helan-Shan Ridge (Ningxia, China). *Zoologicheskiĭ Zhurnal*, 83(8): 995-1007.

Fox J L, Dhondup K, Dorji T. 2009. Tibetan antelope *Pantholops hodgsonii* conservation and new rangeland management policies in the western Chang Tang Nature Reserve, Tibet: is fencing creating an impasse? *Oryx*, 43: 183-190.

Fumagalli L, Taberlet P, Stewart D. 1999. Molecular phylogeny of *Sorex shrews* (Soricidae: Insectivora) inferred from mitochondrial DNA sequence data. *Molecular Phylogeny and Evolution*, 11: 222-235.

Furey N M, Thong V D, Bates P J J, *et al*. 2009. Description of a new species belonging to the *Murina*' *suilla*-Group (Chiroptera: Vespertilionidae: Murininae) from North Vietnam. *Acta Chiropterologica*, 11: 225-236.

Gao X, Xu W, Yang W, *et al*. 2011. Status and distribution of ungulates in Xinjiang, China. *Journal of Arid Land*, 3(1): 49-60.

Gärdenfors U, Hilton-Taylor C, Mace G M, *et al*. 2001. The application of IUCN Red List Criteria at regional levels. *Conservation Biology*, 15(5): 1206-1212.

Garshelis D L, Joshi A R, Smith J L. 1999. Estimating density and relative abundance of sloth bears. *Ursus*, 11: 87-98.

Ge D, Feijó A, Cheng J, *et al*. 2019. Evolutionary history of field mice (Murinae: *Apodemus*), with emphasis on morphological variation among species in China and description of a new species. *Zoological Journal of the Linnean Society*, 187(2): 518-534.

Ge D, Lissovsky A A, Xia L, *et al*. 2012. Reevaluation of several taxa of Chinese lagomorphs (Mammalia: Lagomorpha) described on the basis of pelage phenotype variation. *Mammalian Biology*, 77(2): 113-123.

Ge D, Lu L, Abramov A V, *et al*. 2018. Coalesence Models reveal the rise of the white-bellied rat (*Niviventer confucianus*) following the loss of Asian megafauna. *Journal of Mammalian Evolution*, 26: 423-434.

Ge D, Wen Z, Xia L, *et al*. 2013. Evolutionary history of Lagomorphs in response to global environmental change. *PLoS One*, 8(4): 1-15.

Geissmann T, Lwin N, Aung S S, *et al*. 2011. A new species of snub-nosed monkey, genus *Rhinopithecus* Milne-Edwards, 1872 (Primates, Colobinae), from northern Kachin State, northeastern Myanmar. *American Journal of Primatology*, 73(1): 96-107.

Geissmann T, Momberg F, Whitten T. 2020. Rhinopithecus strykeri. The IUCN Red List of Threatened Species 2020. https://dx.doi.org/10.2305/IUCN.UK.2020-2.RLTS.T13508501A17943490.en. [2020-10-4]

Gippoliti S. 2001. Notes on the taxonomy of *Macaca nemestrina leonina* Blyth 1864 (primates : Cercopithecidae). *Journal of Mammalogy*, 12(1): 51-54.

Giraudoux P, Quéré J P, Delattre P, *et al*. 1998. Distribution of small mammals along a deforestation gradient in southern Gansu, central China. *Acta Theriologica*, 43(4): 349-362.

Giraudoux P, Zhou H, Quéré J P, *et al*. 2008. Small mammal assemblages and habitat distribution in the northern Junggar Basin, Xinjiang, China: a pilot survey. *Mammalia*, 72(4): 309-319.

Goerfoel T, Estok P, Csorba G. 2013. The subspecies of *Myotis montivagus*—taxonomic revision and species limits (Mammalia: Chiroptera: Vespertilionidae). *Acta Zoologica Academiae Scientiarum Hungaricae*, 59(1): 41-59.

Green M J B. 1986. The Distribution, status and conservation of the Himalayan musk deer. *Biological Conservation*, 35(4): 347-375.

Groves C. 2001. Primate Taxonomy. Washington: Smithsonian Institution Press.

Groves C. 2016. Systematics of the Artiodactyla of China in the 21st century. *Zoological Research*, 37(3): 119-125.

Groves C. 1995. Taxonomy of musk-deer, genus *Moschus* (Moschidae, Mammalia). *Acta Theriologica Sinica*, 15(3): 181-197.

Groves C, Grubb P. 2011. Ungulate Taxonomy. Baltimore: Johns Hopkins University Press.

Grueter C C, Jiang X, Konrad R, *et al*. 2009. Are *Hylobates lar* extirpated from China? *International Journal of Primatology*, 30(4): 553-567.

Grueter C C, Li D, Ren B, *et al*. 2012. Food abundance is the main determinant of high-altitude range use in snub-nosed monkeys. *International Journal of Zoology*, DOI: 10.1155/2012/739419.

Grueter C C , Qi X, Zinner D, *et al*. 2020. Multilevel Organisation of Animal Sociality. *Trends in Ecology & Evolution*, 35(9): 834-847.

Guo W, Yu W, Wang X, *et al*. 2017. First record of the collared sprite, *Thainycteris aureocollaris* (Chiroptera, Vespertilionidae) from China. *Mammal Study*, 42(2): 97-103.

Guo Y, Zou X, Chen Y, *et al*. 1997. Sustainability of wildlife use in traditional Chinese medicine. *Conserving China's Biodiversity*, 1: 3.

Gureev A A. 1964. Fauna of the USSR, mammals. Lagomorpha. 3, part 10. Moscow: Nauka. (In Russian)

Han B, Hua P, Gu X, *et al*. 2008a. Isolation and characterization of microsatellite loci in the long-fingered bat *Miniopterus*

fuliginosus. *Molecular Ecology Resources*, 8(6): 799-801.

Han B, Hua P, Gu X, *et al*. 2008b. Isolation and characterization of microsatellite loci in the western long-fingered bat, *Miniopterus magnater*. *Molecular Ecology Resources*, 8(6): 1445-1447.

Han N J, Zhang J S, Reardon T, *et al*. 2010. Revalidation of *Myotis taiwanensis* Arnback-Christie-Linde 1908 and its molecular relationship with *M. adversus* (Horsfield 1824) (Vespertilionidae, Chiroptera). *Acta Chiropterologica*, 12(2): 449-456.

He F, Xiao N, Zhou J. 2015. A new species of *Murina* from China. *Cave Research*, 2: 1-5.

He K, Deng K, Jiang X. 2012a. First record of Van sung's shrew (*Chodsigoa caovansunga*) in China. *Zoological Research*, 33(5): 542-544.

He K, Hu N, Orkin J D, *et al*. 2012b. Molecular phylogeny and divergence time of *Trachypithecus*: with implications for the taxonomy of *T. phayrei*. *Zoological Research*, 33(E5-6): 104-110.

He K, Jiang X. 2015. Mitochondrial phylogeny reveals cryptic genetic diversity in the genus *Niviventer* (Rodentia, Muroidea). *Mitochondrial DNA*, 26(1): 48-55.

He L, García-Perea R, Li M, *et al*. 2004. Distribution and conservation status of the endemic Chinese mountain cat *Felis bieti*. *Oryx*, 38: 55-61.

Heath M E. 1995. *Manis crassicaudata*. *Mammalian Species*, 513: 1-4.

Hedges S, Sagar Baral H, Timmins R J, *et al*. 2008. *Bubalus arnee*. The IUCN Red List of Threatened Species 2008. http://www.iucnredlist.org/details/3129/0. [2016-01-23]

Heideman P D, Heaney L R. 1989. Population biology and estimates of abundance of fruit bats (Pteropodidae) in Philippine submontane rainforest. *Journal of Zoology*, 218(4): 565-586.

Helgen K M. 2005. Order Scandentia. *In*: Wilson D E, Reeder D A. Mammal Species of the World: a Taxonomic and Geographic Reference. Baltimore: Johns Hopkins University Press: 104-109.

Hennig W. 1966. Phylogenetic Systematics. Translated by Davis D D, Zangerl R. Urbana: University of Illinois Press.

Hill J E, Harrison D L. 1986. The baculum in the Vespertilioninae (Chiroptera: Vespertilionidae) with a systematic review, a synopsis of *Pipistrellus* and *Eptesicus*, and the descriptions of a new genus and subgenus. *Bulletin of the British Museum (Natural History) Zoology*, 52: 225-305.

Hjarding A, Tolley K, Burgess N D. 2015. Red list assessments of East African chameleons: a case study of why we need experts. *Oryx*, 49: 653-658.

Hoffmann R S. 1984. A review of the shrew-moles (Genus *Uropsilus*) of China and Burma. *Journal of the Bombay Natural History Society*, 10(2): 69-80.

Hoffmann R S. 1987. A review of the systematics and distribution of Chinese red-toothed shrews (Mammalia: Soricinae). *Acta Theriologica Sinica*, 7(2): 100-139.

Hoffmann R S, Smith A T. 2009. Mammals of Russia and Adjacent Region: Lagomorphs. Translated by Sokolov V E, Ivanitskaya E Y, Gruzdev V V, *et al*. 1994. Smithsonian Institution Libraries. Washington: Amerind Publishing.

Hoffmann R S, Smith A. 2005. Order Lagomorpha. *In*: Wilson D E, Reeder D M. Mammal Species of the World. 3rd ed. Baltimore: Johns Hopkins University Press.

Holt B G, Lessard J P, Borregaard M K, *et al*. 2013. An updated of Wallace's zoogeographic regions of the world. *Science*, 339: 74-78.

Honacki J H, Kinman K E, Koeppl J W. 1982. Mammal species of the world: a taxonomic and geographic reference. Lawrence: Allen Press and the Associated System.

Hong T, Gong Y, Yang J, *et al*. 2011. Partial albino bats of *Miniopterus pusillus* and *Hipposideros pomona* found in Guangdong Province. *Acta Theriologicainica*, 31(3): 320-322.

Horácek I, Hanák V, Gaisler J, *et al*. 2000. Bats of the Palearctic region: a taxonomic and biogeographic review. 11-157. *In*: Woloszyn B W. Proceedings of the VIIIth European Bat Research Symposium. Vol I. Approaches to Biogeography and Ecology of Bats. Krakow: Publication of the Chiropterological Information Center, Institute of Systematics and Evolution of Animals Poland Academy of Science.

Horácek I, Hanák V. 1984. Comments on the systematics and phylogeny of *Myotis nattereri* (Kuhl, 1818). *Myotis*, 21-22: 20-29.

Horwood J. 2018. Sei whale. *In*: Würsig B, Thewissen J G M, Kovacs K M. Encyclopedia of Marine Mammals. 3rd ed. San Diego: Academic Press/ Elsevier: 845-847.

Howell A B. 1928. New Asiatic mammals collected by F. R. WULSIN. *Proceedings of the Biological Society of Washington*, 41:

115-120.

Hu J, Jiang Z, Chen J, *et al*. 2015. Niche divergence accelerates the evolution in Asian endemic *Procapra* gazelles. *Scientific Reports*, 5: 10069

Hu J, Jiang Z, Mallon D. 2013. Metapopulation viability of a globally endangered gazelle on the Northeast Qinghai-Tibetan Plateau. *Biological Conservation*, 166: 23-32.

Hu J, Zhang Y, Yu L. 2012. Summary of Laurasiatheria (Mammalia) Phylogeny. *Zoological Research*, 33(E5-6): 65-74.

Hu Y M, Zhou Z X, Huang Z W, *et al*. 2017. A new record of the capped langur (*Trachypithecus pileatus*) in China. *Zoological Research*, 38(4): 203-205.

Huang C, Li X, Jiang X. 2017. Confirmation of the continued occurrence of Binturong *Arctictis binturong* in China. *Small Carnivore Conservation*, 55: 59-63.

Huang C, Yu W, Xu Z, *et al*. 2014. A cryptic species of the *Tylonycteris pachypus* complex (Chiroptera: Vespertilionidae) and its population genetic structure in Southern China and nearby regions. *International Journal of Biological Sciences*, 10: 200-211.

Huang J C C, Csorba G, Chang H C, *et al*. 2020. *Myotis formosus* of the *IUCN Red List of Threatened Species* 2020. https://dx.doi.org/10.2305/IUCN.UK.2020-2.RLTS.T85736120A95642290.en. [2020-9-28]

Huang S, Karczmarski L, Chen J L, *et al*. 2012. Demography and population trend of the largest population of Indo-Pacific humpback dolphin (*Sousa chinensis*). *Biological Conservation*, 147(1): 234-242.

Hung S, Kyung H. 2010. Genetic distinctness of the Korean hare, *Lepus coreanus* (Mammalia, Lagomorpha), revealed by nuclear thyroglobulin gene and mtDNA control region sequences. *Biochemical Genetics*, 48(7-8): 706-710.

Hunter L. 2015. Wild Cats of the World. London: Bloomsbury Publishing.

Hunter L, Barrett P. 2011. A Field Guide to the Carnivores of the World. London: New Holland Publish.

Hutterer R. 2005. Order Soricomorpha. *In*: Wilson D E, Reeder D M. Mammal Species of the World. Baltimore: Johns Hopkins University Press: 220-311.

International Commission on Zoological Nomenclature. 2003. Opinion 2028 (Case 3073). *Vespertilio pipistrellus* Schreber, 1774 and *V. pygmaeus* Leach, 1825 (currently *Pipistrellus pipistrellus* and *Pipistrellus pygmaeus*; Mammalia, Chiroptera): neotypes designated. *Bulletin of Zoological Nomenclature*, 60: 85-87.

IUCN. 2018. IUCN Red List of Threatened Species. http://www.iucnredlist.org/. [2018-12-1]

IUCN Standards and Petitions Subcommittee. 2017. Guidelines for Using the IUCN Red List Categories and Criteria. Version 13. Prepared by the Standards and Petitions Subcommittee. http://www.iucnredlist.org/documents/RedListGuidelines.pdf. [2018-11-20]

IUCN. 2012. IUCN Red List Categories and Criteria. Version 3.1. 2nd ed. Gland and Cambridge: IUCN. iv, 32.

IUCN. 2020. IUCN Red List of Threatened Species. https://www.iucnredlist.org/. [2020-7-18]

IUCN/SSC Criteria Review Working Group. 2010. Guidelines for Application of IUCN Red List Criteria at Regional and National Levels. Version 4.0. Cambridge: IUCN.

IUCN/SSC Criteria Review Working Group. 1999. IUCN Red List Criteria Review Provisional Report: Draft of the Proposed Changes and Recommendations. Cambridge: IUCN.

Jameson Jr E W, Jones G S. 1977. The Soricidae of Taiwan. *Proceedings of the Biological Society of Washington*, 90: 459-482.

Jefferson T A, Wang J Y. 2011. Revision of the taxonomy of finless porpoises (genus *Neophocaena*): The existence of two species. *Journal of Marine Animals and Their Ecology*, 4: 3-16.

Jiang T L, Liu R, Metzner W, *et al*. 2010a. Geographical and individual variation in echolocation calls of the intermediate leaf-nosed bat, *Hipposideros larvatus*. *Ethology*, 116(8): 691-703.

Jiang T L, Metzner W, You Y, *et al*. 2010b. Variation in the resting frequency of *Rhinolophus pusillus* in Mainland China: Effect of climate and implications for conservation. *Journal of the Acoustical Society of America*, 128(4): 2204-2211.

Jiang T L, Sun K P, Chou C H. 2010c. First record of *Myotis flavus* (Chiroptera: Vespertilionidae) from mainland China and a reassessment of its taxonomic status. *Zootaxa*, 41: 41-45

Jiang T L, Lu G, Sun K, *et al*. 2013. Coexistence of *Rhinolophus affinis* and *Rhinolophus pearsoni* revisited. *Acta theriologica*, 58: 47-53.

Jiang Z. 2013. Re-introduction of Père David's deer "Milu" to Beijing, Dafeng and Shishou, China. *In*: Soorae P S. Global Re-introduction Perspectives: 2013: Further Case Studies from Around the Globe. Gland: IUCN/SSC Re-introduction Specialist

Group and Abu Dhabi, UAE, Environment Agency: xiv, 282.

Jiang Z, Harris R B. 2008. *Elaphurus davidianus*. The IUCN Red List of Threatened Species, version 2015.2. http://www.iucnredlist.org. [2018-11-29]

Jiang Z, Lei F, Zhang C, *et al*. 2015. Biodiversity conservation and its research process. *In*: Li W. Contemporary Ecology Research in China. Springer and Beijing: Verlag Berlin Heidelberg, Higher Education Press: 29-45.

Jiang Z, Ma K. 2014. Scanning the horizon for nascent environmental hazards. *National Science Review*, (3): 330-333.

Jiang Z, Mallon D, Foggin M, *et al*. 2020. A case of reintroducing Saiga highlights the conservation needs of migratory species. doi: 10.20944/preprints202002.0375.v1Preprints . http://www.preprints.org. [2020-2-26]

Jiang Z, Zong H, 2019. Reintroduction of the przewalski's horse in china: status quo and outlook. *Nature Conservation Research*, 4(Suppl. 2): 15-22.

Jing M, Yu H, Wu S, *et al*. 2007. Phylogenetic relationships in genus *Niviventer* (Rodentia: Muridae) in China inferred from complete mitochondrial cytochrome *b* gene. *Molecular Phylogenetics and Evolution*, 44(2): 521-529.

Jones G, Parsons S, Zhang S, *et al*. 2006. Echolocation calls, wing shape, diet and phylogenetic diagnosis of the endemic Chinese bat *Myotis pequinius*. *Acta Chiropterologica*, 8: 451-463.

Jutzeler E, Wu Z, Liu W, *et al*. 2010a. Leopard *Panthera pardus*. *Cat News* (Special Issue), 5: 30-33.

Jutzeler E, Xie Y, Vogt K. 2010b. Fishing cat. *Cat News* (Special Issue), 5: 48-49.

Kaczensky P, Kuehn R, Lhagvasuren B, *et al*. 2011. Connectivity of the Asiatic wild ass population in the Mongolian Gobi. *Biological Conservation*, 144: 920-929.

Kaneko Y. 1987. Skull and dental characters, and skull measurements of *Microtus kikuchii* Kuroda, 1920 from Taiwan. *Journal of the Mammalogical Society of Japan*, 12(1-2): 31-39.

Kaneko Y. 1996. Morphological variation, and latitudinal and altitudinal distribution of *Eothenomys chinensis*, *E. wardi*, *E. custos*, *E. proditor*, and *E. olitor* (Rodentia, Arvicolidae) in China. *Mammal Study*, 21: 89-114.

Kato H, Perrin W F. 2018. Beyde's whale. *In*: Würsig B, Thewissen J G M, Kovacs K M. Encyclopedia of Marine Mammals. 3rd ed. San Diego, CA, USA: Academic Press/ Elsevier: 143-145.

Kawada S, Harada M, Koyasu K, *et al*. 2002. Karyological note on the short-faced mole, *Scaptochirus moschatus* (Insectivora, Talpidae). *Mammal Study*, 27: 91-94.

Kawada S, Shinohara A, Kobayashi S, *et al*. 2007. Revision of the mole genus *Mogera* (Mammalia: Lipotyphla: Talpidae) from Taiwan. *Systematics and Biodiversity*, 5(2): 223-240.

Kawada S, Shinohara A, Yasuda M, *et al*. 2003. The mole of peninsular Malaysia: notes on its identification and ecology. *Mammal Study*, 28: 73-77.

Kawai K, Nikaido M, Harada M, *et al*. 2003. The status of the Japanese and East Asian bats of the genus *Myotis* (Vespertilionidae) based on mitochondrial sequences. *Molecular Phylogenetics and Evolution*, 28: 197-307.

Kemp T S. 2005. The Origin and Evolution of Mammals. Oxford: Oxford University Press.

Kenney D K. 2018. Right whales. *In*: Würsig B, Thewissen J G M, Kovacs K M. Encyclopedia of Marine Mammals. 3rd ed. San Diego: Academic Press/ Elsevier: 817-822.

Koh H S, Lee W J. 1994. Geographic Variation of Morphometric Characters in Five Subspecies of Korean Field Mice, *Apodemus peninsufae* Thomas (Rodentia, Mammalia), in Eastern Asia. *The Korean Journal of Zoology*, 37: 33-39.

Koju N P, He K, Chalise M K, *et al*. 2017. Multilocus approaches reveal underestimated species diversity and inter-specific gene flow in pikas (*Ochotona*) from southwestern China. *Molecular Phylogenetics and Evolution*, 107: 239-245.

Kolleck J, Yang M, Zinner D, *et al*. 2013. Genetic diversity in endangered Guizhou snub-nosed monkeys (*Rhinopithecus brelichi*): contrasting results from microsatellite and mitochondrial DNA data. *PLoS One*, 8: e73647.

Koopman K F. 1993. Order Chiroptera. *In*: Wilson D E, Reeder D M. Mammal Species of the World: A Taxonomic and Geographic Reference. Washington: Smithsonian Institution Press: 137-241.

Kruuk H, Kanchanasaka B, O'Sullivan S, *et al*. 1993. Identification of tracks and other sign of three species of otter *Lutra lutra*, *L. perspicillata* and *Aonyx cinerea* in Thailand. *Natural History Bulletin of the Siam Society*, 41: 23-30.

Kumar B, Cheng J, Ge D, *et al*. 2019. Phylogeography and ecological niche modeling unravel the evolutionary history of the Yarkand hare *Lepus yarkandensis* (Mammalia: Leporidae), through the Quaternary. *BMC Evolutionary Biology*, 19: 113.

Kuo H, Fang Y, Csorba G, *et al*. 2009. Three new species of *Murina* (Chiroptera: Vespertilionidae) from Taiwan. *Journal of Mammalogy*, 90: 980-991.

Kuo H C, Chen S F, Fang Y P, *et al*. 2015. Speciation processes in putative island endemic sister bat species: false impressions from mitochondrial DNA and microsatellite data. *Molecular Ecololgy*, 24: 5910-5926.

Lacy R C, Botbat M, Pollak J P. 2003. Vortex: A Stochastic Simulation of the Extinction Process. Version 9. Chicago: Chicago Zoological Society and Brookfield, IL.

Lai C H, Smith A T. 2003. Keystone status of plateau pikas (*Ochotona curzoniae*): effect of control on biodiversity of native birds. *Biodiversity & Conservation*, 12(9): 1901-1912.

Lau M W N, Fellowes J R, Chan B P L. 2010. Carnivores (Mammalia: Carnivora) in South China: a status review with notes on the commercial trade. *Mammal Review*, 40(4): 247-292.

Lee Y, Kuo Y, Chu W, *et al*. 2007. Chiropteran diversity in different settings of the uplifted coral reef tropical forest of Taiwan. *Journal of Mammalogy*, 88(5): 1239-1247.

Lei R, Jiang Z, Hu Z, *et al*. 2003. Phylogenetic relationships of Chinese antelopes based on mitochondrial Ribosomal RNA gene sequences. *Journal of Zoology*(London), 261(3): 227-237.

Lekagul B, McNeely J A. 1977. Mammals of Thailand. Bangkok: Association for the Conservation of Wildlife, Sahakarnbhat Co.

Leslie D M, Schaller G B. 2009. *Bos grunniens* and *Bos mutus* (Artiodactyla: Bovidae). *Mammalian Species*, (836): 1-17.

Leslie Jr D M, Lee D N, Dolman R W. 2013. *Elaphodus cephalophus* (Artiodactyla: Cervidae). *Mammalian Species*, 45(904): 80-91.

Li C, Jiang Z, Ping X, *et al*. 2012. Current status and conservation of the Endangered Przewalski's gazelle *Procapra przewalskii*, endemic to the Qinghai-Tibetan Plateau, China. *Oryx*, 46: 145-153.

Li C, Zhao C, Fan P. 2015a. White-cheeked macaque (*Macaca leucogenys*): A new macaque species from Modog, southeastern Tibet. *American Journal of Primatology*, 77(7): 753-766.

Li F, Chan B P. 2018. Past and present: the status and distribution of otters (Carnivora: Lutrinae) in China. *Oryx*, 52: 619-626.

Li H, Wu Y. 2011. Study on two species of trematodes in Plagiorchiidae in six bats species from Guangdong Province. *Guangzhou Daxue Xuebao Ziran Kexue Ban*, 10(6): 25-28.

Li J, Song Y, Zeng Z. 2003. Elevational gradients of small mammal diversity on the northern slopes of Mt. Qilian, China. *Global Ecology and Biogeography*, 12(6): 449-460.

Li S, He K, Yu F, *et al*. 2013. Molecular phylogeny and biogeography of *Petaurista* inferred from the Cytochrome *b* gene, with implications for the taxonomic status of *P. caniceps*, *P. marica* and *P. sybilla*. *PLoS One*, 8(7): e70461.

Li S, Sun K P, Lu G J, *et al*. 2015b. Mitochondrial genetic differentiation and morphological difference of *Miniopterus fuliginosus* and *Miniopterus magnater* in China and Vietnam. *Ecology and Evolution*, 5(6): 1214-1223.

Li S, Yu G, Liu S, *et al*. 2019. First record of the ferret-badger *Melogale cucphuongensis* Nadler *et al*., 2011 (Carnivora: Mustelidae), with description of a new subspecies, in southeastern China. *Zoological Research*, 40(6): 575-579.

Li W. 2003. The Comparative research on status of Ili Pika in the past ten years. *Chinese Journal of Zoology*, 38: 64-68.

Li W, Smith A T. 2005. Dramatic decline of the threatened Ili pika *Ochotona iliensis* (Lagomorpha: Ochotonidae) in Xinjiang, China. *Oryx*, 39: 30-34.

Li W, Zhang H, Liu Z. 2006. Brief report on the status of Kozlov's pika, *Ochotona koslowi* (Büchner), in the east Kunlun Mountains of China. *Integrative Zoology*, 1(1): 22-24.

Li Y, Wu Y, Harada M, *et al*. 2008. Karyotypes of three rat species (Mammalia: Rodentia: Muridae) from Hainan Island, China, and the valid specific status of *Niviventer lotipes*. *Zoological Science*, 25(6): 686-692.

Liang R, Dong Y. 1984. Bats from south Anhui. *Acta Theriologica Sinica*, 4: 321-328.

Lin L, Harada M, Moyokawa M, *et al*. 2006. Updating the occurrrence of *Harpiocephalus harpia* (Chiroptera: Vespertilionidae) and its karyology in Taiwan. *Mammalia*, 70(1-2): 170-172.

Lin L, Motokawa M, Harada M, *et al*. 2002a. New record of *Barbastella leucomelas* (Chiroptera: Vespertilionidae) from Taiwan. *Mammalian Biology*, 67(5): 315-319.

Lin L, Motokawa M, Harada M. 2002b. Karyology of ten vespertilionid bats (Chiroptera: Vespertilionidae) from Taiwan. *Zoological Studies*, 41: 347-354.

Lissovsky A A, Ivanova N V, Borisenko A V. 2007. Molecular phylogenetics and taxonomy of the subgenus *Pika* (Ochotona, Lagomorpha). *Journal of Mammalogy*, 88(5): 1195-1204.

Lissovsky A. 2014. Taxonomic revision of pikas *Ochotona* (Lagomorpha, Mammalia) at the species level. *Mammalia*, 78(2): 199-216.

Liu H, Ma H, Cheyne S M, *et al.* 2020. Recovery hopes for the world's rarest primate. *Science*, 368(6495): 1074

Liu J, Du H, Tian G, *et al.* 2008. Community structure and diversity distributions of small mammals in different sample plots in the eastern part of Wuling Mountains. *Zoological Studies*, 29(6): 637-645.

Liu S, He K, Chen S, *et al.* 2018. How many species of *Apodemus* and *Rattus* occur in China? A survey based on mitochondrial Cyt *b* and morphological analyses. *Zoological Research*, 39(5): 309-320.

Liu S, Jin W, Liu Y, *et al.* 2017. Taxonomic position of Chinese voles of the tribe Arvicolini and the description of 2 new species from Xizang, China. *Journal of Mammalogy*, 98(1): 166-182.

Liu S, Liu Y, Guo P, *et al.* 2012a. Phylogeny of Oriental voles (Rodentia: muridae: Arvicolinae): Molecular and morphological evidences. *Zoological Science*, 29(9): 610-622.

Liu S, Sun Z, Liu Y, *et al.* 2012b. A new vole from Xizang, China and the molecular phylogeny of the genus *Neodon* (Cricetidae: Arvicolinae). *Zootaxa*, 3235: 1-22.

Liu S, Sun Z, Zeng Z, *et al.* 2007. A new vole (Muridae: Arvicolinae) from the Liangshan Mountains of Sichuan Province, China. *Journal of Mammalogy*, 88(5): 1170-1178.

Liu W, Wang Y, He X, *et al.* 2011. Distribution and analysis of the importance of underground habitats of cave-dwelling bats in the south of Taihang Mountain. *Acta Theriologica Sinica*, 31: 371-379.

Liu X, Yao Y. 2013. Characterization of 12 polymorphic microsatellite markers in the Chinese tree shrew (*Tupaia belangeri chinensis*). *Zoological Research*, 34(E2): 62-68.

Liu Y, Sun Z, Wang H, *et al.* 2009. 5 new records of small mammals in Tibet, China. *Sichuang Journal of Zoology*, 28(2): 278-279.

Long Y, Momberg F, Ma J, *et al.* 2012. *Rhinopithecus strykeri* Found in China! *American Journal of Primatology*, 74(10): 871-873.

Lu G, Lin A, Luo J, *et al.* 2013. Phylogeography of the Rickett's big-footed bat, *Myotis pilosus* (Chiroptera: Vespertilionidae): a novel pattern of genetic structure of bats in China. *BMC Evolutionary Biology*, 13(1): 241-211.

Lu X. 2011. Habitat use and abundance of the woolly hare *Lepus oiostolus* in the Lhasa mountains, Tibet. *Mammalia*, 75(1): 35-40.

Ludt C J, Schroeder W, Rottmann O, *et al.* 2004. Mitochondrial DNA phylogeography of red deer (*Cervus elaphus*). *Molecular Phylogenetics and Evolution*, 31(3): 1064-1083.

Lunde D P, Musser G G, Son N T. 2003. A survey of small mammals from Mt. Tay Con Linh Ⅱ, Vietnam, with the description of a new species of Chodsigoa (Insectivora: Soricidae). *Mammal Study*, 28: 31-46.

Luo Z, Li C, Tang S, *et al.* 2011. Do Rapoport's rule, the mid-domain effect, land area or environmental factors predict latitudinal range size patterns of terrestrial mammals in China? *PLoS One*, 6(11): e27975.

Luo Z, Tang S, Li C, *et al.* 2012. Environmental Effects on Vertebrate Species Richness: Testing the Energy, Environmental Stability and Habitat Heterogeneity Hypotheses. *PLoS One*, 7(4): e35514.

Ma J, Metzner W, Liang B, *et al.* 2004. Differences in diet and echolocation in four sympatric bat species and their respective ecological niches. *Acta Zoologica Sinica*, 50(2): 145-150.

Ma J, Liang B, Zhang S Y, *et al.* 2008. Dietary composition and echolocation call design of three sympatric insectivorous bat species from China. *Ecological Research*, 23(1): 113-119.

Ma Y Q. 1994. Conservation and utilization of the bear resources in China. *Int. Conf. Bear Res. and Manage*., 9(1): 157-159.

Mace G M, Collar N, Cooke J, *et al.* 1992. The development of new criteria for listing species on the IUCN Red List. *Species*, 19: 16-22.

Mace G M, Collar N J, Gaston K J, *et al.* 2008. Quantification of extinction risk: IUCN's system for classifying threatened species. *Conservation Biology*, 22: 1424-1442.

Mace G M, Cramer W, Diaz S, *et al.* 2010. Biodiversity targets after 2010. *Current Opinion in Environmental Sustainability*, 2(1-2): 3-8.

Mace G M, Lande R. 1991. Assessing extinction threats: toward a reevaluation of IUCN Threatened Species Categories. *Conservation Biology*, 5(2): 148-157.

Macholán M. 1999. Mus musculus. The Atlas of European Mammals. London: Academic Press: 286-297.

Maeda K. 1980. Review on the classification of little tube-nosed bats, *Murina aurata* group. *Mammalia*, 44(4): 531-551.

Malcolm K D, McShea W J, Garshelis D L, *et al.* 2014. Increased stress in Asiatic black bears relates to food limitation, crop raiding, and foraging beyond nature reserve boundaries in China. *Global Ecology and Conservation*, 2(C): 267-276.

Mao X, He G, Zhang J, *et al.* 2013. Lineage divergence and historical gene flow in the Chinese Horseshoe Bat (*Rhinolophus sinicus*). *PLoS One*, 8(2): e56786.

Mao X, Wang J, Su W, *et al.* 2010a. Karyotypic evolution in family Hipposideridae (Chiroptera, Mammalia) revealed by comparative chromosome painting, G- and C-banding. *Zoological Research*, 31: 453-460.

Mao X, Zhu G, Zhang S, *et al.* 2010b. Pleistocene climatic cycling drives intra-specific diversification in the intermediate horseshoe bat (*Rhinolophus affinis*) in Southern China. *Molecular Ecology*, 19(13): 2754-2769.

Marmi J, Lopez-Giraldez F, MacDonald D W, *et al.* 2006. Mitochondrial DNA reveals a strong phylogeographic structure in the badger across Eurasia. *Molecular Ecology*, 15(4): 1007-1020.

Marshall J T. 1977. A synopsis of Asian species of *Mus* (Rodentia, Muridae). *Bulletin of the AMNH*, 158: Article 3.

Masui K, Narita Y, Tanaka S. 1986. Information on the distribution of Formosan monkeys *Macaca cyclopis*. *Primates*, 27: 383-392.

Mattioli S. 2011. Family Cervidae (deer). *In*: Wilson D E, Mittermeier R A. Handbook of the Mammals of the World. Vol. 2. Hoofed Mammals. Barcelona: Lynx Edicions: 350-443.

Mayden R L. 1997. A hierarchy of species concepts: the denoument in the saga of the species problem. *In*: Claridge M F, Dawah H A, Wilson M R. Species: The Units of Diversity. London: Chapman and Hall: 381-423.

Mayer F, Dietz C, Kiefer A. 2007. Molecular species identification boosts bat diversity. *Frontiers in Zoology*, 4: 4.

Mayr E, Kinskey E G, Usinger R L. 1953. Methods and Principles of Systematic Zoology. New York: MaGraw-Hill Book Co. Inc.

Mayr E. 1942. Systematics and the Origin of Species from the Viewpoint of a Zoologist. New York: Columbia University Press.

McCarthy T, Mallon D, Jackson R, *et al.* 2017. *Panthera uncia*. The IUCN Red List of Threatened Species (2017): e.T22732A50664030. doi: 10.2305. Gland: IUCN.

McNeely J A, Miller K, Reid W, *et al.* 1990. Conserving the World's Biodiversity. Gland: IUCN.

Mei Z, Huang S, Hao Y, *et al.* 2012. Accelerating population decline of Yangtze finless porpoise (*Neophocaena asiaeorientalis asiaeorientalis*). *Biological Conservation*, 153: 192-200.

Mei Z, Huang S, Zhao X, *et al.* 2014. The Yangtze finless porpoise: On an accelerating path to extinction? *Biological Conservation*, 172: 117-123.

Meijaard E, Chua M A H, Duckworth J W. 2017. Is the northern chevrotain, *Tragulus williamsoni* Kloss, 1916, a synonym or one of the least-documented mammal species in Asia? *Raffles Bulletin of Zoology*, 65: 506-514.

Meijaard E, Groves C P. 2004. A taxonomic revision of the *Tragulus* mouse-deer (Artiodactyla). *Zoological Journal of the Linnaean Society*, 140(1): 63-102.

Melo-Ferreira J, de Matos A L, Areal H, *et al.* 2017. The phylogeny of pikas (*Ochotona*) inferred from a multilocus coalescent approach. *Molecular Phylogenetics and Evolution*, 84: 240-244.

Miller G S Jr. 1940. Notes on some moles from southeastern Asia. *Journal of Mammalogy*, 21(4): 442-444.

Mittermeier R A, Wilson D E. 2014. Handbook of the Mammals of the World. Volume 4: Sea Mammals. Barcelona: Lynx Edicions.

Moore J C, Tate G H H. 1965. A study of the diurnal squirrels, Sciurinae, of the Indian and Indochinese subregions. *Fieldiana Zoology*, 48: 1-351.

Mootnick A R, Chan B P L, Moisson P, *et al.* 2012. The status of the Hainan gibbon *Nomascus hainanus* and the Eastern black gibbon *Nomascus nasutus*. *International Zoo Yearbook*, 46: 259-264.

Motokawa M. 2004. Phylogenetic relationships within the family Talpidae (Mammalia: Insectivora). *Journal of Zoology* (London), 263(2): 147-157.

Motokawa M, Harada M, Lin L, *et al.* 1997. Karyological study of the gray shrew *Crocidura attenuata* (Mammalia: Insectivora) from Taiwan. *Zoological Studies*, 36: 70-73.

Motokawa M, Harada M, Lin L, *et al.* 1998. Karyological differentiation between two Soriculus (Insectivora: Soricidae) from Taiwan. *Mammalia*, 62: 541-547.

Motokawa M, Harada M, Wu Y, *et al.* 2001. Chromosomal polymorphism in the Gray Shrew *Crocidura attenuata* (Mammalia: Insectivora). *Zoological Science*, 18: 1153-1160.

Motokawa M, Lin L. 2005. Taxonomic status of *Soriculus baileyi* (Insectivora, Soricidae). *Mammal Study*, 30(2): 117-124.

Motokawa M, Lin L. 2002. Geographic variation in the mole-shrew *Anourosorex squamipes*. *Mammal Study*, 27(2): 113-120.

Motokawa M, Suzuki H, Harada M, *et al*. 2000. Phylogenetic relationships among East Asian Crocidura (Mammalia: Insectivora) inferred from mitochondiral cytochrome *b* gene. *Zoological Science*, 17(4): 497-504.

Musser G G, Chiu S. 1979. Notes on taxonomy of *Rattus andersoni* and *R. excelsior*, murids endemic to western China. *Journal of Mammalogy*, 60: 581-592.

Nadler T, Streicher U, Stefen C, *et al*. 2011. A new species of ferret-badger, Genus Melogale, from Vietnam. *Zoologische*, 80: 271-286.

NCBI. 2018. Niviventer huang. http://www.ncbi.nlm.nih.gov/Taxonomy/Browser/wwwtax.cgi?lvl=0&id=979565. [2018-1-15]

Niu H, Wang N, Zhao L, *et al*. 2007. Distribution and underground habitats of cave-dwelling bats in China. *Animal Conservation*, 10(4): 470-477.

Niu K, Tan C L, Yang Y. 2010. Altitudinal movements of Guizhou snub-nosed monkeys (*Rhinopithecus brelichi*) in Fanjingshan National Nature Reserve, China: implications for conservation management of a flagship species. *Folia Primatologica*, 81(4): 233-244.

Niu Y, Wei F, Li M, *et al*. 2004. Phylogeny of pikas (Lagomorpha, *Ochotona*) inferred from mitochondrial cytochrome *b* sequences. *Folia Zoologica-Praha*, 53(2): 141-156.

Nowak R M. 1999. Walker's Mammal of the World. 6th ed. Washington: Johns Hopkins University Press.

Odell D K, McClune K M. 1999. False killer whale *Pseudorca crassidens* (Owen, 1846). *Handbook of Marine Mammals*, 6: 213-243.

Ohdachi S D, Iwasa M A, Nesterenko V A, *et al*. 2004. Molecular phylogenetics of Crocidura shrews (Insectivora) in East and Central Asia. *Journal of Mammals*, 85(3): 396-403.

Ohnishi N, Osawa T. 2014. A difference in the genetic distribution pattern between the sexes in the Asian black bear. *Mammal Study*, 39(1): 11-17.

Ohtaishi N, Gao Y. 1990. A review of the distribution of all species of deer (Tragulidae, Moschidae and Cervidae) in China. *Mammal Review*, 20(2-3): 125-144.

Okada A, Ito T Y, Buuveibaatar B, *et al*. 2012. Genetic structure of Mongolian gazelle (*Procapra gutturosa*): the effect of railroad and demographic change. *Mongolian Journal of Biological Sciences*, 10(1-2): 59-66.

Oldfield S, Lusty C, MacKinven A. 1998. The World List of Threatened Trees. Cambridge: World Conservation Press.

Pan D, Chen J H, Groves C, *et al*. 2007. Mitochondrial control region and population genetic patterns of *Nycticebus bengalensis* and *N. pygmaeus*. *International Journal of Primatology*, 28: 791-799.

Pan R, Oxnard C, Grueter C C, *et al*. 2016. A new conservation strategy for China—A model starting with primates. *American Journal Primatology*, 78(11): 1137-1148.

Patou M L, Wilting A, Gaubert P, *et al*. 2010. Evolutionary history of the *Paradoxurus* palm civets–a new model for Asian biogeography. *Journal of Biogeography*, 37(11): 2077-2097.

Pech R P, Jiebu A A D, Zhang Y, *et al*. 2007. Population dynamics and responses to management of plateau pikas *Ochotona curzoniae*. *Journal of Applied Ecology*, 44(3): 615-624.

Pei K J C, Lai Y C, Corlett R T, *et al*. 2010. The larger mammal fauna of Hong Kong: species survival in a highly degraded landscape. *Zoological Studies*, 49(2): 253-264.

Perrin W F. 2018. Common dolphin. *In*: Würsig B, Thewissen J G M, Kovacs K M. Encyclopedia of Marine Mammals. 3rd ed. San Diego: Academic Press/Elsevier: 205-209.

Perrin W F, Mallette S D, Brownell Jr R L. 2018. Minke whales. *In*: Würsig B, Thewissen J G M, Kovacs K M. Encyclopedia of Marine Mammals. 3rd ed. San Diego: Academic Press/Elsevier: 608-613.

Phillips C J, Wilson N A. 1968. Collection of bats from Hong Kong. *Journal of Mammalogy*, 9: 128-133.

Phillips S J, Dudík M. 2008. Modeling of species distributions with Maxent: New extensions and a comprehensive evaluation. *Ecography*, 31(2): 161-175.

Pieńkowska A, Szczerbal I, Mäkinen A, *et al*. 2002. G/ Q - banded chromosome nomenclature of the Chinese raccoon dog, *Nyctereutes procyonoides procyonoides* Gray. *Hereditas*, 137(1): 75-78.

Poirier F E. 1986. A preliminary study of the Taiwan macaque (*Macaca cyclopis*). *Zoological Research*, 7: 411-422.

Prater S H. 1971. The Book of Indian Animals. 3rd ed. Mumbai and Oxford: Bombay Natural History Society and Oxford University Press.

Qiu Z H. 1987. The Neogene mammalian faunas of Ertemte and Harr Obo in Inner Mongolia (Nei Mongol), China. 6: Hares and pikas–Lagomorpha. Leporidae and Ochotonidae. *Senckenbergiana Lethaea*, 67(5/6): 375-399.

Qureshi B D, Awan M S, Khan A A, *et al*. 2004. Distribution of Himalayan musk deer (*Moschus chrysogaster*) in Neelum Valley, District Muzaffarabad, Azad Jammu and Kashmir. *Journal of Biological Sciences*, 4: 258-261.

Rodrigues A S L, Pilgrim J D, Lamoreux J F, *et al*. 2006. The value of the IUCN Red List for conservation. *Trends in Ecology and Evolution*, 21(2): 71-76.

Rookmaaker L C. 1980. The Distribution of the Rhinoceros in Eastern-India, Bangladesh, China, and the Indo-Chinese Region. *Zoologischer Anzeiger*, 205(3/4): 253-268.

Rossolimo O L, Pavlinov I Y, Hoffmann R S. 1994. Systematics and distribution of the rock voles of the subgenus *Alticola* s. str. in the People's Republic of China (Rodentia, Arvicolinae). *Acta Theriologica Sinica*, 14(2): 86-99.

Ruedi M, Stadelmann B, Gager Y, *et al*. 2013. Molecular phylogenetic reconstructions identify East Asia as the cradle for the evolution of the cosmopolitan genus *Myotis* (Mammalia, Chiroptera). *Molecular Phylogenetics and Evolution*, 69(3): 437-449.

Rydell J. 1993. Eptesicus nilssonii. *Mammalian Species*, 430: 1-7.

Rydell J, Baagøe H J. 1994. Vespertilio murinus. *Mammalian Species*, 467: 1-6.

Saha. 1981. SPECIES *Biswamoyopterus biswasi*. *Bull. Zool. Surv. India*, 4: 333.

Sanborn C C. 1939. Eight new Bats of the genus *Rhinolophus*. *Field Museum Publications Chicago Zoological Series*, 24: 37-43.

Schauer J. 1987. Remarks on the construction of burrows of *Ellobius talpinus*, *Myospalax aspalax* and *Ochotona daurica* in Mongolia and their effect on the soil. *Folia Zoologica*, 36(4): 319-326.

Seim I, Fang X, Xiong Z, *et al*. 2013. Genome analysis reveals insights into physiology and longevity of the Brandt's bat *Myotis brandtii*. *Nature Communications*, doi: 10.1038/ncomms3212.

Shek C T, Chan C S, Wan Y F. 2007. Camera trap survey of Hong Kong terrestrial mammals in 2002-06. *Hong Kong Biodiversity*, 15: 1-11.

Shek C T, Chan C S M. 2006. Mist net survey of bats with three new bat species records for Hong Kong. *Hong Kong Biodiversity*, 11: 1-7.

Shek S T, Lau C T Y. 2006. Echolocation calls of five horseshoe bats of Hong Kong. *Hong Kong Biodiversity*, 13: 9-12.

Shenbrot G I, Sokolov V E, Heptner V G, *et al*. 1995. The Mammals of Russia and Adjacent Regions. Dipodoidea. NH(New Hampshire): Science Publishers.

Shi C, Chen Z, Cheng F, *et al*. 2018. New record of a rodent genus (*Murinae*) in China—*Tonkinomys daovantieni*. *Acta Theriologica Sinica*, 38(3): 309-314.

Silbermayr K, Orozco-ter Wengel P, Charruau D, *et al*. 2010. High mitochondrial differentiation levels between wild and domestic Bactrian camels: a basis for rapid detection of maternal hybridization. *Animal Genetics*, 41: 315-318.

Simmons N B. 2005. Order Chiroptera. *In*: Wilson D E, Reeder D M. Mammal Species of the World. Baltimore: The Johns Hopkins University Press: 312-529.

Sinha A, Datta A, Madhusudan M D, *et al*. 2005. *Macaca munzala*: A new species from western Arunachal Pradesh, Northeastern India. *International Journal of Primatology*, 26(4): 997-989.

Smith A T, Formozov N A, Hoffmann R S. 1990. Pikas. *In*: Chapman J A, Flux J E C. Rabbits, Hares and Pikas: Status Survey and Conservation Action Plan. Gland: IUCN: 14-60.

Smith A T, Xie Y. 2013. *Mammals of China*. Princeton: Princeton University Press.

Smith F A, Boyer A G, Brown J H, *et al*. 2010. The Evolution of maximum body size of terrestrial mammals. *Science*, 330(6008): 1216-1219.

Sokolov V E. 1977. Systematics of mammals. Order Lagomorpha, Rodentia. Moscow: Vysshaya shkola Press. (In Russian)

Son N T, Görfö L T, Francis C M, *et al*. 2013. Description of a new species of *Myotis* (Vespertilionidae) from Vietnam. *Acta Chiropterologica*, 15(2): 473-483.

Song H, Wang H, Chen X, *et al*. 2009. Molecular phylogenetics of nine rhinolophids species (Chiroptera: Rhinolophidae) in Guizhou based on mitochondrial 16S rRNA gene. *Sichuan Journal of Zoology*, 28(6): 816-820.

Sorokin P A, Kiriliuk V E, Lushchekina A A, *et al*. 2005. Genetic diversity of the Mongolian gazelle *Procapra guttorosa* Pallas, 1777. *Russian Journal of Genetics*, 41(10): 1101-1105.

Stone R D. 1995. Eurasian Insectivores and Tree Shrews. IUCN/SSC Insectivore, Tree Shrew and Elephant Shrew Specialist

Group. Gland: IUCN.

Su B, Fu Y, Wang Y, *et al.* 2001. Genetic diversity and population history of the red panda (*Ailurus fulgens*) as inferred from mitochondrial DNA sequence variations. *Molecular Biology and Evolution*, 18(6): 1070-1076.

Subedi A, Aryal A, Koirala R K, *et al.* 2012. Habitat ecology of Himalayan musk deer (*Moschus chrysogaster*) in Manaslu Conservation Area, Nepal. *International Journal of Zoological Research*, 8(2): 81-89.

Sun Z, Liu S, Guo Y, *et al.* 2013. The fauna and distribution of small mammals in Erlang Moutains. *Acta Theriological Sinca*, 33(1): 1-10.

Tanomtong A, Bunjonrat R, Sriphoom A, *et al.* 2005. A Study on karyotype of small-toothed palm civet, *Arctogalidia trivirgata* (Carnivora, Viverridae) by using conventional staining method. *Warasan Songkhla Nakharin* (Sakha Witthayasat lae Technology), 27.

Thomas O. 1912a. On Insectivores and Rodents collected by Mr. F. Kingdon Ward in N. W. Yunnan. *Annals and Magazine of Natural History*, Series 8: 513-519.

Thomas O. 1912b. On a collection of small mammals from the Tsin-ling Mountains, Central China, presented to Mr. G. Fenwick Owen to the National Museum. *Annals and Magazine of Natural History*, Series 8: 395-403.

Thomas O. 1920. Two new Asiatic Bats of the genera Tadarida and Dyacopterus. *Annals Magazine of Natural History*, Series 9: 283-285.

Thorington Jr R W, Hoffmann R S. 2005. Family sciuridae. Mammal Species of the World, A Taxonomic and Geographic Reference. 3rd ed. Washington and London: Smithsonian Institution Press.

Tian Y, Wu J, Wang T, *et al.* 2014. Climate change and landscape fragmentation jeopardize the population viability of the Siberian tiger (*Panthera tigris altaica*). *Landscape Ecology*, 29(4): 621-637.

Tian Z, Jin D. 2012. Study on the gensus *Macronyssus* (Acari: Macronyssidae) with description of a new species, redescription of a known species from the genus *Myotis* (Chiroptera: Vespertilionidae) and a key to the species in China. *International Journal of Acarology*, 38(3): 179-190.

Tilson R, Hu D, Muntifering J, *et al.* 2004. Dramatic decline of wild south China tigers *Panthera tigris amoyensis*: field survey of priority tiger reserves. *Oryx*, 38(1): 40-47.

Tiunov M P, Kruskop S V, Feng J. 2011. A New Mouse-Eared Bat (Mammalia: Chiroptera, Vespertilionidae) from South China. *Acta Chiropterologica*, 13(2): 271-278.

Tu F, Liu S, Liu Y, *et al.* 2014. Complete mitogen me of Chinese shrew mole *Uropsilus soricipes* (Milne-Edwards, 1871) (Mammalia: Talpidae) and genetic structure of the species in the Jiajin Mountains (China). *Journal of Natural History*, 48(23-24): 1467-1483.

Tu F, Tang M, Liu Y, *et al.* 2012. Fauna and species diversity of small mammals in Jiajin Mountains, Sichuan Province, China. *Acta Theriological Sinca*, 32(4): 287-296.

Turghan M, Jiang Z, Groves C P, *et al.* 2013. Subspecies in Przewalski's gazelle *Procapra przewalskii* and its conservation implication. *Chinese Science Bulletin*, 58: 1897-1905.

Turvey S T, Pitman R L, Taylor B L, *et al.* 2007. First human-caused extinction of a cetacean species? *Biology Letters*, 3: 573-540.

Turvey S T, Crees J J, Di Fonzo M M I. 2015. Historical data as a baseline for conservation: reconstructing long-term faunal extinction dynamics in Late Imperial–modern China. *Proceedings to Royal Society*(B), 282: doi.10.1098/rspb.2015.1299.

Van Peenen P F D, Ryan P F, Light R H. 1969. Preliminary Identification Manual for Mammals of South Vietnam. Washington: Smithsonian Institution.

Van Rompaey H. 2001. The crab-eating mongoose. *Herpestes urva. Small Carnivore Conservation*, 25: 12-17.

Vié J, Hilton-Taylor C, Pollock C M, *et al.* 2009. The IUCN Red List: Key Conservation Tool. Gland: IUCN.

Vilà C, Amorim I R, Leonard J A, *et al.* 1999. Mitochondrial DNA phylogeography and population history of the grey wolf *Canis lupus*. *Molecular Ecology*, 8(12): 2089-2103.

Vislobokova I A. 2013. On the origin of Cetartiodactyla: Comparison of data on evolutionary morphology and molecular biology. *Paleontological Journal*, 47(3): 321-334.

Waddell P J, Okada N, Hasegawa M. 1999. Toward resolving the inter-ordinal relationships of placental mammals. *Systematic Biology*, 48(3): 681.

Wallace A R. 1876. The geographic Distribution of Animals. London: McMillan & Co.

Wang J, Frasier T R, Yang S, *et al*. 2008. Detecting recent speciation events: The case of the finless porpoise (genus *Neophocaena*). *Heredity*, 101: 145-155.

Wang L, Wang H, Ou W, *et al*. 2014. Dynamic adjustment of echolocation pulse structure of big-footed myotis (*Myotis macrodactylus*) in response to different habitats. *Journal of the Acoustical Society of America*, 135(2): 928-932.

Wang P, Yao C, Han J, *et al*. 2011. Investigations of stranded and by-caught beaked whales around the coastal waters of Chinese mainland. *Acta Theriologica Sinica*, 31(1): 37-45.

Wang W, Cao L, He B, *et al*. 2013. Molecular Characterization of *Cryptosporidium* in Bats from Yunnan Province, Southwestern China. *Journal of Parasitology*, 99(6): 1148-1150.

Wang X, Guo W, Yu W, *et al*. 2017. First record and phylogenetic position of *Myotis indochinensis* (Chiroptera, Vespertilionidae) from China. *Mammalia*, 81(6): 605-609.

Wang Z, Wang S. 1962. The discovery of a flying fox (*Pteropus giganteus* Brunnich) from Chin-Hai Province, Northwestern China. *Acta Zoologica Sinica*, 4: 494.

Wei F, Feng Z, Wang Z, *et al*. 1999. Current distribution, status and conservation of wild red pandas *Ailurus fulgens* in China. *Biological Conservation*, 89(3): 285-291.

Wei G, Mingxia Z, Liping Z, *et al*. 2017. Large-spotted civet in China: The rediscovery of large-spotted civet *Viverra megaspila* in China. *Small Carnivore Conservation*, 55: 88-90.

Wei L, Flanders J R, Rossiter S J, *et al*. 2010. Phylogeography of the Japanese pipistrelle bat, *Pipistrellus abramus*, in China: the impact of ancient and recent events on population genetic structure. *Biological Journal of the Linnean Society*, 99: 582-594.

Wei L, Wu X, Jiang Z. 2009. The complete mitochondrial genome structure of snow leopard *Panthera uncia*. *Molecular Biology Reports*, 36: 871.

Wilson D E, Lacher T E, Mittermeier R A. 2016. Handbook of the Mammals of the World. Volume 6: Lagomorphs and Rodents I. Barcelona: Lynx Edicions.

Wilson D E, Mittermeier R A, Lacher T E. 2017. Handbook of the Mammals of the World. Volume 7: Rodents II. Barcelona: Lynx Edicions.

Wilson D E, Mittermeier R A. 2009. Handbook of the Mammals of the World. Volume 1: Carnivores. Barcelona: Lynx Edicions.

Wilson D E, Mittermeier R A. 2011. Handbook of the Mammals of the World. Volume 2, Ungulates. Barcelona: Lynx Edicions.

Wilson D E, Mittermeier R A. 2012. Handbook of the Mammals of the World. Volume 3: Primates. Barcelona: Lynx Edicions.

Wilson D E, Mittermeier R A. 2018. Handbook of the Mammals of the World. Volume 8: Insectivores, Sloths and Colugos. Barcelona: Lynx Edicions.

Wilson D E, Mittermeier R A. 2019. Handbook of the Mammals of the World. Volume 9: Bats. Barcelona: Lynx Edicions.

Wilson D E, Reeder D M. 1993. Mammal Species of the World, A Taxonomic and Geographic Reference. 2nd ed. Washington and London: Smithsonian Institution Press.

Wilson D E, Reeder D M. 2005. Mammal Species of the World: A Taxonomic and Geographic Reference. 3rd ed. Baltimore: John Hopkins University Press.

Woodman N. 1993. The Correct Gender of Mammalian Generic Names Ending in *-otis*. *Journal of Mammalogy*, 74(3): 544-546.

Wu C, Li H, Wang Y, *et al*. 2000. Low genetic variation of the Yunnan hare (*Lepus comus* G. Allen 1927) as revealed by mitochondrial cytochrome *b* gene sequences. *Biochemical Genetics*, 38: 147-153.

Wu D, Luo J, Fox B J. 1996. A comparison of ground-dwelling small mammal communities in primary and secondary tropical rainforests in China. *Journal of Tropical Ecology*, 12: 215-230.

Wu S, Wang Y, Feng Q. 2005. A new record of Mammalia in China—*Manis javanica*. *Acta Zootaxonomica Sinica*, 30(2): 440-443.

Wu Y, Harada M, Li Y. 2004. Karyology of seven species bats from Sichuan, China. *Acta Theriologica Sinica*, 24: 30-35.

Wu Y, Harada M, Motokawa M. 2009a. Taxonomy of *Rhinolophus yunanensis* Dobson, 1872 (Chiroptera: Rhinolophidae) with a description of a new species from Thailand. *Acta Chiropterologica*, 11: 237-246.

Wu Y, Harada M. 2005. Karyology of five species of the *Rhinolophus* (Chiroptera: Rhinolophidae) from Guangdong, China. *Acta Theriologica Sinica*, 25: 163-167.

Wu Y, Harada M. 2006. Karyology of seven species of bats (Mammalia: Chiroptera) from Guangdong, China. *Acta Theriologica Sinica*, 26: 403-406.

Wu Y, Li Y C, Lin L K, *et al*. 2012a. New records of *Kerivoula titania* (Chiroptera: Vespertilionidae) from Hainan Island and

Taiwan. *Mammal Study*, 37(1): 69-72.

Wu Y, Motokawa M, Harada M, *et al.* 2012b. Morphometric Variation in the pusillus Group of the Genus *Rhinolophus* (Mammalia: Chiroptera: Rhinolophidae) in East Asia. *Zoological Science* (Tokyo), 29(6): 396-402.

Wu Y, Motokawa M, Harada M. 2008. A new species of the horseshoe bat of the genus *Rhinolophus* from China (Chiroptera: Rhinolophidae). *Zoological Science*, 25(4): 438-443.

Wu Y, Motokawa M, Li Y, *et al.* 2010. Karyotype of Harrison's tube-nosed bat *Murina harrisoni* (Chiroptera: Vespertilionidae: Murininae) Based on the second specimen recorded from Hainan Island, China. *Mammal Study*, 35(4): 277-279.

Wu Y, Motokawa M, Li Y C, *et al.* 2009b. Karyology of eight species of bats (Mammalia: Chiroptera) from Hainan Island, China. *International Journal of Biological Sciences*, 5(7): 659-666.

Wu Y, Thong V D. 2011. A new species of *Rhinolophus* (Chiroptera: Rhinolophidae) from China. *Zoological Science*, 28(3): 235-241.

Xiong Z, Chen M, Zhang E, *et al.* 2013. Molecular phylogeny and taxonomic status of the red goral by Cyto *b* gene analyses. *Folia Zoologica*, 62: 125-130.

Xu A, Jiang Z, Li C, *et al.* 2006. Summer food habits of brown bears in Kekexili Nature Reserve, Qinghai–Tibetan plateau, China. *Ursus*, 17: 132-138.

Xu A, Jiang Z, LI C, *et al.* 2008. Status and conservation of snow leopard in East Burhanbuda Mountain, Kunlun Mountains, China. *Oryx*, 42(3): 460-463.

Xu C, Zhang H, Ma J. 2013. The complete mitochondrial genome of *Martes flavigula*. *Mitochondrial DNA*, 24(3): 240-242.

Xu X, Song J, Zhang Z, *et al.* 2015. The world's second largest population of humpback dolphins in the waters of Zhanjiang deserves the highest conservation priority. *Scientific Reports*, 5: 8147.

Yang C, Xiang C, Zhang X, *et al.* 2013. The complete mitochondrial genome of the Alpine musk deer (*Moschus chrysogaster*). *Mitochondrial DNA*, 24(5): 501-503.

Yang J, Jiang Z. 2011. Genetic diversity, population genetic structure and demographic history of Przewalski's gazelle (*Procapra przewalskii*): implications for conservation. *Conservation Genetics*, 12: 1457-1468.

Yang Q, Meng X, Xia L, *et al.* 2003. Conservation status and causes of decline of musk deer (*Moschus* spp.) in China. *Biological Conservation*, 109(3): 333-342.

Yang T, Hou X, Gu X, *et al.* 2012. The *Hipposideros pomona* in Guizhou Province. *Sichuan Journal of Zoology*, 31: 570-573.

Yang Y, Tian K, Hao J, *et al.* 2004. Biodiversity and biodiversity conservation in Yunnan, China. *Biodiversity and Conservation*, 13: 813-826.

Yao Z, Liu Z, Teng L, *et al.* 2013. Asian badger (*Meles leucurus*, Mustelidae, Carnivora) habitat selection in the Xiaoxing'anling Mountains, Heilongjiang Province, China. *Mammalia*, 77(2): 157-162.

Yoon M H. 1990. Taxonomical study on four *Myotis* (Vespertilionidae) species in Korea. *Korean Journal of Systematic Zoology*, 6: 173-191.

Yoshiyuki M. 1995. A new species of *Plecotus* (Chiroptera, Vespertilionidae) from Taiwan. *Bulletin of the National Science Museum* (Tokyo) (Series A), 17: 189-195.

You Y, Sun K, Xu L, *et al.* 2010. Pleistocene glacial cycle effects on the phylogeography of the Chinese endemic bat species, *Myotis davidii*. *BMC Evolutionary Biology*, 10: 208.

Yu H. 1993. Natural history of small mammals of subtropical montane areas in central Taiwan. *Journal of Zoology* (London), 231(3): 403-422.

Yu H. 1994. Distribution and abundance of small mammals along a subtropical elevational gradient in central Taiwan. *Journal of Zoology* (London), 234(4): 577-600.

Yu H. 1995. Patterns of diversification and genetic population structure of small mammals in Taiwan. *Biological Journal of the Linnean Society*, 55(1): 69-89.

Yu J. 2010. Leopard cat. *Cat News*(Special Issue), 5: 26-29.

Yu N, Zheng C, Shi L. 1997. Variation in mitochondrial DNA and phylogeny of six species of pikas (*Ochotona*). *Journal of Mammalogy*, 78(2): 387-396.

Yu N, Zheng C, Zhang Y P, *et al.* 2000. Molecular systematics of pika (genus *Ochotona*) inferred from mitochondrial DNA sequences. *Molecular Phylogenetic and Evolution*, 16(1): 85-95.

Yu W, Csorba G, Wu Y. 2020. Tube-nosed variations–a new species of the genus *Murina* (Chiroptera: Vespertilionidae) from

China. *Zoological Research*, 41(1): 70-77.

Yuan L, Jiang Z, Cheng Y, *et al.* 2014. Wild camels in the Lop Nur Nature reserve. *Journal of Camel Practice and Research*, 21(2): 137-144.

Yuan S, Jiang X, Li Z, *et al.* 2013. A mitochondrial phylogeny and biogeographical scenario for Asiatic water shrews of the genus *Chimarrogale*: Implications for taxonomy and low-latitude migration routes. *PLoS One*, 8(10): 1-15.

Yuan X, Tian D, Gu X. 2012. Phylogenetics of Rhinolophidae and Hipposideridae based on partial sequences of the nuclear RAG1 gene. *Sichuan Journal of Zoology*, 31: 191-196.

Zahler P, Khan M. 2003. Evidence for dietary specialization on pine needles by the woolly flying squirrel (*Eupetaurus cinereus*). *Journal of Mammalogy*, 84(2): 480-486.

Zeng T, Jin W, Sun Z Y, *et al.* 2013. Taxonomic position of *Eothenomys wardi* (Arvicolinae: Cricetidae) Based on morphological and molecular analyses with a detailed description of the species. *Zootaxa*, 3682: 85-104.

Zeng X, Chen J, Deng H, *et al.* 2018. A New Species of *Murina* from China (Chiroptera: Vespertilionidae). *Ekoloji*, 27: 9-16.

Zhang B, He K, Wan T, *et al.* 2016. Multi-locus phylogeny using topotype specimens sheds light on the systematics of *Niviventer* (Rodentia, Muridae) in China. *BMC Evolutionary Biology*, 16: 261-273.

Zhang C, Zhang M, Stott P. 2013a. Does prey density limit Amur tiger *Panthera tigris altaica* recovery in northeastern China? *Wildlife Biology*, 19: 452-462.

Zhang F, Jiang Z, Xu A, *et al.* 2013b. Recent Geological Events and Intrinsic Behavior Influence the Population Genetic Structure of the Chiru and Tibetan Gazelle on the Tibetan Plateau. *PLoS One*, 8: e60712.

Zhang H, Zhang J, Zhao C, *et al.* 2015. Complete mitochondrial genome of *Canis lupus campestris*. *Mitochondrial DNA*, 26(2): 255-256.

Zhang J, Gareth J, Zhang L, *et al.* 2010. Recent surveys of bats (Mammalia: Chiroptera) from China II. Pteropodidae. *Acta Chiropterologica*, 12(1): 103-116.

Zhang J, Han N, Jones G, *et al.* 2007. A new species of *Barbastella* (Chiroptera: Vespertilionidae) from North China. *Journal of Mammalogy*, 88(6): 1393-1403.

Zhang L. 2011. Current status of Asian elephants in China. *Gajha*, 35: 43-46.

Zhang L, Jones G, Zhang J, *et al.* 2009a. Recent surveys of bats (Mammalia: Chiroptera) from China. I. Rhinolophidae and Hipposideridae. *Acta Chiropterologica*, 11(1): 71-88.

Zhang L, Liu J, Mcsheax W J, *et al.* 2014. The impact of fencing on the distribution of Przewalski's gazelle. *The Journal of Wildlife Management*, 78(2): 255-263.

Zhang L, Zhu G, Jones G, *et al.* 2009b. Conservation of bats in China: Problems and Recommendations. *Oryx*, 43(2): 179-182.

Zhang X, Liu R, Zhao Q, *et al.* 1993. The population of finless porpoise in the middle and lower reaches of Yangtze River. *Acta Theriologica Sinica*, 13(4): 260-270.

Zhang X, Wang D, Liu R, *et al.* 2003. The Yangtze River dolphin or baiji (*Lipotes vexillifer*): population status and conservation issues in the Yangtze River, China. Aquatic Conservation. *Marine and Freshwater Ecosystems*, 13(1): 51-64.

Zhang Z, Tan X, Sun K, *et al.* 2009c. Molecular systematics of the Chinese *Myotis* (Chiroptera, Vespertilionidae) inferred from cytochrome-*b* sequences. *Mammalia*, 73(4): 323-330.

Zhao S, Xu C, Liu G, *et al.* 2013a. Microsatellite and mitochondrial DNA assessment of the genetic diversity of captive Saiga antelopes (*Saiga tatarica*) in China. *Chinese Science Bulletin*, 18: 2163-2167.

Zhao X, Barlowc J, Taylor B L, *et al.* 2008. Abundance and conservation status of the Yangtze finless porpoise in the Yangtze River, China. *Biological Conservation*, 141(12): 3006-3018.

Zhao X, Wang D, Turvey S T, *et al.* 2013b. Distribution patterns of Yangtze finless porpoises in the Yangtze River: implications for reserve management. *Animal Conservation*, 16(5): 509-518.

Zhen X. 1987. A survey of the bats (Chiroptera) from Fujian Province. *Wuyi Science Journal*, 7: 237-242.

Zhong W, Wang G, Zhou Q, *et al.* 2008. Effects of winter food availability on the abundance of Daurian pikas (*Ochotona dauurica*) in Inner Mongolian grasslands. *Journal of Arid Environments*, 72(7): 1383-1387.

Zhou C, Zhou K, Hu J. 2003. The validity of the dwarf bharal (*Pseudois schaeferi*) species status inferred from mitochondial Cyt *b* gene. *Acta Zoologica Sinica*, 49: 578-584.

Zhou K, Li Y. 1989. Status and aspects of the ecology and behaviour of the baiji, *Lipotes vexillifer* in the lower Yangtze River. *In*: Perrin W F, Brownell Jr, Zhou K, *et al.* Biology and Conservation of the River Dolphins. Gland: IUCN Species Survival

Commission Occasional: 86-91.

Zhou K. 2018. Baiji. *In*: Würsig B, Thewissen J G M, Kovacs K M. Encyclopedia of Marine Mammals. 3rd ed. San Diego: Academic Press/ Elsevier: 54-56.

Zhou K, Qian W. 1985. Distribution of dolphins of the genus *Turstops* in the China Seas. *Aquatic Mammals*, 11: 16-19.

Zhou K, Wang X. 1994. Brief review of passive fishing gear and incidental catches of small cetaceans in Chinese waters. *Reports of the International Whaling Commission* (Special Issue), 15: 347-354.

Zhou M. 1964. Evolution of fauna in China during the Quaternary. *Chinese Journal of Zoology*, 24: 19-24.

Zhou X, Xu S, Xu J, *et al*. 2012. Phylogenomic analysis resolves the interordinal relationships and rapid diversification of the Laurasiatherian mammals. *Systematic Biology*, 61(1): 150-164.

Zhou X, Zhou X, Guang X, *et al*. 2018. Population genomics of finless porpoises reveal an incipient cetacean species adapted to freshwater. *Nature Communications*, 9(1): 1276.

Zhou Z, Guillen-Servent A, Lim B, *et al*. 2009. A new species from southwestern China in the Afro-Palearctic Lineage of the Horseshoe Bats (*Rhinolophus*). *Journal of Mammalogy*, 90(1): 57-73.

Zhu X, Shi W, Pan T, *et al*. 2013. Mitochondrial genome of the Anhui musk deer (*Moschus anhuiensis*). *Mitochondrial DNA*, 24(3): 205-207.

物种中文名索引
Index of Species' Chinese Names

E

F

G

H

J

K

L

M

N

P

Q

R

S

T

Z

物种学名索引
Index of Species' Scientific Names

A

B

脊椎动物
Vertebrates

第一卷 哺乳动物（下册）
Volume I, Mammals (III)

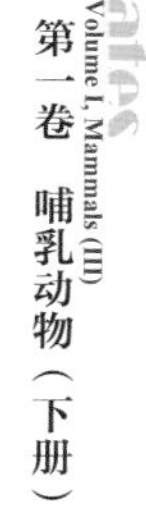

N

O

P

R

S

脊椎动物
Vertebrates
第一卷 哺乳动物（下册）
Volume I, Mammals (III)

后　记

濒危物种红色名录的编研是一项系统工程。大约 25 年前，我开始接触濒危物种红色名录。1996 年和 2008 年，我曾分别应邀参加了《中国濒危动物红皮书·兽类》和《IUCN 受威胁物种红色名录》部分物种濒危等级的研究，并开始探索物种濒危状况评估的方法（蒋志刚和樊恩源，2003）。2008 年，我承担了环境保护部的生物多样性保护专项“中国脊椎动物濒危状况评价”，在完成该项目的过程中，开始了中国哺乳动物的编目与数据库建设，进一步探索物种濒危状况评估的方法。我们提出了区分“生态濒危物种”与“进化濒危物种”，区分不同物种的生活史对策，并引入“经济灭绝”概念等以进行物种等级评估。在 2010 年完成了项目结题报告《中国脊椎动物生存状况》。有关研究结果总结在《物种受威胁状况评估：研究进展与中国的案例》（蒋志刚和罗振华，2012）一文中。

2012 年，我开始主持环境保护部的生物多样性保护专项“中国生物多样性红色名录：脊椎动物卷”，并承担了“中国生物多样性红色名录——哺乳动物卷”的编研。在 2013 ～ 2015 年，我与助手们首先进行了中国哺乳动物的初步编目，而后，开始应用 IUCN 物种红色名录等级标准初步评定了中国哺乳动物濒危等级。我邀请马勇研究员、王应祥研究员、冯祚建研究员、吴毅教授、刘少英研究员等一道审定了中国哺乳动物编目，参加了相关类群的濒危等级评定。在研究中，马勇研究员负责啮齿类，王应祥研究员负责食虫类、灵长类以及云南的哺乳类，冯祚建研究员负责青藏哺乳类，吴毅教授负责翼手目，刘少英研究员承担了兔形目与啮齿目的关键物种的等级评定或审定工作。我们曾通过电子函件邀请了有关专家评审初评的哺乳动物濒危等级，还召开了三次规模不等的会议评审。我们先后完成了《中国哺乳动物多样性》和“中国生物多样性红色名录——哺乳动物卷”，前者在《生物多样性》上发表，后者在 2015 年“生物多样性日”由原环境保护部和中国科学院联合在原环境保护部网站公开发布。此期间，我们还在科学出版社发表了中英双语版本著作《中国哺乳动物多样性与地理分布》。

2016 ～ 2018 年，我们继续更新中国哺乳动物的编目，我邀请了更多的专家参与了中国生物多样性红色名录哺乳动物部分的工作。这些专家包括中国科学院昆明动物研究所蒋学龙研究员、华南濒危动物研究所胡慧建研究员、香港嘉道理农场暨植物园陈辈乐博士等。我们继续更新了中国哺乳动物编目，2017 年发表了《中国哺乳动物多样性》（第 2 版）。我从 2018 年开始写作本书，前后历时两年。承蒙大家的共同努力，现在《中国哺乳动物红色名录：脊椎动物 第一卷 哺乳动物》(2020) 在国家出版基金的资助下，作为中国生物学家的一份生物多样性文件，得以在 2021 年联合国《生物多样性公约》第十五次缔约方大会在中国昆明召开前正式出版。

本书是集体智慧的结晶，本研究团队得到了来自国家野生动物与生物多样性主管部门以及全国及世界有关专家的大力支持。在这一过程中，我们得到原环境保护部、中国科学院的精心指导。在过去的长期野外研究中，我们得到了原林业部、原国家林业局和国家林业与草原局及其有关自然保护区的大力支持和帮助，使得我们能够更新中国哺乳动物分布资料。我们完成的中国哺乳动物编目和红色名录为国家林业与草原局野生动植物保护司修订并发布《国家重点保护野生动物名录》(2021) 提供了参考资料。

我们在此衷心感谢陈宜瑜院士、郑光美院士、张亚平院士、金鉴明院士、马建章院士、曹文宣院士的悉心指导和帮助。我们还感谢原环境保护部李干杰部长、柏成寿副司长、蔡蕾处长，中国科学院动物研究所康乐院士、周琪院士，中国科学院植物研究所马克平先生、覃海宁先生对本书编研工作的关心和帮助。衷心感谢本课题组核心专家组成员马勇研究员、周开亚教授、吴毅教授、刘少英研究员、王应祥研究员、蒋学龙研究员全程参与红色名录编研。衷心感谢《中国生物多样性红色名录：脊椎动物》各卷主持人王跃招研究员、江建平研究员、张鹗研究员、张雁云教授及其团队成员谢锋研究员、张正旺教授、丁平教授、丁长青教授、马志军教授、陈跃英研究员、李家堂研究员、冯祚建研究员、蔡波博士、曹亮博士、董路博士、王斌博士、李成博士等共同承担各卷的工作，并在工作中相互支持与无私帮助。

在本研究中征求了如下专家（按音序为序排列）的意见：鲍伟东教授、鲍毅新教授、毕俊怀教授、初红军教授、范朋飞教授、冯利民博士、Marc Foggin 博士、高行宜教授、葛德燕博士、何锴博士、胡德夫教授、Alice Hughes 博士、花立民教授、黄乘明教授、姜广顺教授、江海声教授、江廷磊博士、金崑研究员、李保国教授、李成先生、李春望博士、李迪强研究员、李明研究员、李晟博士、李松博士、李进华教授、李俊生研究员、李言阔博士、李义明研究员、李玉春教授、李忠秋博士、廖继承教授、林思亮博士、刘丙万博士、刘定震教授、刘伟博士、龙勇诚教授、卢学理博士、罗振华博士、马逸清研究员、毛秀光博士、孟秀祥教授、石红艳博士、时坤教授、宋延龄研究员、苏建平研究员、宛新荣博士、汪松研究员、王丁研究员、王昊研究员、王克雄研究员、王小明教授、王祖望研究员、魏辅文院士、吴诗宝教授、夏霖博士、徐爱春博士、薛达元教授、杨道德教授、杨光教授、杨奇森研究员、杨维康研究员、余文华博士、袁喜才研究员、游章强教授、张劲硕博士、张礼标研究员、张立教授、张明海教授、张林源研究员、张树义教授、张先锋研究员、周江教授、周友兵博士、朱欢兵博士、祝茜教授、宗浩教授，我们的工作得到了他们的大力帮助。我们还感谢美国史密森学会 Don E. Wilson 博士和 Kris Helgen 博士、IUCN 物种生存委员会 Simon Stuart 博士及 IUCN 红色名录工作组（IUCN Red List Committee）David Mallon 博士、伦敦自然历史博物馆 Richard Sabin 馆长及香港嘉道理农场暨植物园陈辈乐博士的支持与帮助。

本书作者还感谢科学出版社李锋总编辑和马俊编辑等为本书的出版所做出的种种努力。此外，我们感谢徐冰冰、李立立、汤宋华帮助整理哺乳动物名录；感谢方红霞高级工程师绘制部分动物插图，丁晨晨、李立立绘制物种分布图，黄元骏校对参考文献。还感谢“西南山地”巫嘉伟先生及其签约摄影师们，中国科学院西双版纳植物园 Alice Hughes 博士，“自然影像中国”谢建国秘书长，香港嘉道理农场暨植物园李松博士，海峡书局曲利明社长、李长青编辑、俞晓佳编辑，王丁研究员、范朋飞教授、袁喜才研究员惠赠照片，Lynx Edicions 出版社特许使用 Wilson D. E. 和 Mittermeier R. A. 所著的 *Handbook of the Mammals of the World* 一书的彩色图版。还感谢武建勇研究员、曾岩博士、平晓鸽博士、臧春鑫博士、李佳琦博士等的大力帮助，感谢中华人民共和国濒危物种科学委员会、中国科学院动物研究所予以大力支持，中国科学院动物研究所野生动物与行为生态研究组的博士后、研究生参与相关工作，在此一并致谢。另外，特别感谢 Plateau Perspectives 的 Marc Foggin 博士校阅文稿并提出宝贵意见，在此谨致真诚谢意。最后，我谨在此再次向你们每一个人表示最诚挚的感谢，感谢你们的帮助。

蒋志刚

2020 年 10 月 5 日于北京中关村

Afterword

The study of the red list of threatened species is a systematic and comprehensive research project. Around a quarter of a century ago, I began to study the conservation status of wildlife species and particularly IUCN red list assessments of endangered species. In 1996 and 2008, I was invited to participate in the studies on the conservation status of certain species for the *China Red Data Book of Endangered Animals: Mammalia* and the *IUCN Red List of Threatened Species*, respectively. Since then, I began to explore the methods for assessing the conservation status of species (Jiang and Fan, 2003). In 2008, I coordinated the key biodiversity conservation project of the Ministry of Environmental Protection, entitled "Assessment of the Endangered Status of Vertebrates in China". In the process of completing this project, my research assistants and I set up a new database and built a preliminary inventory of mammals in China while I continued to explore a variety of methods for assessing the conservation status of species. We proposed to distinguish between "ecologically endangered" and "evolutionarily endangered" species, to distinguish between life history strategies of different taxa, and to introduce the concept of "economic extinction" in assessments of species' status. In 2010, the project concluded with the publication of the report *Living Conditions of Vertebrates in China* and relevant research results are summarized in the article *Assessing species endangerment status: progress in research and an example from China* (Jiang and Luo, 2012).

I presided over the special biodiversity conservation project supported by the Ministry of Environmental Protection in 2012, entitled "China's Red List of Biodiversity: Vertebrates", and I undertook focused research and species assessment on mammals, leading to the compilation of the "China's Red List of Biodiversity: Volume of Mammals". From 2013 to 2015, my assistants and I first renewed the preliminary inventory of mammal species in China, and then we began to assess the conservation status of mammals by applying the IUCN red list criteria for threatened species. I invited Professors Yong Ma, Yingxiang Wang, Zuojian Feng, Yi Wu, Shaoying Liu and others to review the inventory of China's mammals and to participate in the assessment of the conservation status of relevant taxa. Yong Ma reviewed the inventory and helped us assess the status of rodents; Yingxiang Wang reviewed insectivores, primates and mammals of Yunnan Province; Zuojian Feng reviewed mammals on the Qinghai-Tibet (Xizang) Plateau; Yi Wu reviewed the inventory and helped assess the status of bats; and Shaoying Liu reviewed the inventory and helped assess the status of rodents and pika. We also invited relevant experts to review the initial mammal status assessments by e-mails. Three reviewing meetings of varying sizes were held. During the period from 2013 to 2015, we completed the *China's Mammal Diversity* and China's Red List of Biodiversity: Volume of Mammals. The results of the inventory were published in the journal *Biodiversity Science*, and the latter was announced and launched at a special ceremony on "International Biodiversity Day" (2015) hosted by the former Ministry of Environmental Protection and the Chinese Academy of Sciences, with announcement of the event being published on the website of former Ministry of Environmental Protection. During that period, we also published a bilingual edition book entitled *China's Mammal Diversity and Geographic Distribution* in collaboration with Science Press, Beijing.

From 2016 to 2018, we continued to update the mammal inventory for China. I invited more experts to work on the mammal section of the China's red list of biodiversity, including Prof. Xuelong Jiang from the Kunming Institute of Zoology, Chinese Academy of Sciences, Prof. Huijian Hu from the South China Institute of Endangered Animals, and Dr. Bosco P. L. Chan from the Hong Kong Kadoorie Farm and Botanic Garden. We published *China's Mammal Diversity* (2nd edition) in 2017. I started writing the present book in 2018, and this project took two years. Many thanks are hereby extended for the joint efforts of all the people who have contributed to this book. Now, *China's Red List of Biodiversity: Vertebrates, Volume I* (2020) has been officially published with the support of the National Publication Foundation. Developed by Chinese biologists, this national biodiversity document will be officially released prior to the 15th meeting of the Conference of the Parties of the UN *Convention on Biological Diversity* (CBD), which will be held in Kunming, China, in 2021.

This book represents the crystallization and summation of collective wisdom, as the research team has received support from numerous national wildlife and biodiversity management authorities as well as experts from across the country and around the world. During this process, we were guided by the former Ministry of Environmental Protection and the Chinese Academy of Sciences. During this long-term field research, we also were greatly supported by the former Ministry of Forestry, the former National Forestry Administration, the National Forestry and Grassland Administration, and relative nature reserves, enabling us to substantially update mammal distribution data in China. The mammal inventory and the red list of China's mammals have served as critical references for the Department of Wild Fauna and Flora Protection of the National Forestry and Grassland Administration as it sought to update and release the official *List of State Key Protected Wild Animal Species* in 2021.

We sincerely thank Academicians Yiyu Chen, Guangmei Zheng, Yaping Zhang, Jianming Jin, Jianzhang Ma, and Wenxuan Cao for their thoughtful guidance and scholarly help. We also thank the Minister of the former Ministry of Environment Protection, Ganjie Li, former Deputy Director of the Department of Nature, Ministry of Environmental Protection, Mr. Chengshou Bai, and Director Lei Cai, Academician Le Kang, Academician Qi Zhou of the Institute of Zoology, Chinese Academy of Sciences, Prof. Keping Ma and Prof. Haining Qin of the Institute of Botany, Chinese Academy of Sciences, for their kind help in the research about this book. I would like to sincerely thank the members of the core expert group of our research, including Prof. Yong Ma, Prof. Kaiya Zhou, Prof. Yi Wu, Prof. Shaoying Liu, Prof. Yingxiang Wang and Prof. Xuelong Jiang for their full and active participation in the research of this red list. Our heartfelt thanks also are offered to the head of each vertebrate volume of *China's Biodiversity Red List: Vertebrates*, specifically to Prof. Yuezhao Wang, Prof. Jianping Jiang, Prof. E Zhang and Prof. Yanyun Zhang, along with team members Prof. Feng Xie, Prof. Zhengwang Zhang, Prof. Ping Ding, Prof. Changqing Ding, Prof. Zhijun Ma, Prof. Yueying Chen, Prof. Jiatang Li, Prof. Zuojian Feng, Dr. Bo Cai, Dr. Liang Cao, Dr. Lu Dong, Dr. Bin Wang and Dr. Cheng Li for their mutual support and selfless help to each other in the work of editing the books.

During this study, the following experts were also consulted (in Chinese alphabetical order): Prof. Weidong Bao, Prof. Yixing Bao, Prof. Junhuai Bi, Prof. Hongjun Chu, Prof. Pengfei Fan, Dr. Limin Feng, Dr.

Marc Foggin, Prof. Xingyi Gao, Dr. Deyan Ge, Dr. Kai He, Prof. Defu Hu, Dr. Alice Hughes, Prof. Limin Hua, Prof. Chengming Huang, Prof. Guangshun Jiang, Prof. Haisheng Jiang, Dr. Tinglei Jiang, Prof. Kun Jin, Prof. Baoguo Li, Mr. Cheng Li, Dr. Chunwang Li, Prof. Diqiang Li, Prof. Ming Li, Dr. Sheng Li, Dr. Song Li, Prof. Jinhua Li, Prof. Junsheng Li, Dr. Yankuo Li, Prof. Yiming Li, Prof. Yuchun Li, Dr. Zhongqiu Li, Prof. Jicheng Liao, Dr. Siliang Lin, Dr. Bingwan Liu, Prof. Dingzhen Liu, Dr. Wei Liu, Prof. Yongcheng Long, Dr. Xueli Lu, Dr. Zhenhua Luo, Prof. Yiqing Ma, Dr. Xiuguang Mao, Prof. Xiuxiang Meng, Dr. Hongyan Shi, Prof. Kun Shi, Prof. Yanling Song, Prof. Jianping Su, Dr. Xinrong Wan, Prof. Song Wang, Prof. Ding Wang, Dr. Hao Wang, Prof. Kexiong Wang, Prof. Xiaoming Wang, Prof. Zuwang Wang, Academician Fuwen Wei, Prof. Shibao Wu, Dr. Lin Xia, Dr. Aichun Xu, Prof. Dayuan Xue, Prof. Daode Yang, Prof. Guang Yang, Prof. Qisen Yang, Prof. Weikang Yang, Dr. Wenhua Yu, Prof. Xicai Yuan, Prof. Zhangqiang You, Dr. Jinshuo Zhang, Prof. Libiao Zhang, Prof. Li Zhang, Prof. Minghai Zhang, Prof. Linyuan Zhang, Prof. Shuyi Zhang, Prof. Xianfeng Zhang, Prof. Jiang Zhou, Dr. Youbing Zhou, Dr. Huanbing Zhu, Prof. Qian Zhu and Prof. Hao Zong. We received substantial help from all of these colleagues during our work. We would also like to thank Dr. Don E. Wilson and Dr. Kris Helgen from the Smithsonian Institution, USA, Dr. Simon Stuart from the IUCN Species Survival Commission, Dr. David Mallon from the IUCN Red List Committee, Curator Richard Sabin from the Zoological Collections, Natural History Museum, London, and Dr. Bosco P. L. Chan from the Hong Kong Kadoorie Farm and Botanic Garden.

The authors also thank chief editor Feng Li and editor Jun Ma *et al*. of the Science Press, Beijing for their great efforts in publishing this book. In addition, we wish to thank Bingbing Xu, Lili Li and Songhua Tang for compiling the inventory of China's mammal species, Chenchen Ding and Lili Li for drawing the species distribution maps and senior engineer Hongxia Fang for other illustrations, Yuanjun Huang for checking the references list. We also thank Mr. Jiawei Wu and the contracted photographers of "Swild.cn", Dr. Alice Hughes of the Xishuangbanna Tropical Botanical Garden, Chinese Academy of Sciences, Mr. Jianguo Xie, the Secretary-General of "Natural Video China", Dr. Song Li of the Hong Kong Kadoorie Farm and Botanic Garden, president Liming Qu, editors Changqing Li and Xiaojia Yu of the Straits Publishing & Distributing Group, Prof. Ding Wang, Prof. Pengfei Fan and Prof. Xicai Yuan for permission to use their photos, and Lynx Edicions Press for using the color plates from the *Handbook of the Mammals of the World* (edited by D. E. Wilson and R. A. Mittermeier). I would also like to thank Prof. Jianyong Wu, Dr. Yan Zeng, Dr. Xiaoge Ping, Dr. Chunxin Zang and Dr. Jiaqi Li for their great help. I equally wish to extend my thanks to the Endangered Species Scientific Commission, P. R. China as well as to the Institute of Zoology, Chinese Academy of Sciences, for their great support, including all graduate students and postdoctoral fellows who studied or worked in the Group of Wildlife and Behavioral Ecology, Institute of Zoology, Chinese Academy of Sciences. In addition, special thanks also go to Dr. Marc Foggin of Plateau Perspectives for his proof reading as well as invaluable comments on the manuscript. Finally, I here wish to express my most sincere gratitude once again to each and every one of you for all your help.

Zhigang Jiang

Zhongguancun, Beijing, China

October 5, 2020